# REGRESSION ANALYSIS FOR THE SOCIAL SCIENCES

This book provides graduate students in the social sciences with the basic skills that they need to estimate, interpret, present, and publish basic regression models using contemporary standards.

Key features of the book include:

■ interweaving the teaching of statistical concepts with examples developed for the course from publicly available social science data or drawn from the literature;

■ thorough integration of teaching statistical theory with teaching data processing and analysis;

■ teaching of both SAS and Stata "side-by-side" and use of chapter exercises in which students practice programming and interpretation on the same data set and course exercises in which students can choose their own research questions and data set.

**Rachel A. Gordon** is an Associate Professor in the Department of Sociology and the Institute of Government and Public Affairs at the University of Illinois at Chicago. Professor Gordon has multidisciplinary substantive and statistical training and a passion for understanding and teaching applied statistics.

TITLES OF RELATED INTEREST

*Contemporary Social Theory* by Anthony Elliot
*GIS and Spatial Analysis for the Social Sciences* by Robert Nash Parker and Emily K. Asencio
*Social Statistics* by Thomas J. Linneman
*Statistical Modelling for Social Researchers* by Roger Tarling

# REGRESSION ANALYSIS FOR THE SOCIAL SCIENCES

**Rachel A. Gordon**
University of Illinois at Chicago

NEW YORK AND LONDON

First published 2010
by Routledge
270 Madison Ave, New York, NY 10016

Simultaneously published in the UK
by Routledge
2 Park Square, Milton Park, Abingdon, Oxon OX14 4RN

*Routledge is an imprint of the Taylor & Francis Group, an Informa business*

© 2010 Taylor & Francis

Typeset by RefineCatch Limited, Bungay, Suffolk
Printed and bound in the United States of America on acid-free paper by Edward Brothers, Inc.

*Library of Congress Cataloging-in-Publication Data*
Gordon, Rachel A.
Regression analysis for the social sciences / Rachel A. Gordon.—1st ed.
    p. cm.
  1. Social sciences.   2. Regression analysis.   I. Title.
  H61.G578 2010
  519.5′36—dc22

                                     2009050278

ISBN 10: 0-415-99154-4 (hbk)
ISBN 10: 0-203-89249-6 (ebk)

ISBN 13: 978-0-415-99154-4 (hbk)
ISBN 13: 978-0-203-89249-7 (ebk)

List of Trademarks that feature in the text

| | |
|---|---|
| Stata | Microsoft Word |
| SAS | Word Perfect |
| Microsoft Excel | Notepad |
| TextPad | DBMS/Copy |
| UltraEdit | SPSS |
| StatTransfer | R |
| LISREL | Minitab |
| AMOS | S-Plus |
| Mplus | Systat |
| EQS | |

Go to *http://www.routledge.com/textbooks/9780415991544* for an invaluable set of resources associated with *Regression Analysis for the Social Sciences* by Rachel Gordon

# PREFACE

This text is intended for graduate students in the social sciences. The focus is on regression analysis, and we presume prior coursework in basic descriptive and inferential statistics.

Textbooks on regression analysis abound. Even so, there is a gap in the offerings aimed at graduate students in the social sciences, especially given recent developments in substantive and statistical theory and in computing power. We target the social science branches such as sociology, human development, psychology, education, and social work to which students bring a wide range of mathematical skills and have a wide range of methodological affinities. For many of these students, a successful course in regression will not only offer statistical content but will also help them to overcome their fear of statistics and to develop an appreciation for how regression models might answer some of the research questions of interest to them.

To meet these objectives, this book has three distinctive features:

- use of examples of interest to social scientists including both:
  - (i) literature excerpts, drawn from a range of journals and a range of subfields;
  - (ii) examples from real data sets, including two data sets carried throughout the book (the National Survey of Families and Households; the National Organizations Survey);
- thorough integration of teaching statistical theory with teaching data processing and analysis;
- parallel teaching of SAS and Stata.

## THE IMPETUS FOR THIS TEXTBOOK

Over the last few decades, the landscape of quantitative empirical work in the social sciences has changed dramatically, raising the bar on the basic skills that scholars need in order to produce and read quantitatively based publications. One impetus for these changes was the initiation and maturation of a number of large-scale studies of individuals, families, and organizations (for example, the National Longitudinal Survey of Youth began annual interviews in 1979, spawned the Children of the NLSY in 1986 when investigators began to follow the children of mothers from the original cohort, and required an acronym change, to NLSY79, when a new cohort, the NLSY97, was first interviewed). Another impetus for change was the expansion of computing power, allowing these data sets to be readily analyzed on the desktop, and the development of the Internet, which now puts many of these data sets only a click away.

Changing technology has also raised the bar on how the results of regression models are presented. Increasingly, multimedia is integrated into everyday lives. Succinct, clear presentations are required for results to stand out from a flood of information. Clearly organized and presented manuscripts have always been important, but this is increasingly true given demands on reviewers' and editors' time and attention, and pressures to conserve journal space. Strategies for presentation are also important for students and scholars who want to make their work accessible to practitioners, policymakers, and the public, something encouraged by current developments across fields (e.g., public sociology, applied developmental science).

Although many statistics texts exist, none completely meets the needs of core graduate training in the social sciences, instead typically being targeted at a different level, audience, or niche. For example, texts aimed at the undergraduate level often do not meet the goals and coverage of graduate sequences intended to prepare students to understand primary sources and conduct their own publishable research. These texts are sometimes used because they are at the right level for graduate students who are less mathematically inclined, but they do not fully satisfy the needs of graduate students and the faculty. Texts aimed at other disciplines are also problematic because they do not connect with students' substantive training and interests. For example, econometrics texts typically use economic examples and often assume more advanced mathematical understanding than is typical of other social science disciplines.

This text aims to address this current landscape. The goal of the book is to provide graduate students with the basic skills that they need to estimate, interpret, present, and publish regression models using contemporary standards. Key features of the book include:

- interweaving the teaching of statistical concepts with examples developed for the course from publicly available social science data or drawn from the literature;
- thorough integration of teaching statistical theory with teaching data processing and analysis;
- Teaching of both SAS and Stata "side-by-side" and use of chapter exercises in which students practice programming and interpretation on the same data set and of course exercises in which students can choose their own research questions and data set.

## THE AUDIENCE FOR THE BOOK

This book is designed for a semester-long course in graduate-level regression analyses for the social sciences. Such a course typically occurs in the first or second year of graduate study, following a course in basic descriptive and inferential statistics. The skills, motivations, and interests of students vary considerably.

For some students, anxiety is high, and this core sequence comprises the only statistics course that they plan to take. These students will become better engaged in the course if the concepts and skills are taught in a way that recognizes their possible math anxiety, is embedded in substantive examples, connects with the students' research interests, and helps them to feel that they can "do quantitative research." Part of the challenge of connecting with students' research interests, though, is that they are typically just starting their graduate programs when they take their statistics sequence, so the course needs to explicitly make connections to students' budding interests.

Other students in the course are eager to gain a deep understanding of statistical concepts and sophisticated skills in data management and analysis so that they can quickly move on to and excel with advanced techniques. Many of these students will come into their programs believing that quantitative research would be a major part of their career. Some want to use the skills they learn in the course to secure coveted research assistant positions. Many of these students enter the program with solid math skills, prior success in statistics courses, and at least some experience with data management and analysis. For these students, the course will be frustrating and unfulfilling if it doesn't challenge them, build on their existing knowledge and skills, and set them on a path to take advanced courses and learn sophisticated techniques.

Students also vary in their access to resources for learning statistics and statistical packages beyond the core statistics sequence. In some departments, strategies for locating data, organizing a research project, and presenting results in a manuscript are easily learned from mentors and research teams (including through research assistantships) and through informal conversations with fellow students. Some programs also have separate "capstone" courses that put statistics into practice, typically following the core sequence. For other students, there are few such formal and informal opportunities. These students will struggle with implementing the concepts learned in statistics courses without answers to practical questions such as "Where can I find data?" "How do I get the data into SAS (or Stata or SPSS) format?" "How do I interpret a codebook?" "How should I organize my files?" "How do I present my results in my manuscript?" Integrating this practical training within the core statistics sequence meets the needs of students (and faculty) in programs with few formal and informal learning opportunities for such practical skills. We also use this integrated approach in the book to help students practice the statistical concepts they are learning with real data, in order to help reinforce their learning, engage them in the course, and give them confidence in conducting quantitative research.

## THE GOALS OF THE BOOK

The goals of the book are to prepare students to:

(a)  conduct a research project from start to finish using basic regression analyses;
(b)  have the basic tools necessary to be a valuable beginning research assistant;
(c)  have the basic knowledge and skills needed to take advanced courses that build on basic regression models; and
(d)  intelligently and critically read publications in top journals that utilize basic regression models.

We focus especially on concepts and techniques that are needed either to publish basic regression analyses in major journals in the relevant fields (for goals 1–3) or read publications using these models in those journals (for goal 4).

At every stage of the book, we attempt to look through the lens of the social scientist in training: why do I need to know this? How is it useful to me? The book is applied in orientation, and frequently makes concepts concrete through examples based on social science data and excerpts from recent journal publications.

Although the book is applied, we introduce key mathematical concepts aiming to provide sufficient explanation in order to accommodate students with weaker math backgrounds. For example, students are taught to find meaning in equations. Throughout the text, students are shown how to manipulate equations in order to facilitate understanding, with detailed in-text explanations of each step. The goal is to help *all* students feel comfortable reading equations, rather than leaving some to skip over them.

For more advanced students, or students returning to the book later in their careers, we provide references for additional details. We also attempt to present concepts and techniques deeply and slowly, using concrete examples for reinforcement. Our goal is for students to learn an idea or skill well enough that they remember it and how to apply it. This pace and approach allows sufficient time for students who struggle with learning statistics to "really get it" and allows sufficient time for students who learn statistics easily to achieve a more fundamental understanding (including references to more advanced topics/readings).

As part of this approach, we unpack ideas and look at them from multiple angles (again with a goal toward what is needed when preparing a manuscript for publication, or reading a published article). For example, we spend considerable time on understanding how to test and interpret interactions (e.g., plotting predicted values, testing differences between points on the lines, calculating conditional slopes).

We assume that students have had an introductory course in research methods and in descriptive and inferential statistics, although we review concepts typically covered in these courses when we first use them.

## THE CHAPTERS OF THE BOOK

The first part of the book introduces regression analysis through a number of literature excerpts and teaches students how to locate data, use statistical software, and organize a quantitative research project. The second part covers basic ordinary least squares (OLS) regression models in detail. The final chapter pulls together the earlier material, including providing a roadmap of advanced topics and revisiting the examples used in earlier chapters.

### Part 1: Getting Started

Part 1 of the book aims to get students excited about using regression analysis in their own research and to put students on common ground by exposing them to literature excerpts, data sets, statistical packages, and strategies for organizing a quantitative research project. As noted above, this leveling of the playing field is important because students will vary in the prior statistics courses that they have taken and their prior experience analyzing data as well as in opportunities in their program to learn how to put statistical concepts into practice.

■ Chapter 1 introduces the basic ideas of regression analysis using three literature excerpts. By using a range of substantive applications and a range of data sources, a major goal of the excerpts is to get students excited about applying regression analysis to their own work. In this chapter, the examples were also selected because they were completed when the authors were graduate students and published in top journals, thus giving students attainable role models. The examples are also meant to begin to help students read and interpret published regression results (beyond their experiences reading articles that report regression analyses in substantive courses). And, the examples in this chapter were selected to preview some of the central topics to be covered in later chapters (e.g., controlling for confounds, examining mediation, testing for interactions) and others of which will be pointed to in the roadmap in the last chapter of the book (e.g., negative binomial models).

■ Chapter 2 discusses strategies for organizing a research project. Especially with large secondary data sets with numerous variables, it is easy to get lost "playing with the data." We encourage students to keep theoretical ideas and a long-range perspective in mind throughout a project. This chapter directly addresses the variability in formal and informal opportunities for research experiences mentioned above, and attempts to pull together various "words of wisdom" about planning and documenting a project and locating data that some students might otherwise miss. The chapter also exposes students to a breadth of secondary data sets, which can provide the knowledge and comfort needed to access secondary data as their interests develop over the years of graduate study. The chapter teaches students basic skills in understanding documentation for secondary data sources and selecting data sets. The data set carried throughout the in-text examples, the National Survey of Families and Households (NSFH), is introduced in the chapter.

■ Chapter 3 introduces the basic features of data documentation and statistical software. The chapter begins with basic concepts of how data sets are stored in the computer and read by statistical packages. The rationale for using both SAS and Stata is provided, along with the basic types of files used and created by each package. The chapter also covers how to organize files in a project and how to identify relevant variables from large existing data sets. The chapter example uses the data set carried throughout the in-text examples (NSFH). The chapter exercise introduces the data set used for the chapter exercises throughout the remainder of the book (National Organizations Study).

■ Chapter 4 teaches students how to write basic statistical programs. To reach a broad audience of instructors, and to expose students to the flexibility of moving between software packages, SAS and Stata are presented side-by-side. The chapter begins with the basics of the Stata and SAS interfaces and syntax. We then cover how to create new variables and to keep a subset of cases. The chapter ends with recommendations for organizing files (including comments and spacing) and for debugging programs (identifying and fixing errors).

## Part 2: Ordinary Least Squares Regression with Continuous Outcome Variables

■ Chapter 5 covers basic concepts of bivariate regression. Interpretation of the intercept and slope is emphasized through examining the regression line in detail, first generally with algebra and geometry and then concretely with examples drawn from the literature and developed for the course. We look at the formulae for the slope coefficient and its standard error in detail, emphasizing what factors affect the size of the standard error. We discuss hypothesis testing and confidence intervals for testing statistical significance and rescaling and effect sizes for evaluating substantive significance.

■ Chapter 6 covers basic concepts of multiple regression. We look in detail at a model with two predictors, using algebra, geometry, and concrete examples to offer insights into interpretation. We look at how the formulae for the slope coefficients and their standard errors differ from the single predictor variable context, emphasizing how correlations among the predictors affect the size of the standard error. We cover joint hypothesis testing and introduce the general linear $F$-test. We again use algebra, illustrations, and examples to reinforce a full understanding of the $F$-test, including its specific uses for an overall model $F$-test and a partial $F$-test. We re-emphasize statistical and substantive significance and introduce the concepts of $R$-squared and Information Criteria.

■ Chapter 7 covers dummy variable, predictors in detail, starting with a model with a single dummy predictor and extending to (a) models with multiple dummies that represent one multicategory variable, and (b) models with multiple dummies that represent two multicategory variables. We look in detail at why dummy variables are needed, how they are constructed, and how they are interpreted. We present three approaches for testing differences among included categories.

■ Chapter 8 covers interactions in detail, including an interaction between two dummy variables, between a dummy and interval variable, and between two interval variables. We present the Chow test and fully interacted regression model. We look in detail at how to interpret and present results, building on the three approaches for testing among included categories presented in Chapter 7.

■ Chapter 9 covers nonlinear relationships between the predictor and outcome. We discuss how to specify several common forms of nonlinear relationships between an interval predictor and outcome variable using the quadratic function and logarithmic transformation. We discuss how these various forms might be expected by conceptual models and how to compare them empirically. We also show how to calculate and plot predictions to illustrate the estimated forms of the relationships. And, we also discuss how to use dummy variables to estimate a flexible relationship between a predictor and the outcome.

- Chapter 10 examines how adding variables to a multiple regression model affects the coefficients and their standard errors. We cover basic concepts of path analysis, including total, direct, and indirect effects. We relate these ideas to the concept of omitted variable bias, and discuss how to contemplate the direction of bias from omitted variables. We discuss the challenge of distinguishing between mediators and confounds in cross-sectional data.

- Chapter 11 encompasses outliers, heteroskedasticity, and multicollinearity. We cover numerical and graphical techniques for identifying outliers and influential observations. We also cover the detection of heteroskedasticity, implications of violations of the homoskedasticity assumption, and calculation of robust standard errors. Finally, we discuss three strategies for detecting multicollinearity: (a) variance inflation factors, (b) significant model $F$ but no significant individual coefficients, and (c) rising standard errors in models with controls. And, we discuss strategies for addressing multicollinearity based on answers to two questions: Are the variables indicators of the same or different constructs? How strongly do we believe the two variables are correlated in the population versus our sample (and why)?

## Part 3: Wrapping Up

The final chapter provides a roadmap of topics that students may want to pursue in the future to build on the foundation of regression analysis taught in this book. The chapter organizes a range of advanced topics and briefly mentions their key features and when they might be used (but does not teach how to use those techniques). Students are presented with ideas about how to learn these topics as well as gaining more skill with SAS and Stata (e.g., searching at their own or other local universities; using summer or other short course opportunities). The chapter also revisits the Literature Excerpts featured in the first chapter of the book.

## SOME WAYS TO USE THE BOOK

The author has used the complete textbook in a 15-week semester with two 75-minute lectures and a weekly lab session. Typically, chapters can be covered in a week, although a bit less time is needed for Part 1 (usually accomplished in the first two to three weeks) and extra time is often taken with the earliest chapters in Part 2 (two weeks each on the basics of bivariate regression, the basics of multiple regression, dummy variables, and interactions).

Depending on the local resources and the expertise of the instructor and teaching assistant, however, the book can be used to teach only SAS, only Stata, or both. To meet the needs of a heterogeneous pool of instructors and students, we provide comparable SAS and Stata commands for the same task throughout the book. We also hope that putting the commands side-by-side helps students to see the similarities (and differences) between the two languages and helps to prepare students for the varied and often unexpected directions their interests may take them. Even when instructors use one package, the side-by-side presentation allows the most motivated students to implement both (during the course or after the course is completed).

With a few exceptions, each chapter has a common set of materials at the end: key terms, review questions, review exercises, chapter exercises, and a course exercise.

- **Key Terms** are in bold within the chapter and defined in the glossary index.
- **Review Questions** allow students to demonstrate their broad understanding of the major concepts introduced in the chapter.
- **Review Exercises** allow students to practice the concepts introduced in the chapter by working through short, standalone problems.
- **Chapter Exercises** allow students to practice the applied skills of analyzing data and interpreting the results. The chapter exercises carry one example throughout the book (using the 1996–7 National Organizations Survey) allowing students to ask questions easily of one another, the teaching assistant, and instructor as they all work with the same data set. The goal of the chapter exercises is to give students confidence in working with real data, which may encourage them to continue to do so to complement whatever other research approaches they use in the future.
- The **Course Exercise** allows students to select a data set to apply the concepts learned in each chapter to a research question of interest to them. The author has used this option with students who have more prior experience than the average student, who ask for extra practice because they know that they want to go on to advanced courses, or who are retaking the course as they begin to work on a masters or dissertation. The course exercises help students to gain confidence in working independently with their own data.

Answers to the review questions, review exercises, and chapter exercises, including the batch programs and results for the chapter exercises, are available on the textbook web site (*http://www.routledge.com/textbooks/9780415991544*). The data sets, programs, and results from the in-text examples are also available on the textbook web site.

# ACKNOWLEDGMENTS

This book reflects many individuals' early nurturing and continued support of my own study of statistics, beginning at Penn State University, in the psychology, statistics, computer science and human development departments, and continuing at the University of Chicago, in the schools of public policy and business, in the departments of statistics, sociology, economics and education, and at the social sciences and public policy computing center. I benefited from exposure to numerous faculty and peers who shared my passion for statistics, and particularly its application to examining research questions in the social sciences.

UIC's sociology department, in the College of Liberal Arts and Sciences, was similarly flush with colleagues engaged in quantitative social science research when I joined the department in 1999 and has provided me with the opportunity to teach graduate statistics over the last decade. This book grew out of my lecture notes for that course, and benefits from numerous interactions with students and colleagues over the years that I have taught it. Kathy Crittenden deserves special thanks, as she planted the idea of this book and connected me with my publisher. I also have benefitted from interacting with my colleagues at the University of Illinois' Institute of Government and Public Affairs, especially Robert Kaestner, as I continued to study statistics from multiple disciplinary vantage points.

My publisher, Steve Rutter, was instrumental in taking me over the final hurdle in deciding to write this book and has been immensely supportive throughout the process. He has ably provided advice and identified excellent reviewers for input as the book took shape. The reviewers' comments also importantly improved the book, including early reviews of the proposal by Peter Marsden at Harvard University, Timothy Hilton at North Michigan University, Robert Kaufman at Ohio State, Alan Acock at Oregon State, Sarah Mustillo at Purdue University, Evan Schofer at the

University of California, Irvine, François Nielsen at the University of North Carolina, Chapel Hill, and Thomas Pullum at the University of Texas at Austin; first chapter reviews by Gretchen Cusick at the University of Chicago, Scott Long at the University of Indiana, Erin Leahey at the University of Arizona and Tom Linneman at William and Mary; and special thanks to John Allen Logan at the University of Wisconsin, Madison who provided excellent comments on the first and revised draft of the full manuscript. I also want to express my appreciation to Stata for their Author Support Program, and especially thank Bill Rising for providing comments on Chapters 3 and 4 of the book. I also want to thank all of the staff at Routledge who helped produce the book, especially Leah Babb-Rosenfeld and Mhairi Baxter. Any remaining errors or confusions in the book are my own.

I also thank my husband, Kevin, and daughter, Ashley, for their support and for enduring the intense periods of work on the book.

Every effort has been made to trace and contact copyright holders. The publishers would be pleased to hear from any copyright holders not acknowledged here, so that this acknowledgement page may be amended at the earliest opportunity.

# TABLE OF CONTENTS IN BRIEF

## PART 3: WRAPPING UP

# TABLE OF CONTENTS IN DETAIL

## PART 2: THE REGRESSION MODEL

## PART 3: WRAPPING UP

## APPENDICES

*Part 1*

# GETTING STARTED

*Chapter 1*

# EXAMPLES OF SOCIAL SCIENCE RESEARCH USING REGRESSION ANALYSIS

## CHAPTER 1: EXAMPLES OF SOCIAL SCIENCE RESEARCH USING REGRESSION ANALYSIS

"Statistics present us with a series of techniques that transform raw data into a form that is easier to understand and to communicate or, to put it differently, that make it easy for the data to tell their story."

Jan de Leeuw and Richard Berk (2004)
Introduction to the Series
Advanced Quantitative Techniques in the Social Sciences

**Regression analysis,**\* a subfield of statistics, is a means to an end for most social scientists. Social scientists use regression analysis to explore new research questions and to test hypotheses. This statement may seem obvious, but it is easy to get sidetracked in the details of the theory and practice of the method, and lose sight of this bigger picture (especially in introductory statistics courses). To help keep the big picture in sight, this chapter provides excerpts from the social science literature.

These excerpts also help to focus attention on how regression analyses are used in journal articles, consistent with two of the major reasons graduate students learn about regression analysis: (a) to be able to read the literature, and (b) to be able to contribute to the literature. You have probably already read at least some articles and books that report the results of regression analyses (if not, you likely will be doing so in the coming weeks in your substantive courses). Examining such literature excerpts in the context of a course on regression analysis provides a new perspective, with an eye toward how the technique facilitates exploring the question or testing the hypotheses at hand, what choices the researchers make in order to implement the model, and how the results are interpreted. At this point, you do not have the skills to understand fully the regression results presented in the excerpts (otherwise this book would not be needed!), so the purpose of this chapter is to present these features in such a way that they overview what later chapters will cover, and why. We will revisit these excerpts in the final chapter of the book, to help to reinforce what we have covered (and what advanced topics you might still want to pursue).

We have purposefully chosen excerpts in this chapter to hopefully appeal to you because they were written by young scholars who published work that they completed as graduate students (in later chapters, we use shorter excerpts, including some from more senior scholars). To appeal to a broad array of interests, we have also selected the examples from a range of subfields and with different levels of analysis (person, community, informal organization). All of the articles use existing data, but they

\* **Terms in color** in the text are defined in the glossary/index.

illustrate a wide variety of sources of such data (including ethnographies). Although using existing data presents some limitations (discussed for each excerpt), doing so can let you answer questions of interest to you in less time and cost, and with a larger and more representative sample, than you could have achieved with new data collection. Chapter 2 helps you to identify similar data sources for your own research.

## 1.1: WHAT IS REGRESSION ANALYSIS?

Later chapters will develop the statistical details of regression analysis. But, in order to provide some guideposts to the features we will examine in the literature excerpts, it is helpful first to briefly consider what regression analysis is conceptually, and what are some of its key elements.

Why is it called regression, and why is it so widely used in the social sciences? The term regression is attributed to Francis Galton (Stigler 1986). Galton was interested in heredity and gathered data sets to understand better how traits are passed down across generations. The data he gathered ranged from measures of parents' and children's heights to assessments of sweet peas grown from seeds of varying size. His calculations showed that the height or size of the second generation was closer to the sample average than the height or size of the first generation (i.e., it reverted, or regressed, to the mean). His later insights identified how a certain constant factor (such as exposure to sunlight) might affect average size, with dispersion around that group average, even as the entire sample followed a normal distribution around the overall mean. Later scientists formalized these concepts mathematically and showed their wide applicability. Ultimately, regression analysis provided the breakthrough that social scientists needed in order to study social phenomena when randomized experiments were not possible. As Stephen Stigler puts it in his *History of Statistics* "beginning in the 1880s . . . a series of remarkable men constructed an empirical and conceptual methodology that provided a surrogate for experimental control and in effect dissipated the fog that had impeded progress for a century" (Stigler 1986, 265).

Regression analysis allows scientists to quantify how the average of one variable systematically varies according to the levels of another variable. The former variable is often called a **dependent variable** or **outcome variable** and the latter an **independent variable, predictor variable**, or **explanatory variable**. For example, a social scientist might use regression analysis to estimate the size of the gender wage gap (how different are the mean wages between women and men?), where wage is the dependent variable and gender the independent variable. Or, a scientist might test for an expected amount of returns to education in adults' incomes, looking for a regular increment in average income (outcome) with each additional year of schooling (predictor). When little prior research has addressed a topic, the regression analyses may be exploratory, but these variables are ideally identified through theories and concepts applied to particular phenomena. Indeed, throughout the text, we encourage forward thinking and conceptual grounding of your regression models. Not only is this most consistent with the statistical basis of hypothesis testing, but thinking ahead (especially based on theory) can facilitate timely completion of a project, easier interpretation of the output, and stronger contributions to the literature.

An important advantage of regression analysis over other techniques (such as bivariate *t*-tests or correlations) is that additional variables can be introduced into the model to help to determine if a relationship is genuine or spurious. If the relationship is spurious, then a third variable (a confounder, common cause, or extraneous variable) causes both the predictor and outcome; and, adjusting for the third variable in a regression model should reduce the association between the predictor and outcome to near zero. In some cases, the association may not be erased completely, and the predictor may still have an association with the outcome, but part of the initial association may be due to the third variable. For example, in initial models, teenage mothers may appear to attain fewer years of schooling than women who do not have a child until adulthood. If family socioeconomic status leads to both teenage motherhood and academic achievement, then adjusting for the family of origin's education, occupation, and income should substantially reduce the average difference in school attainment between teenage and adult mothers. In Chapters 6 and 10, we will discuss how these adjustments are accomplished, and their limitations.

Such statistical adjustments for confounding variables are needed in the social sciences when randomized experiments cannot be conducted due to ethical and cost concerns. For example, if we randomly assigned some teenagers to have a child and others not to, then we could be assured that the two groups were statistically equivalent except for their status as teenage mothers. But, of course, doing so is not ethical. Although used less often than in the physical sciences to test basic research questions, experiments are more frequently used in certain social science subfields (e.g., social psychology) and applications (e.g., evaluations of social programs). When experiments are not possible, social scientists rely on statistical adjustments to **observational data** (data in which people were not assigned experimentally to treatment and control groups, such as population surveys or program records). Each literature excerpt we show in this chapter provides examples of using control variables in observational studies in an attempt to adjust for such confounding variables.

Regression models also allow scientists to examine the **mechanisms** that their theories and ideas suggest explain the association between a particular predictor variable and an outcome (often referred to as mediation). For example, how much of the wage gap between men and women is due to discriminatory practices on the part of employers and how much is due to differences in family responsibilities of men and women (such as child-rearing responsibilities)? If the mediators—discriminatory practices and family responsibilities—are measured, regression models can be used to examine the extent to which they help to explain the association between the predictor of interest—in this case, gender—and the outcome—wages. In Chapter 10, we will discuss how these mechanisms can be identified in regression models and some of the challenges that arise in interpreting them. Some of the literature excerpts we discuss below illustrate the use of mediators.

Social scientists also use regression models to examine whether two predictors jointly associate with an outcome variable (often referred to as moderation or interaction). For example, is the gender wage gap larger for African Americans than for whites? Are the returns to education larger for workers who grew up in lower versus higher income families? In Chapter 8, we will discuss how such research questions can be examined with regression models, and below we illustrate their use in the literature.

Regression models rest on a number of assumptions, which we will discuss in detail in later chapters. It is important for social scientists to understand these assumptions so that they can test whether they are met in their own work and know how to recognize the signs that they are violated when reading others' work. Although the basic regression model is linear, it is possible to model nonlinear relationships (as we discuss in Chapter 9) and important to check for cases that have a great effect on the slope of the regression line, referred to as outliers and influential observations (as we discuss in Chapter 11). Other technical terms that you may have seen in reading about regression models in the past, and that we will unpackage in Chapter 11, include the problems of heteroskedasticity and multicollinearity. Although these terms undoubtedly merely sound like foreign jargon now, we will spend considerable time discussing what these terms mean and how to test for and correct for them, such that they become familiar.

It is also helpful from the outset to recognize that this book will help you to understand a substantial fraction, but by no means all, of what you will later read (or potentially need to do) with regression models. Like all social scientists, you may need to take additional courses, engage in independent study, or seek out consultation when the research questions you pose require you to go beyond the content of this textbook. An important skill for a developing social scientist is to be able to distinguish what you know from what you need to know, and find out how to learn what you still need to know. We use the literature excerpts in this chapter, and the excerpts in future chapters, to begin to help you to distinguish between the two. One of our goals is to provide a solid foundation that adequately prepares you for such advanced study and help seeking. Chapter 12 provides a roadmap of advanced topics and ideas about how to locate relate courses and resources.

In the interest of space and to meet our objectives, we extract only certain details from each article in the extracts below. Reading the entire article is required for a full picture of the conceptual framework, method and measures, results and interpretations.

## 1.2: LITERATURE EXCERPT 1

Lyons, Christopher J. 2007. "Community (Dis)Organization and Racially Motivated Crime." *American Journal of Sociology*, 113(3): 815–63.

A paper by Christopher Lyons published in the *American Journal of Sociology* illustrates the innovative use of several existing data sources to examine the correlates of hate crimes. Lyons (2007, 816) summarizes his research questions this way: "Are racial hate crimes the product of socially disorganized communities low in economic and social capital? Or are racially motivated crimes more likely in communities with substantial resources to exclude outsiders?" Lyons creatively used multiple existing data sources to examine these questions. He took advantage of the Project on Human Development in Chicago Neighborhoods (PHDCN) which had gathered data that precisely measured concepts about social organization for the same time period in which he was able to obtain Chicago Police Department statistics on hate crimes for the same communities. The availability of these data provided a unique opportunity to go beyond other community measures that would be less close to the concept of social organization (e.g., measures of the poverty rate or mobility rate in the area from the US Census Bureau).

Bringing together these various data sources is an important innovation of the study. On the other hand, using existing data sources means the fit with the research questions is not as precise as it could be with newly gathered data (e.g., as Lyons discusses, the community areas might ideally be smaller in size than the 77 Chicago community areas, which average some 40,000 people, and the police districts and PHDCN data had to be aggregated to that level with some imprecision; pp. 829, 833–4). But, the creative combination of existing data sources allowed a novel set of research questions to be tested with moderate time and cost by a graduate student. The results raise some questions that might be best answered with in-depth methods targeted at the problem at hand. Yet, the existing data provide an important first look at the questions, with the results suggesting a number of next steps that could productively be pursued by any scholar, including a post-doc or young assistant professor building a body of work.

Lyons identifies three specific hypotheses that are motivated by theory and are testable with regression analyses. Importantly, the third hypothesis produces expectations different from the first two, increasing intellectual interest in the results. If hate crimes are similar to general crime, then Lyons (p. 818) expects that two traditional models would apply:

1. *Social disorganization theory* predicts more crimes in disadvantaged areas with low levels of social capital.
2. *Resource competition theories* specify that crimes are most likely when competition between racial groups increases, especially during economic downturns when resources are scarce.

The third model is unique to hate crimes, and differs from the others in its predictions:

3. The *defended community perspective* implies that interracial antagonism is most likely in economically and socially *organized* communities that are able to use these resources to exclude racial outsiders.

Lyons also posed a subquestion that is an example of a statistical interaction (a situation in which the relationship between the predictor and outcome variables differs for certain subgroups). In particular, Lyons expected that the defended community result will be particularly likely in a subset of communities: "Social cohesion and social control may be leveraged for hate crime particularly (or perhaps exclusively) in racially homogenous communities that are threatened by racial invasion, perhaps in the form of recent in-migration of racial outgroups." (p. 825).

Lyons separately coded antiblack and antiwhite hate crimes because of theoretically different expectations for each and because antiwhite hate crimes have been understudied. Literature Excerpt 1.1a provides the results for the regression analyses of antiblack hate crime (Table 6 from the article).

The table may seem overwhelming at first glance, but as you learn regression analyses you will find that it follows a common pattern for reporting regression results. The key predictor variables from the research questions and other variables that adjust for possible confounders are listed in the leftmost column. The outcome variable is indicated in the title ("antiblack hate crimes"). Several models are run, and presented in different columns, to allow us to see how the results change when different variables are included or excluded. In the final chapter of the book, we will return to this table to link all the pieces of the table to what we have learned (such as the term "unstandardized coefficients") or what is left to be learned, in our roadmap of advanced topics (such as the term "overdispersion"). For now, we will consider the major evidence related to Lyons' research questions based on the significance (indicated by asterisks) and sign (positive or negative) of the relationship of key predictor variables.

# Literature Excerpt 1.1a

Table 6. Negative Binomial Regressions: Community Characteristics and "Bonafide" Antiblack Hate Crimes, 1997–2002

| | Model 1 | Model 2 | Model 3 | Model 4 | Model 5 | Model 6 | Model 7 | Model 8 | Model 9 | Model 10 | Model 11 |
|---|---|---|---|---|---|---|---|---|---|---|---|
| Constant | -4.01* | -6.36** | -6.29** | -6.38** | -5.64** | -14.78*** | -11.15*** | -10.11*** | -11.10** | -10.18** | -12.67** |
| | (2.32) | (2.39) | (2.39) | (2.66) | (2.60) | (3.71) | (4.05) | (4.12) | (4.76) | (4.79) | (5.75) |
| Ln population 1990 | .46** | .67** | .67*** | .60** | .67*** | .88** | .67*** | .70** | .68*** | .75** | .78** |
| | (.22) | (.23) | (.23) | (.25) | (.26) | (.25) | (.26) | (.26) | (.26) | (.26) | (.24) |
| Spatial proximity | .67** | .31 | .30 | .25 | .26 | .24 | .20 | .22 | .19 | .21 | -.15 |
| | (.34) | (.31) | (.31) | (.30) | (.30) | (.32) | (.29) | (.29) | (.29) | (.29) | (.28) |
| Disadvantage | | -.65*** | -.60** | -.22 | -.05 | | | | .03 | .19 | .17 |
| | | (.18) | (.20) | (.34) | (.32) | | | | (.37) | (.37) | (.35) |
| Stability | | .11 | .13 | .17 | .24 | | | | .03 | .11 | .16 |
| | | (.15) | (.15) | (.15) | (.16) | | | | (.18) | (.18) | (.17) |
| White unemployment | | | -.02 | | | | | | | | |
| | | | (.02) | | | | | | | | |
| %white 1990 | | | | .013 | | | .01* | | .013 | | .03 |
| | | | | (.01) | | | (.006) | | (.01) | | (.06) |
| %black 1990 | | | | | -.018** | | | -.01** | | -.018** | |
| | | | | | (.01) | | | (.006) | | (.01) | |
| %Hispanic 1990 | | | | .01 | -.007 | | .01 | -.002 | -.012 | -.005 | .02** |
| | | | | (.01) | (.01) | | (.01) | (.01) | (.01) | (.01) | (.01) |
| Informal social control | | | | | | 1.50** | 1.34** | 1.35** | 1.31** | 1.29** | 1.48** |
| | | | | | | (.60) | (.59) | (.58) | (.67) | (.66) | (.90) |
| Social cohesion | | | | | | -.22 | -.23 | -.26 | -.24 | -.26 | -.43 |
| | | | | | | (.47) | (.56) | (.53) | (.59) | (.57) | (.50) |

| | | | | | | | | | | | %change in black population, 1990–2000 |
|---|---|---|---|---|---|---|---|---|---|---|---|
| **%change in black population, 1990–2000** | | | | | | | | | | | −.52 (.44) |
| **Informal social control × %white** | | | | | | | | | | | −.002 (.02) |
| **Informal social control × %change in black population** | | | | | | | | | | | .14 (.13) |
| **%white 1990 × %change in black population** | | | | | | | | | | | −.02* (.01) |
| **Informal social control × %white × %change in black** | | | | | | | | | | | .005** (.002) |
| **Overdispersion** | 1.23 (31.00) | .87 (.25) | .87 (.25) | .80 (.24) | .76 (.23) | .84 (.25) | .71 (.23) | .70 (.22) | .71 (.23) | .67 (.22) | .31 (−.16) |
| **Log likelihood** | 146.17 | 139.10 | −138.87 | 137.73 | −136.46 | −139.27 | −136.19 | −135.21 | −36.17 | −134.90 | −126.23 |

Note.—$N = 77$ Chicago community areas; unstandardized coefficients; SEs are in parentheses.
* $P < .10$.
** $P < .05$.
*** $P < .001$.

---

▨ **Literature Excerpt 1.1a—continued**

Table 7.   Predicted Antiblack Hate Crime, Chicago Communities, 1997–2002

| | White Communities (85% White) | | Nonwhite Communities (10% White) | |
|---|---|---|---|---|
| | Threat[a] | No Threat[b] | Threat[a] | No Threat[b] |
| High informal social control | 34.7 | 3.9 | 3.1 | 2.1 |
| Low informal social control | .1 | .9 | .1 | .3 |

*Note.*—See table 6, model 11. Except for %white, informal social control, and change in %black, all variables held at mean values. High informal social control: 1 SD above mean; low informal social control: 1 SD below mean.
[a] Black in-migration 15%.
[b] Black in-migration 0%.

**Source:** Lyons, Christopher J. (2007). "Community (Dis)Organization and Racially Motivated Crime." *American Journal of Sociology*, 113(3): 815–63.

---

In Lyons' Table 6, a red circle encloses the coefficient estimate of a three-way interaction between social control, percentage white in 1990, and percentage change in black between 1990 and 2000 which is relevant to the defended communities hypothesis. The asterisks indicate that the interaction is significant. It is difficult to interpret this interaction based only on the results in Lyons' Table 6. His Table 7 (also reproduced in Literature Excerpt 1.1a) provides additional results from the model that aid in interpretation of the interaction results (we will detail how to calculate such results in Part 2).

Each cell value in Lyons' Table 7 is the number of antiblack hate crimes that the model predicts for a cell with the listed row and column characteristics. The results show that when white communities are under threat (have experienced black in-migration of 15 percent or higher) and are high in social control, they have substantially more antiblack hate crimes (predicted number of 34.7, circled in red) than any other communities. This is consistent with the author's expectation about which communities would be most likely to conform to the defended communities perspective: racially homogeneous areas with recent in-migration of racial outgroups. Although less extreme, it is also the case that other communities conform to the defended communities perspective: within each column, communities with high social control have more predicted antiblack hate crimes than those low in social control (i.e., two to four versus fewer than one antiblack hate crime predicted by the model).

Literature Excerpt 1.1b shows the regression results for antiwhite crime, Table 8 in Lyons' article.

A key finding here is that the basic results differ substantially from those seen for antiblack crime (compare results enclosed by black circles in Literature Excerpts 1.1a

# Literature Excerpt 1.1b

Table 8. Negative Binomial Regressions: Community Characteristics and "Bonafide" Antiwhite Hate Crime, 1997–2002

| | All Chicago[a] | | Excluding Outliers[b] | | | | | | | | |
|---|---|---|---|---|---|---|---|---|---|---|---|
| | Model 1 | Model 2 | Model 2 | Model 3 | Model 4 | Model 5 | Model 6 | Model 7 | Model 8 | Model 9 | Model 10 |
| Constant | −6.12** | −5.34** | −5.82** | −5.85 | −6.24** | −5.97** | −2.31 | −3.95 | −3.63 | −7.83** | −5.59* |
| | (1.84) | (1.87) | (1.83) | (1.84) | (1.89) | (1.97) | (2.68) | (2.96) | (3.17) | (3.77) | (3.13) |
| Ln population 1990 | .64*** | .56** | .59** | .59 | .64** | .63** | .58** | .68*** | .66*** | .69** | .69** |
| | (.18) | (.18) | (.17) | (.17) | (.19) | (.18) | (.18) | (.19) | (.19) | (.20) | (.20) |
| Spatial proximity | .69** | .73** | .57** | .58 | .54** | .53** | .59** | .46** | .49** | .46** | .45* |
| | (.22) | (.24) | (.24) | (.24) | (.24) | (.24) | (.21) | (.23) | (.24) | (.24) | (.24) |
| Economic disadvantage | | .13 | .30** | .29 | .26 | .20 | | | | .33 | .28 |
| | | (.12) | (.12) | (.14) | (.21) | (.21) | | | | (.25) | (.25) |
| Residential stability | | −.15 | −.26** | −.25 | −.26** | −.27** | | | | −.28** | −.30** |
| | | (.12) | (.12) | (.12) | (.13) | (.13) | | | | (.13) | (.14) |
| Black unemployment 1910 | | | | .01 | | | | | | | |
| | | | | (.01) | | | | | | | |
| % black 1990 | | | | | .001 | | | | .002 | .004 | |
| | | | | | (.01) | | | | (.00) | (.01) | |
| % white 1990 | | | | | | −.003 | | −.005 | | | −.01 |
| | | | | | | (.005) | | (.01) | | | (.01) |
| % Hispanic 1990 | | | | | −.004 | −.005 | | −.007 | −.005 | .000 | −.004 |
| | | | | | (.01) | (.01) | | (.01) | (.01) | (.01) | (.01) |
| Informal social control | | | | | | | −1.10* | −1.11* | −1.10* | −.54 | −.56 |
| | | | | | | | (.58) | (.58) | (.58) | (.61) | (.61) |
| Social cohesion | | | | | | | .18 | .44 | .30 | .77 | .85 |
| | | | | | | | (.40) | (.50) | (.30) | (.51) | (.52) |
| Overdispersion | .36 | .34 | .24 | .24 | .23 | .24 | .24 | .21 | .21 | .18 | .18 |
| | (.16) | (.16) | (.14) | (.14) | (.14) | (.14) | (.15) | (.14) | (.15) | (.13) | (.13) |
| Log likelihood | −130.39 | −129.06 | −116.68 | −116.67 | −116.31 | −116.21 | −118.32 | −117.63 | −117.91 | −115.13 | −114.80 |

*Note.*—Unstandardized coefficients, SEs are in parentheses.
[a] $N = 77$ Chicago community areas.
[b] $N = 74$ Chicago community areas.
* $P < .10$.
** $P < .05$.
*** $P < .001$.

**Source:** Lyons, Christopher J. (2007). "Community (Dis)Organization and Racially Motivated Crime." *American Journal of Sociology*, 113(3): 815–63.

and 1.1b). For antiwhite crime, the results are consistent with social disorganization theories: in Model 2, economic disadvantage associates with more antiwhite hate crimes (asterisks and positive sign) and residential stability associates with less antiwhite hate crimes (asterisks and negative sign; see again the black circle in Literature Excerpt 1.1b). In contrast, for antiblack crime, economic disadvantage is associated with less antiblack hate crime (asterisks and negative sign) and residential stability is not associated with antiblack hate crime (no asterisks; see again the black circle in Literature Excerpt 1.1a).

## 1.3: LITERATURE EXCERPT 2

Vaisey, Stephen. 2007. "Structure, Culture, and Community: The Search for Belonging in 50 Urban Communes." *American Sociological Review*, 72: 851–73.

Stephen Vaisey provides an example that applies regression techniques to a collection of ethnographies, illustrating one approach to combining quantitative and qualitative methods. In particular, Vaisey reanalyzed dozens of ethnographies of urban communes using regression analyses to test several research questions about how a sense of collective belonging develops in groups; his particular interest was in how structural and cultural mechanisms influence the development of collective belonging.

The data that Vaisey analyzes were originally gathered in the mid-1970s. As discussed further in Chapter 2, these data have been made publicly available, allowing researchers such as Vaisey to reanalyze them. In the original study, communes were drawn systematically from a known sampling frame (list of all members of the population) making the results more generalizable than they would be if the ethnographies were conducted with a convenience sample of communes. Specifically, fieldworkers developed lists of communes in six US metropolitan areas (Atlanta, Boston, Houston, Los Angeles, New York, and the Twin Cities). Ten communes were selected from the list in each of the six areas, with a goal of a good representation of communes associated with key constructs (ideology, size, longevity). To be listed, a group had to have at least five members and, so as to exclude monasteries and convents, the commune had to have at least one member of each sex (or resident children). Participant observers gathered data from each commune. They completed standardized forms based on their interactions and experiences in the commune, and they asked commune members to fill out surveys about their attitudes, beliefs and relationships.

Using existing data has disadvantages as well as advantages for Vaisey's research questions. For example, Vaisey notes that communes are not representative of all groups that develop collective belonging, his broader conceptual interest. The studied communes' members were "whiter, younger, and more educated" than the general population (p. 855). On the other hand, communes are "bounded" groups, making it easier to examine their social interactions than would be the case with larger groups with less clearly defined boundaries. And, because of the sampling frame, the included communes are more diverse than the stereotypical commune (e.g., the majority of studied communes are religious, political, or countercultural, although some more simply serve alternative family or cooperative living functions) and have considerable variation on the constructs of interest.

Vaisey drew on both reports from the participant observers and the ratings from the members to construct variables representing the concepts of interest to him. The outcome, referred to as *gemeinschaft*, captures a "sense of we-feeling, a sense of collective self, or the feeling of natural belonging" (p. 852). It is operationalized using a scale comprised of six measures, including the ethnographer's overall rating of the "feeling of community" in the commune and members' reports of whether they see the other members as their true family, as people who care about others in general and the respondent in particular, and whether they expect to be in the commune 10 years in the future and would leave if offered $10,000 to do so (reversed).

Other variables capture the major processes that Vaisey expects explain the outcome, including structural processes (properties of the organization) and substantive processes (cultural meaning of the group). The structural processes suggest that belongingness is a by-product of frequent interaction among group members, similarity (homophily) of group members, required investments from group members, and strict leadership (authority). The substantive theories suggest that structural factors are not sufficient to produce belongingness; rather, shared moral culture allows groups to "withstand centrifugal forces" (p. 854). The moral culture was assessed by the ethnographer's rating of the consensus among members on ideology, values, and beliefs and the importance of ideology, values, and beliefs in the commune's life as well as members' responses to questions about having clear beliefs of a "right" and "wrong" way to live. Vaisey also introduces variables to the model to adjust for confounders, which might be correlated with the predictors and outcome (e.g., group size—number of members, age—number of years in existence, and origin—whether arose from a prior group).

Vaisey uses bivariate correlations and regression analyses (as well as a recently developed technique called "fuzzy set analysis" beyond the scope of this book). He sees the unique advantage of regression analyses as allowing him to identify the "proximate mechanism" leading to belongingness (*gemeinschaft*). Indeed, nearly all variables are correlated with belongingness in the expected direction (see Literature Excerpt 1.2a, which has asterisks on all of the theoretically motivated variables except the three measures of homogeneity and the measure of density—persons per room—circled in red). But, in the regression analysis shown in Literature Excerpt 1.2b, when all variables are entered together in the model, only three variables remain significantly associated with the outcome: authority, investment, and moral order having asterisks. Importantly, the sign for authority reverses when all variables are included together in the regression model. It was positive in the bivariate correlations and now is negative in the regression model (see black circles in Literature excerpts 1.2a and 1.2b; we will have more to say about why such sign reversals may be observed, and how to interpret them, in Chapter 10).

---

■ **Literature Excerpt 1.2a**

Table 1.   Correlations between *Gemeinschaft* Scale and All Predictor Variables

| | ρ | s.e. | | ρ | s.e. |
|---|---|---|---|---|---|
| SPATIOTEMPORAL | *.355* | *.155*** | MORAL ORDER | *.713* | *.094**** |
| Meetings | .352 | .122** | Ideological unity | .707 | .084*** |
| Eating together | .456 | .126*** | Importance of ideology | .693 | .069*** |
| Density | .124 | (133) | Role certainty | .580 | .116*** |
| | | | "How to live" | .564 | .119*** |
| AUTHORITY | *.379* | (.124**) | | | |
| Authoritarian governance | .337 | .125** | TYPE OF GROUP | n/a | |
| Extent of authority | .364 | .145* | Eastern religious | .330 | .202 |
| Number of rules | .469 | .112*** | Christian | .688 | .153*** |
| | | | Political | .053 | .208 |
| INVESTMENT | *.543* | *.122*** | Counter cultural (hippie) | −.362 | .155* |
| Time spent | .301 | .136* | Alternative family | −.169 | .225 |
| Communism | .630 | .081*** | Household | −.489 | .157** |
| Bar to entry | .441 | .172* | Personal growth | −.256 | .117* |
| Assigned chores | .714 | .111*** | | | |
| | | | CONTROLS | n/a | |
| HOMOGENEITY | n/a | | Size of group | −.033 | .134 |
| Age | .044 | .141 | Age of group | .123 | .119 |
| Education | .178 | .139 | Evolved from previous | .538 | .152*** |
| Class | −.193 | .139 | | | |

*Notes:* Italicized statistics are for scale measurements. Categorical variables use polychoric correlations. Other variables use Pearson's *r*.
* *p* < .05;** *p* < .01;*** *p* < .001 (two-tailed).

**Source:** Vaisey, Stephen. 2007. "Structure, Culture, and Community: The Search for Belonging in 50 Urban Communes." *American Sociological Review*, 72: 851–73.

▨ **Literature Excerpt 1.2b**

Table 2. OLS Regression of *Gemeinschaft* Scale on Independent Variables

| Mechanisms | b | β | t |
|---|---|---|---|
| Spatiotemporal Interaction | −.281 | −.192 | −1.300 |
| Authority | −.665 | −.595 | −3.170** |
| Investment | .485 | .374 | 2.460** |
| Strength of Moral Order | 1.033 | .936 | 4.360*** |
| Group Types | | | |
| Eastern religious | (reference) | | |
| Christian | .403 | .153 | 1.470 |
| Political | .315 | .106 | .780 |
| Counter cultural | −.095 | −.034 | −.220 |
| Alternative family | .655 | .203 | 1.410 |
| Household | −.050 | −.018 | −.110 |
| Personal growth | .250 | .061 | .490 |
| Controls | | | |
| Size of group | −.017 | −.155 | −1.310 |
| Age of group | .118 | .191 | 2.010 |
| Evolved from previous | .457 | .220 | 1.790 |
| Constant | −.464 | | −1.630 |
| N | | | 50 |
| $R^2$ | | | .754 |
| Adjusted $R^2$ | | | .665 |

$**p < .01$; $***p < .001$ (two-tailed).

**Source:** Vaisey, Stephen. 2007. "Structure, Culture, and Community: The Search for Belonging in 50 Urban Communes." *American Sociological Review*, 72: 851–73.

Vaisey interprets the results as indicating that, with the exception of investment, the structural constructs are not as important as the theory suggests; and, "in fact, there is evidence here that the direct effects of authority can be alienating." The latter interpretation reflects the negative sign of authority, with the other variables controlled. In contrast, the author tells us in a footnote that the association between moral order and belongingness increases with the other variables controlled (again a result we will discuss further in Chapter 10). The author concludes that, in contrast to prior theorizing, "*reciprocity and trustworthiness do* not *simply 'arise' from social networks, except, perhaps, as that interaction is either animated by or productive of shared moral understandings*" (p. 866, italics in original). In other words, structure alone does not produce collective belonging; moral order is essential.

## 1.4: LITERATURE EXCERPT 3

Li-Grining, Christine P. 2007. "Effortful Control Among Low-Income Preschoolers in Three Cities: Stability, Change, and Individual Differences." *Developmental Psychology*, 43(1): 208–21.

As we discuss in Chapter 2, numerous large-scale data sets are now available that follow individuals over time. These data sets are particularly well-suited to regression analyses because they are drawn to represent a population (often with a stratified, clustered design; we will consider some of the implications of these designs in Chapter 11). They often have large enough sample sizes to allow the estimation of complicated models with numerous conceptually important variables and control variables. And, more recent data sets are frequently designed to oversample subgroups of interest to social scientists, such as racial-ethnic groups or single-parent households.

Christine Li-Grining uses one of these data sets—Welfare, Children, and Families: A Three-City Study—to examine how preschoolers develop the ability to control impulses. This capacity, referred to by Li-Grining as *effortful control*, is an important predictor of children's later academic achievement (Duncan et al. 2007). Thus, understanding how this skill varies among low-income children, and what predicts better effortful control, is important.

Li-Grining's central research goals are:

- describing typical patterns of stability and change over time in effortful control among low-income preschoolers;
- examining to what degree characteristics of children and stressors associated with poverty explain variation in effortful control among low-income children;
- testing whether results differ by race-ethnicity and gender.

The fact that the Three-City Study investigators took well-established methods typically used to measure effortful control in small psychology labs and adapted them for use in a large-scale study allowed Li-Grining to address these goals. Doing so was also facilitated by the large size of the study overall and within the African American and Latino subgroups. Of course, the data set also has limitations. For example, it was drawn to represent three cities (Boston, Chicago, and San Antonio) rather than the nation. And, it focused on low-income children, preventing comparisons with higher income children. However, the data provide an opportunity for her to begin to address her research questions and to contribute to the growing literature on effortful control.

Li-Grining considers two separate but interrelated aspects of effortful control: the ability for children to inhibit impulses (delayed gratification) and the ability of children to focus attention (executive control). Both are assessed with rigorous procedures, such as coding whether children peak at presents while they are being wrapped and whether they can stay in the lines while tracing a turtle's versus a rabbit's path home.

Literature Excerpt 1.3 shows the results of regression models examining her second goal: predicting each outcome based on child characteristics and poverty-related risk factors. Among the child factors, race was not significantly related to either type of effortful control (no asterisks), but gender was significantly related to delayed gratification (see asterisks in red circle). The negative sign indicates that boys score lower than girls on this outcome, consistent with other literature that shows boys' poorer performance on behavioral than on cognitive control. Among the poverty risk factors, low birth weight is significantly related to less effortful control of both types (negative

---

### ■ Literature Excerpt 1.3

Table 3.   Regressions: Child Characteristics, Risk Factors, and Child–Mother Interaction Predicting Delayed Gratification and Executive Control at Wave 2

| Variable | Delayed Gratification | | | | | Executive Control | | | | |
|---|---|---|---|---|---|---|---|---|---|---|
| | $R^2$ | $\Delta R^2$ | B | SE B | β | $R^2$ | $\Delta R^2$ | B | SE B | β |
| Child characteristics | .19 | | | | | .42 | | | | |
| Age | | | .35*** | .05 | .42 | | | .55*** | .05 | .62 |
| Gender | | | −.27** | .08 | −.19 | | | −.04 | .08 | −.03 |
| Negative emotionality | | | .00 | .01 | .00 | | | −.01 | .01 | −.03 |
| Race | | | | | | | | | | |
| European American (omitted) | | | | | | | | | | |
| African American | | | .08 | .19 | .06 | | | .34 | .25 | .22 |
| Latino | | | .00 | .18 | .00 | | | .43† | .25 | .28 |
| Other | | | .27 | .24 | .07 | | | .35 | .29 | .09 |
| Risk factors and child–mother interaction | | 0.04** | | | | | 0.07*** | | | |
| Low birth weight | | | −.27* | 0.13 | −.09 | | | −.48** | .17 | −.16 |
| Psychosocial risk | | | .02 | 0.04 | .03 | | | .04 | .04 | .05 |
| Sociodemographic risk | | | −.01 | 0.03 | −.01 | | | −.08** | .03 | −.15 |
| Residential risk | | | −.01 | 0.05 | −.01 | | | −.16* | .07 | −.16 |
| Dyadic connectedness | | | .21** | 0.06 | .19 | | | .07 | .06 | .06 |

*$p < .05$.   **$p < .01$.   ***$p < .001$.   †$p < .10$.

**Source:** Li-Grining, Christine P. 2007. "Effortful Control Among Low-Income Preschoolers in Three Cities: Stability, Change, and Individual Differences." *Developmental Psychology*, 43(1): 208–21.

sign and asterisks circled in black). Tests for interactions reported in the text indicate that gender also moderates the association for delayed gratification: low birth weight is associated with delayed gratification only for boys. Several other poverty-related risks are related to only one type of effortful control (asterisks only in one, rather than both, columns).

## 1.5: SUMMARY

SUMMARY

1

The three literature excerpts in this chapter cover a diverse range of applications of regression analysis to examine interesting research questions. All represent research completed when the authors were graduate students with existing data sources.

Lyons' research on hate crimes shows that a seemingly positive aspect of community—social capital—can work to achieve ingroup goals at the cost of outgroup members. Socially organized areas are observed to have more antiblack hate crime. This is especially evident in primarily white communities that have experienced high levels of black in-migration. A different process applies for antiwhite hate crimes, which correlate with characteristics of neighborhoods as emphasized by general theories of crime, especially residential instability.

Vaisey's research shows that all communes are not the same. Frequent interactions among members and investments in the commune are not sufficient to produce a sense of collective belonging; rather, interactions and investments need to occur in the context of shared values and beliefs to produce belonging. Vaisey also unexpectedly finds that once these values, and aspects of structure, are adjusted, strict authority can alienate members.

Li-Grining's research contributes to the growing literature on young children's ability to control impulses, an important factor in later school success. She found that low-income boys are generally less able than low-income girls to delay behavioral impulses and this is particularly true for boys born with low birthweight. Her large data set allowed her better to test for differences by race than prior studies, and she found evidence of similarity in impulse control among low-income children from different racial backgrounds.

In future chapters we will dig deeper to understand questions these excerpts may have raised for you. For example: how do we interpret the size (not just the significance and sign) of coefficients? Why do coefficients change when other variables are controlled? How exactly does the regression model test for an interaction? We hope that these excerpts spark enthusiasm for answering these questions and learning how to implement regression models in your own work.

## KEY TERMS

Dependent Variable (or Outcome Variable)

Independent Variable (Predictor Variable or Explanatory Variable)

Mechanisms

Observational Data

Regression Analysis

## CHAPTER EXERCISES

**1.1.** Locate a journal article using regression analyses, preferably on a topic of interest to you. You can use an article you already have or draw something from the syllabus of one of your substantive classes or search a bibliographic database. Read (or reread) the article, paying particular attention to the research questions or hypothesis statements and how these are examined in the regression models. Examine at least one table of regression results in detail, and pull out some of the features that were discussed in this chapter (e.g., significance and sign of coefficients relevant to the study's research questions and hypotheses). Write a short paragraph discussing what is easiest and most difficult to understand about the article's regression analysis. Keep a copy of this article to revisit in the chapter exercise in the final chapter of the book.

**1.2.** Answer the following questions to help your instructor get to know you better and help you think about your goals for the course. Some questions may be difficult to answer, especially if your research interests and empirical approach are still developing, so answer as best you can. In addition to helping your instructor tailor lectures and interactions, this exercise can also help you to think about how to get the most out of the course.

(a) Have you taken previous statistics courses? (If yes, what/when? How well do you feel you mastered the material?)

(b) Have you worked, or are you currently working, as a research assistant on any research projects (qualitative or quantitative)?

(c) Have you collected your own data or analyzed existing data for a prior paper, including for an undergraduate senior thesis or a graduate master's thesis (qualitative or quantitative)?

(d) What do you hope to get out of the current class?

(e) Do you see yourself more as a qualitative researcher, a quantitative researcher, both, neither, or are you unsure? Elaborate as desired.

(f) What is the substantive area of your research (or what major topics do you think you would like to study as a graduate student)?

(g) Provide an example of a regression-type relationship from your area of research; that is, list an "outcome" (dependent variable) with one or more "predictors" (independent variables).

**COURSE EXERCISES**

**1**

## COURSE EXERCISE

Write three research questions in your area of interest. Be sure to identify clearly the outcome (dependent variable) and one or more predictors of interest (independent variables) for each question. Ideally, your outcome could be measured continuously for at least one research question so it will be appropriate for the techniques learned in Part 2 of the book.

If you are able to write your research question ("Do earnings differ by gender?") as a hypothesis statement (e.g., "Women earn less than men.") you should do so, but if prior research and theory do not offer enough insights for a directional hypothesis statement, a nondirectional research question is fine at this stage.

If your interests are still developing, you may want to get ideas by scanning a number of articles, chapters or books from one of your substantive classes, perusing the table of contents of recent top journals in your field, or by conducting a few bibliographic literature searches with key terms of broad interest to you.

The research questions should be of interest to you and something that you would like to examine as you move forward with the course. They could be already well studied in your field (i.e., you do not need to identify a dissertation-type question that makes a novel contribution to the literature). You may modify and refine them as you progress, especially so that you can apply the techniques learned in future chapters (e.g., modeling different types of variable, testing for mechanisms, etc.).

*Chapter 2*

# PLANNING A QUANTITATIVE RESEARCH PROJECT WITH EXISTING DATA

## CHAPTER 2: PLANNING A QUANTITATIVE RESEARCH PROJECT WITH EXISTING DATA

A primary goal of this chapter is to help you to plan for a quantitative research project. Whether you collect your own data, or use existing data, careful planning will produce a more efficient project. You will be able to complete it more quickly and to interpret the results more easily than you could otherwise. At the planning stage you should think through questions such as: what variables are needed to operationalize my concepts? Can I predict the direction of association between each predictor variable and the outcome? Do I anticipate that any predictor variables' associations with the outcome may change when confounding variables are included in the models? Do I anticipate that any predictor variables' associations with the outcome will differ for various subgroups? Doing so can help you to choose from among existing data sets (selecting the one that has the right variables and sample coverage). Forward thinking should save you time in the long run, because you will be less likely to have to backtrack to create additional variables or to rerun earlier analyses. Forward thinking will also help you to avoid being unable to examine particular research questions because subsamples are too small or constructs were not measured.

We focus on existing data in this book. We expose you to numerous data sets that are readily available on the Web. Knowing about these data sets should make it easier for you to engage in using regression analyses to examine research questions of interest to you. This chapter also contributes to our goal of leveling the playing field by offering information about where and how to look for such existing data, since not all students will have access to this information through mentors and peers.

Generally, existing data sources also have three distinct advantages over collecting new data:

1. They often provide designs and sample sizes that support fuller regression models and hypothesis tests.
2. They allow research questions to be addressed when resources are limited.
3. Their use provides a greater return on public investment in large-scale data collection.

There are instances where regression analysis might be used on non-random samples (Berk 2004). But, in order to support statistical tests based on regression analyses—and thus test hypotheses or answer research questions—a sample of adequate size (e.g., often 100 or more cases) should be drawn randomly from a known population. More complicated models, with more variables and interactions among variables, require even larger sample sizes, overall and within subgroups. Yet, gathering large-scale data sets drawn systematically from populations is expensive. If an existing data set has the relevant measures to answer a social scientist's research question, then the research can be conducted more quickly and with less cost than if the researcher attempted to gather new data. Turning first to existing data also best utilizes scarce resources for funding new studies. And, first testing a research question on existing data can provide the preliminary answers needed to modify the research questions and to demonstrate competence for seeking funding for new data collection.

That said, even though we focus on existing data in this chapter, the general strategies about planning for the analysis at the data-gathering stage also apply generally for your own data collection (e.g., being sure you ask all the right questions and oversample subgroups when needed). You should, of course, consult with mentors and, where needed, take advanced courses to prepare to implement your own data collection. We assume that you have taken, or will take, a basic research methods class, which covers a range of data collection methodologies.

## 2.1: SOURCES OF EXISTING DATA

Existing data come from a number of sources, including:

- multi-data set **archives**;
- single-data set archives;
- individual researchers.

Table 2.1 provides links to the resources we discuss below in each of these categories. Before turning to that discussion, we first want to emphasize the importance of checking with your local Institutional Review Board (IRB) regarding necessary IRB approval if you plan to use one of these sources for research purposes (versus for educational purposes only, such as course exercises). It may seem that IRB approval is not needed since some existing data sets are prepared for **public release**, for example with careful removal of all identifying information, and can be downloaded from the Web. Indeed, some local IRBs have determined that publicly available data sets accessed from preapproved data archives do not involve human subjects and thus do not require IRB review. But, current human subjects' policy requires researchers to verify this with their local IRB, rather than making such determinations independently (see Box 2.1 for an example). Some existing data sets also require special data use or data security agreements, for

> **■ Box 2.1**
>
> The University of Chicago's Social and Behavioral Sciences Institutional Review Board provides a nice summary of its policies and procedures for four types of existing data sets: *http://humansubjects. uchicago.edu/sbsirb/publicpolicy.html.*

▨ **Table 2.1: Examples of Data Archives**

**General Multi-data Set Archives**

| | |
|---|---|
| Inter-University Consortium for Political and Social Research (ICPSR) | http://www.icpsr.org/ |
| Henry A. Murray Research Archive at Harvard University | http://www.murray.harvard.edu/ |
| Sociometrics | http://www.socio.com/ |

**Government Multi-data Set Archives**

| | |
|---|---|
| US Census Bureau | |
|   American FactFinder | http://factfinder.census.gov/ |
|   Census Data Products | http://www.census.gov/mp/www/cat/ |
|   Integrated Public Use Microdata Series (IPUMS) | http://www.ipums.umn.edu/ |
| NCHS Public Use Data | http://www.cdc.gov/nchs/datawh/ftpserv/ftpdata/ftpdata.htm |
| NCES Surveys and Programs | http://nces.ed.gov/pubsearch/surveylist.asp |
| NCHS Research Data Center | http://www.cdc.gov/nchs/r&d/rdc.htm |
| Census Research Data Centers | http://www.ces.census.gov/ |

**Single-Data Set Archives**

| | |
|---|---|
| National Survey of Families and Households | http://www.ssc.wisc.edu/nsfh/ |
| Urban Communes Data Set | http://sociology.rutgers.edu/Ucds/ucds.htm |
| Three City Study | http://www.jhu.edu/welfare |

Notes: NCHS = National Center for Health Statistics; NCES = National Center for Education Statistics.

example that spell out special precautions for limiting access to the data. If you want to use these data for research, you should build in time for completing these agreements, which may require signatures from your adviser and administrators at your university, and for review by your local IRB. Some **restricted-use data sets** cannot be sent to you, but can be analyzed in special secure locations. These data sets require the greatest time lags for approvals, including local IRB review.

## 2.1.1: Multi-Data Set Archives

One of the oldest and largest multi-data set archives is the Inter-University Consortium for Political and Social Research (**ICPSR**; see Table 2.1). ICPSR began in 1962 to archive data from computer-based studies in political science. In the mid-1970s, the name "social" was added to the title to reflect the broader set of disciplines that were archiving data (Vavra 2002). The founders recognized the need to store centrally the growing amounts of quantitative data being collected by political scientists across the country. This allowed other scientists not only to replicate the findings of the original scholar, but also to tap the data for additional purposes, beyond those within the interests and time limits of the original scholar (Vavra 2002; see Box 2.2). The archive is housed at the Institute for Social Research at the University of Michigan and over 600 universities from across the world are institutional members, giving their faculty and students access to tens of thousands of data sets.

Data sharing is common today in part because funders often require it. For example, the National Institutes of Health (NIH) now requires that all proposals requesting $500,000 or more in direct costs in any single year must include a plan for sharing the data (with appropriate removal of identifying information to protect confidentiality). In its policy, NIH notes the intention to give broader access to the data, once the original researchers have had a chance to pursue their main research objectives: "initial investigators may benefit from first and continuing use but not from prolonged exclusive use" (NIH 2003). Thus, archives have become important to scholars, as investigators can turn to such archives to deposit their data and meet such requirements.

> ■ **Box 2.2**
>
> Today's new user of ICPSR is used to having a wealth of data at her fingertips over the Internet. The remarkable achievement underlying this ease of access to data sets spanning centuries is easy to overlook. Especially in recent decades, the speed with which hardware and software advanced made data sets vulnerable to being inaccessible, even if their files were stored. In other words, data sets stored in formats written for now obsolete operating systems or statistical packages are not directly readable on today's computers. Fortunately, the archives had the foresight to convert these data sets to more general formats, making them still usable to today's researchers (Vavra 2002).

ICPSR's coverage is the broadest among today's major data archives. At the time of this writing, ICPSR used 19 thematic categories covering broad topics including wars, economic behavior, leadership, geography, mass political behavior, health care, social institutions, and organizational behavior. The subtopic of family and gender, within social institutions, returns 123 results, ranging from one-time polls to multiyear surveys. A bibliography of publications using the archived data sets includes over 45,000 citations.

Two other data archives often used by social scientists are the Henry A. Murray Research Archive at Harvard University and the Sociometrics Archive (see again Table 2.1). The Murray Archive focuses on data sets that study human development across the lifespan from a range of disciplines. The Sociometrics archive focuses on nine topics: teen sexuality and pregnancy, the American family, social gerontology, disability, maternal drug abuse, HIV/AIDS, contextual influences on behavior, child well-being, and complementary and alternative medicine.

Federal government agencies also provide direct access to data that they gather. Census data, aggregated to geographic areas such as states, counties, and census tracts, are available freely from the Web or for purchase on DVD. Most students are familiar with the decennial census of population, but data are also available from the economic censuses, which survey business establishments every five years. Individual level data from the Decennial Census of Population are also available through the Integrated Public Use Microdata Series (IPUMS).[1] Similarly, the National Center for Health Statistics and the National Center for Education Statistics also make numerous data sets available, in public use and restricted formats.

The costs of accessing these data range from zero to several hundreds of dollars. Data in the ICPSR archive are free to faculty and students at ICPSR member institutions. In most cases, individual faculty and students can create an account using their institutional affiliation, and download data and documentation directly from the ICPSR web site.[2] Some of the other data resources are also freely available on the Web (e.g., Census American FactFinder, IPUMS); others have data set-specific use agreements and fees. Sometimes, multiple archives contain the same data set; at other times, a data set is available in only one archive, or is available in more detail in one archive. So, it is worth searching several archives when initially exploring data sets for a topic.

In addition to these resources, which allow a copy of the data to be analyzed at the researcher's place of work (sometimes with specific requirements to limit others' access to the data), the US Census Bureau and National Center for Health Statistics also make data available to researchers at several Research Data Centers across the country (the Web addresses listed in Table 2.1 include descriptions of the data sets available at the centers). Currently, there are Census Research Data Centers located in Berkeley and Los Angeles, CA; Chicago, IL, and Ann Arbor, MI; Boston, MA; Ithaca and New York, NY; Washington DC; and Durham, NC. The NCHS has one Research Data Center, located in Hyattsville, MD, but the NCHS data sets recently became available through the nine Census centers as well. At these sites researchers can access data that are not included in public releases. For example, researchers can identify individuals at small levels of geography and match the individuals to information about these contexts. In addition, firm-level data from the economic censuses can be linked across time to study the births and deaths of organizations. Although the time to get approval is longer for these restricted-use data than with publicly available data sets, and the fees for using the center can raise the cost of a project, these centers allow scholars to pursue unique and innovative projects.

### 2.1.2: Single-Data Set Archives

Some investigators archive their data individually. Sometimes these data sets are also available through multi-data archives. For example, the National Survey of Families and Households (**NSFH**), which we will use for examples throughout the book, maintains its own web site from which data and documentation can be freely downloaded. The first two waves of this data set are also available through the ICPSR and Sociometrics archives.

The data sets used in the literature excerpts in Chapter 1 also illustrate the range of access points for some data sets. The Project on Human Development in Chicago Neighborhoods, used in the first excerpt in Chapter 1, does not maintain its own public archive, but is available through ICPSR and the Murray Archive. The Urban Communes Data Set, which was used in the second excerpt highlighted in Chapter 1, is available freely on its own web site (once researchers establish a login and password) but is not available through any archives. The Three City Study, used in the third excerpt from Chapter 1, is available on its own web site as well as in the Sociometrics and ICPSR archives. Table 2.1 provides links to each of the individual data set web sites.

## 2.1.3: Individual Researchers

It is also sometimes possible to access data through proximity to or a relationship with an individual researcher. Obvious cases are graduate students analyzing data collected by their advisers (or by other members of their department or university). Such data may not be generally publicly available, but accessible through local access or personal relationships. One example of reanalysis of locally available existing data is Robert Sampson and John Laub's use of data on juvenile delinquents originally gathered by Sheldon and Eleanor Glueck. As the authors write in their book *Crime in the Making* (1995, 1):

> Eight years ago we stumbled across . . . dusty cartons of data in the basement of the Harvard Law School Library . . . These data, along with the Gluecks' eighteen-year follow-up of the 1,000 subjects . . . were given to the Harvard Law School Library in 1972 . . . The papers and other items were sorted and fully cataloged as part of the Glueck archive. The cartons of data were simply stored in the sub-basement of the library.

The data had been gathered decades earlier, and were stored in boxes, rather than electronically. Thus, the first step toward analyzing these data involved extensive recoding and entering of the data. These investments paid off. Applying modern data analytic techniques to the data provided new insights into the correlates of juvenile crime and the factors that lead to stability and change in criminal behavior into adulthood (for example, the book won the annual book award from the American Society of Criminology). The data are now in the Murray Archive. You may similarly be able to access data through local depositories or personal networks.

## 2.2: THINKING FORWARD

Whether selecting an existing data set or planning for new data collection, thinking forward will improve the chances that your data can address your research interests and will reduce time spent on correcting course. Again, we focus on planning a project with existing data, because the details of planning and executing new data collection are beyond the scope of this textbook and are often taught in research methods courses.

Some key questions that can help to guide your selection of an existing data set include:

1. How was the sample identified?
2. What was the response rate?
3. How can I operationalize my key constructs with these data (including mechanisms through which a key variable associates with the outcome)?
4. Does the data set measure important confounding variables?
5. Do any of my research questions suggest that an association of a key predictor with the outcome differs by subgroups? If so, are the sample sizes large enough for these subgroups in the data set?

When multiple data sources exist that might address your research interests, the strengths and limitations of each data set across these questions can help you to choose among them (for example, all else equal, a data set with a higher response rate and larger sample would be preferred).

In choosing among data sets, it is also useful to consider constraints on time, money, and other resources, such as:

- What is your deadline?
- How much time and money can you devote to obtaining and analyzing the data by that deadline?
- How experienced are you already with statistical programming?

Clearly, planning for a course project differs from planning a thesis or dissertation. And, achieving goals by a deadline is easier if it covers a period with few other time demands (a summer semester funded by a fellowship versus a fall semester juggling other classes and teaching or research assistantships). In general, researchers tend to underestimate the time needed to complete a project using existing data. The time frame seems shorter because the data have already been gathered. Yet, time needs to be built in for locating the data set, understanding its design, extracting and coding variables from the raw data, running descriptive analyses, estimating regression models, and interpreting the results. Throughout, time is also needed to adequately double-check and document your work. More time is needed for each of these tasks when you are new to quantitative research and whenever using a new data set. When you have other concurrent demands and a short time frame (such as when writing a course paper), you may want to choose a data set with a simpler design (e.g., **cross-sectional** versus **longitudinal**) and to simplify your research questions where possible (e.g., replicate or modestly extend an existing study). For projects with longer timelines and higher expectations (such as theses or dissertations), you will likely want to take advantage of data sets with more complex designs and to address more nuanced questions.

As with other aspects of your research, you should draw on as much information as possible to identify the best existing data set for your particular research interests. This might include conversations with mentors and peers, looking at what data sets are used in published books and articles in your area, and searching archives on your topic. A review article on the topic of interest, for example in the *Annual Reviews*, may be a helpful starting point. Using the *Web of Science* to search forward and backward from the citations of key articles will likely be useful as well.[3] You may find it helpful to move back and forth among these strategies (talk to colleagues, read articles, search online) as you identify possible data sets and home in on those most relevant to your questions.

Because the ICPSR archive covers the largest number of data sets and broadest range of subjects, we will use it as an example. We will illustrate how to search for a data set relevant to a topic that we will then use as an example in the remainder of the book.

## 2.3: EXAMPLE RESEARCH QUESTIONS

Because we are using a research question for pedagogical purposes, we chose a fairly straightforward topic with a good body of literature. The data set we chose allows for more complicated analyses, but for instructional purposes we will focus on a small set of straightforward variables taken from one time point of the longitudinal study.

The research question involves the distance adults live from their mothers. The topic is important because at the same time that there may be benefits from moves away from family (such as access to new labor markets or a particular job) there may also be costs, both for an adult-child who moves away (such as less access to help with childcare) and an adult-parent who is left behind (such as less access to help during old age; Mulder 2007). Although some aspects of emotional and instrumental support can occur at a distance, direct instrumental support, such as assistance with childcare or household chores, require physical proximity. Prior studies have confirmed that such practical assistance is more likely when adult children live close to parents (Cooney and Uhlenberg

1992) and are less mobile (Magdol and Bessel 2003). There is also empirical support for a relationship between living close to family and frequency of contact (Lawton, Silverstein, and Bengtson 1994), and more frequent contact has in turn been associated with less loneliness (Pinquart 2003). One set of geographers argue that:

> Progress in understanding the nature and effects of kin proximity is of paramount importance if we are to be well prepared for the planning and policy decision that will accompany the arrival of the baby boom generation into its elderly years (Rogerson, Weng, and Lin 1993).

Prior literature has examined a number of correlates of geographic proximity, suggesting that characteristics of both the adult child and their parent will be relevant. For example, studies have found that higher education is associated with long-distance moves in general (Clark and Withers 2007) and distance from parents in particular (Rogerson, Weng, and Lin 1993). Having siblings, particularly sisters, has been found to reduce an individual's time spent on care for aging parents (Wolf, Freedman, and Soldo 1997), and the presence of siblings and sisters has been found to relate to greater distance from parents, although not always in expected ways (Michielin and Mulder 2007). Some research also suggests that women and minorities live closer to parents (Roan and Raley 1996). We will focus specifically on the following questions as we learn about the basic regression model in Part 2:

- ▬ Do higher education and earnings predict greater distance?
- ▬ How do the number of brothers and sisters associate with distance?
- ▬ How does proximity to the mother vary by the respondent's gender and race-ethnicity?

We will also adjust for the age of the respondent and mother in our models.

This example will raise a number of issues that we will resolve in later chapters. For example, what is the best approach to examining an outcome that is highly skewed (addressed in Chapter 9)? How can we best capture the effect of count and categorical predictor variables (such as number of siblings), especially when we anticipate that associations are not linear (addressed in Chapters 7 and 9)? In earlier chapters, we will look at simpler models that associate the continuous outcome (ignoring skewness) with continuous predictors. In later chapters, we will introduce other approaches for modeling these outcomes and predictors as well as strategies for thinking about which models are most conceptually appropriate and how to test between them empirically. In the remainder of this chapter, we illustrate how to locate a data set to examine these questions. We will identify, extract, and prepare the relevant variables in Chapters 3 and 4.

## 2.4: EXAMPLE OF LOCATING STUDIES IN ICPSR

There are three main ways to identify data sets through the ICPSR archive:

(a) browsing the data holdings;
(b) searching the data holdings;
(c) searching the bibliography.

We will discuss each strategy in general and then demonstrate it with our research questions. Our research questions illustrate how useful it can be to use multiple strategies to search for data. In our case, the third strategy (searching the bibliography) works the best. For other topics, another strategy may work best. In most cases, using all three strategies will help you either to assure you that you have found the best data for your question (because it turns up through all three methods) or to help you to identify a data set (when two of the three come up empty).

### 2.4.1: Browsing the Data Holdings

Browsing the ICPSR archives may take some time, but will give you the broadest overview of holdings in your area of interest. Such browsing may be especially helpful to stimulate and refine your research ideas, since it will expose you to a wide array of data sets available in the ICPSR archives.

Display C.2.1 shows the result after selecting the thematic area of "Social Institutions and Behavior" (found from the main page in Data/Browse/Data By Subject) and then choosing "H. Family and Gender" and sorting by relevance. The results illustrate the broad array of studies available in the ICPSR archive, ranging from national surveys gathered by the National Center for Health Statistics to several studies gathered by individuals or groups of investigators, including a study from three cities, a study from Los Angeles, and a study from Chicago. The topics range from welfare to immigration to intimate partner violence.

The number at the left of each study title is the ICPSR Study Number. It is helpful to take note of these numbers as you browse and identify studies of potential interest. You can use these easily to locate a study again in the archive (as discussed below).

Beyond its general archive, ICPSR also collaborates with other groups on special topic archives, such as on aging, childcare, crime, demography, education, health and medicine, and psychiatric epidemiology (found under Data/Archives on the main page). For example, if your topic was on aging, you might go to the National Archive of Computerized Data on Aging and then use its thematic areas to narrow the studies for browsing.

## 2.4.2: Searching the Data Holdings

Searching may also be useful, depending on the topic. Display C.2.2 shows the Advanced Search screen which allows for searching across multiple fields. The search utility can be useful if you already know something about a data source (such as a portion of the title or the name of the scholar who led the data collection effort). If you have already identified a data set, the highlighted search screen is also an easy way to find its details using the ICPSR study number.

You can also use the search utility shown in Display C.2.2 to search across the database fields (including the study summary) for key terms related to your research interests. The thesaurus (found under Data/Browse from the main page) can help you to identify key terms relevant to your topic, which can also be searched in the "subject term" field. Browsing the thesaurus may be particularly helpful as you initially formulate a research question. For our research questions, several possible key terms were not listed in the thesaurus (such as distance, proximity, or miles) so we tried the broader term "adult children." Although a search for this "phrase" in the "summary" was not productive for our research questions, we used this key term in the bibliographic search (discussed below) which was more successful. For other research questions, key terms may be more productive (e.g., key terms such as "income" and "child abuse" return more results).

The National Archive of Computerized Data on Aging also provides access to a variable-level search utility, within the "Survey Documentation and Analysis" or SDA tool (this tool more broadly allows online analysis of certain data sets).[4] Display C.2.3 shows a search for "miles" within the data sets. Numerous results returned. Many identified variables were not relevant to our research questions because they measured driving distances, but one hit was specific to the distance between a child and parent in the National Longitudinal Study of Adolescent Health (Display C.2.4). But, clicking on the Study Description from the initial results page reveals that "The Add Health cohort has been followed into young adulthood with three in-home interviews, the most recent in 2001–2, when the sample was aged 18–26."[5] Thus, the data do not cover the full adult age group, and are too young to be relevant for our research questions. And, clicking on the variable shows that most respondents legitimately skipped this question (Display C.2.5).

## 2.4.3: Searching the Bibliography

Because large public use data sets are so widely available, and used by many researchers, they typically produce numerous publications. The ICPSR bibliography contains many publications based on archived data sets (found from the main page in

Data/Publications Based on our Data). For example, the data set we will use (the NSFH) has nearly 700 publications in the database.

The bibliography's Advanced Search utility (shown in Display C.2.6) provides another tool for identifying existing data sets relevant for your research questions. For our topic, searching the "subject terms" for "proximity" and "miles" produced no results and the terms "distance" and "adult child" each produced five or fewer results. Searching in the title for these terms produced one result for "miles" but several dozen each for "distance" and "proximity," and 74 results for "adult child." We used the titles of the results to exclude those clearly irrelevant to our research questions (keeping any that specifically mentioned proximity, as well as those more broadly on intergenerational contact, relationships, and coresidence). Across these results, several data sets occurred several times, including:

■ NSFH (40 times);
■ National Long Term Care Survey (5 times);
■ Americans Changing Lives (3 times);
■ Health and Retirement Study (3 times);
■ Panel Study of Income Dynamics (3 times).

The NSFH clearly came up the most and several results had titles that specifically dealt with parent-adult child proximity. This data set also looks promising for our research questions because its description page indicates a large sample (over 13,000) covering the full range of US adults, ages 19 and older, and because minorities were oversampled.

Of the other data sets, the study description pages indicate that:

■ The National Long Term Care Survey is more restrictive in its coverage (adults ages 65 and older with the disabled over-represented).
■ The Americans Changing Lives study covers the population ages 25 and over, and oversamples blacks, although the total sample size is smaller than the NSFH (less than 4,000).
■ The Health and Retirement Study is a large survey of over 22,000 individuals, but is limited to adults ages 50 and over.
■ The Panel Study of Income Dynamics had nearly 3,000 adults in its original sample and close to 7,000 in its follow-ups, with a unique design that follows original household members when they form a new household.

The titles of publications using these data sets show that only one, based on the Panel Study of Income Dynamics (PSID), specifically addressed distance between adults and their parents. Thus, the PSID seems the most promising alternative to the NSFH in terms of coverage of a broad age range of adults and ability to examine proximity.

The Panel Study of Income Dynamics maintains its own web site (see psidonline.isr.umich.edu) where variable-level searches can be conducted, allowing us to examine farther its fit for our research questions. A search for the word "miles" on the PSID page identified four variables on miles to father and mother (see Display C.2.7). Clicking on the first of these variables showed codebook details of the question and response structure, verifying its relevance (Display C.2.8). Additional searches on the PSID and NSFH web sites, suggest that although either could be used for our research questions, the NSFH will be simpler to start with and offer large sample sizes within race-ethnicity. However, the design of the PSID has unique possibility for our research questions, because distances between parents and adult children could be tracked from the time the young adult first leaves the parental household. Thinking over a career as a graduate student and young scholar, the NSFH might be used for a first project and the PSID the next project (e.g., masters and dissertation, dissertation and post-doc, etc.).

### 2.4.4: Putting It All Together

Again, for our research questions, the third strategy produced the best results. But, in other cases, the first or second might be most productive. Thus, it is useful to use multiple strategies to identify possible data sets. Especially for a project that will result in a thesis, dissertation, or published work, once you have identified a set of candidate data sets, it is helpful to complete a broader search for publications on your topic, especially those that have used the identified data set(s). The ICPSR bibliography is a good starting point (You can use the search utility to search by ICPSR study number, in our case #6041). You can use the Web of Science to move forward and backward to studies that are cited by or cite these references. And, you can use bibliographic databases from your library, such as JSTOR ("Journal Storage"), ERIC ("Educational Resources Information Center"), PsycINFO (the electronic version of the older "Psychological Abstracts"), and Sociological Abstracts, to search abstracts and/or full text for key terms for publications you may have missed (e.g., for our question, we searched for "NSFH" or "National Study of Families and Households" combined with "proximity," "distance," or "miles").

With existing data, it is possible that other students or scholars have already used the data set to examine questions similar to yours. Don't get immediately discouraged, though, if you find work on your topic. Be sure to read the publications, and determine if and how you can make a contribution. And, as we have noted, replicating and modestly extending prior studies is often a good starting point, especially for a course paper.

## 2.5: SUMMARY

In this chapter, we introduced various strategies you can use to locate existing data relevant to your research interests. We covered archives that encompass a single data set or multiple data sets (summarized in Table 2.1) and strategies for locating data through your professional network and local resources. We introduced one set of research questions that we will use as examples in Part 2, and illustrated how to locate data sets in the ICPSR to address these research questions. We emphasized the importance of using multiple strategies to locate relevant data, and for choosing among identified data sets.

## KEY TERMS

Archive

Cross-Sectional

ICPSR

Longitudinal

NSFH

Public Release (Data Set)

Restricted-Use (Data Set)

## REVIEW EXERCISE

Use the ICPSR to search for a data set that might address a specific hypothesis or research question (either a topic assigned by your instructor or voted on by your class, or a research question of interest to you). Follow each of the three strategies discussed in the chapter (browsing and searching the data holdings; searching the bibliography) to locate data sets. Write a short paragraph discussing what you take away from this process (were you surprised by the number of data sets that you located? Which approach did you find most useful?).

## CHAPTER EXERCISE

Locate the NSFH data in ICPSR. Read the study description. Search the bibliography for publications using this data set. Search at least one other bibliographic data base for publications using the NSFH. Write a short paragraph about what you found (for example, were you surprised by the number of publications? Would the NSFH be relevant to your research interests?).

## COURSE EXERCISE

Use the ICPSR to locate possible data sets for each of your hypotheses or research questions. Use the procedures discussed in the chapter to scan the documentation on key study features, including the sampling design, coverage of relevant subpopulations, measures of conceptual constructs, and historical timing of data collection. Use what you find to focus or refine your research interests. For example, are no data sets available for some research questions? Are several data sets available for others? When multiple data sets are available, what are the advantages and limitations of each (e.g., in terms of sampling design, coverage of relevant subpopulations, measures of conceptual constructs, recency of collection, etc.)? If possible, you can extend your search for relevant data by talking to mentors and peers, your instructor and classmates, and through general searches of the literature on your topic. Putting together the results from your search of the ICPSR archive with these additional sources will help you to identify the best data for your research interests (recognizing that for the course, you may want to choose a simple question or data set given the limited time frame and your newness to programming).

*Chapter 3*

# BASIC FEATURES OF STATISTICAL PACKAGES AND DATA DOCUMENTATION

## CHAPTER 3: BASIC FEATURES OF STATISTICAL PACKAGES AND DATA DOCUMENTATION

Once you have gathered your data or identified an existing data set, you have the raw material needed to examine your research questions. As a scientist, your job is to put those raw materials together in a creative way to make a novel product. To do this, you need to understand those raw materials in detail—what information do you have, exactly, and where are they stored? And, you need a **statistical package** to serve as your tool for implementing your ideas about how to put together those raw materials. In this chapter, we offer the basic strategies you need to follow in order to understand what is contained in a raw data file and its documentation and what the advantages are of using various statistical packages. In the next chapter, we will learn how to use those statistical packages to access the data and create some of the variables we will need for our regression models.

Typically, at this stage, we want to jump forward to see the statistical answer to our research questions. Yet, once the raw data are identified, much work remains to understand it, prepare it, and analyze it. Doing so in a steady, systematic fashion avoids later backtracking to correct errors and allows the documentation needed for replication. Although it may seem tedious, seasoned researchers know that time spent at this stage of a project is essential and will pay off many times over in the long run.

## 3.1: HOW ARE OUR DATA STORED IN THE COMPUTER?

To students new to quantitative data analysis, a data file may seem mysterious at first. In reality, the file is a simple transcription of the data. Quantitative data are stored in a matrix of rows and columns, in which each cell represents a specific data point. A **data point** captures how a particular respondent answered a particular question. Typically, your project will begin with what is called a **raw data file**. The *raw data file* contains the participants' exact responses to the study's instruments. By convention, each row represents a study participant. The participants could be people, or they could be other entities such as organizations, schools, states, or countries (they are often more generally referred to as **cases**). Each column represents a variable.

Variables

...Data points...

Cases

Typically, the first column contains a variable that identifies the participant (or case), often called a *caseid* for short. With public use data sets these **identifiers** are numbers and/or letters with no intrinsic meaning. The study investigators maintain a list that links the *caseid* to meaningful identifiers (the individual's first and last name or the name of the organization or geographic area). The process of removing all meaningful identifiers is called **de-identification**. This de-identification is a critical part of the process of preparing data for public release. It allows other researchers to analyze a data set while protecting the identity and preserving the confidentiality of the study participants (see Box 3.1). Although you will not use the *caseid* in basic statistical analysis, it is good practice to keep it in your data file. For example, you will need to if you ever need to link to other study information about participants (e.g., if the study's data are stored in multiple files or the study follows participants across multiple waves; See Box 3.2).

Table 3.1 illustrates how data might be arranged for some variables relevant to our example research questions, with the *caseid* in the first column and then the mother's education and age (in years) and distance from the respondent (in miles).

The first respondent in this hypothetical data set was given the *caseid* of 9,332 by the study and reported that her mother had 13 years of schooling, was 73 years old, and lived 4 miles away. The seventh respondent in the datafile

**Table 3.1:**

| caseid | momeduc | momage | mommiles | ... |
|--------|---------|--------|----------|-----|
| 9,332 | 13 | 73 | 4 | ... |
| 8,454 | 11 | 55 | 6 | ... |
| 3,743 | 15 | 42 | 1,202 | ... |
| 2,581 | 2 | 65 | 13 | ... |
| 5,617 | 6 | 56 | 287 | ... |
| 4,668 | 4 | 57 | 348 | ... |
| 1,533 | 10 | 50 | 12 | ... |
| . | . | . | . | . |
| . | . | . | . | . |
| . | . | . | . | . |

**■ Box 3.1**

If you are analyzing data that you gathered, de-identification is still a good practice. When your project went through IRB approval, you were likely asked how you would protect the confidentiality of respondents. For example, you may keep a master list linking *caseids* to names in a locked file cabinet. Any data sets kept on your computer would include only the *caseid* as a variable.

**■ Box 3.2**

In more complicated designs, a data set may have more than one identifier. For example, your data set may have information about multiple members of a family. In this case, each family would have an identifier (perhaps called a *famid*). And, each member would have a unique identifier (perhaps called the *personid*). All members of the same family would have the same *famid*. But, each family member would have her own *personid*. Or, your data set might have information about multiple students in a classroom (in which case you would have separate identifiers for classes and for students). Or, your data set might have information about multiple families within neighborhoods, and separate identifiers for neighborhoods and families.

was given the *caseid* of 1,533 and reported that his mother had 10 years of schooling, was 50 years old, and lived 12 miles away. Typically, the order of the variables follows the order of administration in the survey (the first variable in the first column, second in the second, and so on).

In the past, the raw data file was typically first stored in an *ascii* or **plain text** file, with just rows and columns of numbers. This plain text file could be viewed through any software program on any computer, including a text editor like Notepad. For example, the data from Table 3.1 above might have been represented as follows and saved as plain text:

```
9332 13 73 4
8454 11 55 6
3743 15 42 1202
2581 2 65 13
5617 6 56 287
4668 4 57 348
1533 10 50 12

. . .
```

On its own, this data file is simply a series of numbers. To interpret these numbers requires information about what each value represents. The column labels in Table 3.1 gave us some information, but the information needed to interpret the numbers fully would more generally be found in the data's documentation. To prepare the data for analysis by a statistical package, we would need to instruct the software on what each value means. For example, we would tell the software which variables are located in which position (e.g., the case identifier, then the mother's education, then the mother's age, then the distance away the mother lives, separated by spaces). And, we would choose variable names for each variable (such as *caseid, momeduc, momage,* and *mommiles*). We might want to label the variables to remind ourselves about important details found in the documentation (e.g., that *momeduc* and *momage* are recorded in years and *mommiles* is recorded in miles). Our instructions would convert these numbers into a data file format that the statistical package recognizes. Once converted, the data would be ready for analysis by that statistical package, and could be saved in that package's format (see Box 3.3).

▪ **Box 3.3**

You probably have experience with such software-specific files in word processing. A Word Perfect file is in a particular format. A Microsoft Word file is in another format. One format cannot be read directly by the other software, although nowadays the packages often have built-in conversion utilities (accessed by selecting the type of file from the File/Open menu). We will discuss some similar conversion utilities for statistical software below and in Chapter 4.

In the past, scientists received data in plain text format and needed to write instructions such as these to "read" the data. This process was time-consuming and error-prone. As we will see in our NSFH example, these days, the raw data file can often be obtained directly from an archive in a format ready to be directly understood by the statistical package. But, it still can be viewed in a row and column format, similar to Table 3.1.

## 3.2: WHY LEARN BOTH SAS AND STATA?

Many basic tools for working with the raw data—statistical packages—exist, and a first step in our project is to choose which tool to use. General purpose statistical software can accomplish a wide array of functions, ranging from **data management** (e.g., defining new variables) to **data analysis** (e.g., calculating statistics). You have likely heard of some of the most common statistical packages, such as SPSS, SAS, and Stata. Others exist as well (e.g., R, Minitab, S-Plus, Systat). Special purpose software typically has fewer data management utilities, and focuses on a particular statistical technique (often advanced techniques with names like "structural equation modeling" or "hierarchical linear modeling"). We include two general purpose packages—SAS and Stata—in this book. It is worth taking a moment to explain why, since many students in the social sciences find it tedious and frustrating to learn one statistical package, let alone two.

### 3.2.1: Social Scientists Commonly Use Multiple Statistical Packages

Many scholars today who regularly conduct quantitative research move back and forth among statistical packages. They are not "SAS users" or "SPSS users" or "Stata users." Rather, they choose one software package over another for a particular research project because one has a unique feature or capability needed for that project. Or, they may use a software package because another researcher or group of researchers in a collaboration uses that package. Some researchers use one package to manage the data and create variables (e.g., SAS) and another to run regression analyses (e.g., Stata). This is possible because data can now be easily moved between formats using **file conversion software**, such as StatTransfer and DBMS/Copy, which have been available since the mid-1990s (Hilbe 1996). With file conversion software, the researcher need only click a couple of buttons to tell the conversion software where the original file is located and its format and where to store the new file and in what new format (e.g., SAS, Stata, or SPSS). Depending on the size of the file, it typically takes only seconds to convert from one format to another. So, today, it is easy to use whichever software tool is best suited to the particular task at hand or particular collaborative project.

### 3.2.2: Different Packages Fit the Needs of Different Students and Instructors

Including both SAS and Stata in parallel fashion in the book also offers flexibility to the mixture of instructors and students teaching and taking graduate statistics courses in the social sciences. Your instructor may choose to focus on SAS or Stata because it fits her expertise or the local resources better (i.e., your access to SAS or Stata). Your instructor may also choose to focus on one, rather than teach both, because the majority of students in the class are apprehensive about learning statistics and are uncertain about using quantitative methods. In such cases, the book meets the needs of the subset of students who are eager to learn statistical computing. If you fit in that group, you can work through the examples using the second language on your own. Other students may realize after finishing the course that they need to know the other package in order to meet the needs of their thesis, dissertation, or research project. If you need to do so, you can easily go back and rework examples or revisit the parallel commands in the other package within the book after completing the course.

Including both packages, and allowing students to learn the second package on their own during the course or afterward also meets the book's goal of preparing students for advanced courses and research assistantships. Sometimes, instructors in advanced courses require students to use a particular package, use multiple packages, or use a specialized package targeted at the technique being taught. Similarly, some faculties use a particular statistical package across projects or for a given project, and want their research assistants to use that package. In either case, at least being exposed to (if not learning) both SAS and Stata offers you an advantage because: (a) one or the other may be the package used in the course or by the faculty member, or (b) when you do not feel wedded to one package (you have seen the parallel language of another package), you will feel more comfortable and find it easier to learn an additional package (including others beyond SAS and Stata, such as SPSS).

### 3.2.3: Each Package has Relative Strengths

As you prepare to learn SAS and Stata, it is also helpful to know something about their general comparative advantages, as well as those of SPSS (which is used within or alongside many introductory statistics textbooks).

Many researchers view SAS as having a comparative advantage in data management. As Alan Acock (2005, 1093) wrote in a review of SAS, Stata, and SPSS: "SAS is the first choice of many power users . . . The most complex data sets we use hardly tap the capability of SAS for data management." On the other hand, SAS is often seen as relatively difficult to learn. As Acock put it (2005, 1093–4): "SAS is a long way down the list on ease of use . . . Graduate instruction on methods gets bogged down on how to use SAS rather than on methodological and statistical issues." The documentation for SAS is extensive, with details that an advanced user will appreciate but that make it hard to find answers to the new user's common questions. We address these challenges of learning SAS by teaching selective aspects of its extensive capabilities and by doing so in small doses that are connected to the statistical techniques and substantive examples of each chapter of the book.

In contrast, SPSS's comparative advantage is ease of use, but with lesser capabilities for data management or statistical analysis. As Acock (2005, 1094) writes:

> SPSS is the first choice for the *occasional* user who is doing basic data management and statistical analysis . . . [However] looking to the future, it is fair to say that it will be the weakest of the three packages in the scope of statistical procedures it offers. It can manage complex data sets, but often relies on brute force in the code required to do this.

Its Windows-based "point and click" interface makes it easy for users to open a data set and conduct analyses by choosing a procedure and variables from the menus. But, for social scientists, we argue that it is easier to organize and document a research project, even a seemingly small and simple project, using batch programming code. Although it is possible to write in SPSS code (or ask SPSS to generate the code written "behind the scenes" of the Windows interface), the actual code in SPSS is no easier to write and understand than the code in Stata or SAS. Finally, as Acock puts it "SPSS is the easiest of the three packages to grow out of" (Acock 2005, 1094). Because you cannot predict where your interests will take you, we see teaching in SAS and Stata as preferable to SPSS. We try to make this easier by using strategies that simplify teaching each language.

Stata is usually viewed as somewhere in between SAS and SPSS on these fronts, easier to learn and use than SAS but with stronger data management and statistical analysis capabilities than SPSS. Stata also stays at the forefront of statistical analysis because it is easy for users to write code for things they would like to do but are not yet in the official release (Kolenikov 2001). Scholars regularly share such code, and user-written code is generally easy to install from the Internet. Not only does some such code allow users to conduct cutting-edge statistical analyses, but some code is also written explicitly to help users interpret and present results (e.g., code for interpreting regression models written by Long and Freese, see Long 2007; code for presenting results written by Ben Jann, see Jann 2007). As Acock writes (2005, 1094–5) "Although Stata has the smallest development team, all their efforts are focused on the statistical needs of scholars. Looking to the future, Stata may have the strongest collection of advanced statistical features." Stata's code is also especially easy to pick up because it is very consistent across commands. Once you know how to write the command for one type of regression model in Stata, it is easy to write the command to run most types of regression model. Like SPSS's Windows-based interface, it is also easy to sit down and conduct analyses interactively with Stata's command line. This allows you to run a quick command or explore the data while staying within the straightforward and consistent Stata command language. Such interactive sessions can be tracked, and the code can be edited to the final desired results and saved for project documentation.

In short, we use both SAS and Stata because they are best suited to the needs of social scientists. Our goal is to present them in a parallel fashion so that it is easy for students to see the similarities (and differences) between the languages. And, we use a slow and steady building of knowledge and skills together in statistics and statistical programming. Let's get started!

## 3.3: GETTING STARTED WITH A QUANTITATIVE RESEARCH PROJECT

In the remainder of this chapter we will use the National Survey of Families and Households and the research questions presented in Chapter 2 as an example of getting started with a quantitative research project. We will cover the following topics:

- What are the basic SAS and Stata files?
- What are some good ways to organize these files?
- How do we identify the variables needed for our research question from a larger existing data set?

In Chapter 4, we will learn how actually to use the software to prepare the data for our regression models. To simplify what we need to teach at the outset, we save some basic data management tasks for later chapters (e.g., creating indicators for categorical variables), although in a real project you would implement them at this stage also.

### 3.3.1: What are the Basic SAS and Stata Files?

SAS and Stata are sometimes referred to as languages as well as statistical packages. Each has its own syntax which we use to tell the computer how to perform a particular task. Usually, both packages can achieve the same objective. Often, the syntax is similar in each language, but with slightly different conventions. There are some resulting benefits and costs of these similarities and differences, as with spoken languages: it is easier to learn a second language once you've learned one; but, when learning a second language, you can get confused, sometimes merging the two languages when you speak.

Throughout the book, we set the syntax for the two packages side-by-side, making these similarities and differences very clear (and offering a ready reference to look up the details when forgotten).

The similarities and differences between the languages are also evident in the basic files used by each package. Display A.3 lists the three basic kinds of file used and created by each statistical package: data, batch programs, and results. Each has a standard **extension**, also listed in Display A.3. (As you probably know, an extension consists of the letters following the "." after a file name.)

It is best to stick with conventions, and use the file extensions listed in Display A.3. Even if, for example, using an extension like *.prg* would help you initially remember that a file contains a batch program, in the long run you will find it easier if you use the conventions *.sas* and *.do* for SAS and Stata batch programs, respectively, especially as you interact with others who use each package. Your computer also uses such extensions to determine which software to use to open a file, known as **file associations** (e.g., *.wpd* is usually associated with Word Perfect files; *.doc* is usually associated with Microsoft Word files). With a standard installation, your computer will also associate the standard SAS and Stata extensions with the respective software, making it easy to launch the software by double-clicking an associated file. It is helpful to keep in mind, though, that there is nothing inherent in the extension that alters the content of the file (if you changed an extension from *.wpd* to *.doc* without converting the contents, the contents would remain in their original Word Perfect format). Some extensions will be assigned by default when SAS and Stata create files (e.g., when saving data files in the package's format). Some you will assign (e.g., when you write commands in batch programs).

### Data Files

We already discussed above the contents of a raw data file. Display A.3 shows us that the typical extensions for raw data files are *.dat, .raw*, and *.txt* for the plain text versions and *.sas7bdat* and *.dta* for the SAS and Stata versions. The extensions *.sas7bdat* and *.dta* are also used for other data files that you create with the respective software. If your computer has SAS or Stata installed on it, and you (or your university's computer staff) allowed the default file associations with the installation, then the computer should recognize files with these extensions as SAS and Stata files, respectively (thus double clicking on the file will launch the associated software).

Below, we provide examples of two hypothetical data files in the folder *c:\data*. Notice in the Name column that each file has the extension relevant for each package (*.dta* for Stata; *.sas7bdat* for SAS). In the Type column we see that the computer recognizes the *.dta* file as a Stata dataset and the *.sas7bdat file* as a SAS data set. Thus, double clicking on either file would launch the appropriate software.

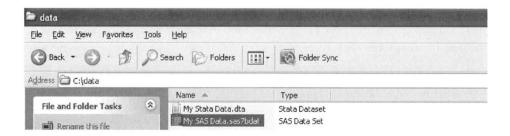

## Batch Programs

A **batch program** is simply a list of instructions to modify or analyze your data, written with SAS or Stata syntax. We emphasize the use of such batch programs (rather than interactive variable creation and analysis). The word "batch" simply means that the series of instructions, in SAS or Stata syntax language, are gathered together so that the computer can run them one after another without prompting you for what to do next. When you gather your work together in such batch programs, you have a record of every step in your project, from raw data to final regression results. As we will emphasize in the next section, learning to organize your files is a critical piece of learning to conduct a quantitative data analysis project. Even the best and most insightful findings will be lost if you cannot find or recreate them. And, the ability of others to replicate your results will increase your credibility (to yourself and others; Freese 2007).

The extensions for these batch programs (*.sas* in SAS and *.do* in Stata) will also be associated with their respective software by the default installation, meaning that SAS or Stata will be launched when you double click on files with these extensions. When we discuss the interfaces used by SAS and Stata in Chapter 4, we will talk about what editors to use to create these files (they are always stored in plain text format).

For example, below we show two hypothetical batch programs with the appropriate extensions (*.do* for Stata and *.sas* for SAS). The computer recognizes these files as a Stata do-file and SAS program, and would launch the appropriate software if we double-clicked on the file. The word "**do-file**" reflects the fact that the extension *.do* is so commonly used for Stata batch programs that a Stata batch program is called a "do-file."

## Results

The final files we discuss contain the output from your batch programs, which could range from basic means and tabulations of variables to sophisticated regression analyses. These files also contain error messages which are useful for "**debugging**"—removing the errors from—your batch program (if you make a severe enough syntax mistake in either language, the software may stop until you fix it, just as if you make too serious an error in a spoken language you may need to try again until the proficient speaker you are trying to communicate with can understand you). In SAS the output and the error messages are contained in two separate files (the *.rtf* and *.log* files, respectively, in Display A.3). In Stata, they are combined together in a single file (the *.log* or *.smcl* file in Display A.3).

Below are examples of the SAS and Stata files with the *.log* extension. Depending on the order and choices made during installation, *.log* files may be interpreted by the computer as being associated with SAS or Stata. On our computer, for example, they are associated with SAS.

As noted above, this doesn't mean that the contents of the Stata *.log* file are now in SAS format; but we cannot double-click the *.log* file to open it in Stata. Because our computer has associated *.log* files with SAS, double-clicking will launch SAS. We would need to do something else (such as use the File/Open menu from within the software interface) to open the Stata results. We will have more to say about the other versions of the results files in Chapter 4 *(.rtf* in SAS and *.smcl* in Stata).

## 3.3.2: What are Some Good Ways to Organize these Files?

Social science research relies on replication (Freese 2007; Long 2009). By first replicating and then extending others' studies, new studies best contribute to a growing body of knowledge on a topic. To facilitate such replication, it is important for you to document your work so that others could repeat your steps. If you publish your results, another student or scholar may ask you to share your batch programs so they can replicate your results before extending them, especially when you use a publicly available data set. Some journals require researchers to make their data and batch programs available to others if their results are published in the journal. Being open to such sharing benefits the research community, and ensures that researchers take extra care with their own work.

Good organization is also important simply to be sure you can trace the steps of your project from start to finish, and can confirm that your calculations are correct. As we will see in Chapter 4, it is possible to analyze your data "interactively," especially in Stata; that is, you can type a command to create a new variable, examine that new variable's distribution, and then use that variable in an analysis, all typing commands one by one that simply scroll by on the computer screen (and aren't permanently saved).[1] You can also use menus to make it even easier to complete this kind of interactive variable creation and analysis (without remembering the syntax). Although most researchers do some of this sometimes, it is best to move quickly toward defining variables and analyzing them through batch programs, otherwise it is easy to forget exactly how a variable was defined (especially if you experiment with several coding decisions). Writing in batch programs and organizing these programs helps you to avoid these problems and allows you to document and check your work easily.

### Project Folders

We recommend that when you start a new project you create a folder on your computer to store your data, batch program, and results files (see Box 3.4). It is useful to come up with a convention for naming your files (and variables) that makes it easy to remember their purpose and contents. The convention is up to you—it should be something that makes sense to you, making it easy for you to organize and locate your files. Two general suggestions for folder names are:

■ Use words and abbreviations that convey the meaning of the folder contents. For example, you might use *Chapter 4* as a folder name for the files created in the

---

■ **Box 3.4**

Especially if you are using a restricted rather than public use data file, you may want to store the raw data file separately from your other files. Often, these data files can only be analyzed on a single computer with particular security safeguards (e.g., disconnected from the Internet). The data should not be removed from that computer, so often for these types of project it is easier to store your programs and results separately from the data, so that you can copy the program/output files to a flash drive for backup, summarizing, and writing. Sometimes such policies also require that you delete your analytic data files at regular intervals, and in these cases you may also want to store your analytic files in a third, separate location to make it easy to locate them for such deletion (and leave untouched in other folders the raw data, programs, and results).

exercises of the next chapter. Or, you might put together a data set name (e.g., *nsfh*) and research topic (e.g., *distance*) to create a compound name such as *nsfhdistance*. Some researchers find it hard to distinguish these separate words, and use underscores or capitalization to make them stand out (*nsfh_distance* or *NsfhDistance*). Others find the underscores and capitals distracting or difficult to type, and stick with the original compound word.

■ Use words and abbreviations for the date that you create the folder or files (e.g., 2008jan15 or 20080115), again potentially with symbols or capitalization to increase readability (e.g., 2008_Jan15 or 2008_01_15). If you use this date strategy, it can be helpful to keep a separate log, on paper or electronically, to make notes about what you did on that date. In electronic form, such a log can be stored in the same project folder as the data set, batch programs, and results, so it is easy to find.

These strategies can be combined as well. We might create a folder called *nsfh_distance* for a larger project and then use dated folders within, such as *2008_Jan15*, to organize our work. For this course, think a little about which of these might suit your style, and try it (or try another strategy that you come up with). You can even experiment with different styles throughout the course to help choose one that works best for you (or realize that certain styles work better for certain tasks or projects; see Box 3.5).

> ■ **Box 3.5**
>
> In general it is a good idea to avoid spaces in names. If you do include spaces, then you will need to put them in quotes in your program (e.g., "nsfh distance"). The same holds for folder names on your computer. If you are working in a public lab, and the folder where your files are stored includes a space, then you need to enclose it in quotes (e.g., "c:\Documents and Settings\Rachel Gordon\My Documents").

### Analytic Data Files

The public use version of a data file can be very large. The NSFH data set we will use as an example contains 4,355 variables from 13,007 participants. Data sets intended for public release, like the NSFH, typically cover a broad range of subjects so that they are of great utility to a broad research community, especially when funded by federal agencies. If you were conducting new data collection for your targeted research question, your data set would likely include fewer variables (and, you would likely not have the resources to achieve such a large sample size). For any project, but especially with such large public data sets, it is useful to create a smaller ***analytic data file*** which contains only the variables (and cases) needed for your particular project. Even though today's computing power often makes it possible to manipulate the large raw data file quickly, it will be easier to review the contents of a smaller file for completeness and accuracy. And, even today, some software may not be able to read a very large raw data file in its entirety.

For our example research question, we will focus on a subset of cases and a subset of variables from the NSFH. We will exclude adults whose mother is no longer living (since distance is irrelevant for these people). We will also exclude those who are coresiding with their mother (since decisions about whether to live in the same household may be explained by somewhat different constructs than the closeness of residences among those already living apart). And, we will exclude adults whose mothers live outside the USA. We also focused our research questions in Chapter 2 on a subset of constructs, including:

- distance between respondent and mother's residences;
- respondent gender, race-ethnicity, age, earnings, and number of brothers and sisters;
- mother's education and age.

In Chapter 4 we will show you how to create an analytic data file that contains just the relevant cases and variables.

You can, and should, use names that make sense to you for your analytic data files (we'll offer some suggestions in Chapter 4). *But, do not under any circumstances make changes to the raw data file (such as keeping a subset of variables or a subset of cases) and "overwrite" the original raw data file (i.e., do not save the revised data set under the original raw data file's name).* Later confusion is almost guaranteed if you overwrite the original raw data file! We also recommend that you *do not rename the original raw data file*, even though its name may make little sense to you because it was named by other scholars or archivers to fit their naming conventions (or what made sense to them). For example, the Stata version of the raw data file for our NSFH example is named *d1all004.NSDstat_F1.dta*. If you leave it with its original name, it will be easy to trace it back to the source (for example, if you need to call the archivers or original study investigators with questions, they will recognize their raw data file name, not what you renamed it). You need only refer to the original name once—in the file where you create your analytic data file. You can then save your analytic data file with a name that makes sense to you, and use that name from that point forward!

### 3.3.3: How Do We Identify the Variables Needed for Our Research Questions from a Larger Existing Data Set?

We are nearly ready to access the raw data and prepare it for analysis. The last step before doing so is to gain a thorough understanding of the data's design and documentation, including understanding which variables you need to extract and how those variables are coded. In Chapter 2, we relied on basic study descriptions and publications to identify appropriate candidates for our research questions. At this stage, we need to drill down into the study's documentation so we understand exactly how the data were collected and exactly what questions were asked (see Box 3.6).

For our example, we will download the National Survey of Families and Households (NSFH) data set and the subset of relevant variables. The NSFH began in 1987–8 (Wave I), with two follow ups (Wave II in 1992–4 and Wave III in 2001–3). One adult household member was randomly selected for an in-person interview at Wave I, and data are available from many spouses or cohabiting partners and children of these primary respondents. As an example of basic regression techniques, we will focus on the Wave I data from the primary respondent.

We will take advantage of the NSFH's own web site to access the data and documentation rather than relying on ICPSR. The data can be downloaded in their original ascii format from ICPSR (and the main NSFH page). A utility called BADGIR that is accessible from the NSFH web site (see Display C.3.1) makes it easy to download the data in a variety of formats (e.g., directly in Stata format) and to access the documentation files (see Box 3.7).

Data documentation usually includes the following:

- **Measure/Variable Lists.** Data sets often include a summary list of the study's measures (sometimes called **instruments**) and variables. These provide a useful overview of the content when you are first learning about a data set (these lists can be helpful in selecting a data set at the planning stage). They can also help you to identify where a variable occurs in a datafile (near the beginning, middle, or end).
- **Questionnaires.** The actual questions, as asked of the respondents, are contained in the questionnaires. When questionnaires are administered in paper-and-pencil, these are straightforward records of the interviewing process. When questionnaires are administered by computer (Computer Assisted Telephone Interviews, CATI, or Computer Assisted Personal Interviews, CAPI), there may not be a straightforward document containing the questions. This is especially true if the questionnaire includes complex skip structures, where only a subset of respondents are asked to respond to some questions.

■ **Box 3.6**

In reality, you would likely identify two or three candidate data sets, and choose among them by looking at the details or their designs and questionnaires to identify their relative strengths for your research questions. This step is a critical piece of demonstrating the feasibility of your project, to your thesis or dissertation committee early in your career or to a funding agency later in your career (i.e., can you demonstrate that the design and measures are well-suited to your research questions?)

■ **Box 3.7**

It is not uncommon for investigators to maintain their own web sites like these for publicly available data sets, in addition to the data being deposited in archives. It is a good idea to check for study web sites and see if they have additional options for downloading data before accessing them through ICPSR.

▪ **Skip Maps.** Skip instructions are usually listed in questionnaires. Sometimes data documentation also includes separate skip maps to help users to understand which subgroups of respondents should have been asked which questions. These skip maps are very helpful in understanding response structures, especially if you see in the data tabulations that a large group of respondents are missing responses to a variable or series of variables. Respondents who skip over a question are typically coded as *not applicable* on skipped questions. It is always worth understanding the skip structures yourself, and carefully comparing the distribution of responses in your downloaded data to the skip patterns. This is especially important when the survey was administered in paper-and-pencil format where it is more likely that some respondents may have been asked questions that they should have skipped (or not asked questions that they should not have skipped) due to interviewers incorrectly applying the skip rules.

▪ **Codebooks.** Codebooks describe the contents of the raw data files. Usually, they include the name and description of the variables (typically following the same order as the questionnaires). Often, the codebooks include a tabulation of the number of study participants who chose each response for each variable (e.g., for a variable that captured the gender of study participants, the number who chose *male*, the number who chose *female*, and the number who did not respond might be listed). These tabulations can be extremely helpful at the planning stage of a project (to assure that there is enough variation on key variables and large enough subsample size for the data set to be a good candidate for your research interests). These tabulations can be used to check the accuracy of your downloaded data (this is especially important if you have to write a batch program to read the data from plaint text format, although it is worth checking these distributions as we will illustrate in Chapter 4 even if you download the data in SAS or Stata format; it is always better to catch errors earlier rather than later!). Codebooks will also indicate how missing data are coded (often, missing data are captured with multiple codes, such as *refused, not applicable*, and *don't know*; sometimes these are already coded in a format that SAS or Stata will recognize as missing, although often they are stored with numbers in the raw data file and you must recode them so SAS and Stata recognize them as missing as we will show how to do in Chapter 4).

### Using the NSFH Documentation

The NSFH data includes all of these kinds of documentation—measure lists, questionnaires, codebooks, and skip maps. We will illustrate each as we identify the variables relevant to our research questions. In the interest of space, we provide details for just a few questions to demonstrate the process you would follow in your own work.

The NSFH documentation is all accessible through the BADGIR utility. Usually, it takes several passes through a study's documentation to become familiar with it.

Do not be surprised if you have to go back through materials a few times to understand the "big picture" and to locate the specific files you need. It is often helpful to approach learning a data set as you would detective work—as a process of iterative discovery.

*Variable Lists*

A helpful variable list is found in the first working paper written by the study investigators, titled *The Design and Content of the NSFH*, which is available within the *Metadata* section of BADGIR. This working paper provides relevant general background about the data set, generally worth reading to help understand the data. For our current purposes, we will focus on *Section VIII*, beginning on p. 37, which provides an *Outline of the Content of the National Survey of Families and Households* with enough detail to allow us to pinpoint the location of relevant variables.

Display C.3.2 provides excerpts from this outline. Topics related to our research questions are found in two major sections of the outline: (1) Interview with Primary Respondent, and (2) Self-Administered Questionnaire: Primary Respondent. Within the Interview with Primary Respondent, the section of questions on Household Composition include age and sex of household members (including the respondent) and the section of questions on Social and Economic Characteristics include race and parent's education. Within the section on Parents, Relatives, and General Attitudes in the Self-Enumerated Questionnaire Number 13 "SE-13" we find the mother's age and current residence and the respondent's number of brothers and sisters.

Browsing the documentation also reveals that the study investigators created some variables based on the raw data, and include these in the data file. Such created variables can be useful for replicating prior studies and saving time and potential errors in recoding complicated questionnaire sections (e.g., with extensive skip patterns). We use one such variable which the study investigators created to capture the respondent's total earnings, from wages, salary, and self-employment. The details of their coding are found in Appendix I (*Instructions for creating income variables and poverty status*) of the "Other Documentation" on BADGIR.

*Skip Maps and Questionnaires*

Display C.3.3 shows where the relevant questionnaires and skip maps are found in BADGIR. Based on the sections that we identified in the variable lists, we check for the questions we needed in the primary respondent's main interview and self-administered questionnaires. In this section, we will look at the skip maps and questionnaires for the mother's age and distance as examples.

The NSFH skip maps provide an easy overview of this section of questions. The skips among the questions about the respondent's mother are found in the file that opens when we click on *Skipmap: Wave 1 Self-Administered Questionnaire* in BADGIR (see again Display C.3.3). The relevant section of the skip map is duplicated in Display C.3.4. After reading through this section, we identify several items that will be relevant to our project. One question will help us to select the correct subgroup of adults whose mothers are still alive from the full data set: a skip from Question 1 directs subsets of respondents to relevant questions, depending on whether their mother is currently alive or deceased. Two items that we will want to include in our analyses are also found on this page. The second question captures the mother's age and the fifth question the distance she lives from the respondent, if she is alive.

After identifying the relevant items in the skip map, we click on *Questionnaire: Primary Respondent Self-Enumerated Schedule* in BADGIR (see again Display C.3.3) to see the corresponding questions as completed by the respondent. The relevant section is reproduced in Display C.3.5. Notice that this page is laid out for self-administration. The questions to be answered by respondents whose mothers are living are in the left column. The questions to be answered by respondents whose mothers are deceased are in the right column. Thus, the skip pattern is clear in this relatively simple section of the questionnaire, with its straightforward layout. Our relevant questions about the mother's age (*2. How old is she? __ Years old*) and the mother's location (*5. About how far away does she live? ___ Miles . . .*) should have been answered only by those respondents whose mother was living (although in paper and pencil format it is possible that some respondents may have incorrectly followed the instructions, so we may want to check for this after downloading the data). The questionnaire also alerts us that we should expect to see in the data file that the mother's age is recorded in years and the distance in miles. The notes on Question 5 also clue us that respondents who coresided with their mother should skip to the next page, suggesting that there may be a code that will help us to select adults who do not coreside with their mothers from the full sample. The particular wording of questions and associated notes are also important for understanding the data, even when not used directly in writing our batch programs (e.g., it appears that the study investigators calculated distances for respondents who did not know the distance, based on their report of the city and state where their mother lived).

### Codebooks

Now that we understand the skip structure and questions for the items, we can use the codebook to see exactly how the responses were recorded in the data. We will rely on the BADGIR *Variable Description* section to access the codebook. The original codebooks are also available on the NSFH web site, but these tabulations reflect the data as they were released in the late 1980s, rather than the latest release. In contrast, BADGIR's codebook tabulations are based on the current version of the data set, which is the version we will download. (Like all researchers, investigators who gather data that is later made publicly available identify errors after its release; in fact, sometimes public users identify problems the initial investigators missed.) BADGIR is also easy to browse to locate items.

Display C.3.6 provides an example of the codebook from BADGIR for Question 1 from the self-administered questionnaire regarding whether the mother is still living or deceased. Notice first that the name of this variable is E1301. This follows the study's naming convention. E13 represents the 13<sup>th</sup> self-administered questionnaire (see again top of Display C.3.4 with the title SE—13). And, 01 represents Question 1 in this section. This is the name we will use to refer to this variable in our batch program in Chapter 4. We will show below how to create new variables that we intend to use in our analysis, and we will use names that are meaningful to us for these new created variables. Like raw data files, we recommend not renaming or recoding the raw variables themselves (otherwise, later confusion will result!).

The *Values* column of the codebook shows us which values correspond to which categories (e.g., 1 = *Still Living* and 2 = *Deceased*). In the *N* column we see that over one third of respondents report that their mother is deceased, whereas close to two thirds report that she is alive. Even after excluding respondents whose mother is deceased, over 8,000 cases will remain. The tabulation in the codebook also shows us the values that the study investigators used to indicate types of missing data. As expected, based on the skip map, *Not Applicable* is not included as a type of missing data for this question, since all respondents should have answered Question 1 of this section (see note at top of Display C.3.4). But, over 250 cases are missing data because the respondent refused (one case), didn't know (13 cases), or didn't answer (252 cases).[2] These types of missing data are recorded with the values 7, 8, and 9, respectively.

The next example, in Display C.3.7, shows the codebook results for Question 2, on the age of the respondent's mother. Only the smallest and largest values are shown, since viewing all of the values requires scrolling across multiple screens online. Notice first that the file-naming convention is again followed for this variable, called E1302 (second question on the 13th self-administered questionnaire). The tabulation shows us that one respondent reported that her mother's age was 29, one reported 30, and three reported 33 and so on down to seven respondents reporting that she was 94 years old and seventeen respondents reporting that she was 95 years old. Notice also that *inapplicable* occurs as a missing value for this variable, recorded with the value of 96 for 4,466 respondents (who should be valid skips because their mother was not alive).[3] Ten cases are also recorded as *refusals*, 44 as *don't know*, and 305 as *no answer*, with the values 97, 98, and 99, respectively.

The third example, in Display C.3.8, shows the responses for Question 5, on the distance away from the respondent that the mother lives. Note again that the variable name follows the study's convention (E1305 for the fifth question on the 13th self-administered questionnaire). The distribution of values shows that a large number of respondents (1,020) reported that their mother lived just one mile away. In contrast, just a few report very large distances (two respondents report a distance of 7,000 miles, five report 8,000 miles, and three report 9,000 miles). In addition to the missing data codes we have seen already, this question has two specific codes that we can use for keeping the desired subset of cases: (1) 9995 indicates that the mother lived with the respondent (recall that these respondents were instructed to skip this question), and (2) 9994 indicates that the mother lived in a foreign country.[4] An additional 4,464 respondents are coded as *inapplicable* (9996), two respondents *refused* (9997), 22 said *don't know* (9998), and 321 provided *no answer* (9999).

## Naming Conventions for Created Variables in Analytic Data Files

Just as with the naming of our analytic data files, we want to think forward to a naming convention for new variables that we create. Many programmers use names that signal the content of the variable (e.g., *miles* rather than E1305 for the distance the mother lives from the respondent). In the past, variable names were restricted to eight characters, forcing researchers to be quite creative in naming variables. These restrictions have been lifted in recent software releases, and currently, variable names in SAS and Stata can be up to 32 characters. However, most researchers still limit the length of variable names because shorter names are easier to read and better for display (some output will abbreviate long variable names). We recommend variable names of 12 characters or less because we have found that they display best in results output.

Let's plan ahead to think about what to name the variables we will create for our analytic data files. Display B.3.1 summarizes the variables we will include in the analytic data file. The original variable name and description from the codebook are included. Based on the frequency distribution in the codebook, the table also has notes about the codes used for missing data for each variable and relevant notes for recoding. The table also includes a name that we chose for the variable we will create in the analytic data file (students are encouraged to think about other possible names that make sense to them, but if you use our naming conventions to follow along with the examples, it will be easiest to replicate our results). In our naming convention, we started the variable with either *g1* (to designate the Generation 1 mother) or *g2* (to designate the Generation 2 adult child respondent). We then chose a word or abbreviation to represent the content of the variable. We do not include variable names for variables we will use only to subset the data (M497A, E1301), nor for categorical variables that we will learn how to recode in Chapter 6 (M2DP01, M484). And, we will keep the case identifier as named in the raw data file, MCASEID.

As a final step for preparing to write our SAS and Stata batch programs in Chapter 4, let's walk through the notes in Display B.3.1 for each variable so that we can anticipate how they will be used in the batch program.

■ The variables M497A, E1301, and E1305 will be used to keep only a subset of cases. Distance is irrelevant unless the mother is known to be still alive, so we will only retain cases in which E1301 is coded as a "1" (still living). In addition, we treat coresidence and immigration as conceptually distinct events, and focus on distances when respondents live apart from their mother but both live in the USA. These are captured in the codes for E1305. Because the survey was completed in paper-and-pencil, some errors may have occurred (some distances are quite large, over 5,000 miles) and we use the respondent's country of birth (M497A) as an additional restriction to help exclude distances to foreign countries.

■ For age (M2BP01 and E1302) and distance (E1305), we simply need to recode missing data values. If we left the numeric values of missing data, such as 96–99 for age or 9996–9999 for distance, our results would be inaccurate. Means would be drawn higher by the high value of these missing data codes. For example, nearly 100 cases are coded 96–99 on E1302, and would be treated erroneously as quite old rather than of unknown age if not recoded. Both SAS and Stata use the "." symbol to represent missing data (this can be read as "**dot missing**;" see Box 3.8). By default, SAS and Stata will exclude cases with missing values from statistical calculations, which is preferable to including them in calculations with their arbitrary missing data values.[5]

■ For the mother's years of schooling and respondent's number of siblings, we need to recode some valid responses, in addition to recoding missing values. For M502, a GED is represented by the value 25. We will treat GEDs as equivalent to 12 years of schooling, and recode any cases coded 25 on M502 to 12 on our new *g1yrschl* variable. For E1332A and E1332B, respondents skipped the questions, and were coded *inapplicable* (96) if they said on an earlier question that they had no living siblings. Thus, cases with values of 96 on these original variables (E1332A and E1332B) will be coded as "0" on our new variables, *g2numbro* and *g2numsis*.

> ■ **Box 3.8**
>
> As we will see in later chapters, SAS stores '.' missing as the lowest possible value and Stata stores '.' missing as the highest possible value. We will discuss in later chapters how this is important in appropriately accounting for missing data, especially when creating new variables. SAS and Stata both also allow for extended or special missing values. Both allow a letter to be placed after the . to designate different types of missing value (e.g., .a could designate not applicable, .b could designate refusal, and .c don't know). In Stata, excluding these extended missing values could be accomplished with <. and in SAS with >. in the if qualifiers discussed in Chapter 4. We focus on the . "system missing" for simplicity in the textbook, which can be addressed with ~=. in the if qualifiers discussed in Chapter 4.

▪ For earnings, the respondents reported their wages, salaries, and self-employment income from 1986. Because of inflation, these levels of earnings will be difficult for contemporary readers to interpret substantively. Researchers commonly use the Consumer Price Index (CPI) to convert such values to more recent dollars, for which readers have ready benchmarks in mind. The conversion is the ratio of the annual CPI in the more recent year to the annual CPI in the earlier year. For 2007 to 1986, this ratio is 207.342/109.6 (Bureau of Labor Statistics 2008).

## 3.4: SUMMARY

This chapter covered the process of planning for a quantitative research project once you have located an appropriate data set. We introduced the basic types of file used by software packages (data, batch programs, and results files) and helped you to think forward to how you can name and organize these files. We also introduced the basic types of file you will encounter in the documentation for most data sets (variable lists, skip maps, questionnaires, and codebooks). Using the NSFH as an example, we modeled the process of locating the variables needed for an analysis in such documentation. We discussed how to develop conventions for naming the new variables we will create based on these basic raw variables, again modeling the process of thinking forward to the next task (writing the commands to extract the variables, keeping the relevant subgroups from the full data, and creating the new variables needed for our analysis).

## KEY TERMS

Analytic Data File

Batch Program

Case

Codebook

Data Analysis

Data Management

Data Point

De-Identification

Debugging

"Do-File"

Dot Missing

Extension

File Associations

File Conversion Software

Identifiers

Instruments

Measure/Variable Lists

Plain Text (Data File) (Ascii (Data File))

Project Folder

Questionnaire

Raw Data File

Skip Map

Statistical Package

## REVIEW QUESTIONS

REVIEW
QUESTIONS
3

3.1. What are the conventional extensions for SAS and for Stata data files, batch program files, and results files?

3.2. What are some conventions you might use to name a project folder and to name your analytic data file, your batch program, and your results files?

3.3. What are some conventions you might use to name variables?

3.4. Why is a skip map useful?

## REVIEW EXERCISE

Look in BADGIR for the documentation related to some of the variables listed in Display B.3.1, beyond E1301, E1302, and E1305 which we covered in the text. Locate the relevant sections in the variable lists and then find the corresponding sections in the skip maps, questionnaires, and codebooks.

## CHAPTER EXERCISE

Throughout the book, we will use the National Organizations Survey for end-of-chapter exercises.

Three separate cross-sectional National Organizations Surveys have been conducted, in 1991, 1996–7, and 2002; Kalleberg, Knoke, Marsden, and Spaeth 1994; Kalleberg, Knoke, and Marsden 2001; Smith, Kalleberg, and Marsden 2005. The design and content of each survey differs slightly. The first and third surveys both identified a representative sample of US businesses through the places of work of respondents to that year's General Social Survey. The second survey was drawn from Dun and Bradstreet's list of business establishments.

Each of the three surveys is available through the ICPSR archive. We use the 1996–7 study for the examples. Informants from over 1,000 establishments provided information through telephone and mail surveys. The 1996–7 survey gathered a range of organizational constructs, with a particular focus on how employers interact with other organizations to obtain and train new workers.

In our Chapter Exercises, we will examine some correlates of the percentage of managers who are female in the organization. Prior studies examined similar questions with the NOS 1991 (Reskin and McBrier 2000), although not all the same constructs were measured in the 1996 NOS and we look at a simplified model.

For this exercise, use the documentation for the NOS 1996–7 from ICPSR to identify the following variables. You can locate the data set in ICPSR by searching for the names of the investigators or the study name.

1.  Identifier for organization.

2.  When was the organization founded?

3.  Is the organization independent, or part of a larger organization?

4. Does the organization have to report the demographic composition (gender, race-ethnicity, etc.) of its workforce to the government?

5. Is the organization in the service sector?

6. The number of employees in the organization. Look for separate questions about the total number of: (a) all full-time employees, (b) all full-time employees in the company's core occupation, and (c) all managers. Watch for a variable that will tell you whether the core occupation is a managerial position (because in these cases respondents will skip the questions about managers).

7. The percentage of women in each of the three categories of employees (all full-time, core, and managers). Watch for how this variable was coded (respondents could report either the percentage directly, or the number of women from which the percentage can be calculated; we will need all the relevant variables).

8. Questions about the level of formalization of rules and procedures (for example, are there formal dispute resolution procedures, are there written records and documentation of human resource decisions, are there employment contracts, etc.)? You should be able to locate six variables along these lines.

9. How was the interview completed (by mail or telephone?)

10. When was the interview completed?

Create a table similar to Display B.3.1 with the name of the original variable, its description, its missing data code(s), and any notes about the variables. Don't worry about coming up with created variable names yet. In the exercise for Chapter 4, we will think about how to create and name several variables.

## COURSE EXERCISE

Choose one of the data sets that you identified in Chapter 2 for one of your research questions or hypothesis statements. Use the data archive or study web site to access its documentation. Locate relevant lists of variables, questionnaires and skip maps, and codebooks. Use these to identify the variables you will need for your regression model, including at least the outcome variable (ideally a continuous variable) and at least two predictor variables (ideally a categorical and a continuous variable to use across later chapters). Also, identify the case identifier and any variables you will need to subset your data to the relevant subsample needed for your research questions.

**COURSE
EXERCISE
3**

*Chapter 4*

# BASICS OF WRITING BATCH PROGRAMS WITH STATISTICAL PACKAGES

## CHAPTER 4: BASICS OF WRITING BATCH PROGRAMS WITH STATISTICAL PACKAGES

We are now ready to "roll up our sleeves and get our hands dirty with the raw data." We know what variables we want to extract from the raw NSFH data file, what subset of the cases we want to keep, and how we want to create and name our new variables. We are poised to write our batch programs to accomplish these tasks, and in this chapter we will learn how to do so. We will start by illustrating the SAS and Stata interfaces and syntax with a few of the NSFH variables. We will write a small batch program to read these variables from the raw data set, verify that their values are what we expect based on the documentation, and save the results and an analytic data set. We will then learn how to create new variables, keep a subset of cases, add comments, check for errors, and some general finishing touches. Display A.4.2 provides a summary of the syntax we will learn throughout the chapter.

## 4.1: GETTING STARTED WITH SAS AND STATA

Like most people these days, you probably use lots of software, including for word processing, emailing, playing media, etc. The latest versions of SAS and Stata have windows and menu structures that should be generally familiar to you based on these experiences (see Box 4.1). As with most software, you can begin to use each statistical package and accomplish many needed tasks by learning just some of its basic features.

---

■ **Box 4.1**

Like most software, various versions of SAS and Stata are available. This book is written with Version 10 of Stata and Version 9.2 of SAS. Stata released Version 11 as this manuscript was being copy-edited. We make comments in the text about some new features in Version 11 and will post additional information about using the textbook with Stata 11 on the textbook web site (*http://www.routledge.com/textbooks/9780415991544*). Stata also comes in different formats, each with its own restrictions on the number of variables and number of observations it can analyze. We assume that you have access to at least the IC, SE, or MP formats (see Stata 2009a for more information about these formats). We also assume that you are using a Windows operating system, although both SAS and Stata are available for other operating systems. Your instructor will be able to guide you to local labs and resources to access the software, and your local computing support center should be able to help you to find out whether your university has licensing agreements that provide you with student discounts if you want to install the software on your own computer.

As you need to accomplish more complicated tasks, you may be motivated to learn some of the "bells and whistles" of the software. But, like most users, you will probably never need to learn all of the features of the software, similar to what is likely the case with software you use daily to read email or write papers.

Our goal in this book is to help you to feel as comfortable turning to SAS and Stata to accomplish *basic* data management tasks and estimate *basic* regression models as you feel turning to whichever word processing software you use. But, that means we set aside many of the sophisticated capabilities and nuances of each software, including approaches for accomplishing some tasks that are unique to that software package. Indeed, if you explore the menus and help files, you may feel overwhelmed by the possibilities. (We will show some examples of each software's help files in Appendix F.) If that happens, remember that you will be able to accomplish what you need for a basic regression analysis just with the syntax we introduce in each chapter of this book. When you are ready or desire to go beyond these basics, Chapter 12 provides numerous resources for learning more, including strategies for finding opportunities at your local university, on the Web, and across the country.

### 4.1.1: Locating Batch Program and Results Files

We will now take a look at the basic menus and windows in SAS and Stata, and see where our basic data, batch program, and results files are located. We will first connect each window to the types of file and organizational strategy that we discussed in Chapter 3. Then, we will show you more specifically how to enter the example batch program in your computer to duplicate our results.

▪ **Box 4.2**

You can change which software program is associated with an extension, even after installation. One easy way to do this is to "right click" on a file, choose "properties", and then click on the "change" button next to "opens with" on the "general" tab. However, if you are working on a computer in a public lab at your university you may not be allowed to use this function.

To see the menus and windows, you first need to start the software. Like most software, you can accomplish this with SAS and Stata in multiple ways. You should be able to find SAS and/or Stata from the list of software programs in your computer's Start menu. There may be an icon for the software program on your desktop. And, as noted in Chapter 3, if your computer has a standard installation of SAS and Stata and you use standard extensions for your files (as shown in Appendix A.3), you can double-click on a file to launch its associated software (see Box 4.2).

Once you start the software program, you are in a **session**. The session continues until you close the software. Keeping this in mind is important because, as we will discuss below, changes you make during a session are not always saved permanently. If you want changes to remain in the next session (the next time you start the software program), then you need to save them explicitly (as we will discuss below).

Let's now take a look at some of the basic features of the menus and windows that we see when we launch SAS and Stata.

## Stata Menus and Windows

Display B.4.1 provides examples of the Stata windows for editing a batch program (panel a) and viewing the results on screen (panel b). When you first launch Stata, you will see something like the screenshot shown in Display B.4.1b. Notice that there are four main windows visible: (1) The large window with black background which contains results and three smaller windows with white background labeled, (2) Review, (3) Variables, and (4) Command (see Box 4.3).

> ■ **Box 4.3**
>
> It is possible to change the background colors, and other features of the display (such as the size and color of text). From the Edit menu, choose Preferences (or right click on the large black results window, and choose Preferences).

The small white box labeled Command is referred to as the **command window**. We can use the command window to work with our data interactively in Stata. We can type a command into this box, hit return, and view the results. Then, we can type another command, hit return, and view the results. And so on. Many users like to use Stata's command window to try out some commands easily before adding them to their batch programs.

The command window also offers us one way to open the window where we will write our batch program: by typing the word "doedit" in the command window, and hitting return (see Box 4.4). Doing so opens the **Do-File Editor**, Stata's built-in text editor (shown in Display B.4.1a). When it first opens, the main screen is blank (similar to any text editor or word processor). We can type text into this main white screen to write our batch program. Display B.4.1a shows the text we typed to create an example batch program based on the NSFH data (we'll have more to say about the contents of this batch program below). Once we have

> ■ **Box 4.4**
>
> Typically, there are several ways to accomplish a task in SAS and Stata software, like most software. To open the Do-File Editor, you can also click the icon circled in red in Display B.4.1b. You can also choose the Do-File Editor from the Window menu.

written our batch program, we want to save it so that we can ask Stata to run its contents (and so that we have a record of our work, which we can save and share with others). You can save the file in the same way as most windows software (File/Save, Ctrl-S, or the "save" button). We recommend saving your batch program frequently as you write, to avoid lost work. We saved our example batch program with the name

*ReadAFewVariables.do* (shown in bold in Display B.4.1a). Stata will automatically give the extension *.do* when you save the batch programs you write in the Do-File Editor.

Once you have written a batch program, you are ready to ask Stata to run its commands, from beginning to end. One way to do this is to choose the icon circled in black in Display B.4.1a. We can also run the batch program by typing `do ReadAFewVariables` in the Stata command window (in Display B.4.1b). The word `do` is a Stata command that asks Stata to run all of the commands contained in the listed file (in our case, *ReadAFewVariables.do*). We do not need to put the extension *.do* explicitly at the end of the filename, because Stata assumes that we used that extension for our batch program.

We asked Stata to run the *ReadAFewVariables* batch program by using the `do` command on the Command Window. The end result is what is shown in Display B.4.1b. Again, we'll examine the results more carefully below, but let's look for now at the Review and Variables windows. The Review window contains a list of the commands we have typed on the command window. Our Review window in Display B.4.1b lists just two things, the `doedit` command we used to open the Do-File Editor and the `do ReadAFewVariables` command we used to execute our batch program. A convenience of the Review window is that you can double-click on any item in the list to execute that command again. For example, if we double-clicked on `do ReadAFewVariables` in the list, it would run our batch program again. This feature is especially convenient as you write and debug a batch program (we will discuss how to do this in Section 4.5.2). The Variables window lists the four variables that our batch program read from the larger NSFH raw data file. You can double-click on variables in this list to include them in your command window, again a useful feature for interactive work.

### SAS Menus and Windows

Display B.4.2 provides similar examples of the basic SAS windows and menus.

Display B.4.2a shows the windows that will be visible when you first open SAS. The window labeled "ReadAFewVariables.sas" contains our batch program. This window will be blank when we first open SAS, and here we can type the commands for our batch program. Again, we recommend saving frequently to avoid lost work (using File/Save, Ctrl-S, or the "save" icon). SAS will give the batch program the *.sas* extension by default, which is what we want to use.

When we are ready to run our batch program in SAS, we can choose the "run" icon (circled in red in Display B.4.2a) or choose Run/Submit from the menus. After SAS executes the commands, the Log window contains messages. As we will discuss in Section 4.5.2, the log window can be especially useful for debugging (locating and fixing errors). The Explorer window allows us to view our data (which we will do in the next section).

The Output window contains results, such as frequency distributions and regression estimates. Although this window is not visible initially, we can move to it by clicking the tabs at the bottom of the screen (see "Output—(Untitled)" the dotted circle in Display B.4.2a) or by choosing Window/Output from the menus. Again, we will discuss the results further below.

## 4.1.2: Viewing Data

Both SAS and Stata allow you to view the data in a spreadsheet format, similar to the layout we showed in Table 3.1. Although these spreadsheets can be used to modify the data, we do not recommend that you use them in that way as you work the exercises that go along with this textbook. But, the spreadsheet views can be useful for helping you get a "feel" for the data, and they can be a useful aid to verifying variable creation, as we will discuss in Section 4.3.2.

In Stata, the **Data Browser** can be used to view the data by typing "browse" in the command window (or by choosing Data/Data Browser from the menus or by clicking on the Data Browser icon, circled in black in Display B.4.1). Doing so opens a spreadsheet view of the data, as shown in the Display B.4.3a. (The Data Editor in Stata allows the data to be edited as well as viewed).

In SAS, the **Table Editor** can be opened by double-clicking on the data set, located in one of the file drawers in the Explorer window. Our data set *ReadAFewVariables.sas7bdat* is located in the Library file drawer (circled in black in Display B.4.2a). Double-clicking the Library file drawer reveals a number of icons, such as this:

■ **Box 4.5**

The data are actually displayed with their formats by default. This means that each value is associated with a value label. For example a 1 for E1301 is displayed as "STILL LIVING." Such formats are useful because they allow you to link the meaning of values from the codebook to values in the data set. We will generally not rely on formats in this book, but we mention them in Appendix D as we cover the process of reading the raw data.

■ **Box 4.6**

If you are using a computer in your school's computer lab, and cannot store to the c drive, you can choose the appropriate letter and folder name where you can store your personal files (e.g., *h:\rgordon\nsfh_distance* if we have a personal folder called *h:\rgordon*).

Each of the two icons on the left represents a data set, the original raw NSFH data set begins with *D1all004_* and our small data set with four variables is called *ReadAFewVariables*. Double-clicking on *ReadAFewVariables* opens the spreadsheet view shown in Display B.4.3b. In both the SAS and Stata views, we show the first 15 cases in the data set in Display B.4.3. The values are the same in each view, as expected. For example, the first case has MCASEID of 3 with E1301 of 1, E1302 of 75, and E1305 of 2. We can look back to Display B.3.1 to remind ourselves that these data mean that the mother is still living (E1301 = 1), is age 75, and lives 2 miles away (see Box 4.5).

### 4.1.3: Organizing Files

In general, we find it easiest to keep our work organized if we use the same filename, but different extensions, for a batch program and the results and analytic data files that it creates. In our case, we used the same filename—ReadAFewVariables—with different extensions, for our batch program, results, and analytic data files so that it is easy to link them together. We chose the name using the convention of conveying the meaning of the task ("read a few variables") with capitalization to increase readability of the single word. We saved all of the files in our project folders (*c:\nsfh_distance\Stata* and *c:\nsfh_distance\SAS*; see Box 4.6).

We will take a look at screenshots of the files in these directories, since it is often helpful for students to see these concretely as they start to work with the various files. For Stata, we have a set of four files in our project folder (*c:\nsfh_distance\Stata*).

The first is the original raw data file (*d1all004.NSDstat_F1.dta*), which we discuss how to access in Appendix D. The remainder are the three Stata files: the Stata batch program (*ReadAFewVariables.do*), the analytic data set (*ReadAFewVariables.dta*), and the results file (*ReadAFewVariables.log*) that the batch program creates. Notice that the analytic data file is substantially smaller in size than the original raw data file (104 kbyte versus 70,342 kbyte), because our analytic data file keeps just four variables.

For SAS, we have five similar files in our project directory (*c:\nsfh_distance\SAS*).

The original raw data file is in SAS format (*d1all004_nsdstat_f1.sas7bdat*) and its associated formats file (*formats.sas7bdat*) as well as the SAS batch program (*ReadAFewVariables.sas*), the analytic data file (*ReadAFewVariables.sas7bdat*), and results file (*ReadAFewVariables.rtf*) that the batch program creates.

As we have noted, project directories such as these help us to keep our files organized, and doing so is important for checking our own work and sharing our work with others. The batch program provides a ready record of all of the steps from the raw data to the created analytic data file. Retaining this file allows you later to verify that the steps were correct, easily make modifications as needed, and readily share your work with others.

Your project directory is also important because you use it to tell SAS and Stata where to look for files and where to save files. When SAS and Stata were installed on the computer, default locations where they would look for and save information were defined. Some of the files that are created during a session are only saved temporarily. They are erased when you exit the session (close the software). Both SAS and Stata provide ways to save changes permanently and shortcuts that allow you more easily to save files in your project directory.

Stata uses a concept called the **working directory**. The working directory is set by the default installation of Stata, usually as a file folder location, usually something like *c:\data*. You can override the default location when you double-click on a Stata file to launch Stata. When you do so, Stata uses that file's location as the working directory. We did this when we started Stata and thus the working directory is our project directory, *c:\nsfh_distance\Stata.* This working directory is shown in the lower left of the Stata windows (the dotted circle in Display B.4.1b). You can also change from the default working directory to your project directory after launching Stata by typing:

```
cd c:\nsfh_distance\Stata
```

in the command window (cd stands for "change directory"). This can also be accomplished by selecting "File\Change Working Directory" from the menus, which allows browsing to the desired directory. Changing the working directory to your project directory is convenient because Stata will look for and store files in this location; thus, if we change our working directory to our project directory, we do not need to type the entire path when we want to refer to a file in our batch program (e.g., we can type *ReadAFewVariables* rather than *c:\nsfh_distance\Stata\ReadAFewVariables*).

SAS uses a concept called the **libname** ("library name") as a shortcut. You can use the libname to assign a shortcut name for a longer project directory pathname. For example, the batch program shown in the window in Display B.4.2a includes the following statement:

```
libname LIBRARY 'c:\nsfh_distance\SAS';
```

This libname associates the name LIBRARY with the path c:\nsfh_distance\SAS. It is because we had defined this libname that we were able to find our data files in the library file drawer icon (circled in black in Display B.4.2a) when we used the spreadsheet data view above.

We could use this *libname* more extensively in our batch program whenever we want to refer to files stored in our project directory, although we will generally not do this in this book because students sometimes find the necessary syntax too abstract. For example, to refer to the file:

```
c:\nsfh_distance\SAS\ReadAFewVariables
```

we would need to use the following syntax:

```
LIBRARY.ReadAFewVariables
```

*Libnames* are convenient shorthands if you move on to writing more complicated SAS batch programs in your future work; but, in this book, to keep our SAS batch programs less abstract, we will use full paths to refer to the location of files in SAS batch programs such as

```
c:\nsfh_distance\SAS\ReadAFewVariables
```

## 4.2: WRITING A SIMPLE BATCH PROGRAM

Display B.4.4 contains the contents of the *ReadAFewVariables* batch programs and a portion of the results. These simple batch programs contain the basic syntax we need to read a data file, save an analytic data file, obtain a variable's frequency distribution, and save the resulting frequency distribution. We will discuss each command in the text. They are also summarized in Display A.4.2. Appendix F provides information about using each software's help files to learn about more features for these commands (although what we provide in the text is sufficient for what we need to accomplish in this chapter).

### 4.2.1: Basic Syntax to Read and Save Data Files

In both SAS and Stata, we must first open a data file before we can analyze its contents, often referred to as "reading the data into memory" or "using the data file." If our batch program modifies a raw data set (for example, we read only a portion of its cases or variables into memory) we may also want to save this smaller analytic data set (so that we can share it with others or use it in future batch programs).

**Reading a Stata Data File**

The command we will use to read Stata data into memory is:

```
use <variable list> using <data filename>
```

Since this is the first command we will discuss in detail, we will comment on the conventions we will follow throughout the book. The commands will be shown in a `different font`. Items you will supply are in angle brackets < >. Optional items are shaded.

In our case, the <data filename> would be the raw NSFH data file `d1all004.NSDstat_F1`. Stata assumes the file name has the extension *.dta*, so we do not need to list the extension explicitly (although if your file name had spaces, you would need to put it within quotes `use "my file.dta"`). We also only want to use four of the 4,355 variables in the raw data file, so we will use the optional shaded portion of the command and write the names of the desired four variables, separating each name with a space, in place of <variable list>. This is the typical structure of a variable list (write the names of one or more variables, separated by spaces). The resulting command would be:

```
use MCASEID E1301 E1302 E1305 using d1all004.NSDstat_F1
```

In Stata's language, this command asks Stata to read four variables (`use MCASEID E1301 E1302 E1305`) from the NSFH raw data file (`using d1all004.NSDstat_F1`). If we wanted to read additional variables (as we will below) we can simply add them to the list. Notice that we do not include the path to the data filename because we assume you have changed the working directory to your project directory. Otherwise, we would need to type:

```
use MCASEID E1301 E1302 E1305 using c:\nsfh_distance\Stata\d1all004.NSDstat_F1
```

Since this is the first time we have written a Stata command, it is also helpful to point out that Stata is **case-sensitive**, meaning that if a variable name includes a capital letter, we must capitalize it and if a command name does not include a capital letter, then we must write it in lowercase. In other words, `USE` and `Use` are not equivalent to `use`, and, `mcaseid` and `Mcaseid` are not equivalent to `MCASEID`. The NSFH raw variable names were capitalized in the codebook and are capitalized in the raw data file, so we must capitalize them when we refer to them in our batch program.

### Saving a Stata Data File

When we create a modified data file (e.g., reading just four of a data set's variables) that modified data file is available during the current session (until we close the software). We can always recreate the modified data file in a future session by re-running the batch program that created it; but if we want to save the modified data file, so that it is easier to use in the future or so that we can share it with others, we need to save it permanently. For example, when we expand our batch program below to define new variables, we will save an analytic data file with all the variables we need for regression analyses in future chapters.

Stata's command to save a data file is called `save`. The general syntax for this command is:

```
save <data filename>, replace
```

As emphasized in Chapter 3, we want to save such modified analytic data files with a different name from the original raw data file. We chose the name *ReadAFewVariables*. Stata automatically adds the *.dta* extension. So, our command would be:

```
save ReadAFewVariables, replace
```

The word `replace` after the comma is referred to as an **option**. (In Stata's syntax, words that follow commas are options.) We recommend using this option with the `save` command in your batch programs. You will often need to run a batch program multiple times. After the first run, when the data file already exists (e.g., once *ReadAFewVariables.dta* exists in the project directory), Stata would stop with an error message such as `file ReadAFewVariables.dta already exists` unless you use the option `replace`. Adding the option `replace` to the `save` command in your batch programs will avoid this error.

## Reading and Saving a SAS Data File

In SAS, we use one basic command, referred to as the **DATA Step**, to read and save data files (see Box 4.7). The basic syntax is:

```
data <path and data filename>;
  set <path and data filename>
  (keep= <variable list>);
run;
```

The <path and data filename> following the word `data` is the name of the file to be saved; in our case it will be *c:\nsfh_distance\SAS\ReadAFewVariables* (see Box 4.8). The <path and data filename> following the word `set` is the name of the file we want to start with, in our case the raw NSFH data file *c:\nsfh_distance\SAS\d1all004_nsdstat_f1*. Although students often find it counterintuitive that the original data file is listed on the second line of SAS's DATA step, we can think of it as starting with (*setting*) our original data (in this case the raw data file) and creating the new data file (in this case our analytic data file) listed in the first line.

The variable list is the same as we used in our Stata command above `MCASEID E1301 E1302 E1305`. Thus, substituting into the general syntax would give the following:

```
data "c:\nsfh_distance\SAS\ReadAFewVariables";
  set "c:\nsfh_distance\SAS\d1all004_nsdstat_f1"
      (keep = MCASEID E1301 E1302 E1305);
run;
```

Note that SAS assumes the extension *.sas7bdat* for our data files, so we do not need to list the extensions explicitly.

> ■ **Box 4.7**
>
> SAS separates its programs into two major sections, the DATA step and the PROC steps. The DATA step reads data into memory. Any data manipulation must occur in the DATA Step, such as creating new variables, keeping variables, or keeping a portion of the data. The PROC steps calculate statistics. The programs we write in this book typically have one DATA step followed by one or more PROCs.

> ■ **Box 4.8**
>
> If the path is omitted from the filename following the word `data`, then the new data file is created temporarily. It is available during the current SAS session, but deleted when the session ends. In contrast, if a path is specified, the data set is saved in that location, and available in future SAS sessions. The advantage of the temporary file is to avoid cluttering the computer with data files (since the temporary file can always be recreated by re-executing the program). As we have noted, a permanent file is useful once you have finalized all of your data manipulation and are ready for analysis, especially if the raw data file is sizable, and when you need to share an analytic file with others. In Stata, we can similarly only create the analytic data file temporarily if we do not include a `save` command in our program.

There are two general differences between SAS and Stata syntax that we will pause to point out. First, you may have noticed that each line of SAS syntax ends with a semi-colon. The semicolon tells SAS where the command ends. Because SAS keeps reading until it reaches the semicolon, a command can extend over multiple lines. Breaking a command over multiple lines can be useful in organizing your batch program (as we discuss in Section 4.5.1). In fact, above we put the `keep` portion of the `set` statement on a second line to make the command easier to read. We will learn later how to achieve this result in Stata, which by default requires commands to appear on only one line. In addition, SAS (unlike Stata) is *not case-sensitive*. Your commands and your data file names and variable names are interpreted in the same way by SAS if written in lowercase, uppercase, or a mixture of both. In other words, `Data` and `DATA` are interpreted as the same command as `data`. In fact, the SAS documentation often capitalizes key command names when referring to them, as we will sometimes do in the text.

There is also one unique feature of SAS that we'll note here. The final line in our command is `run`; The RUN statement tells SAS to submit the DATA command for processing. We will see that most of the commands we use in SAS have a matching RUN statement at the end. Although technically there are many situations in which your batch program will successfully execute without these paired `run`; statements (and thus you may see that some peers or collaborators omit them from their batch programs) it is best to get into the habit of including the RUN statements (to avoid the problem of omitting them when they are essential).

### 4.2.2: Basic Syntax to Check Variables Against the Codebook

In the long run, it is worth spending the time to check that your downloaded variables are correct before moving forward. As an example of this process, we will check the frequency distribution of one of the NSFH variables, E1301, against the frequency distributions shown in the BADGIR codebook.

### Checking a Variable's Distribution in Stata

In both SAS and Stata there are multiple ways to obtain a frequency distribution. We will use the following Stata command which provides useful information about the variable, including its frequency distribution:

```
codebook <variable list>, tabulate(400)
```

The option `tabulate(400)` is useful if some variables have many values (by default the frequency distribution is only shown for up to nine values; the tabulate option requests that it be shown for more values, for example we changed from the default to 400). As an example, we will just include a single variable, E1301, in the variable list:

```
codebook E1301, tabulate(400)
```

The results, shown in Display B.4.4, can be compared to the results from BADGIR, shown in Display C.3.6. Notice that the values and labels match BADGIR exactly, as does the number of cases listed for each value (column labeled *Freq* in Stata and labeled *N* in BADGIR). For example, both show that 8,307 respondents reported that their mothers were still living.

### Checking a Variable's Distribution in SAS

We will use the SAS command statement PROC FREQ similarly to request a frequency distribution. The general syntax is:

```
proc freq; tables <variable list>; run;
```

Notice that like the DATA statement, the PROC FREQ statement has a matching RUN statement, which we recommend always to include in your batch program. In our case, to request the frequency distribution for E1301 we would type:

```
proc freq; tables E1301; run;
```

As shown in Display B.4.4, the results match those seen for Stata and the original BADGIR codebook (Display C.3.6).

### 4.2.3: Basic Syntax to Save Results

By default, results appear only on the screen in the SAS Output Window and the Stata Results Window. Typically, we would like to save these to use when we write about our results in a paper (or when we answer homework questions for a class).

In Stata, we accomplish this with the command:

```
log using <filename>, replace text
```

Stata will add the *.log* extension to the saved results file. The option `replace` asks Stata to replace the file, even if it already exists, similar to the `replace` option on the `save` command. The option `text` asks Stata to store the results in plain text format. The default is a special format called `.smcl` which can only be read by Stata (whereas the plain text file can be read by any text editor or word processor). Our examples of Stata output in this book come from the plain text format (we change the font to `Courier New` of font size 9 to make it easier to read in a word processor). In our case, we want to use the same name for our log file as our batch program file, so we would type:

```
log using ReadAFewVariables, replace text
```

Each `log using` command should be paired with a `log close` command. We usually put the `log using` command near the start of our batch program, and put the `log close` command in the last line of our batch program. Then, all results between the two commands are saved in the *.log* file. Display B.4.4 provides an example of the full batch program.

Recent releases of SAS allow results to be saved in Rich Text Format (RTF). This format is convenient because it can be opened by any word processor, and retains special formatting, such as table gridlines. The command to create the .rtf file is

```
ods rtf body=<"path and filename.rtf">;
```

In our case, the command would be:

```
ods rtf body="c:\nsfh_distance\SAS\ReadAFewVariables.rtf";
```

| ■ **Box 4.9** |
| --- |
| SAS will prompt you to open the .rtf file with the program associated with .rtf on your computer (often Microsoft Word). If you open the file, be sure to close it before rerunning the program. |

This command should be paired with the `ods rtf close;` statement. As with Stata, we typically put these statements at the beginning and end of the batch program, respectively. Note that we placed the filename in quotes and explicitly put the *.rtf* extension in the filename used in the `ods` command. We have found that this improves performance (see Box 4.9).

## 4.3: EXPANDING THE BATCH PROGRAM TO CREATE NEW VARIABLES

The data file we created above included just four of the variables needed for our research questions (see again Display B.3.1). We now will expand the batch program to include the remaining variables and create the new variables. We also kept all of the cases in the data file above, although we planned to make some exclusions for our research question (exclude respondents whose mothers are not alive, respondents who coreside with their mothers, and respondents whose mothers live outside the USA). We will add the command to keep only a subset of cases now as well. First, we will learn the use of expressions, a central part of all of these tasks.

### 4.3.1: Expressions

We use **expressions** to write with SAS and Stata syntax (using variable names, values, and mathematical symbols) how we would like to define a variable or subset the data.

For example, to subset our data we might say in plain English "I would like to keep only respondents whose mothers are still living." Or, we could refer to the variable names and values, but still say this in English, "I would like to keep respondents who have the value 1 on the variable E1301." Similarly, we listed in Display B.3.1 a number of ways we would like to recode some of the values of some variables. We could again say this in plain English. For example, "I would like to treat respondents who completed a GED the same as those with a high school diploma." Or, referring to the variable names and values, we might say "If a respondent is coded a 25 on M502, I would like to replace her value with a 12." How do we communicate these requests to SAS and Stata? In each case, we use an expression—a combination of variable names, numbers, and mathematical symbols—to do so. The mathematical symbols are referred to as **operators** and common symbols used in expressions are listed in Display A.4.1.

The notes under Display A.4.1 tell us where on the keyboard to find some symbols that may be unfamiliar. SAS can use either the letter abbreviations or symbols. Although the letters may be easier for you to interpret at first, using the symbols will be more parallel to Stata and become easier to interpret over time. Most symbols are similar between SAS and Stata, with the exception of the highlighted cells in the top row. Stata uses the double equal symbol to denote equality in expressions, whereas SAS uses the single equal sign. For example, `M502=25` in SAS means the same thing as `M502==25` in Stata.

## 4.3.2: Creating and Modifying New Variables

It is worth reiterating here that we highly recommend not altering the variables in the raw data file. Instead, create a new variable for your analytic data file. Not only does this allow you to give your analytic variable a name that is meaningful to you, but it also preserves the original variables so that you can double-check your work and easily trace back to the raw data.

### Stata Syntax for Creating and Modifying New Variables

The basic commands for creating and modifying variables in Stata are:

```
generate <variable name> =<expression> if <expression>
replace <variable name> =<expression> if <expression>
```

For example, referring to our summary in Display B.3.1 (in Chapter 3), for the respondent's age we might type:

```
generate g2age = M2BP01
```

Or, for earnings converted to 2007 from 1986 dollars, we might type

```
generate g2earn=IREARN*207.342/109.6
```

Importantly, this code has not yet addressed the missing data values. We can use Stata's optional `if` qualifier to do so. The `if` qualifier asks Stata only to run the command for the subset of cases that are consistent with its expression. For example, `if E1301==1` would restrict a command only to cases in which the mother is living. The `if` qualifier can be used when creating new variables to ask that the command be processed only for cases that have valid values on the original variable. Any case that is not consistent with the `if` qualifier on the generate command will be coded a '.' missing on the new variable.

For example:

```
generate g2age = M2BP01                if M2BP01<97
generate g2earn= IREARN*207.342/109.6  if IREARN<9999997
```

will result in cases with missing value codes on each variable to be '.' missing on the new variable. We determined what value to use in the expression (97 versus 9999997) based on our notes in Display B.3.1.

Notice that we used the *less than* operator in the `if` qualifier. As with writing in English, programming syntax can be expressed in many different equivalent ways. Any that achieves the objective of conveying which cases to run the command for is fine, although some may be shorter than others. For example, based on Display B.3.1, we could write `if M2BP01<97` equivalently as `if M2BP01~=97 & M2BP01~=98`. Some students may find `if M2BP01~=97 & M2BP01~=98` easier to connect back to the two missing value codes for the variable M2BP01 summarized in Display B.3.1. You should feel free to use an expression that makes sense to you, as long as it conveys the right meaning (i.e., in this case the meaning in words is "if the respondent's age is not missing").

We can also use the `if` qualifier, along with the replace command, to address the other coding modifications in our notes to Display B.3.1. For example, in order to recode the GEDs when we create the *g1yrschl* variable, we could use the following syntax:

```
generate g1yrschl=M502 if M502<98
replace g1yrschl=12 if M502==25
```

The first command creates the new variable, `g1yrschl`, carrying over all the original values except missing data codes which are translated to '.' missing. The second command replaces the value of 25 with the value of 12 for respondents who report that their mother received a GED. Notice that in the expression of the `if` qualifier we use the double-equal sign to denote equality (see again Display A.4.1), but the equals sign is used in conjunction with the `replace` and `generate` commands (e.g., `replace g1yrschl=12`). In Stata, we use the double-equal sign within expressions, generally as part of an `if` qualifier, where the computer will check whether or not the expression is true (does M502 take on the value of 25 for this case?).

## Checking Created Variables in Stata

Similar to our check that the frequency distributions of the downloaded data were consistent with the study's codebooks, it is important to check that newly created variables recode the original variable in the way we intend. As you become more comfortable with programming, you may be tempted to skip this checking step, feeling you can look at the code to see that it is correct. We strongly recommend that you always check newly created variables.

One concrete way to spot check created variables is to use the spreadsheet views we introduced in Section 4.1.2. After running a command that adds a new variable to the data set, that variable will appear as a new column in the spreadsheet view. This approach is particularly useful for variables that take on many values. For example, this approach might be helpful as a check of our `g2earn` variable. For variables that take on only a few variables, we can get a more complete view of all of the cases by cross-tabulating the created variable against the original variable. This might be helpful for a variable such as the mother's education, in which we want to verify that the GED code of 25 on the original variables gets appropriately converted to a 12 on the new variable.

In Stata, the syntax for cross-tabulating two variables is:

```
tabulate <var1> <var2>, missing
```

We typically include the `missing` option. By default, Stata omits cases that are coded `'.'` missing on either variable from the cross-tabulation. But, we would like to verify that cases with missing value codes on the original variable are appropriately converted to `'.'` missing on the new variable, so we want to include them in the cross-tabulation.

In our case, for the education variable, we would type:

```
tabulate g1yrschl M502, missing
```

Display B.4.5 provides the commands and results for a short batch program that creates the new earnings and education variables. In the interest of space, we present just a portion of the cross-tabulated results between the created and original earnings variables. We also show a portion of the spreadsheet view of the new data.

In the cross-tabulation, we see that cases generally retain their value from the original variable to the new variable, as we expected based on the `generate g1yrschl=M502` portion of our command. For example, 290 cases were coded a *0* on the original variable and are coded a *0* on the new variable; and, 27 cases were coded a *1* on the original variable and are coded a *1* on the new variable. We can also see that our *if* qualifier, `if M502<98`, accomplished our goal of converting cases coded 98 or 99 on the original variable to '.' missing on the new variable (values in red circle in Display B.4.5). Finally, our `replace` command, `replace g1yrschl=12 if M502==25`, also worked as we desired. The value of 4 circled in black shows that four cases were originally coded a 25 on M502 and are now coded a 12 on *g1yrschl*.

The spreadsheet view also allows us to verify some values for the education variable (although no values of 25 on the original variable are visible in the cases shown in the screenshot, illustrating why the cross-tabulation is more helpful for this variable). The spreadsheet view is particularly useful for the earnings variable. The value of 1014.008 highlighted in the seventh row of the spreadsheet view allows us to spot check the results to see if `generate g2earn=IREARN*207.342/109.6` worked properly for this case. For the seventh case, we need to substitute the original value of IREARN into the formula `IREARN*207.342/109.6` In the seventh row, IREARN is 536. Plugging into the formula gives us, `536*207.342/109.6=1014.0083` which matches the value of the new variable *g2earn* that is highlighted in the seventh row. Based on the spreadsheet, we can also see that the *if* qualifier, `if IREARN<9999997`, appropriately converted the visible cases larger than 9,999,997 on IREARN to '.' missing in the first and sixth rows.

### SAS Syntax for Creating and Modifying New Variables

In SAS, there is not a command devoted to creating a new variable. We simply type:

```
<variable name> =expression;
```

For example,

```
g2age = M2BP01;
g2earn=IREARN*207.342/109.6;
```

Importantly, though, this command must occur within the DATA step (between the word DATA and the RUN statement). This means that we must anticipate all of the new variables that we will need in our batch program, and create them all in the DATA step or remember to put them between DATA and RUN when we later add them to the batch program. For example, we cannot create a new variable outside of the DATA step, right before a PROC FREQ where we want to use it. (In contrast, Stata does allow variables to be created anywhere within a batch program, although we generally find it easier for proofing purposes to group variable creation together at the beginning of a batch program.)

We can also use an `if` statement to address missing data in SAS, as in Stata. However, in SAS, the `if` qualifier comes at the *beginning* of the command and is paired with the word `then`. The general syntax is:

```
if <expression> then <variable name> = <expression>;
```

For example,

```
if M2BP01<97 then g2age = M2BP01;
if IREARN<9999997 then g2earn=IREARN*207.342/109.6;
```

### Checking Created Variables in SAS

We can use both the spreadsheet view and the cross-tabulation to check our new variables in SAS, as we did in Stata. In SAS, we can use the PROC FREQ command to request crosstabulations. The syntax is:

```
proc freq; tables <var1>*<var2> /missing; run;
```

The `/missing` portion of the command is a SAS option (in SAS, options follow a slash rather than a comma). As with Stata, the option requests that missing values be shown in the cross-tabulation.

For the education variable that we examined in Stata, we would type:

```
proc freq; tables g1yrschl*M502
/missing; run;
```

Display B.4.6 shows a short batch program in SAS that accomplishes the same tasks as the short Stata batch program shown in Display B.4.5. The SAS cross-tabulation and spreadsheet show the same results as did Stata (although in SAS, the '.' missing value is treated as the lowest rather than highest possible value, so it is listed in the first rather than the last row of the table; see Box 4.10).

> **■ Box 4.10**
>
> In the interest of space, we used additional options in SAS so that only the number in each cell was shown (rather than also row, column, and total percentages). So, our actual command was `proc freq; tables g1yrschl*M502 /nocum nocol norow nopercent missing; run;`

## 4.4: EXPANDING THE BATCH PROGRAM TO KEEP A SUBSET OF CASES

We would like to restrict our entire analytic file using the variables capturing whether the mother is alive (E1301), whether the respondent was born in the USA (M497A), and whether the mother doesn't coreside with the respondent and doesn't live outside the USA (E1305). We can translate these plain English statements into SAS or Stata syntax expressions, similarly to what we did above in creating new variables, and then use `if` qualifiers to keep just a subset of the data.

Based on the documentation in Display B.3.1 and the operators in Display A.4.1, we can write the corresponding syntax expressions for these statements:

| In Words | In Stata Syntax |
| --- | --- |
| Mother is alive | E1301==1 |
| Respondent was born in the USA | M497A<=51 \| M497A==990 \| M497A==996 |
| Mother doesn't coreside with the respondent and doesn't live outside the USA | E1305~=9994 and E1305~=9995 |

These three expressions can then be combined together with additional `&` operators to create an `if` qualifier:

```
if E1301==1 & (M497A<=51 | M497A==990 | M497A==996) & E1305~=9994 & E1305~=9995
```

Note that we put the second expression, which contains symbols for "or," in parentheses to help us proof it and assure that it is properly interpreted by SAS and Stata. In Stata, we can place this qualifier at the end of our `use` command. Then, Stata will only read into memory the subset of cases that meet the conditions of these expressions.[1] In SAS, we put a similar `if` statement in the DATA step (although the double equals must be changed to single equals for SAS; see again shaded text in Display A.4.1). We added these `if` qualifiers, and show in Display B.4.7 the full batch program, including all of the new variable creation listed in Display B.3.1. We also added some finishing touches to the batch program, which we will turn to in the next section.

## 4.5: SOME FINISHING TOUCHES

We end the chapter by discussing some finishing touches for a batch program—adding comments and spacing to make the program easier to understand and proof, and checking the program for errors.

### 4.5.1: Comments and Spacing

Looking at Display B.4.7, you may realize that you have forgotten what some variables represent, what some code accomplishes, or why some decisions were made. Adding comments to your batch program helps you to remember such intents and decisions, especially when you come back to code that you wrote days, weeks, months, or even years earlier. These comments also help others, including your adviser, peers, or collaborators, to understand your batch program.

How can we add a comment? SAS and Stata cannot directly distinguish notes that we might type into our batch program from commands meant to manipulate or analyze the data. However, both packages provide special symbols that can be used to tell the software that what follows is not a command to be run.

- In either SAS or Stata, comments may be added to batch programs using pairs of /* and */. These symbols can be used to "comment out" any section of the batch program. The software ignores whatever comes between the symbols. These comment symbols can be useful for lengthy comments as well as for excluding a subsection of the batch program from running (useful when debugging—locating and fixing errors).
- Both SAS and Stata also allow single lines or commands to be "commented out" by placing an asterisk (*) at the beginning of the line or command.
- Stata also allows the double slash // to be used to comment out anything from the double slash to the end of the line and the triple slash /// to be used to comment out the end-of-line delimiter allowing for commands to extend over one line.

Spacing can also be helpful to assist in understanding and debugging a batch program. You can add white space to your batch program simply by adding tabs or lines. For example, commands do not have to begin in the first position of a line, but may be tabbed over. We used such tabbing in our SAS code in the left column of Display B.4.7. We also added spacing within lines to make it easier to scan and check them (e.g., lining up the word then). We added a comment to remind ourselves about the factor multiplied by IREARN. We included both what the factor did and where we obtained the CPI values online, so that it would be easy to return to the source if needed in the future.

Stata's default requirement that a command cannot extend over more than one line is sometimes limiting for adding spacing, and very long lines can be difficult to read. Stata does offer several ways around this default, though. The preferred method used by Stata programmers is to put three forward slashes (///) at the end of each continuing line (see Display B.4.7 for an example).[2]

## 4.5.2: Debugging

Students usually find it frustrating when their batch program ends with an error message, but even experienced programmers expect to spend some time finding and fixing errors—debugging—their batch program. Because the statistical software packages lack human intelligence, they cannot forgive our syntax mistakes as someone might if we were just learning a spoken language. Many typos will cause SAS and Stata to end abruptly and wait for us to fix the error and re-submit the batch program.

Common errors include misspelling commands, misspelling file or variable names, and (in SAS) forgetting the semicolon at the end of lines. SAS's **enhanced editor** is useful for finding many such errors. For example, it uses red font to show syntax it cannot understand as you type (even before you try running the syntax).[3] Display B.4.8 shows an example in which we misspelled `if` and `proc`. On your screen, these error messages would stand out in red font (circled in the Display). We also left the semicolon off of the end of one line. This error would not be displayed by SAS in red font, so it would not be quite as easy to see, but the `if` following the missing semicolon would be in black font rather than the blue font of the other `if` statements.

**■ Box 4.11**

We like to use the shortcut keys F7, F6, F5 on your keyboard to move among the windows when we are debugging a program. The key stroke Ctrl-E can also be used to erase the Log and Output windows before rerunning a program. Try typing F7, Ctrl-E, F6, Ctrl-E, F5, F8 (where F8 submits the program).

If you do not catch errors before submitting the batch program, then SAS will issue an error message in the Log window. Examples are shown in Display B.4.9. The errors again would stand out on your screen because they would be in red font. A message accompanies the ERROR note for each of our three errors. Some of these messages make it easy to diagnose the problem, while others do not. The middle message clearly tells us that we have forgotten a semicolon. But, the first simply says the statement is "out of order" and the last that there is "no default data file." In these cases, we have to scrutinize the syntax around the underlined code to look for errors (see Box 4.11).

Unlike SAS, Stata's editing window does not highlight errors with colors. However, Stata stops running the batch program when an error occurs, requiring you to fix the error before it moves forward in the code. In Display B.4.10 we show Stata's message when we create a spelling error in the command `generate`. Stata is also very careful about overwriting data and log files. As we noted above, unless we explicitly tell it to replace an existing data file or log file, Stata will stop with an error message. Thus, it is useful to place the option, `replace` at the end of `log` and `save` statements, as we suggested above.

Stata's caution about helping us to avoid losing our work can also result in errors when we rerun a batch program repeatedly, as we write it in stages or debug it. If a data set is already in memory or a log file is already open from the first run of the batch program, we will get an error message when we try to reopen the data or re-open the log in a second run of the batch program (e.g., `no; data in memory would be lost`; `log file already open`). We can avoid these problems by starting our batch program with the commands `capture drop _all` (which drops any data that may be in memory) and `capture log close` (which closes any logs that may be open). When we put these at the beginning of a batch program, they allow us to "start fresh" every time (knowing that our batch program will later use a data set and open a log).

It is also useful to start all of your Stata batch programs with two additional commands. The command `version 10` (or whichever version you are using) tells Stata what version to use when interpreting the batch program. This assures that your batch program will run, even if changes are made to future releases of Stata. We also like to type `set more off` at the beginning of a batch program, so that we don't have to press a key to move to the next screen as the results appear (instead, we will open the log file to view the entire set of results once the batch program has successfully run without errors).

Like naming files and variables, and writing expressions, there are many different ways to lay out and annotate a batch program. You should experiment with what works best for you, with the ultimate goal of making your batch programs easy for you (and others) to understand and proof.

## 4.6: SUMMARY

SUMMARY
**4**

In this chapter, we learned the basics of writing batch programs in SAS and Stata. We learned how to work with each software package and where to write our batch programs, how to save and locate our results, and how to save and locate our analytic data files. We learned some basic syntax and wrote a short batch program to read data, check variables from the raw data file against the documentation, and check the creation of new variables for our analytic data file. In learning how to create new variables and keep a subgroup of the full sample, we introduced how to use expressions to tell SAS and Stata what we want to accomplish. We also reinforced the importance of thoughtfully naming our files and storing them in project folders and of using comments and spacing to help us check and document our work.

## KEY TERMS

Case-Sensitive

Command Window

DATA Step

Data Browser

Do-File Editor

Enhanced Editor

Expression

Libname

Operators

Option

Session

Table Editor

Working Directory

## REVIEW QUESTIONS

**4.1.** Where in the batch program can you create new variables in SAS and Stata?

**4.2.** How do you tell SAS and Stata where a line ends, by default?

**4.3.** How can you add a comment to SAS and to Stata?

**4.4.** What are some instances in which we can use expressions?

**4.5.** What are key differences in operators between SAS and Stata?

## REVIEW EXERCISES

**4.1.** Follow the steps in Appendix D to create the SAS and/or Stata raw data files for the NSFH data set.

**4.2.** Replicate the ReadAFewVariables example in SAS and/or Stata from Section 4.1.1; that is, launch the software, open the editor, and type the code to match the batch program shown in Display B.4.4. Be careful to check for typos (which can lead to error messages). Run the batch program and view the results. If necessary, use the suggestions from Section 4.5.2 to identify and fix errors.

**4.3.** Request the frequency distributions for E1302 and E1305 from SAS and/or Stata and compare them to the results in BADGIR (shown in Display C.3.7 and Display C.3.8) to confirm that these variables were correctly downloaded.

**4.4.** Write and run the CreateData example shown in Display B.4.7 in SAS and/or Stata. Use the techniques described in Section 4.3.2 to verify that the variables are created correctly.

## CHAPTER EXERCISE

Download the NOS 1996–7 data file from ICPSR. Create a SAS and Stata version of the data file (see Appendix E for suggestions). Extract all of the raw variables that you identified in the chapter exercise to Chapter 3. Create three new variables for use in Chapters 5 and 6.

| Created Variable Name | Variable Coding (In Words) |
| --- | --- |
| *age* | The difference between the year the interview was conducted (*mdoc*) and the year the organization started operations (*a2*). The study documentation (and tabulating the downloaded variable) shows us that *mdoc* is stored as a 3–4 digit value capturing the MonthDay of the interview. Interviews were conducted between June 10, 1996 and June 13, 1997, so we can logically use 1996 for the interview year if *mdoc* is between 610 and 1230 (June 10 and December 31) and use 1997 for the interview year if *mdoc* is bweteen 101 and 609.[4] The *status* variable tells us whether the interview was conducted by mail or by phone. If we tabulate *mdoc* for those completed by mail, we see that the year was not captured for |

these cases. We will treat them as 1997.[5] Be sure to pay attention to missing data codes on *a2* and *mdoc* variables in your code to create *age*.

*ftsize*

Use the variable *a7* to code the total number of full-time employees. Be sure to pay attention to missing data codes on the *a7* variable itself when you create your new variable.

In addition, because the questionnaire was created in paper-and-pencil there are some inconsistencies with the other key variables, including the number of full-time employees in the *core* and *management* occupations. Although we will preview in Chapter 12 additional more sophisticated ways to deal with the resulting inconsistencies and missing data, for now use the number of core (*f3*) and management (*f10*) employees to deal with the worst inconsistencies. To simplify the coding for now, first replace *ftsize* with *f3* if *ftsize* is less than *f3* and *f3* is not missing and is not zero and then replace *ftsize* with *f10* if it is still less than *f10* and *f10* is not missing and is not zero.

*mngpctfem*

Create the percentage of managers who are female using the *f10* and *f11* variables. Remember, *f10* is the total number of full-time managers and *f11* tells us whether female representation was reported as a number or percentage. If *f11* indicates that a number was reported, then *f11_n* contains that number. If *f11* indicates that a percentage was reported, then *f11_p* contains that percentage. When a number was reported, we need to divide by *f10* to calculate the percentage. We have to be careful about missing data on all the variables as we make these calculations.

We also need to calculate a similar variable for core employees (*corepctfem* using the *f3* and *f4* variables), since if the core job was a management position, then respondents skipped the section about managers. *f7_int* tells us that about 128 organizations had a management position as their core job. For these 128 cases, *mngpctfem* should be replaced with *corepctfem*. Because of the paper and pencil administration of the survey, it is possible to get calculated values of *mngpctfem* that exceed 100. Again, there are various ways we might deal with these cases. For the purpose of our chapter exercises, we will replace these few cases with observed values that exceed 100 with 100.

## COURSE EXERCISE

Choose one of the data sets that you identified in Chapter 2 for one of your research questions. Download the raw data. Create an analytic data set that contains at least three variables: the case identifier, a dependent variable, and an independent variable. Recode these variables as needed (e.g., to convert missing data codes to `. missing`).

If you identified multiple candidate data sets, you may want to focus on the simplest (e.g., a data set that can be downloaded directly in SAS or Stata format as a single file). Or, you can utilize one of the example data sets used in the textbook (NSFH or NOS).

*Part 2*

# THE
# REGRESSION
# MODEL

*Chapter 5*

# BASIC CONCEPTS OF BIVARIATE REGRESSION

## CHAPTER 5: BASIC CONCEPTS OF BIVARIATE REGRESSION

With our analytic data file now ready, we are poised to use regression analysis to examine our research questions. We could open our analytic file in SAS or Stata and easily estimate a basic regression model right now (possibly simply using the pull down menus). If we were thoughtful in examining the results, however, doing so would likely lead to many questions and considerable confusion. Indeed, much work lies ahead of us in learning how to appropriately implement and interpret a regression analysis. We will begin in this chapter with basic concepts of **bivariate regression** with one continuous dependent variable and one continuous independent variable.

Fundamentally, regression modeling involves the algebra and geometry of a function, starting with a straight line. Understanding this basic math aids interpretation of our results. Returning to this basic math will be quite helpful in understanding more complicated models as we move forward, and we will repeatedly revisit the strategies for interpretation that we introduce in this chapter.

Whereas a mathematical straight line function is deterministic, statistical models contain a systematic (straight line) and probabilistic component. Because of this, in regression, there are three basic parameters of the model: the intercept and slope of the straight line and the conditional variance of the distribution of values around that straight line. We will see that this is true in both the population and the sample.

To test social science research questions, we need a reliable strategy for estimating the parameters of the model (the intercept, slope, and conditional variance). Ordinary least squares (OLS) is the most commonly used approach. We look in detail in this chapter at the least squares estimators and their standard errors and how to answer our research questions and evaluate our hypotheses using these estimates. We then discuss a number of strategies for evaluating the substantive size of statistically significant effects (see Box 5.1).

We will use two examples in this chapter, both taken from the NSFH. One uses the datafile we created in Chapter 4. We will learn the syntax to estimate a regression model in SAS and Stata based on *g1miles* and *g2earn* from that datafile. We will also use a second example taken from the NSFH, in which we examine how the hours of chores that employed women complete each week is predicted by their number of children and by their hours of paid work. The batch programs for creating this datafile from the NSFH and reproducing the in-text examples are provided on the textbook web site (*http://www.routledge.com/textbooks/9780415991544*). In this chapter, we use three versions of this "hours of chores" datafile:

(a) the *full sample* of 3,116 women who were currently employed;
(b) a *stylized sample* of 50 women that we created to illustrate the conditional regression equation;
(c) a *hypothetical* population and 5,000 *samples* drawn from it that we created to illustrate the sampling distribution of the slope.

We will also use this hours of chores example in future chapters, when it illustrates a point better than our distance example.

## 5.1: ALGEBRAIC AND GEOMETRIC REPRESENTATIONS OF BIVARIATE REGRESSION

We start with basic concepts of a straight line, first in a graph and then in an equation, using geometry and algebra to reinforce the basic concepts. These basic mathematical concepts give us the building blocks we need in order to understand the regression model.

Regression models with a single predictor variable are typically referred to as **simple regression** or **bivariate regression** (bivariate because two variables are involved—the outcome variable, which we often call $Y$, and the predictor variable, which we often call $X$). At the heart of bivariate regression models is the geometry of a straight line. A straight line is defined by an **intercept**, which denotes the point at which the line intersects the $Y$ axis, and a **slope**, which measures the amount that $Y$ changes when $X$ increases by 1. We can write this algebraically as $Y = \beta_0 + \beta_1 X$ and display this geometrically as:

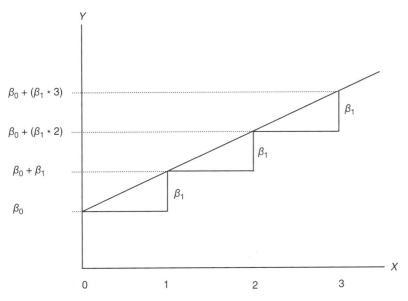

**Figure 5.1**

The graph of the line makes clear that the slope, $\beta_1$, measures the amount of **linear association** between $Y$ and $X$ (see Box 5.2).

If $\beta_1 = 0$ then $Y$ would not change when $X$ increased. This result can be seen both algebraically and geometrically.

Geometrically, when $\beta_1 = 0$ our straight line plot becomes horizontal (see Fig. 5.2). Clearly, $Y$ remains the same regardless of the values of $X$ in this situation. In this "flat line" context, $X$ and $Y$ are unrelated.

Algebraically, if $\beta_1 = 0$ then everything but the intercept "drops out" of the equation. That is, $Y = \beta_0 + \beta_1 X = \beta_0 + 0 * X = \beta_0$. No matter what the value of $X$, $Y = \beta_0$.

The regression equation defines the values along the line. We can substitute any $X$ value into the equation to calculate its associated $Y$ value. With any two pairs of $X$ and $Y$ values we can plot the straight line.

■ **Box 5.2**

We follow the convention of using Greek letters to denote population parameters. We use subscripts to indicate associated variables for the regression coefficients. In this chapter, we represent the intercept with the subscript zero and the single predictor with the subscript 1. In subsequent chapters, numbers 2 and above will designate additional predictor variables. Below, we will put "hats" on the Greek letters to denote sample estimates of these population parameters, such as $\hat{\beta}_0$. Keep in mind that other books and articles may use different notation (e.g., lowercase Roman letters, such as $a$ and $b$, for the sample coefficients).

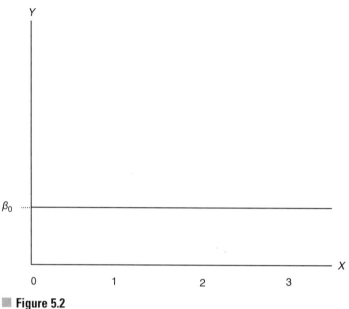

■ Figure 5.2

Suppose we have an equation with an intercept of 3 and a slope of 0.5. We would write this algebraically as $Y = 3 + 0.5X$. Substituting in the values 0, 1, 2, and 3 into the equation results in the following values of $Y$:

| $X$ | $Y = 3 + 0.5X$ | $Y$ |
|---|---|---|
| 0 | $Y = 3 + 0.5 * 0 = 3 + 0.0$ | 3.0 |
| 1 | $Y = 3 + 0.5 * 1 = 3 + 0.5$ | 3.5 |
| 2 | $Y = 3 + 0.5 * 2 = 3 + 1.0$ | 4.0 |
| 3 | $Y = 3 + 0.5 * 3 = 3 + 1.5$ | 4.5 |

We can draw a simple graph of these values to show the line visually (see opposite page).

### 5.1.1: Interpretation of the Intercept

The interpretation of the intercept and slope are also made clear from the algebraic and geometric forms of the regression line.

In regression analysis, the focus is on the $Y$ intercept, referred to simply as the *intercept* for short. The intercept provides the value of $Y$ when $X = 0$. This interpretation always holds mathematically.

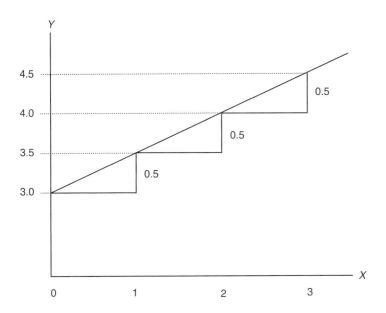

Although the intercept can always be calculated and interpreted in this way, and will always be shown in our SAS and Stata output, it may or may not make sense to interpret it substantively. In social science applications, the intercept is of little substantive interest when zero is not in the range of values for the $X$ variable, either by definition or within the sample. For example:

| | |
|---|---|
| Years of Education | With mandatory enrollment through middle or junior high school, the lower bound may be 6–9 years. |
| Legal Wages | With a minimum wage law, legal wages in many occupations cannot fall below a given level, such as $7.25 h. |
| SAT Scores | SAT total scores are designed to range from 900 to 2,400. |
| Age in a Sample of Adults | Many surveys include age restrictions. Some include only adult respondents (e.g., sampling only persons over, say, 18 years of age). Others focus on older persons (e.g., sampling persons over, say, 65 years of age). |

On the other hand, it is also helpful to keep in mind that zero need not be the minimum value for either $Y$ or $X$. Although most introductions to the simple regression model use a graph like the one above in which only positive values of $X$ and $Y$ are depicted, the $X$s and $Y$s can be less than zero. Consider, for example, a variable constructed to measure the difference in ages between two partners in a couple such as $age_{male} - age_{female}$. This difference will be positive when the male is older than the female and negative when the male is younger than the female. Perhaps we would want to relate these differences in ages across generations, for example with $Y$ capturing the age difference in the younger generation and $X$ capturing the age differences in the older generation.

The graph below depicts a hypothetical model in which the couples' age differentials would be reproduced exactly from generation to generation, i.e., $Y = X$. Here the intercept is zero and the slope is 1. Thus, $Y = \beta_0 + \beta_1 X = 0 + 1X = X$. The mechanical interpretation of the intercept in this case is "$Y$ is zero when $X$ is zero." In this case, the intercept would also have substantive meaning, telling us that when parents in the older generation are of the same age, then the partners in the younger generation are also of the same age.

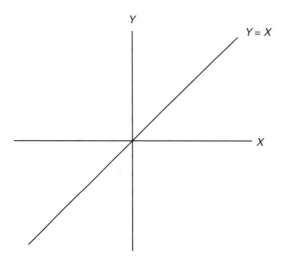

Other examples of variables including negative values might be measures of family income or business profits, in which debts and losses are allowed, or standardized values of a variable constructed using the formula you likely learned in your introductory statistics class:

$$Z = \frac{X_i - \overline{X}}{\hat{\sigma}_X}$$

(5.1)

Notice that $Z$ takes on the value of zero when $X$ is at its mean. We will see in Chapter 8 how subtracting the mean (or another value) from a predictor can be useful for interpretation.

### 5.1.2: Interpretation of the Slope

The slope measures how much the outcome variable, $Y$, changes for each one unit increase in $X$. We assume initially that the relationship between $Y$ and $X$ is a straight line. (We will allow for nonlinear relationships in Chapter 9.) *Because we are modeling a straight line, the amount of change on Y for a one unit increase in X does not depend on the starting value of X.* This is a simple but critical concept. We will come back to it several times later in the textbook so let's explore it in more detail in order to reinforce it.

We can see in Figure 5.1 that the change in $Y$ when $X$ goes from zero to 1 is equivalent to the change in $Y$ when $X$ goes from 1 to 2 which is equivalent to the change in $Y$ when $X$ goes from 2 to 3. We show this algebraically in the table below, explicitly writing the result of substituting each value of $X$ into the equation. Then, the change in $Y$ for a one-unit increase in $X$ is the difference in the resulting value of $Y$ for the current row and prior row:

| | $Y = \beta_0 + \beta_1 X$ | | Change in $Y$ when $X$ increased by 1 | | |
|---|---|---|---|---|---|
| When $X = 0$ | $Y = \beta_0 + \beta_1 * 0$ | $= \beta_0$ | — | | |
| When $X = 1$ | $Y = \beta_0 + \beta_1 * 1$ | $= \beta_0 + \beta_1$ | $(\beta_0 + \beta_1) - \beta_0$ | $= \beta_0 - \beta_0 + \beta_1$ | $= \beta_1$ |
| When $X = 2$ | $Y = \beta_0 + \beta_1 * 2$ | $= \beta_0 + 2\beta_1$ | $(\beta_0 + 2\beta_1) - (\beta_0 + \beta_1)$ | $= (\beta_0 - \beta_0) + (2\beta_1 - \beta_1)$ | $= \beta_1$ |
| When $X = 3$ | $Y = \beta_0 + \beta_1 * 3$ | $= \beta_0 + 3\beta_1$ | $(\beta_0 + 3\beta_1) - (\beta_0 + 2\beta_1)$ | $= (\beta_0 - \beta_0) + (3\beta_1 - 2\beta_1)$ | $= \beta_1$ |

Although the above result may seem obvious, we will use this technique of calculating and substituting values of $Y$ as we work with more complicated models in future chapters. Thus, it is helpful to have this simpler case very clearly in mind.

This constant slope reflects the way in which we have written these models to be linear in $X$. When the model is linear, it allows for a very concise interpretation of the slope: "$Y$ changes by $\beta_1$ units when $X$ increases by 1." We will see in Chapter 9 that nonlinear models require lengthier explanations and interpretations.

In our linear model, what if we changed $X$ by 2 units instead of 1 unit? How much would $Y$ change then? We can use a similar table structure to work out this result. Now, we subtract from the result for $Y$ in the current row the value of $Y$ two rows prior (e.g., $Y$ when $X = 2$ minus $Y$ when $X = 0$, $Y$ when $X = 3$ minus $Y$ when $X = 1$).

| | $Y = \beta_0 + \beta_1 X$ | | Change in $Y$ when $X$ increased by 2 | | |
|---|---|---|---|---|---|
| When $X = 0$ | $Y = \beta_0 + \beta_1 * 0$ | $= \beta_0$ | — | | |
| When $X = 1$ | $Y = \beta_0 + \beta_1 * 1$ | $= \beta_0 + \beta_1$ | — | | |
| When $X = 2$ | $Y = \beta_0 + \beta_1 * 2$ | $= \beta_0 + 2\beta_1$ | $(\beta_0 + 2\beta_1) - (\beta_0)$ | $= (\beta_0 - \beta_0) + (2\beta_1)$ | $= 2\beta_1$ |
| When $X = 3$ | $Y = \beta_0 + \beta_1 * 3$ | $= \beta_0 + 3\beta_1$ | $(\beta_0 + 3\beta_1) - (\beta_0 + \beta_1)$ | $= (\beta_0 - \beta_0) + (3\beta_1 - \beta_1)$ | $= 2\beta_1$ |

More generally, if we increase $X$ by $i$ units, then $Y$ will change by $i\beta_1$ units. We will come back to this result near the end of the chapter.

## 5.2: THE POPULATION REGRESSION LINE

The algebraic and graphic straight lines we've been discussing so far are **deterministic**. $Y$ is predicted exactly from the intercept and slope coefficients and the value of $X$. This model actually captures only one portion of the statistical regression model, what is often called the **systematic component**. The other portion of the statistical regression model is an error term. Adding this error term produces the statistical or **probabilistic** regression model:

$$Y_i = \beta_0 + \beta_1 X_i + \varepsilon_i \tag{5.2}$$

where $\beta_0 + \beta_1 X_i$ is the systematic component and $\varepsilon_i$ is the **nonsystematic** (also known as **random** or **stochastic**) **component**. This model is probabilistic rather than deterministic in that the exact value of $Y$ that we observe depends not only on the fixed value calculated by plugging in the values of the intercept, slope, and $X$, but it also depends on the value of the error term. Later, we will make assumptions about the probability distribution of this error term. We refer to Equation 5.2 as the **population regression line**, assuming that we know or can calculate exactly the values of the intercept and slope (i.e., we know the values of $Y$ and $X$ for every member of the population). Later, we will learn how to estimate these parameters using a sample from the population. The sample regression model, like the population regression model, has both a systematic and stochastic component.

To help understand these systematic and random components of the regression model, we will examine them in a graph. The values predicted from the systematic portion of the equation are the points that plot the regression line. This portion of the population regression line is also known as the **expected value** of $Y$ given $X$ which is denoted as $E(Y|X_i)$. The expected value is essentially another name for the mean and as we shall see $E(Y|X_i)$ is the **conditional mean** of $Y$ for a given level of $X$ (see Box 5.3). It is calculated from the systematic portion of the regression line: $E(Y|X_i) = \beta_0 + \beta_1 X_i$. For example, $E(Y|X_i = 1) = \beta_0 + \beta_1 * 1 = \beta_0 + \beta_1$.

The expected value of $Y$ calculated from the systematic portion of the regression model is the same for each member of the population with the same value of $X$. The error term allows population members with the same value of $X$ to vary on $Y$. It represents the distribution of the observed $Y_i$ around the conditional expected values at each level of $X$. It is important to note that this variation exists in the *population*. The population regression line is probabilistic. The error terms can be thought of as capturing many various forces that explain $Y$ in addition to the predictor, $X$, that is in the regression equation. These might include variables that have not been considered by a particular theory, that are not available in the data, or that represent "pure" randomness in human behavior.

■ **Box 5.3**

The concept of "expected value" is intuitive. It is the value that we expect given the probability distribution of the variable. In calculating the expected value, we weight each possible value of the variable by the probability of observing that value. A value that is highly probable gets a higher weight (is more expected). A value that has a lower probability gets a lower weight (is less expected). This concept is easiest to illustrate with a discrete variable. Consider a hypothetical variable that can take on five values: zero with a probability of 50 percent, 1 with probability of 30 percent, 2 with a probability of 10 percent, 3 with a probability of 7 percent, 4 with a probability of 3 percent. Zero is most likely in this distribution, and so should be weighted highest in calculating the expected value. Four is least likely, and thus should receive the lowest weight. The expected value would be $0 * 0.50 + 1 * 0.30 + 2 * 0.10 + 3 * 0.07 + 4 * 0.03 = 0.83$. The expected value is similarly calculated for continuous variables, but we integrate over the continuous values of the variable, each weighted by its probability, rather than summing.

If these error terms were distributed approximately normally within each level of $X$, the model might look something like the graph shown below.

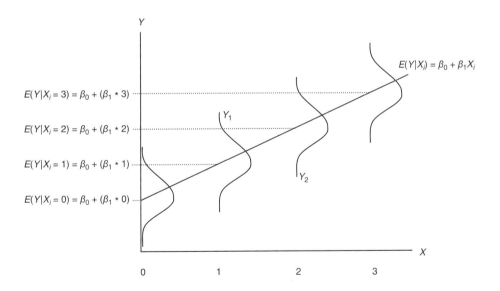

The figure depicts four **conditional distributions** of $Y$ within four levels of $X$. Each conditional distribution is centered on a conditional mean which is determined by the systematic portion of the regression line. Each conditional distribution varies around that conditional mean, with the variability determined by the random portion of the regression line. For example, when $X = 0$, the distribution is centered around the conditional mean of $E(Y|X_i = 0) = \beta_0 + (\beta_1 * 0)$. And, when $X = 3$, the conditional distributed is centered around conditional mean of $E(Y|X_i = 3) = \beta_0 + (\beta_1 * 3)$.

The line is drawn to represent a positive slope. Thus, the conditional means increase as $X$ increases. The systematic portion of the regression model captures the fact that *on average* the values of $Y$ move linearly upward with increasing $X$. However, the random component of the regression model allows for variability around these conditional averages. Because of the random component, it is not the case that *all* $Y$s at a higher level of $X$ will have higher values than *all* $Y$s at a lower level of $X$. In fact, any two observed $Y$s, one at the higher level of $X$ and one at the lower level of $X$, may reflect a reverse relationship from the conditional averages (i.e., higher $Y$ at the smaller rather than the larger $X$ value). The points $Y_1$ and $Y_2$ in the figure depict this situation. $Y_1$ is in the upper tail of the conditional distribution at $X = 1$ and $Y_2$ is in the lower tail of the conditional distribution at $X = 2$. Thus, $Y_1 > Y_2$ even though $E(Y|X_i = 1) < E(Y|X_i = 2)$.

These conditional distributions of the $Y$s at various levels of $X$ are fundamental to the standard regression model. The systematic and stochastic components of the regression model represent, respectively, the mean and variance of these conditional distributions; that is, the mean of each of these conditional distributions is $(Y|X_i) = \beta_0 + \beta_1 X_i$. The variance of each of these conditional distributions is the variance of $\varepsilon_i$ which is typically denoted simply $\sigma^2$. You likely saw this Greek symbol "sigma" used for the variance in general in your prior statistics classes. In the regression context, it is conventional to denote the **conditional variance** as $\sigma^2$ and to use explicit subscripts to denote other variances. For example, the unconditional (or marginal) variance of $Y$ is typically denoted $\sigma_Y^2$.

It is important to keep in mind both the systematic and the probabilistic components of the regression model when we estimate and interpret the slope coefficients. It is easy to slip into deterministic language (e.g., "Earnings are \$10,000 higher for every additional year of schooling") when really we should use language that reminds the reader that the relationship is probabilistic, and our slopes reflect the conditional means (e.g., "Earnings are \$10,000 higher, on average, for every additional year of schooling").

## 5.3: THE SAMPLE REGRESSION LINE

In the social sciences, we can rarely observe the values of the desired variables for all members of a population. Instead, we draw a sample from a population and we apply statistics to describe a sample regression model and to make inferences from that sample regression model to the population regression model. In particular, the sample regression model is:

$$Y_i = \hat{\beta}_0 + \hat{\beta}_1 X_i + \hat{\varepsilon}_i \tag{5.3}$$

The ^ symbols or "hats" are used to distinguish the **population parameters** (e.g., $\beta_1$) from **sample estimates** (e.g., $\hat{\beta}_1$). The systematic or deterministic portion of this sample regression model is:

$$\hat{Y}_i = \hat{\beta}_0 + \hat{\beta}_1 X_i \tag{5.4}$$

These values are often referred to as the **fitted values, predicted values**, or **Y-hats**, and the equation is called a **prediction equation**. We can substitute a value of $X$ into this prediction equation, multiply by the estimated slope, and add the estimated intercept to obtain a predicted value for $Y$. The estimated error, $\hat{\varepsilon}_i$, is the difference between the observed sample value, $Y_i$, and the fitted value based on the sample regression, $\hat{Y}_i$, or $\hat{\varepsilon}_i = Y_i - \hat{Y}_i$. This can be seen by substituting Equation 5.4 into Equation 5.3 and rearranging the terms.[1] The variance of the errors—the conditional variance or $\hat{\sigma}^2$—is also estimated based on the sample. In Section 5.4, we will define the formula for calculating the estimate.

Figure 5.3 shows an example of the association between hours of chores and number of children, using the stylized sample that we created based on the employed women

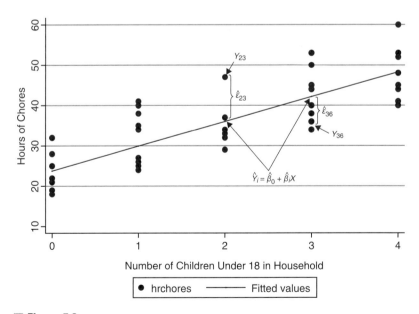

■ **Figure 5.3**

*Source*: Stylized sample from National Survey of Families and Households, Wave 1.

in Wave 1 of the NSFH.[2] The example shows 10 observations within each level of number of children, illustrating the distribution of observed values around the conditional means. Two errors are shown, one above and one below the sample regression line (for case $i = 23$ and case $i = 36$ in the stylized sample). These two observed sample points also illustrate the fact that, although the average hours of chores is higher among households with more children in the household, Case 23 with two children reported more chores than Case 36 with three children.

The reason why the observed value for Case 23 is higher than predicted and the observed value for Case 36 is lower than predicted may simply be pure random variation in human behavior. Or, there may be something that we could measure about each of these women that would help explain their deviations from the conditional means (for example, perhaps Case 36 works very long hours, and has less time left over to devote to chores than the average women with three children; and perhaps Case 23 works only a few hours per week, and has more time left over to devote to chores than the average woman with two children). In multiple regression, we can put additional predictors (such as hours of paid work) in the model that may explain some of this remaining conditional variation.

### 5.3.1: Sampling Distributions

Recall that an **estimator** is a formula that tells us how to calculate our best guess of the population parameter based on the information available in our sample from that population. The **estimate** is the value calculated from the formula for a particular sample.

Like all estimators, estimators for the population intercept, slope, and conditional standard deviation of a regression model have **sampling distributions**. This means that estimates of the population parameters of the regression line based on any sample that we draw randomly from the population ($\hat{\beta}_0 \, \hat{\beta}_1 \, \hat{\sigma}^2$) will differ to some degree from the actual population parameters ($\beta_0 \, \beta_1 \, \sigma^2$). Differences in estimates between repeated random samples, drawn in the same way from the same population, reflect sampling fluctuations (that is, most samples contain primarily the most typical values of the population, but some samples may happen to draw mostly unusual values of the population). Thus, you probably learned in your introductory statistics class that an estimate of the sample mean based on the estimator $\frac{\sum_{i=1}^{n} Y_i}{n}$ will be close to but not exactly equal to the population mean. Similarly, the estimates of the intercept, slope, and conditional standard deviation will be close to but not necessarily exactly equal to the population parameters (with the expected closeness measured by the standard deviation of the sampling distribution for the estimator, whose formula we will examine below).

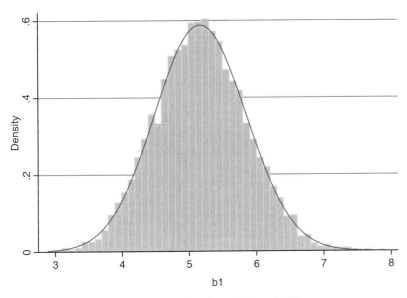

**Figure 5.4:** Sampling distribution for slope with $n = 1,000$.

Figure 5.4 illustrates this sampling variation using a hypothetical population created based on our NSFH "hours of chores" example. To produce this example, we used the NSFH observations as a starting point to generate a simulated population of nearly 700,000. We calculated the intercept and slope for this population. (We will discuss below the formulae for making these calculations; for now, we just rely on the computer to make these calculations.) We then drew a sample of 1,000 from the original population, and estimated the intercept and slope for this sample. We then returned those 1,000 cases, and drew a fresh sample of 1,000 from the original population, and estimated the intercept and slope for this new sample. We repeated this sampling 5,000 times. This produced 5,000 estimates of the intercept and 5,000 estimates of the slope, all based on samples of size 1,000 and all drawn from the same population. Figure 5.4 plots these 5,000 estimates to help us to visualize the sampling distribution of the slope. The mean of this sampling distribution should be close to the population value. And, the standard deviation of this sampling distribution (known as the standard error) estimates how close any particular sample estimate would be to this population value.

We also calculated the average of the 5,000 estimates of the slope and the average of the 5,000 estimates of the intercept to compare with the population value of the slope and intercept. And, we calculated the standard deviation of the 5,000 estimates of the slope and the standard deviation of the 5,000 estimates of the intercept to evaluate the standard error.

|  | Population Value ($n$ = 699,765) | Average of 5,000 Sample Estimates (each $n$ = 1,000) | Standard Deviation of 5,000 Sample Estimates (each $n$ = 1,000) |
|---|---|---|---|
| Intercept | 27.96 | 27.95 | 0.8421 |
| Slope | 5.16 | 5.17 | 0.6777 |

The population value of the slope is 5.16. So, with each additional child in the household, the average hours of chores per week increases by about 5. The slope calculated in each sample of $n$ = 1,000 is an estimate of this population slope. Figure 5.4 shows the sampling fluctuation in our estimate of the slope across the 5,000 random samples. This is a simulation of the sampling distribution of the estimator for the regression slope. Across our 5,000 samples, the slopes range from about 3 to 7.5 with a mean of 5.17, nearly exactly the population value of 5.16. The standard deviation of the sampling distribution of slopes is 0.68. We will discuss later assumptions about the distribution of the slope estimator, but for now we have superimposed a normal distribution line on the histogram. The sampling distribution of the slopes is quite close to this normal distribution. So, based on the empirical rule, about 95 percent of slopes should fall between $5.17 - 1.96 * 0.6777 =$ 3.84 and $5.17 + 1.96 * 0.6777 = 6.50$, which appears to be the case in Figure 5.4.[3]

When we have only one sample, as we normally do, we cannot know if we have drawn a typical sample, whose estimate falls in the center close to the population parameter, or an unusual sample, whose estimate falls in the tails far from the population parameter. But, when the standard deviation of the sampling distribution is smaller (all else equal), then every sample's estimate will be close to the population parameter.

We will now turn to the formulae we used to estimate the slope and its standard error (i.e., the standard deviation of this sampling distribution) as well as the formulae for the intercept and conditional variance.

## 5.4: ORDINARY LEAST SQUARES ESTIMATORS

How do we calculate the value of the intercept and slope from a collection of data points? A standard approach is **OLS regression**. In OLS, the estimators provide the sample estimates of the intercept and slope that minimize the sum of the squared errors, $\hat{\varepsilon}_i$, which as we discussed above measure the distances between the observed values of the dependent variable, $Y_i$, and the estimated conditional means, $\hat{Y}_i$.

### 5.4.1: The Least Squares Approach

Today, least squares is so commonly used that it is easy to lose sight of the incremental process and major breakthrough leading to the discovery of this approach around the end of the eighteenth century. Imagine looking at the sample points from Figure 5.3 without the regression line superimposed. How might you calculate the intercept and slope for a line that best summarizes the systematic linear trend that is clear from the data points?

The intercept and slope, calculated for the stylized sample in Figure 5.3, are 23.78 and 6.13, respectively (see Box 5.4). Figure 5.5 shows one guess we might draw by freehand, rounding the intercept and slope to 24 and 6, respectively. How might we choose between these two sets of values? Figure 5.5 indicates the same two observed values as we showed in Figure 5.3. The errors, representing the difference between these observed values and the new fitted values, are now labeled $\hat{\mu}_{23}$ and $\hat{\mu}_{36}$. Perhaps we could use the errors to choose which line (the one in Figure 5.3 or in Figure 5.5) comes closest to the observed values? An intuitive choice might be to minimize the sum of these errors (which is smaller $\Sigma\hat{\varepsilon}_i$ or $\Sigma\hat{\mu}_i$?). Unfortunately, such a choice does not produce a unique solution.

■ **Box 5.4**

Note that the intercept and slope for the stylized sample differ from those for the hypothetical population, shown on the previous page, because of the different ways each was created from the NSFH full sample of 3,116 employed women. The intercept and slope also differ in the full sample of 3,116 than in either the stylized sample or hypothetical population. (As we will see later, the intercept is 28.69 and the slope is 6.38 in the full sample of 3,116.)

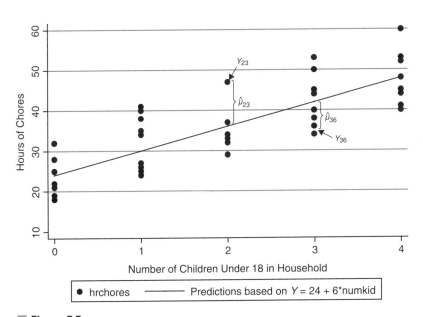

■ **Figure 5.5**

*Source*: Stylized sample from National Survey of Families and Households, Wave 1.

During the eighteenth century, astronomers confronted a similar problem. In 1805, mathematician Adrien Marie Legendre published the method of least squares, which solved this problem. As quoted by noted statistician and historian Stephen Stigler (1986, 13), Legendre saw:

> an element of arbitrariness in any way of, as he put it, "distributing the errors among the equations," but that did not stop him from dramatically proposing a single best solution: "Of all the principles that can be proposed for this purpose, I think there is none more general, more exact, or easier to apply, than what which we have used in this work; it consists of making the sum of the squares of the errors a *minimum*. By this method, a kind of equilibrium is established among the errors which, since it prevents the extremes from dominating, is appropriate for revealing the state of the system which most nearly approaches the truth (Legendre 1805, 72–3).

The least squares approach has numerous advantages, including those mentioned in Legendre's quote as well as offering a single solution for the estimates of the intercept, slope, and conditional variance whose formulae can be derived in a straightforward manner by mathematicians.

We will not reproduce the derivations here. However, we will note conceptually that it starts with the formula for the squared errors:

$$\sum \hat{\varepsilon}_i^2 = \sum \left(Y_i - \hat{Y}_i\right)^2 = \sum (Y_i - (\hat{\beta}_0 + \hat{\beta}_1 X_i))^2$$

This formula has two unknown values, $\hat{\beta}_0$ and $\hat{\beta}_1$. Given our set of data, with observed values for $X$ and $Y$, we would like to find the values of $\hat{\beta}_0$ and $\hat{\beta}_1$ that minimize these squared residuals. Using calculus, it is possible to differentiate these equations with respect to each unknown, set the result to zero, and solve for the unknown (for details, see Fox 2008; Kutner, Nachtsheim, and Neter 2004; Wooldridge 2009).

### 5.4.2: Point Estimators of the Intercept and Slope

Following are these closed form solutions for the intercept and slope in the bivariate regression model:

$$\hat{\beta}_1 = \frac{\sum (X_i - \bar{X})(Y_i - \bar{Y})}{\sum (X_i - \bar{X})^2} \tag{5.5}$$

$$\hat{\beta}_0 = \bar{Y} - \hat{\beta}_1 \bar{X} \tag{5.6}$$

Through the calculus derivation of the least squares estimators it can be shown that the estimated errors sum to zero. That is, $\sum \hat{\varepsilon}_i = \sum (Y_i - \hat{Y}_i) = 0$. Note that this applies to the errors, *not* the squared errors. The nonzero quantity of squared errors is what we minimize to estimate the intercept and slope. But, since the errors themselves sum to zero, across all sample points positive errors (above the regression line) balance out with negative errors (below the regression line).

It is also the case that if we substitute the sample mean of the predictor variable, $\bar{X}$, into the prediction equation, $\hat{Y}_i = \hat{\beta}_0 + \hat{\beta}_1 X$, the resulting $\hat{Y}_i$ will equal the sample mean of the outcome variable, $\bar{Y}$.[4] This is referred to as the sample regression line passing through the sample means.

### 5.4.3: Point Estimator of the Conditional Variance

The conditional variance is estimated based on the sample errors. For the bivariate regression model with one predictor the point estimator for the conditional variance is:

$$\hat{\sigma}^2 = \frac{\Sigma \hat{\varepsilon}_i^2}{n-2} = \frac{\Sigma \left(Y_i - \hat{Y}_i\right)^2}{n-2} = \frac{\Sigma \left(Y_i - (\hat{\beta}_0 + \hat{\beta}_1 X)\right)^2}{n-2} \tag{5.7}$$

This value is also referred to as the **Mean Square Error** (MSE). Notice that the value in the numerator is the sum of the squared errors, which is what we minimize in least squares. This sum is divided by the sample size, less two, so the result is very nearly the simple average of the squared errors (hence the term Mean Square Error). The denominator is known as the **degrees of freedom**. The subtraction of 2 from the sample size accounts for the two estimates—the estimated intercept and the estimated slope—used in the calculation of $\hat{Y}_i$ for the numerator of Equation 5.7. As we discussed above, $\hat{\sigma}^2$ is an estimate of the conditional variance of the $Y$s at each level of $X$. The square root of this value is the estimate of the **conditional standard deviation**, commonly referred to as the **Root Mean Square Error** or **Root MSE**.

Recall that the error is the stochastic portion of the regression equation. Based on the other (systematic) portion of the sample regression equation, all observations at a given $X$ level have the same fitted value estimated by:

$$\hat{Y}_i = \hat{\beta}_0 + \hat{\beta}_1 X_i$$

Thus, the conditional variation in $Y$ is determined by the variation in the error term. Consequently, the Root MSE is an estimate of both the conditional standard deviation of the error term and the conditional standard deviation of the dependent variable.

If $Y$ and $X$ are strongly associated, then the standard deviation of the conditional distribution of the $Y$s given $X$ will be considerably less than the standard deviation in the **unconditional distribution** of $Y$ ($\hat{\sigma} < \hat{\sigma}_Y$). In other words, if there is a linear relationship between $Y$ and $X$ then, across observations, the observed values fall closer to the regression line than to the overall sample mean. The figure below depicts such conditional distributions. These distributions are centered on the estimated conditional means. And, their estimated conditional standard deviation is smaller than the unconditional standard deviation ($\hat{\sigma}_Y$) shown to the left.

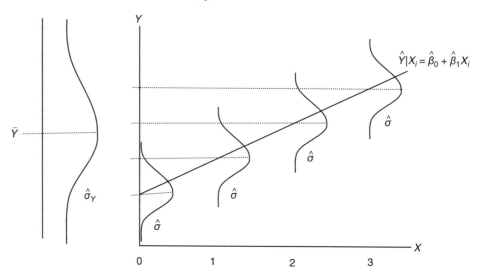

Below, we will ask SAS and Stata to calculate the conditional and unconditional standard deviations using our example of the distance the mother lives from her adult child. First, let's take a look at the results for our stylized example based on 50 cases created from the NSFH. Using the formula above, the prediction equation is estimated to be: $\hat{Y}_i = 23.78 + 6.13X$. We might also write this as $\widehat{hrchores}_i = 23.78 + 6.13 *$ *numkid* to make the variables in our example explicit. We can use what we learned above about the algebra and geometry of a straight line to say in words what these values represent. The intercept estimates that when a household contains no children, a woman will spend nearly 24 hours per week on chores, on average. With each additional child in the household, the woman spends about six additional hours per week on chores, on average.

Although we will rely on the computer to make these calculations throughout the textbook, it can be helpful to see an explicit set of calculations to help demystify the process. Appendix G provides a spreadsheet of these calculations for our stylized example. (The spreadsheet is available on the textbook web site *http://www.routledge.com/textbooks/9780415991544*.) The first two columns contain the observed values of the $Y$ and $X$ variables, *hrchores* and *numkid*. The rows circled in red show the two observations highlighted in Figure 5.3. The third and fourth columns calculate the deviations of each of these observed values from their respective means (the means are shown at the bottom of columns 1 and 2). The fourth and fifth columns provide the product of these deviations, across $Y$ and $X$ and between $X$ and itself. The sum of these products are shown below columns 5 and 6, and these are entered into the equations to make the calculations labeled *Slope* and *Intercept*.

$$\hat{\beta}_1 = \frac{\sum(X_i - \bar{X})(Y_i - \bar{Y})}{\sum(X_i - \bar{X})^2} = \frac{613}{100} = 6.13$$

$$\hat{\beta}_0 = \bar{Y} - \hat{\beta}_1\bar{X} = 36.04 - (6.13*2) = 23.78$$

The seventh column then uses the intercept and slope to predict the $Y$s based on the observed $X$s (e.g., for case 23, the calculation is $23.78 + 6.13 * 2 = 36.04$; for case 36, the calculation is $23.78 + 6.13 * 3 = 42.17$). Notice that, as expected, within levels of number of children, the predicted value of $Y$ is the same. These values represent the systematic portion of the regression line (i.e., the conditional mean). The stochastic portion of the model is estimated by the errors in the next column, calculated by subtracting this predicted $\hat{Y}_i$ from the observed $Y_i$. The square of these errors is in the final column, and the sum of the errors and the sum of their squares appear at the bottom of these columns. Notice that, as expected, the sum of the errors is zero, but the sum of the squared errors is a positive value. It is difficult to interpret this sum of squared errors in absolute terms, but in relative terms we know that it represents the smallest possible sum of squared errors for a straight line drawn for these 50 observations. The RMSE value divides this sum of squared errors by the degrees of freedom and then takes the square root.

$$\hat{\sigma} = \sqrt{\frac{\sum\left(Y_i - \hat{Y}_i\right)^2}{n-2}} = \sqrt{\frac{1614.2}{50-2}} = \sqrt{33.63} = 5.80$$

The resulting estimate of the conditional standard deviation, 5.80, is smaller than the unconditional standard deviation, 10.47.[5]

## 5.4.4: Standard Error of the Estimators of the Intercept and Slope

In social science applications, our ultimate goal is generally to answer our research questions or evaluate our hypotheses. We will see below that an essential ingredient for doing so is the standard error of our estimators, which allow us to calculate test statistics and confidence intervals. As discussed above, the standard error is the standard deviation of the sampling distribution for the estimator. A relatively large standard error means that in any given sample our estimate is more likely to fall further from the population value. A relatively small standard error means that in any given sample our estimate is more likely to fall closer to the population value.

Under least squares, the standard errors of the estimators of the intercept and slope are calculated with the following formulae (for details on the derivations, see Fox 2008; Kutner, Nachtsheim, and Neter 2004; Wooldridge 2009).

$$\sigma_{\hat{\beta}_1} = \frac{\sigma}{\sqrt{\Sigma(X_i - \bar{X})^2}} \qquad\qquad (5.8)$$

$$\sigma_{\hat{\beta}_0} = \sigma\sqrt{\frac{\Sigma(X_i)^2}{n\Sigma(X_i - \bar{X})^2}}$$

Note that these formulae depend on the unknown conditional standard deviation, $\sigma$. By substituting the estimate of the conditional standard deviation, $\hat{\sigma} = \sqrt{\frac{\Sigma(Y_i - \hat{Y}_i)^2}{(n-2)}}$, we can estimate the standard errors for the intercept and slope.

$$\hat{\sigma}_{\hat{\beta}_1} = \frac{\hat{\sigma}}{\sqrt{\Sigma(X_i - \bar{X})^2}}$$

$$\hat{\sigma}_{\hat{\beta}_0} = \hat{\sigma}\sqrt{\frac{\Sigma(X_i)^2}{n\Sigma(X_i - \bar{X})^2}}$$

Our research questions typically involve hypothesis tests about the slope, so we are particularly concerned with how close our estimate of the slope is to the population value.

What does the formula for the standard error of the slope shown in Equation 5.8 tell us about the key components that affect its size?

- The standard error of the slope is directly proportional to the conditional standard deviation. So, with all else held constant, *a larger conditional standard deviation would correspond to a larger standard error for the slope and thus less precision in our estimate of the slope.* In other words, to the extent that the observed $Y_i$ do not cluster closely around the regression line, then our estimates of the slope will vary more from sample to sample.
- The standard error of the slope is inversely proportional to the sum of the squared deviations of the $X_i$ from their mean $\bar{X}$. Thus, all else equal, *greater variation in the observed values of the predictor variable will correspond to a smaller standard error for the estimates of the slope.* In other words, we can more precisely estimate the slope when we have good variability in the $X$ values.
- We can also see in the denominator of the equation for the standard error of the slope estimate that *as the sample size increases the standard error of the estimate of the slope decreases, all else equal.* This can be seen since each additional

observation in the sample adds another squared difference between $X_i$ and $\bar{X}$ to be summed in the denominator (i.e., the denominator contains the *variation* in the $X$s not the *variance* of the $X$s).

It is useful to keep these factors in mind as you plan a research project and interpret your results. You will be better able to estimate the association between your predictor of interest and the outcome if you have: (a) a larger sample size, (b) more variation on the predictor variable, and (c) a better fit around the regression line. The first two of these can be built into your study design. Using existing data sources is one way to obtain a larger sample size. Collaboration with peers and mentors or external funding may help you to increase the sample size if you collect your own data. Variation on the predictor variable can also be increased (or at least not unwittingly limited) through thoughtful study design. Using existing data sets may again be helpful. For example, national designs may provide the greatest variation on some variables, such as neighborhood income, which would be constrained in localized studies. Thinking carefully may also help you to avoid decisions that limit variation (e.g., design decisions such as drawing a sample from particular organizations or geographic areas, or restricting the age of sample members, may affect the variation of a predictor of interest). The closeness that the observed data fall to your regression line may seem harder to control, although reducing measurement error in your outcome variable and predictor, and relying on prior studies and solid theory to identify your predictors of interest, are good strategies.

### 5.4.5: The Standard OLS Assumptions

If our goal were description—finding the best fitting line for the sample data—then we could stop with the formulae for the **point estimates** of the slope and intercept. But, to draw inferences about the population parameters, as we typically want to do in the social sciences, we need some additional assumptions. We will examine these assumptions in greater detail in later chapters, especially Chapter 11. Here, we preview a few of them. If the OLS assumptions hold, it can be shown that the OLS estimators are unbiased and have the minimum standard error of all unbiased linear estimators (they are called the **Best Linear Unbiased Estimators, "BLUE"**).

One important assumption of OLS is that we are modeling the conditional means of the outcome variable in relation to the predictor of interest (i.e., is the average value of $Y$ larger [or smaller] within higher levels of $X$ than within lower levels of $X$)? We allow for variation around these conditional means, so that any given individual at a particular level of $X$ may have an observed value of $Y$ that is higher or lower than the conditional mean. Still, it is important to be careful in discussing our findings to emphasize the distinction between prediction of averages and individual variation around those averages (i.e., the estimates should not be interpreted as though they reflect a deterministic relationship between $X$ and $Y$).

The model is also assumed to be linear in the parameters, meaning that we do not raise our coefficients to a power or multiply coefficients together. We can model nonlinear associations between $X$ and $Y$, however, and we will learn how to do so in Chapter 9. For example, we might expect a positive but diminishing association between children's reading achievement and family income: the difference in reading achievement for children from families with $10,000 and $20,000 of income may be larger than the difference in reading achievement for children from families with $110,000 and $120,000 in income. Or, we might expect that the association between age and verbal acuity is curvilinear, rising from childhood to middle adulthood, and then falling in older age. Incorrectly modeling such associations as though they are linear is not a problem inherent to regression analysis. Rather, it is the job of the researcher to conceptualize the relationship appropriately between the predictor and outcome, as linear or nonlinear (or check for nonlinear associations during model estimation) and to specify correctly the model to identify nonlinear relationships, if needed. We will learn how to do this in Chapter 9.

Our OLS model also assumes that the conditional variance is the same across levels of $X$. This has been implicit above, for example in the lack of a subscript on $\sigma^2$. Constant conditional variance is referred to as homoskedasticity. Nonconstant conditional variance is called heteroskedasticity and is often indicated by subscripting the conditional variance as $\sigma_i^2$. We will discuss how to identify and address heteroskedasticity in Chapter 11.

We will also deal with other assumptions of the model in later chapters. For example, when we add additional predictors to the model in multiple regression in Chapter 6 we assume that the predictors are not collinear (not perfectly correlated). We also assume that the error terms are not correlated across individuals, and in Chapter 12 touch on ways to address potential correlation and refer to additional advanced techniques for doing so (see Box 5.5). More broadly, we assume that we have correctly specified the model, including the issue of nonlinear relationships noted above, but also in other ways (such as including the correct set of variables in the model).

■ **Box 5.5**

For example, errors might not be independent if we include multiple members of the same family, multiple students from the same class, or multiple companies from the same city, rather than drawing samples of individuals, students, and companies completely randomly. Clustering is sometimes used to reduce the costs of a study and can allow for better testing of conceptual ideas about contextual effects, but such clustering should be addressed by the statistical models. We briefly preview techniques for doing so in the roadmap in Chapter 12.

## 5.4.6: The Normality Assumption

We do not need to assume any particular form for the distribution of the outcome variable, and the residuals, in order for the OLS estimators to be BLUE. However, the normality assumption is commonly used to justify hypothesis testing. We will discuss problem of outliers and influential observations that often go along with non-normality in Chapter 11.

We are able to calculate confidence intervals and test hypotheses in the standard OLS regression model because we assume that the errors, $\varepsilon_i$, follow a normal distribution. The error term is the stochastic component in the regression equation. This means that assumptions about the distribution of the error term carry over into assumptions about the distribution of the outcome variable and assumptions about the distribution of the estimators of the intercept and the slope (because the $Y$s enter the formulae for the intercept and the slope, and the error is the stochastic portion of the $Y$s; the $X$s are assumed given).

The normality assumption is sometimes summarized with notation such as the following:

$$\varepsilon_i \sim idN(0,\ \sigma^2)$$

$$Y_i | X \sim idN(\beta_0 + \beta_1 X_i,\ \sigma^2)$$

In words, this simply means that the errors are distributed ($\sim$) independently ($id$) and normally ($N$), with a mean of zero and variance of $\sigma^2$. We already noted above the assumptions of constant variance and independence. The zero mean of the errors means that positive errors balance out negative errors, so the systematic portion of the regression model accurately captures the conditional mean of $Y$. Because of this, and the fact that the errors comprise the stochastic portion of the regression model, the $Y$s are also distributed independently and normally with a variance of $\sigma^2$, but their mean is the conditional mean, $\beta_0 + \beta_1 X_i$, the systematic portion of the regression model.

The assumption of normal errors can be motivated by the conceptualization of the error term that we discussed earlier. Namely, if we think of the error term as capturing the sum of many excluded factors that impact $Y$, then the central limit theorem can provide a justification for the normality assumption.[6]

## 5.4.7: Hypothesis Testing about the Slope

Given our normality assumption, we can use standard techniques for hypothesis testing based on the normal distribution. We will describe these in detail below for the slope, but first let's review the general steps of hypothesis testing.

### Review of Basic Concepts

As you know from your basic statistics course, to conduct a hypothesis test, you follow four basic steps:

1. List the null and alternative hypotheses about the value that you expect for the parameter being estimated (in regression models, typically the slope).
2. Calculate a test statistic (typically, the point estimate of the parameter minus the hypothesized value, divided by the estimated standard error).
3. Determine the critical value given the test statistic's distribution (e.g., $Z$- or $t$-distribution) and a pre-specified alpha level (typically 0.05).
4. Reject the null hypothesis if the magnitude of the calculated test statistic is larger than the critical value.

Alternatively, at steps 3 and 4 you could calculate the $p$-value, given the test statistic's distribution, and reject the null hypothesis if the $p$-value is smaller than the alpha level.

You may also recall that hypothesis tests can be one-sided or two-sided. In a **two-sided hypothesis test**, the null hypothesis states that the parameter equals a particular value (the **hypothesized value**) and the alternative hypothesis states that the parameter does not equal that value. In a **one-sided hypothesis test**, the null hypothesis states that the parameter is either "less than or equal to" or "greater than or equal to" a particular value (the hypothesized value) and the alternative hypothesis states that the parameter is "greater than" or "less than" a particular value. The hypothesized value is typically zero in regression. In the two-sided context this means that the alternative hypothesis typically states that the relationship is expected to not equal zero ($\neq 0$). In the one-sided context this means that the alternative hypotheses typically state that the relationship is expected to be positive ($>0$) or negative ($<0$).

Let's take a look at the standard normal distribution to reinforce these ideas. Figure 5.6 shows a hypothetical example of a two-sided hypothesis test based on the $Z$ distribution. Recall that with a two-sided test we split the alpha value between the two tails, as shown in the shaded areas in the figure. For the conventional $\alpha = 0.05$ the critical $Z$-value for a two-sided test is $|1.96|$. We show the negative value explicitly on the left in Figure 5.6. The probability to the left of that $Z$-value is 0.025. So, 2.5 percent of the $Z$-values are more negative than $-1.96$.

The $Z$-distribution is symmetric, and the shaded area on the right side of the figure similarly depicts the 2.5 percent of $Z$-values that are larger than 1.96.

In Figure 5.6, we also show a hypothetical calculated $Z$-value of $-1.40$. For a two-sided test, remember that the $p$-value is the proportion of $Z$-values that are more extreme

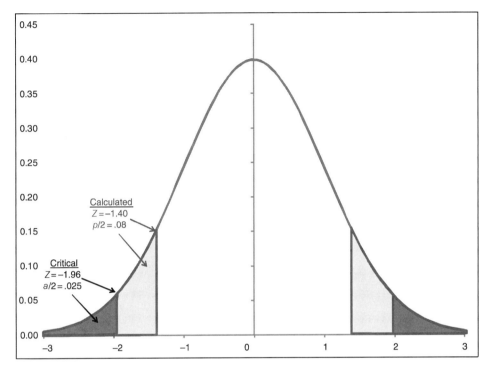

■ **Figure 5.6**

than the absolute value of our $Z$-value. On the left, we show that 8 percent of $Z$-values are more negative than $-1.40$ (the sum of the light and dark gray areas on the left of the figure). The value 0.08 is half of the $p$-value. The other half is represented by the light and dark gray areas on the right side of the figure: an additional 8 percent of the $Z$-values are more positive than 1.40. So the $p$-value in our hypothetical example is $0.08+0.08 = 0.16$.

To make a conclusion about our hypothesis test in this hypothetical case, we would either compare the calculated $Z$-value to the critical $Z$-value or compare the calculated $p$-value to the alpha level. Each comparison will lead us to the same conclusion:

1.  The absolute value of the calculated $Z$-value $|-1.40|$ is smaller than the absolute value of the critical $Z$-value $|-1.96|$ thus we fail to reject.
2.  The calculated $p$-value of 0.16 is larger than the alpha of 0.05 thus we fail to reject.

Figure 5.7 shows an example of a one-sided test. To illuminate the differences between the one-sided and two-sided cases, we assume that we have the same calculated

$Z$-value of $-1.40$. In the one-sided case, we do not double the percentage of $Z$-values more extreme than our calculated $Z$-value. Rather, we concentrate on one tail. In our case, let's assume that our alternative hypothesis was that our point estimate would be negative. (We will have more to say about this below.) So, the $p$-value is the probability of being more negative than the calculated $Z$-value, which we saw above is 0.08. Notice that there is nothing shaded in the right tail of Figure 5.7. Thus, in this one-sided case, the $p$-value is 0.08.

In the one-sided case, we also do not split the alpha level between the two tails. We place it all in one tail. Again, in this case, since we assume that our alternative hypothesis was that our point estimate would be negative, we place the $p$-value in the left tail. The $Z$-value is now $-1.645$ because 5 percent of $Z$-values are more negative than $-1.645$.

In our hypothetical example, our conclusion about the test would be the same as in the two-sided case. We fail to reject the null hypothesis because $|-1.40| < |-1.645|$ and $0.08 > 0.05$. We will provide an example below of a case in which the one-sided and two-sided tests would lead to difference conclusions for the same point estimate.

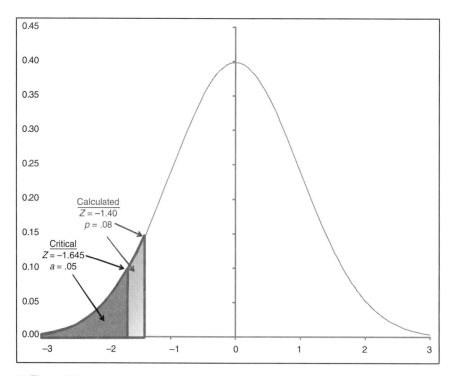

▩ **Figure 5.7**

Alternatively, we might calculate a **confidence interval** by adding and subtracting a multiple of the estimated standard error from the point estimate. The confidence interval approach leads to the same conclusion as a two-sided hypothesis test, with the same alpha level. In our case, we can test our null hypothesis by seeing whether the hypothesized value falls within these lower and upper bounds of the confidence interval. If the confidence interval contains the hypothesized value, then we fail to reject the null hypothesis. If we use $Z = 1.96$ for the multiplier of the estimated standard error, our decision will be the same as for the test statistic approach with a two-sided alternative and $\alpha = 0.05$. The resulting confidence interval is referred to as a 95 percent confidence interval, because it is based on the $Z$ value for an alpha of 5 percent. For a one-sided test, we must rely on the test statistic approach.

Generally, for the normal distribution, we could use the standard normal table to look up a critical $Z$-value for a given alpha level or to look up a $p$-value for a calculated $Z$-value. In regression analysis, we use the $t$ distribution instead of the $Z$-distribution, however, for tests about the slope and the intercept. Statistically, the $t$ distribution is needed because we used the estimated standard error of the slope given the conditional standard deviation was unknown. The degrees of freedom for the $t$ statistic is $n - 2$ in the bivariate regression case. However, as you learned in your basic statistics course, the $t$ becomes identical to the $Z$ for large samples. Computer output will typically provide the exact $t$-values and associated $p$-values for the relevant degrees of freedom. If the sample is large ($n$ greater than 60), we also know that $t$-values greater than about 2 will be significant at a two-sided 5 percent Type I error level, based on the empirical rule. This rough cutoff is useful when reviewing $t$-statistics that do not have $p$-values listed (e.g., in a book or journal article).

### Confidence Interval Approach for the Slope

More precisely, the confidence interval for the slope can be constructed as $\hat{\beta}_1 \pm t_{\alpha/2} \hat{\sigma}_{\hat{\beta}_1}$ where $t_{\alpha/2} = 1.96$ for $\alpha = 0.05$ and a large sample size.

As noted above, to test a null hypothesis about a population parameter against a two-sided alternative, we check whether the constructed confidence interval contains the hypothesized value of the population parameter. For the slope, we would proceed as follows.

$$H_0: \beta_1 = \beta_1^*$$
$$H_a: \beta_1 \neq \beta_1^*$$

Decision rule:

If $\beta_1^*$ is contained within the bounds of the confidence interval, then do not reject the null hypothesis.

If $\beta_1^*$ is not contained within the bounds of the confidence interval, then reject the null hypothesis.

Recall that we do not accept the null hypothesis when we fail to reject. In either case of rejecting or failing to reject our null hypothesis, the range of slopes within the lower and upper bounds of the confidence interval provide our best estimates of the true population slope parameter.

This decision rule should make intuitive sense. If the hypothesized slope value is within the estimated range of values, then we cannot reject the hypothesis that the parameter takes on that null value. However, if the hypothesized slope value falls outside of the estimated range of values, then we can reject the null hypothesis that the parameter takes on the hypothesized value.

### Test Statistic Approach for the Slope

For the test statistic approach, we define:

$$t = \frac{\hat{\beta}_1 - \beta_1^*}{\hat{\sigma}_{\beta_1}}$$

where $\beta_1^*$ is the hypothesized value of the slope under the null hypothesis (often zero) and $df = n - 2$ in bivariate regression.

#### Two-Sided Hypothesis Test

For a two-sided test, we then either contrast this $t$-statistic to the critical $t$-value for the given $\alpha$ and degrees of freedom (with $\alpha$ divided by two to include both tails) or we look up the $p$-value for our $t$-statistic and degrees of freedom (with the probability that the $t$-value being more extreme than our $t$-value being doubled to encompass both tails) and compare the calculated $p$-value to the selected $\alpha$ value. It is conventional to use $\alpha = 0.05$ and $t_{\alpha/2} = 1.96$ as the cutoffs.

As reviewed above, if the absolute value of the $t$-statistic is larger than the critical value (e.g., 1.96) or the $p$-value is less than the alpha value (e.g., .05) then we reject the null hypothesis; otherwise, we fail to reject the null hypothesis. In the latter case, we do not accept the null hypothesis. Under the test statistic approach, our best guess of the true population slope parameter is the point estimate $\hat{\beta}_1$.

As with the confidence interval, the decision rules for the hypothesis test should be intuitive. If our $t$-statistic is highly unusual (far in the tails of the sampling distribution constructed under the null hypothesis) then our findings are not consistent with the null hypothesis.

## One-Sided Hypothesis Test

One-sided significance tests proceed similarly. Although we do not divide alpha into two tails.

| Two-sided | One-sided (right tail) | One-sided (left tail) |
|---|---|---|
| $H_0: \beta_1 = \beta_1^*$ <br> $H_a: \beta_1 \neq \beta_1^*$ | $H_0: \beta_1 \leq \beta_1^*$ <br> $H_a: \beta_1 > \beta_1^*$ | $H_0: \beta_1 \geq \beta_1^*$ <br> $H_a: \beta_1 < \beta_1^*$ |
| Calculate $p$-value using the area to the right of the positive value of the calculated $t$-statistic and to the left of the negative value of the $t$-statistic. | Calculate $p$-value using the area to the right of the calculated $t$-statistic. | Calculate $p$-value using the area to the left of the calculated $t$-statistic. |

When $\beta_1^* = 0$, then the one-sided "right tail" hypothesis corresponds to the situation where our theory suggests that our predictor and outcome variables are positively associated (i.e., $H_a: \beta_1 > 0$). Similarly, when $\beta_1^* = 0$, then the one-sided "left tail" hypothesis corresponds to the situation where our theory suggests that our predictor and outcome variable are negatively associated (i.e., $H_a: \beta_1 < 0$).

We will see below how to use SAS and Stata output to conduct a one-sided hypothesis test.

## Zero as the Hypothesized Value

The typical hypothesis test in OLS regression is a two-sided test of the null hypothesis that $H_0: \beta_1 = 0$ versus the alternative that $H_a: \beta_1 \neq 0$.

The $t$-value and $p$-value provided in standard SAS and Stata output correspond to this test. This "straw man" test has a null hypothesis that $Y$ is not linearly predicted by $X$ and an alternative hypothesis that $Y$ is linearly predicted by $Y$ (either positively or negatively). Referring back to Figure 5.2, when the slope is zero, the relationship between $X$ and $Y$ is graphed as a horizontal line; in this situation, the conditional mean of $Y$ is the same regardless of the level of $X$.

Sometimes enough prior research or theoretical clarity exists to specify in advance the expected direction of the relationship between the predictor and outcome. As noted above, if a positive association is expected, then the null hypothesis would be $H_0: \beta_1 \leq 0$ and the alternative hypothesis would be $H_a: \beta_1 > 0$. If a negative association is expected, then the null hypothesis would be $H_0: \beta_1 \geq 0$ and the alternative hypothesis would be $H_a: \beta_1 < 0$.

In these cases, if the estimated coefficient is consistent with the alternative hypothesis (the "correct sign"), we can calculate the one-sided $p$-value for a null hypothesized value of zero by dividing the *p*-value listed in SAS's output in half. A correct sign would be a positive estimate of the coefficient if our alternative hypothesis is $H_a$: $\beta_1 > 0$ and a negative estimate of the coefficient if our alternative hypothesis is $H_a$: $\beta_1 < 0$. If the estimated coefficient is not consistent with the alternative hypothesis ("wrong sign"), then we should subtract half of the listed *p*-value from one.

### Example of Hypothesis Testing

Let's look at our NSFH distance example to put these approaches into practice. In Chapter 2, we initially stated our research questions in a nondirectional fashion: how does an adult's earnings relate to how near she lives to her mother? In this case, we would use a nondirectional or two-sided alternative.

> $H_0$: Adults' earnings do not linearly predict the distance they live from their mothers.
> $H_a$: Adults' earnings linearly predict the distance they live from their mothers.

In symbols, this is our standard hypothesis test, using the straw man hypothesized value of zero to represent no linear relationship:

$$H_0: \beta_1 = 0$$
$$H_a: \beta_1 \neq 0$$

### *SAS and Stata Syntax*

We can ask SAS and Stata to estimate the slope and its standard error and to calculate the associated *t*-value and *p*-value using their basic regression commands.

| | SAS | Stata |
|---|---|---|
| Basic Syntax | proc reg;<br>  model <depvar>=<indepvar>;<br>run; | regress <depvar> <indepvar> |

Notice the similar structure of the commands, mirroring our regression equation. The dependent variable is listed first (*depvar* on the left) followed by the independent variable (*indepvar* on the right). In SAS, the equals sign is included to separate the dependent from the independent variable, further mimicking how we write the equation. Both use a short word to reflect regression (reg or regress).[7] Following SAS convention, REG is a procedure (PROC). The command name should be paired with the word RUN; and, all variables must have been previously defined in a DATA step. SAS allows multiple model statements to be included in a single PROC REG, and we will use this approach in some cases. For example, we might type:

```
proc reg;
    model <depvar>=<indepvar1>;
    model <depvar>=<indepvar2>;
run;
```

Both SAS and Stata allow many more options, some of which we will learn later in this chapter and in future chapters (to access the documentation, type "REG procedure" in the SAS help index and type `help regress` in the Stata command window).

### Prediction Equation

Display B.5.1 shows the commands for the regression of *g1miles* on *g2earn* and the resulting output. We show the full set of output, but we will focus for now on the elements circled in red. Each package shows the same calculated values, but in different positions. Near the top of both sets of output we see the sample size of 6,350. And, in the middle, the estimated Root MSE is 635.1 (rounded to the first decimal). Both then have tables which show the sample estimates of the intercept and slope (labeled `Coef.` in Stata and `Parameter Estimate` in SAS). Although SAS lists the intercept in the first row, and Stata in the last (labeled `_cons`), the estimated value is the same: 249.81 (rounded to two decimals).[8] The row for the slope is labeled with the predictor variable's name, *g2earn*. The estimate is 0.00113 (rounded to five decimals).[9]

Putting these estimates together we can write the prediction equation:

$$\hat{Y} = 249.81 + .00113 * X$$

or using the variable names

$$\widehat{g1miles} = 249.81 + .00113 * g2earn$$

The slope is very small in size because one unit of change on the predictor—the adult child's earnings—represents a dollar change on annual earnings. As discussed above, we can multiply the coefficient estimate by a factor to convert to a more meaningful amount of change. For example, if we set $i = \$10,000$ as the amount of increase on $X$, then the amount of change on $Y$ would be $0.0011333 * 10,000 = 11.333$. We can interpret this rescaled slope as: "The distance a mother lives from her adult child increases by over 11 miles, on average, with each additional \$10,000 the child earns." We will discuss below additional strategies for making substantively meaningful interpretations (but, as we will see, doing so will not affect our decision on the hypothesis test).

## Two-Sided Hypothesis Test

Can we reject the null hypothesis that the slope is equal to zero? The SAS and Stata output list the estimated standard error to allow us to complete this test. Each also calculates a *t*-value and *p*-value based on the basic straw man hypothesis test (a null of no linear relationship against an alternative of some linear relationship, without specification of direction). In this case, $t = (\hat{\beta}_1 - 0)/\hat{\sigma}_{\hat{\beta}_1} = .0011333 - 0/.00021852 = 5.19$ with $n-2 = 6{,}350-2 = 6{,}348$ degrees of freedom.[10] The computer has done the work of looking up the *p*-value for this test statistic. Just based on the *t*-value, though, we should recognize that we can reject the null hypothesis. The *t*-distribution for our sample size of over 6,000 is very close to the normal and $t_{\alpha/2} \doteq 1.96$ in this large sample. Our *t*-value of 5.19 is larger than this critical value. Likewise, the *p*-value is listed as less than .0001 in both sets of output. So, we can reject the null hypothesis. We do find evidence of a linear relationship between the respondent's earnings and the distance his or her mother lives away.

Notice that SAS labels the column containing the *p*-value "`Pr > |t|`" and Stata labels it "`P>|t|`." In both cases, placing the letter *t* within the two vertical lines represents the absolute value of *t*. This reminds us that SAS and Stata list the *p*-values for two-sided tests—the probability of a *t*-value larger than a *t* of the calculated magnitude with either positive or negative sign. In our case, it is the probability in a *t*-distribution with 6,348 degrees of freedom of a value larger than 5.19 or smaller than −5.19. This **two-sided p-value** is appropriate given the way we stated the alternative hypothesis above, without direction ($H_a$: $\beta_1 \neq 0$).

## One-Sided Hypothesis Test

For our research question, it might have been possible to draw on conceptual ideas and prior empirical work to specify instead a directional alternative hypothesis. For example, children who earn more have more financial resources to allow a distant move and may be in higher status jobs where opportunities are available in distant locations.[11] Thus, we might have written the following research hypotheses:

> $H_0$: Adults' earnings are unrelated to or negatively associated with the distance they live from their mothers.
> $H_a$: Adults' earnings are positively associated with the distance they live from their mothers.

In symbols:

> $H_0$: $\beta_1 \leq 0$
> $H_a$: $\beta_1 > 0$

The alternative hypothesis denotes that positive estimates for the slope would be consistent with our expectations (a positive estimate would be of the "right sign"—greater than zero as in our alternative hypothesis). The null hypothesis indicates that an estimate of zero or an estimated negative value for the slope would not be consistent with our expectations (a negative estimate would be of the "wrong sign"—less than or equal to zero as in our null hypothesis).

Because SAS and Stata list two-sided $p$-values, we need to do some additional work to calculate a **one-sided p-value** and make a decision about such one-sided tests. This diagram illustrates the decision-making process for a one-sided test, based on statistical output.

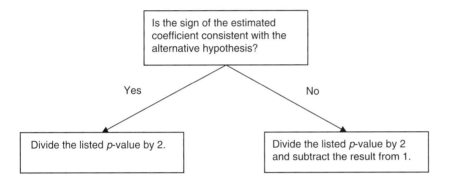

The resulting $p$-value is then compared to our alpha (typically 0.05) to make a decision about the one-sided null hypothesis. In our case, the sign is consistent with our alternative hypothesis (0.00113 is positive in sign), so we can divide the listed $p$-value in half. Doing so is unimportant in our example because the two-sided $p$-value is already quite small.

In cases of marginal significance, dividing the $p$-value in half can have greater importance. For example, if the listed two-sided $p$-value were 0.06, then it would not be significant at a conventional 5 percent alpha level. But, dividing it in half (to 0.03) would lead to rejection of the one-sided null hypothesis at a conventional 5 percent alpha level. Thus:

■ If you are able to state your hypothesis in a directional manner before examining the association in your data, then you should use a one-sided test.
■ If you are not able to state your hypothesis in a directional manner before examining the association in your data, you should use a two-sided test.

What if you specify a directional hypothesis a priori, and the estimated coefficient is the "wrong sign" (inconsistent with your alternative hypothesis) and shown to be statistically significant by SAS and Stata's two-sided test? In this case, if you use the decision rules above, you should convert this significant two-sided $p$-value to a nonsignificant one-sided $p$-value. For example, suppose we had an estimated coefficient that was negative in sign, our alternative hypothesis expected a positive sign, and the listed two-sided $p$-value was 0.04? Then, we should calculate the one-sided $p$-value to be $1 - (0.04/2) = 1 - 0.02 = 0.98$. This $p$-value indicates that we should fail to reject our original null hypothesis. In fact, the negative sign of the coefficient suggests that we were quite wrong in our alternative hypothesis, especially given its large magnitude relative to the estimated standard error (i.e., the fact that the estimate was in the wrong direction with a small two-sided $p$-value). This might be noted in discussing the findings, although it is not consistent with statistical methods to revise your original hypothesis to be consistent with the unexpected finding.

### Confidence Interval Approach

Stata always includes a 95 percent confidence interval for the intercept and slope in its default output. We can ask SAS to report a confidence interval by adding the option /clb (Confidence Limits for the parameter estimates, sometimes also known as Betas). Display B.5.2 implements this option in SAS. The results from both packages are circled in red, and are the same within rounding. By hand, we can calculate them as follows:

$$\hat{\beta}_1 \pm t_{\alpha/2} \hat{\sigma}_{\hat{\beta}_1} = 0.0011333 \pm 1.96 * 0.00021852 = (0.000705, 0.001562)$$

Within rounding to four significant digits, our results match the packages.[12] Both packages and our hand calculations produce a lower bound of 0.000705 and an upper bound of 0.001562. As expected, this result does not contain the null value of zero and thus our decision is consistent with the two-sided hypothesis test (reject the null hypothesis). The results allow us to estimate that the coefficient falls between 0.000705 and 0.001562 with 95 percent confidence. Whether it actually contains the parameter in this sample is unknown, and the upper and lower limits of the confidence interval would differ from sample to sample.

### 5.4.8: Inference About the Intercept

Test statistics and confidence intervals for the intercept are constructed similarly.

As we saw at the beginning of this chapter, the intercept measures the point at which the regression line crosses the $Y$-axis (i.e., the conditional mean of $Y$ when $X = 0$).

In our example, Display B.5.1 showed that the intercept is estimated to be 249.8116. Thus, adults with no earnings are estimated to live about 250 miles away from their mothers, on average.

As we discussed above, when $X = 0$ does not fall within the range of the $X$ values, inference about the intercept will not be substantively meaningful. Similarly, the default hypothesis about the intercept included in computer printout tests $H_o$: $\beta_0 = 0$ against the alternative that $H_a$: $\beta_0 \neq 0$ may not be substantively meaningful in all applications (even though the SAS and Stata output will always conduct and present the results of this test in the default output). Even if $X = 0$ falls within the range of $X$s, testing whether the conditional mean of $Y$ when $X = 0$ is significantly above or below zero may not be substantively meaningful if zero is not a value in the range of $Y$ (e.g., if the outcome was SAT scores). In other words, while testing whether the *slope* differs from zero is an important straw man test of $X$ significantly linearly predicting $Y$, the substantive importance of testing whether the *intercept* equals zero will vary across applications.

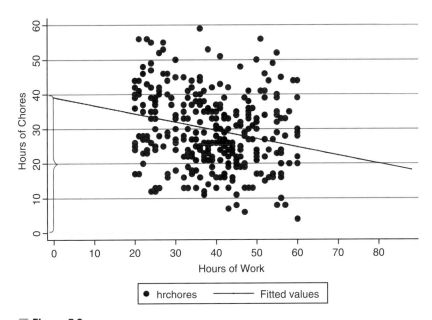

**■ Figure 5.8**

*Source:* Stylized sample from National Survey of Families and Households, Wave 1.

For example, suppose we hypothesized that hours of paid work was negatively related to hours of chores. This raises two questions in relation to the interpretation of the intercept. First, is $X = 0$ a valid value? In our example, we can ask whether it makes sense to include those working no hours in the sample (i.e., can we think of a continuous progression from no work to increasing hours of work? Or, should we view the movement from nonemployment to employment as a distinct process from the hours worked once employed)? In the stylized example in Figure 5.8, we excluded women who worked fewer than 20 or more than 60 hours per week. Thus, $X = 0$ is not valid in this case (the sample points fall between 20 and 60 on the $X$ axis). But, we allowed the sample line to extend to the point where the line would meet the $Y$ axis, showing how a prediction could be made from the regression line when $X = 0$, even if no such values exist in the sample.

Second, even if zero is a valid value for the predictor, does it make sense to test whether the conditional mean is zero? Figure 5.9 illustrates the negative association between hours of work and hours of chores. Here, we included those who worked fewer than 20 hours per week (including the small number reporting zero hours in the survey week). Given our conceptualization of a negative association between hours of chores and hours of work, and our focus on women, does it make sense to test a null hypothesis of no hours spent on chores? The dashed line in Figure 5.9 shows a line with the sample slope as the estimated regression line, but shifted down so that it begins with an intercept of zero. The estimated negative slope would predict that women who worked more hours engaged in a negative level of chores, clearly not sensible. And, in fact, in the NSFH sample, only 1 percent of women report spending no hours per week on chores.

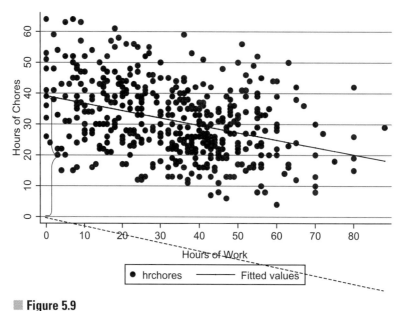

**▨ Figure 5.9**

*Source:* Stylized sample from National Survey of Families and Households, Wave 1.

### 5.4.9: Substantive Interpretation

It is important to consider the substantive importance of the point estimates and confidence intervals for our coefficients, along with the results of our decisions about our null and alternative hypotheses.

Recall that when the sample size is larger, the standard error is smaller, and thus it is especially possible in larger samples to reject null hypotheses for substantively small associations. As a consequence, it is important to ensure that findings that reject the null are substantively meaningful, especially when our sample sizes are large. In other words, for a substantively meaningful increase in $X$, is the estimated change in $Y$ based on the point estimate or confidence interval of "real world" importance?

### Reinterpreting for a Different Amount of Change

We discussed at the beginning of this chapter how to translate the slope into a more meaningful value if a one unit change in $X$ is not substantively meaningful. We saw that the effect of $X$ on $Y$ for changes in $X$ other than from 1 can be easily calculated using the following rule: if we increase $X$ by $i$ units, then $Y$ will change by $i\hat{\beta}_1$ units. In this case, we are reinterpreting the slope coefficient after estimating a regression model. When moving between different $X$ and $Y$ scalings, the significance of the effect of $X$ on $Y$ remains the same, but the ease with which the substantive relevance is seen may be clearer in some units of measurement than others.

### Rescaling Before Estimation

We can also rescale $X$ and/or $Y$ before running the regression model to obtain the desired results directly from SAS and Stata. Although this does not affect our decisions about hypothesis tests, it can make the output easier to read. To rescale *g2earn* into units of $10,000 we could create a new variable that is equal to *g2earn10000* = *g2earn*/10000. An increase of 1 on this new variable represents a $10,000 change in earnings. The following table shows some example values to help make the correspondence between the values of the old and new variable concrete.

| g2earn | g2earn10000=g2earn/10000 |
|---|---|
| 0 | 0.0 |
| 1,000 | 0.1 |
| 5,000 | 0.5 |
| 10,000 | 1.0 |
| 20,000 | 2.0 |
| 30,000 | 3.0 |
| 50,000 | 5.0 |
| 100,000 | 10.0 |

Comparing the fourth with the first rows shows that when *g2earn* changes from 0 to 10,000, *g2earn10000* changes from zero to 1. Similarly, when *g2earn* changes from 20,000 to 30,000 *g2earn10000* changes from 2 to 3.

▨ **Box 5.6**

More explicitly, notice that if we multiply both the numerator and denominator by 10,000 in the original calculation of the *t*-value, the 10,000 cancels out:

$$t = \frac{\hat{\beta}_1 - 0}{\hat{\sigma}_{\hat{\beta}_1}} = \frac{.00113 * 10000 - 0}{.0002185 * 10000}$$

$$= \frac{.00113 * 10000}{.0002185 * 10000}$$

$$= \frac{.00113}{.0002185}.$$

If we had a nonzero hypothesized value, rather than zero, we would need to rescale it equivalently by 10,000, so the 10,000 would still cancel out of the numerator and denominator, even with a nonzero hypothesized value.

Display B.5.3 shows the results of re-estimating our models with such a rescaled variable. The estimated slope coefficient based on the regression of *g1miles* on *g2earn10000* reports directly the result 11.333 that we calculated above with our $i\hat{\beta}_1$ rule. Notice that in Display B.5.3, the estimated standard error is now also equivalent to the original estimate of the standard error multiplied by 10,000 (0.0002185 * 10,000 = 2.185). Because both the estimated coefficient and standard error are multiplied by the same factor, the *t*-value remains the same (5.19; see Box 5.6). Thus, our conclusion regarding our hypotheses remains unchanged by rescaling the coefficient estimate. Notice also that the intercept and its standard error remain the same as in our original model (compare Display B.5.3 to Display B.5.1). This result will hold in general. Unless a variable is involved in an interaction (as discussed in Chapter 8), rescaling one variable will affect only its own estimated coefficient and standard error, not the estimate of the intercept or estimates for other variables in the model.

In our case, the range of values on *g2earn* was quite large, making a one unit change very small relative to its overall variation, and thus the original coefficient estimate was tiny. It is also important to pay attention to cases in which a one unit change is very large relative to the variation on the predictor. For example, percentages are sometimes used to define contextual variables (e.g., percentage of neighbors who are the same race as the respondent, percentage of students receiving free lunch). If these variables are recorded as decimals, a one unit change will represent the maximal possible change on the variable and may exceed the actual observed range of the variable (i.e., zero to 1 represents 0 percent to 100 percent). In this case, estimates can look substantively larger than they actually are if they are not appropriately rescaled. For example, if we had a variable *psamerace* that captured the proportion of neighbors who are the same race as the respondent in decimals, we could create a new variable *psamerace10* = *psamerace* * 10 that would allow us to interpret the estimated coefficient as the average change in the outcome for a 10 percentage point increase on the predictor.

| psamerace | psamerace10=psamerace*10 |
|-----------|--------------------------|
| 0.0 | 0 |
| 0.1 | 1 |
| 0.2 | 2 |
| 0.3 | 3 |
| 0.4 | 4 |
| 0.5 | 5 |
| 0.6 | 6 |
| 0.7 | 7 |
| 0.8 | 8 |
| 0.9 | 9 |
| 1.0 | 10 |

We can see that a change of zero to 1 on the new variable *psamerace10* represents a change from zero to 0.10 on the original variable *psamerace*.

### The Algebra of Rescaling

There is another general way to figure out the rule that we deduced by examining the regression line in Section 5.1.2 (if we increase $X$ by $i$ units, then $Y$ will change by $i\hat{\beta}_1$ units). If we define the following scale factors:

$$newY_i = rescaleY * Y_i$$
$$newX_i = rescaleX * X_i$$

The following relationships hold for the estimated slope coefficient and its standard error:

$$new\hat{\beta}_1 = \frac{rescaleY}{rescaleX} * \hat{\beta}_1$$

$$new\hat{\sigma}_{\hat{\beta}_1} = \frac{rescaleY}{rescaleX} * \hat{\sigma}_{\hat{\beta}_1}$$

The new $X$ variable, *g2earn10000*, that we created above is created from the original $X$ variable *g2earn* by multiplying by $1/10,000$. That is, *g2earn10000* = $1/10,000$ * *g2earn*. Thus, *rescaleX*=$1/10,000$. We did not rescale the $Y$ value in this case, thus *rescaleY* = 1 (i.e., *newY*=1*$Y$*). Plugging into the formulae we get:

$$new\hat{\beta}_1 = \frac{1}{\frac{1}{10,000}}\hat{\beta}_1 = 10,000 * \hat{\beta}_1$$

$$new\hat{\sigma}_{\hat{\beta}_1} = \frac{1}{\frac{1}{10,000}}\hat{\sigma}_{\hat{\beta}_1} = 10,000 * \hat{\sigma}_{\hat{\beta}_1}$$

■ **Box 5.7**

Professional and journal style guides in the social sciences often use the Greek symbol $\beta$ to designate a standardized coefficient. We have used the symbol $\beta$ to indicate the unstandardized regression coefficient in the population. You should consult style guides, such as the American Sociological Association Style Guide (American Sociological Association 2007) and Publication Manual of the American Psychological Association (American Psychological Association 2009), and journal submission guidelines to choose the preferred conventions in your field or for a particular journal. We will provide some examples of these varying conventions in the literature excerpts in Chapters 6 and 7.

Thus, by multiplying the listed coefficient estimate and its standard error for *g2earn* by 10,000 we calculate the results that we would obtain if we ran the regression model using *g2earn10000*. Knowing this simple formula makes it easy to convert results so that they can be interpreted for different units of change, if other units are more substantively meaningful.

### Standardized Slope Coefficient

The formula in Equation 5.5 is sometimes referred to as an **unstandardized regression coefficient**. A **completely standardized regression coefficient** would result if we rescaled both the $X$ and the $Y$ variables by dividing by their respective standard deviations (see Box 5.7). Such rescaled variables have standard deviation of 1. Variables whose values have been converted to $Z$- or standard normal-scores are examples of such rescaled variables and are sometimes referred to as standardized variables (see Equation 5.1). If we regressed the standardized $Y$ value on the standardized $X$ value, we call the resulting slope a standardized coefficient estimate. Such a standardized estimate of the slope may be especially useful for interpretation purposes if the original variables do not have substantively meaningful natural units (e.g., a sum of variables rated on scales such as 1 = *never*, 2 = *sometimes*, 3 = *often*, and 4 = *always*). In bivariate regression, the standardized slope is equivalent to the Pearson correlation. Recall that the *Pearson correlation* or *r* is a unitless measure of linear association between two variables, such as $Y$ and $X$.

Based on our discussion above, it is straightforward to produce the formula that relates the standardized slope to the unstandardized slope. In this case, $newY_i = (1/\hat{\sigma}_Y)Y_i$ and $newX_i = (1/\hat{\sigma}_X)X_i$.[13] Thus, $rescaleX = (1/\hat{\sigma}_X)$ and $rescaleY = 1/\hat{\sigma}_Y$ and we can calculate the new slope as follows:

$$new\hat{\beta}_1 = \frac{rescaleY}{rescaleX} * \hat{\beta}_1 = \frac{\dfrac{1}{\hat{\sigma}_Y}}{\dfrac{1}{\hat{\sigma}_X}} \hat{\beta}_1 = \frac{\hat{\sigma}_X}{\hat{\sigma}_Y} \hat{\beta}_1$$

This new (rescaled) estimate of the slope is the estimate of the standardized slope coefficient.

Both SAS and Stata have commands to create standardized variables and calculate the estimates of the standardized regression coefficients. We will see how to request estimates of the standardized regression coefficients below.

We will first calculate them directly so we can see how to use the rescale formula. To obtain the standard deviations, we use Stata's command `summarize <varlist>` and the SAS statement `proc means; var <varlist>; run;` Display B.5.4 shows the results of running these commands for SAS and Stata to obtain the standard deviations of the variables *g1miles* and *g2earn*.

We can then plug these values into the formula to calculate by hand an estimate of the new slope (the standardized slope coefficient):

$$new\hat{\beta}_1 = \frac{\hat{\sigma}_X}{\hat{\sigma}_Y}\hat{\beta}_1 = \frac{36475.81}{636.394} * .0011333 = .06496$$

Let's now replicate this process in SAS and Stata, by creating new rescaled variables that divide each variable by its respective standard deviation and estimate a regression model with the new standardized variables. Display B.5.5 shows our use of an expression with the division operator to create these new variables in SAS and Stata (see again Display A.4.1 and A.4.2) and our estimation of the regression model using these new variables. The estimate matches our hand calculation, within rounding error (0.06496 in SAS and 0.0649594 in Stata). Notice that the *t*-value is the same as before (5.19) as expected because the estimated standard error is rescaled by the same factor of $\hat{\sigma}_X/\hat{\sigma}_Y$ as the slope.

How do we interpret this new standardized estimate of the slope? Recall above that when we divided *g2earn* by 10,000 a one unit change on the new variable, *g2earn10000*, represented a $10,000 change on the original variable. Similarly, when we divide a predictor by its standard deviation, a one-unit change represents a one-standard deviation change. For *g2earn*, the following table shows how values on *g2earnSD* relate to values on *g2earn* and *g2earn10000*.

| g2earn | g2earn10000=<br>g2earn/10000 | g2earnSD=<br>g2earn/36475.81 |
|---|---|---|
| 0 | 0.0 | 0.000 |
| 1,000 | 0.1 | 0.027 |
| 5,000 | 0.5 | 0.137 |
| 10,000 | 1.0 | 0.274 |
| 20,000 | 2.0 | 0.548 |
| 30,000 | 3.0 | 0.822 |
| 36,475.81 | 3.6 | 1.000 |
| 50,000 | 5.0 | 1.371 |
| 72,951.62 | 7.3 | 2.000 |
| 100,000 | 10.0 | 2.742 |
| 109,427.43 | 10.9 | 3.000 |
| . . . | . . . | . . . |

Thus, a change from 0 to 1 on *g2earnSD* corresponds to a change of one standard deviation (from zero to 36,475.81) on *g2earn*. And, a change from 1 to 2 on *g2earnSD* also corresponds to a change of one standard deviation (36,475.81 to 72,951.62) on *g2earn*.

Because we have also rescaled the outcome variable by dividing by its standard deviation, the slope estimate reflects by what fraction of a standard deviation the outcome variable will change when the predictor variable increases by one standard deviation. In our case, we might say "For each one standard deviation more that adults earn, they live about .065 of a standard deviation further away from their mothers."

### Semistandardized Slope Coefficient
Sometimes, it is also useful to standardize only the outcome or only the predictor variable. This is done especially when one but not both of the variables is measured in units that have no intrinsic meaning (e.g., based on a rating scale or a sum of rating scale items). The results are referred to as semi-standardized coefficient estimates (Stavig 1977).

We can use our formula to calculate the new coefficient estimate we would obtain if we only standardized the predictor or the outcome prior to re-estimating the regression. If we rescale $X$ by dividing by its standard deviation but leave $Y$ untouched, then we have *rescaleX*=1/36475.81 and *rescaleY*=1. Substituting into the equation *new* $\hat{\beta}_1 = (rescaleY/rescaleX)\hat{\beta}_1$ with our original slope of $\hat{\beta}_1 = .0011333$ gives us:

$$new\hat{\beta}_1 = \frac{1}{1/36475.81} * .0011333$$

$$= \frac{36,475.81}{1} * .0011333$$

$$= 41.34$$

This would be interpreted as "For each standard deviation more that adults earn they live about 41 miles further from their mothers, on average." If we rescale $Y$ by dividing by its standard deviation but leave $X$ untouched, then we have *rescaleY*=1/636.394 and *rescaleX* = 1. Substituting again, into the equation *new* $\hat{\beta}_1 = (rescaleY/rescaleX)\hat{\beta}_1$ gives us:

$$new\hat{\beta}_1 = \frac{(1/636.394)}{1} * .0011333$$

$$= \frac{1}{636.394} * .0011333$$

$$= 0.000001781$$

This number is tiny because it represents the amount of change on the outcome, in standard deviation units, for a one dollar increase in annual earnings. We might instead rescale earnings to $10,000 units while rescaling distance to standard deviation units. To do so, we use $rescaleY=1/636.394$ and $rescaleX=1/10,000$. Substituting again, into the equation $new\hat{\beta}_1 = (rescaleY/rescaleX)\hat{\beta}_1$ gives us:

$$
\begin{aligned}
new\,\hat{\beta}_1 &= \frac{(1/636.394)}{(1/10,000)} * .0011333 \\
&= \frac{10,000}{636.394} * .0011333 \\
&= 0.01781
\end{aligned}
$$

This would be interpreted as "For each $10,000 more that adults earn they live nearly 0.02 standard deviations further from their mothers, on average."

Although it is instructive to see how to hand-calculate the standardized variable and coefficient estimate, once you understand the process, you can use commands in SAS and Stata to obtain them directly. Stata has an option, `beta` which can be added to any regression command and SAS has the option `/stb` which can be added to any model statement. Stata's term "beta" reflects the fact that in some subfields it is conventional to use the greek symbol $\beta$ or corresponding word *Beta* to denote the standardized slope. The SAS option can be remembered as shorthand for "standardized beta." Display B.5.6 shows these options implemented in SAS and Stata, and the results (0.06496 in SAS and 0.0649594 in Stata) match our earlier results (compare results of Display B.5.6 and Display B.5.5).

Finally, we can use a command to obtain the Pearson correlation in SAS and Stata to reinforce the fact that, in bivariate regression, the standardized coefficient estimate is equivalent to the Pearson correlation. In SAS, the relevant syntax is: `proc corr; var <varlist>; run;` in Stata, the command is `correlate <varlist>`. The results, shown in Display B.5.7 again match the standardized estimates of the slope coefficient, within rounding (0.06496 in SAS and 0.0650 in Stata).

### Reversing the Predictor and Outcome

As we interpret regression results, it is also important to keep in mind the asymmetric nature of the regression model. In regression modeling, we consider one variable to be the outcome variable and another variable (or set of variables in multiple regression) to be the predictor variable. The shorthand expression **"left-hand side"** and **"right-hand side"** of the equation reflects this asymmetry.

$$Y = \beta_0 + \beta_1 X + \varepsilon_i$$

Left-hand side      Right-hand side

The theoretical and conceptual undergirding of our model provide the basis for setting up, or specifying, this asymmetric relationship. It is these theories and conceptions that move us from a symmetric correlation relationship to an asymmetric regression model.

The asymmetric nature of regression, in contrast to the symmetric correlation coefficient, is evident if we reverse the predictor and outcome in our example (i.e., we regress *g2earn* on *g1miles*). The regression coefficient differs when we do so. It now reflects the amount of change in *g2earn* when *g1miles* increases by 1. Display B.5.8 shows the result of this regression. The coefficient estimate for *g1miles* can be interpreted as: "For each mile further the mother lives from her child, the child earns $3.72 more dollars per year, on average." Note that the *t*-value remains 5.19, so our conclusion based on our hypothesis test is the same. The coefficient estimate is quite different in Display B.5.8 than in Display B.5.1 (3.72324 versus 0.00113) because of the difference in standard deviations between the two variables (about 636 for miles and 36,476 for earnings). But, if we instead standardize both variables so that one unit of change represents one standard deviation for that variable, the coefficient is the same regardless of the order of the variables (and is equal to the Pearson correlation coefficient; see Display B.5.9 and compare slope to Display B.5.5). Likewise, if we reverse the order of the two variables in the variable lists asking SAS and Stata to calculate the correlation, the result is unchanged.[14]

These results bring out three major points:

1. We must always keep in mind the units of measurement of our variables when interpreting the substantive size of a coefficient estimate.
2. Which variable we choose to be the outcome will alter the size of the estimated slope coefficient (but not the statistical strength of the underlying relationship).
3. The basic OLS regression model cannot help us to decide which variable belongs on the left-hand versus the right-hand side.

As we move forward in Chapter 6, we will add more variables to the right-hand side, but keep the single outcome on the left. Our choices about which variable is the outcome and which are the predictors are up to us. Ideally, they are based on a specific theory or conceptual framework. But, ultimately, another researcher might reverse their order (and more sophisticated techniques exist to try to examine causal ordering; see roadmap in Chapter 12).

## Effect Size

We saw different ways to rescale the slope estimates to aid interpretation, but we still did not answer the bottom line question of whether the association is big or small, substantively. For statistical significance, we have an exact critical value which is conventionally used to determine whether or not a $t$-statistic is statistically significant. What kinds of criterion can we use to determine if the size of the association is substantively meaningful? Although there is no hard-and-fast rule for doing so, the concept of **effect size** can be used to help us make such conclusions.

A quick scan of books and articles devoted to effect size will reveal numerous formulae which might appear unfamiliar on first blush, but actually are intimately related to the standardized and semistandardized slopes we have already discussed. In fact, effect size formulae are often grouped into those that capture linear relationships and those that capture group differences (Huberty 2002; McCartney and Rosenthal 2000). The Pearson correlation coefficient is the basic effect size measure in the linear relationship set. A measure conventionally referred to as Cohen's $d$ is the basic effect size measure in the group difference set. It is defined as:

$$\frac{\overline{Y}_{Group_1} - \overline{Y}_{Group_2}}{\sigma_Y} \tag{5.9}$$

Notice that this formula is nothing more than the difference in means on the outcome variable between two groups divided by the standard deviation of the outcome variable. We will use this formula directly in Chapter 7 when we introduce dummy variables to model predictor variables that indicate two or more groups. With continuous predictor variables, we can use this measure as an analogy for the semistandardized coefficient in which we keep the $X$ variable in its natural units and standardize the $Y$ variable. As we will emphasize below, it is important to think still about rescaling the $X$ variable so that a one unit increase on the rescaled variable represents a meaningful amount of real world change when doing so.

Using effect sizes for interpretation has the same benefit as standardizing, discussed above. Effect sizes convert the size of the association into standard deviation units (see Box 5.8). Especially when the units of measure of the outcome do not have intrinsic meaning, this conversion can help you interpret the size of the effect. And, it can allow you more easily to compare associations across different variables that have different standard deviations. Whether to use the completely standardized or semistandardized measure will depend on your application. If your predictor variable does not have meaningful natural units, then the completely standardized measure may be preferred. If your predictor variable does have meaningful natural units, then the semistandardized measure may be helpful. We present an example of each below.

■ **Box 5.8**

In the multiple regression context which we will consider in the next chapter, effect size may also be calculated in other ways, including based on the value of *R*-squared. But, the analogy to Equation 5.9 is intuitive and as we show can be related to the semistandardized coefficient (and the dummy variable predictors we will discuss in Chapter 7). We encourage you to think carefully about how to characterize the size of effects in your particular applications, either in terms of the natural units of the variables, in comparison to real world benchmarks or in terms of the standard deviation of the variables.

First, though, let's address the question of what size of an effect is meaningful. Recall that the Pearson correlation coefficient is bounded between zero and 1, in absolute value. Because the completely standardized coefficient in bivariate regression is equivalent to the Pearson correlation coefficient, it is also bounded between zero and 1, in absolute value. The semistandardized coefficient could be larger than 1 if a one-unit increase on the predictor variable is associated with more than a one standard deviation change on the outcome variable. But, the semi-standardized effect is still interpreted as the fraction of a standard deviation that the outcome variable increases when the predictor variable increases by one (e.g., a semistandardized coefficient estimate of 1.5 would indicate that the outcome variable increase by one and a half standard deviations when the predictor variable increases by 1).

What is a large value for a completely standardized or semistandardized coefficient estimate? Guidelines developed by statistician Jacob Cohen (1969) are probably the most widely applied. Cohen recommended considering effect sizes of about 0.20 as small, 0.50 as medium, and 0.80 as large (Cohen 1992). This rule is widely used, particularly in some subfields, although some have argued that it is overused, especially given the limited empirical basis on which Cohen first developed the cutoffs (Huberty 2002; McCartney and Rosenthanl 2000). Another alternative to using absolute cutoffs like Cohen's is to place effect sizes in the context of related literature on the outcome and predictor of interest or to compare the effect size for one predictor and outcome to effect sizes for other predictors. Researchers might also consider the real world costs and benefits of the predictor and outcome in interpreting the results. For example, McCartney and Rosenthal (2000) provide an example of a very small effect size that was interpreted as of practical importance because the outcome was mortality and the predictor was an inexpensive method for preventing heart attacks (daily aspirin use).

How do our examples stack up using these guidelines? Above, we calculated the standardized coefficient estimate for our distance example to be 0.0649594. The semistandardized coefficient estimate was tiny for a $1 increase in annual earnings. It was larger, although still small, when we rescaled $X$ so that a one-unit increase represented $10,000 in annual earnings (0.01781). Thus, in all of these cases (for a one standard deviation, a $1, and a $10,000 increase in annual earnings) the distance that the adult lives from his or her mother increases by less than one tenth of a standard deviation. All are small, based on Cohen's cutoffs. In Chapter 6 we will demonstrate how to place these effect sizes for earnings in the context of the size of effects of other predictor variables.

Both our predictor and our outcome in this distance example have meaningful natural units. Thus, we can also use real world benchmarks to evaluate the substantive importance of the estimated association. Recall that our unstandardized coefficient estimate was 11.33 when we rescaled earnings so that a one-unit increase reflected an increase of $10,000. Does a change of about 11 miles of distance for a $10,000 increase in annual earnings seem large? To answer this question, we might think about the practical consequence of 11 miles of distance for interactions between family members. Most of us would likely come to a similar conclusion based on these natural units of $Y$ as we did with Cohen's cutoffs and the semistandardized slope: the association between earnings and distance, although statistically significant, seems substantively small (i.e., a change of 11 miles in average distance for a $10,000 increase in earnings is not large).

We can also convert the lower and upper bounds of our confidence intervals into standardized or semistandardized form to help us to assess the substantive size of an effect. The 95 percent confidence interval we calculated above was (0.000705, 0.001562). We can convert these bounds to units of $10,000 change in earnings by multiplying both numbers by 10,000. This results in a confidence interval of (7.05, 15.62). Thus, we might say that we have 95 percent confidence that the amount of increase in distance associated with a $10,000 increase in the adult's earnings falls between about 7 and 16 miles. We could also convert the bounds to semistandardized values by dividing by the standard deviation of distance (636.394). Since $7.05/636.394 = .011078$ and $15.62/636.394 = .02454$ we could say that we have 95 percent confidence that the amount of increase in distance associated with a $10,000 increase in the adult's earnings falls between about 0.01 and 0.03 of a standard deviation of distance. Completely standardizing the confidence interval bounds would involve multiplying the original values by the standard deviation of earnings and dividing by the standard deviation of distance. Since $0.000705*36475.81/636.394 = .0404$ and $0.001562*36475.81/636.394 = .0895$ we could say that we have 95 percent confidence that the amount of increase in distance associated with a standard deviation increase in the adult's earnings falls between about 0.04 and 0.09 of a standard deviation of distance. As with the point estimates, the confidence intervals suggest a fairly small effect size.

The association we examined between number of children in the household and hours of chores, in contrast, seems larger in substantive size. Recall that in our stylized example of 50 cases above, we estimated that the women completed about six more hours of chores per week when one more child was in the household, on average. Across the seven days of the week this would represent nearly one more hour of chores per day. This seems a substantively larger amount of real world change than we saw in our distance example. For example, for women employed full-time, and allocating 8 hours to sleeping and 10 hours to work and commuting leaves six hours in a day; spending one more of those six remaining hours on chores seems a substantial change.

What about the effect size of the association between number of children and hours of chores? As noted above, the standard deviation of work hours is 10.47 in our stylized sample of 50. Thus, the semistandardized group difference between those having one fewer versus one more child is also sizable on Cohen's cutoffs. $6.13/10.47 = 0.58$, somewhat larger than his criteria for a moderate association. The standard deviation of number of children is 1.43, thus the completely standardized coefficient estimate would be $6.13*1.43/10.47 = 0.84$, a large effect by Cohen's cutoffs. Thus, in our stylized hours of chores example, we come to a similar conclusions based on examining the real world importance of the effect in its natural units and the relation to Cohen's cutoffs of the effect in its standardized units: the effect is moderate to large in size.[15]

**SUMMARY 5**

## 5.5: SUMMARY

In this chapter, we introduced the basic concepts of regression in the bivariate context (with one outcome variable and one predictor variable). We presented the population regression line with algebra and graphs, emphasizing how each approach helps us to understand the interpretation of the intercept and slope. We also introduced the idea of conditional distributions, emphasizing that the population regression line has two components. The systematic component is the prediction equation determined by the intercept and slope and provides the mean of the conditional distributions. The random component is the error term and determines the variance of the conditional distributions. These error terms allow individual $Y$s within a given level of $X$ to fall above or below the regression line (i.e., above or below the conditional means).

The sample regression line provides estimates of the conditional mean (the intercept and slope of the prediction equation) and conditional variance. Like all estimates of population parameters, these estimates have sampling distributions. In any one sample the estimate is unlikely to be exactly equal to the population parameter. The standard error is the standard deviation of the sampling distribution. The smaller the standard error, the closer a sample estimate is expected to fall to the population parameter. Because most of our research questions and hypothesis statements in the social sciences involve slopes, we looked in detail at the factors that affect the standard error of the estimator for the slope. These include the sample size, variation on the predictor variable, and strength of the relationship between the predictor and outcome.

We also considered how to evaluate our research questions (two-tailed tests) or hypothesis statements (one-tailed tests) using $t$-statistics and confidence intervals. For a two-tailed test, the confidence interval and $t$-statistic provide equivalent results. If the magnitude of the test statistic is larger than the critical value, then the confidence interval will not contain the hypothesized value. If the magnitude of the test statistic is smaller than the critical value, then the confidence interval will contain the hypothesized value. For a one-tailed test, we rely on the $t$-statistic, and reject the null hypothesis if the test statistic is more extreme than the critical value.

Regardless of our decision about rejecting or failing to reject the null hypothesis, the point estimate and confidence intervals provide our best guesses of the value of the population parameter. The confidence interval is superior to the point estimate in that it captures information about the uncertainty of our guess (the standard error of our estimate). For either point estimates or confidence intervals it is also important to consider the substantive size of the effect. We can rescale the predictor and/or outcome variables to aid in such interpretation. If the predictor and outcome variables have meaningful natural units, then we can rescale the predictor to an amount of change that has real world importance (e.g., \$10,000 rather than \$1 increase in annual earnings). Especially if the predictor and/or outcome variables do not have meaningful units, we can rescale based on their respective standard deviations. When we rescale both the predictor and outcome based on their standard deviations, we calculate a completely standardized coefficient estimate. When we rescale only one (the predictor or the outcome) we have a semistandardized coefficient estimate. The completely standardized and semistandardized coefficient estimates can be interpreted using the concept of effect size. They can be compared to cutoffs introduced by Jacob Cohen and, preferably, will also be considered in relation to other effect sizes in the literature and/or the effect sizes for other variables in a study.

## KEY TERMS

Best Linear Unbiased Estimators (BLUE)

Bivariate Regression

Completely Standardized Regression Coefficient

Conditional Distribution
(also Conditional Mean, Conditional Standard Deviation, Conditional Variance)

Confidence Interval

Degrees Of Freedom

Deterministic

Effect Size

Estimate

Estimator

Expected Value

Fitted Value

Hypothesized Value

Intercept

Left-Hand Side/Right-Hand Side

Linear Association

Mean Square Error

Nonsystematic Component

One-Sided Hypothesis Test

One-Sided *p*-Value

Ordinary Least Squares (OLS) Regression

Point Estimate

Population Parameters

Population Regression Line

Predicted Values

Prediction Equation

Probabilistic

Random

Root Mean Square Error (or Root MSE)

Sample Estimates

Sampling Distributions

Semistandardized Regression Coefficient

Simple Regression

Slope

Stochastic

Systematic Component

Two-Sided Hypothesis Test

Two-Sided $p$-Value

Unconditional Distribution
(also Unconditional Mean, Unconditional Standard Deviation,
Unconditional Variance)

Unstandardized Regression Coefficient

$Y$-Hat

## REVIEW QUESTIONS

REVIEW
QUESTIONS
5

**5.1** What does it mean, algebraically and geometrically, if the slope is equal to zero?

**5.2** We can always mechanically interpret the intercept. What should we consider to determine if the intercept is substantively meaningful?

**5.3** Write the general statement that we use to interpret the slope from a bivariate regression model (i.e., do so in the "general" case of $Y$ as the outcome variable, $X$ as the predictor variable and $\beta_1$ as the slope).

**5.4**   Write a general bivariate population regression equation and bivariate sample regression equation, including the error terms. Indicate which is the systematic and which the random component of each. Discuss what the systematic and random components represent in the conditional distributions associated with each regression equation.

**5.5**   Write the formulae for the point estimate of the slope, conditional variance, and standard error of the slope, in bivariate regression.

**5.6**   Will the conditional standard deviation be larger than, smaller than, or equal to, the standard deviation of the unconditional distribution of the outcome variable if the outcome and predictor variable are strongly associated? Why?

**5.7**   What are the three major factors that affect the size of the standard error of the slope in bivariate regression?

**5.8**   Write the formula for the test statistic of the slope in bivariate regression. What is the most common hypothesized value for the slope? Why? What are the null and alternative hypotheses about the slope tested in the default SAS and Stata output?

**5.9**   Write the formula for a 95 percent confidence interval about the slope. If you reject the null hypothesis in a two-sided hypothesis test, what do you expect to see in relation to the confidence interval?

**5.10**  What are the three ways discussed in the chapter to "change units" in order to interpret a slope in a more substantively meaningful way (e.g., in units of 10,000s of dollars rather than units of dollars, for example).

**5.11**  What are some strategies you can use to determine if a significant effect is substantively important (large or small)?

**5.12**  Define the standardized slope using the standard deviation of $X$, standard deviation of $Y$, and the unstandardized slope.

**5.13**  How do the standardized slope, unstandardized slope, and Pearson correlation change if we reverse which variable we use as the outcome and which variable we use as the predictor in bivariate regression?

## REVIEW EXERCISES

**5.1**  Suppose that you estimate a bivariate regression of earnings on years of schooling and obtain the following estimates: $\hat{\beta}_0 = \$20{,}000$ and $\hat{\beta}_1 = \$5{,}000$.

  (a)  Interpret the intercept.

  (b)  Interpret the slope.

**5.2**  Suppose that you regress children's reading achievement scores on the number of books in the household and obtain the following estimates: $\hat{\beta}_0 = 70$ and $\hat{\beta}_1 = 0.60$.

  (a)  Interpret the intercept.

  (b)  Interpret the slope.

  (c)  Reinterpret the slope for an increase of 10 books.

**5.3**  Show how to calculate the unstandardized regression coefficient based on the Pearson correlation coefficient, standard deviation of $X$, and standard deviation of $Y$.

**5.4**  Suppose you estimate an unstandardized slope coefficient of 0.5. If $\hat{\sigma}_Y = 4$ and $\hat{\sigma}_X = 2$, what is the completely standardized slope?

**5.5**  Suppose that $\hat{\beta}_1 = 0.035$ when we regress $Y$ on $X$. Now, suppose we create a rescaled $X$ variable, which is equal to the original $X$ variable divided by 5. What would be the slope coefficient from the regression of $Y$ on the new variable?

**5.6**  Write the formula for the standard error of the slope in bivariate regression and discuss what you might do as a researcher in planning a study in order to try to achieve a relatively small standard error.

**5.7**  What would happen to Equation 5.8 in the extreme case of no variation in the $X$s? (That is, all $X$ have the same value, and thus all $X_i$ are equal to $\bar{X}$).

**5.8**  Does a positive slope coefficient from a simple regression model of $Y$ on $X$ mean that all observed $Y$ values in the sample at $X = 11$ are larger than all observed $Y$ values in the sample at $X = 10$, that the conditional mean predicted from the regression model at $X = 11$ is larger than the conditional mean predicted from the regression model at $X = 10$, or both? Why?

**5.9** Consider a simple regression of $Y$ on $X$ based on a large sample size in which $\hat{\beta}_1 = 5$ and $\hat{\sigma}_{\hat{\beta}_1} = 2$.

(a) Construct a 95 percent confidence interval for the slope.

(b) Write a one-sentence interpretation of the confidence interval.

**5.10** Suppose a colleague sends you a data set that contains two variables, a measure of mental health, and a measure of marital satisfaction. How would you determine which variable should be the dependent variable and which the independent variable?

**5.11** Many social scientists would probably order the variables in our distance example as we have, predicting distance from the mother based on the adult's earnings. Can you think of some reasons for this order? Can you think of reasons for reversing the order (how might distance affect the adult's earnings)?

**5.12** Consider a regression model with prediction equation $\hat{Y} = 2.32 + 4.59X$, and Root Mean Square = 2.32. Based on these results, what are the estimated mean and standard deviation for the conditional distribution of $Y$ when $X = 0$ and when $X = 10$?

CHAPTER
EXERCISES
5

## CHAPTER EXERCISES

In these Chapter Exercises, you will write a SAS and a Stata batch program to create a new variable, calculate some descriptive statistics, and estimate several bivariate regression models ("SAS/Stata Tasks"). You will want to have Displays A.4.1, A.4.2, and A.5 handy as you write your batch programs. You will use the results to answer some questions related to the substance of what we learned in this chapter ("Write-Up Tasks").

**To begin,** prepare the shell of a batch program, including the commands to save your output and helpful initial commands (e.g., In SAS, use a libname to assign the SAS library; in Stata, turn *more* off and close open logs; see Display A.4.2). Use the NOS 1996 dataset that you created in Chapter 4, but use an *if expression* to only keep the organizations that were founded after 1950 (be careful about missing values on founding year) and that do not have missing values on the *mngpctfem, age,* and *ftsize* variables (see Display A.4.1 and A.4.2).

## 5.1  Interpretation and Hypothesis Testing in Bivariate Regression

(a)  SAS/Stata tasks

   (i)  Regress *mngpctfem* on *age*.

   (ii)  In SAS, use the `/clb` option to request the 95 percent confidence intervals (recall that Stata calculates the confidence intervals by default).

(b)  Write-up task

   (i)  Write the null and alternative hypotheses about the *slope* that are being tested by the *t*-value and *p*-value listed in the output. Why is this null hypothesis interesting? Write the steps for making a decision about these hypotheses (use alpha = .05). Interpret the point estimate of the slope.

   (ii)  Calculate by hand, and interpret, a 95 percent confidence interval for the slope. Note whether and how the confidence interval results relate to the *t*-test examined in Question 5.1bi.

   (iii)  Repeat the hypothesis test from Question 5.1bi, but use a one-sided alternative, based on the expectation that younger organizations have a higher percentage of female managers. Be sure to write the null and alternative hypotheses and indicate the *t*-value and *p*-value for this test.

   (iv)  Write the null and alternative hypotheses about the intercept that are being tested by the *t*-value and *p*-value listed in the output. You do not need to go through the steps of making a decision about this null hypothesis, but you should comment on whether this test is meaningful in this situation.

## 5.2  The Conditional Nature of Regression

(a)  SAS/Stata tasks

   (i)  Refer to the output from Question 5.1.

   (ii)  Calculate the unconditional standard deviation of *mngpctfem*.

(b)  Write-up tasks

   (i)  Write algebraically and present graphically the regression model estimated in Question 5.1. Calculate at least two predicted values for your graph (you can graph them roughly in freehand; the plot does not need to be exact).

(ii) What are the estimated mean and standard deviation for the conditional distribution of *mngpctfem* when *age* is 5? What about when *age* is 15? And when *age* is 25?

(iii) Compare the unconditional and conditional standard deviations of *mngpctfem* (Which is larger? What do their relative sizes suggest about the strength of the relationship between *mngpctfem* and *age*?)

## 5.3 Changing the Units of *X*

(a) SAS/Stata tasks

(i) Compute a new variable defined as age divided by 5. (Remember: in SAS this new variable creation must happen in the DATA step!)

(ii) Regress *mngpctfem* on this new variable.

(b) Write-up tasks

(i) Based only on the output from Question 5.1, interpret the slope coefficient for a five-year increase in age. Comment on the potential substantive meaning of this result. What other amounts of change might be substantively interesting?

(ii) Compare the slope from the regression of *mngpctfem* on the new variable, age divided by 5, to the slope you calculated in response to Question 5.3bi.

## 5.4 The Asymmetric Nature of Regression

(a) SAS/Stata tasks

(i) Correlate *mngpctfem* with *age*.

(ii) Regress *age* on *mngpctfem*.

(iii) Calculate the unconditional standard deviation of *age*

(b) Write-up tasks

(i) Show how to calculate the slope in the output from Question 5.1 and the slope in the output from this question (i.e., the new regression of *age* on *mngpctfem*) based on the Pearson correlation.

(ii) Briefly, discuss how you would decide which variable (*mngpctfem* or *age*) would be the dependent variable if you were writing a paper.

(iii) Under what circumstances would the slope based on regressing *mngpctfem* on *age* equal the slope based on regressing *age* on *mngpctfem*?

## COURSE EXERCISE

Using the dataset you created in the course exercise to Chapter 4, regress one of your continuous outcomes on one of your continuous predictors. Before running the model, make your best guess of the expected direction of the association and justify your choice of outcome variable (can you think of a conceptual rational or an argument another scholar might make about reversing the order of the variables, so that a predictor becomes the outcome?).

Conduct a hypothesis test for the slope, using both a two-sided test for any linear relationship and a one-sided test using the direction you expect for the association as the alternative. Write a one-sentence interpretation of the slope estimate. If the outcome and predictor have natural units that are meaningful, consider rescaling them to other amounts of change (both by reinterpreting the original estimates and rescaling before a new estimation). Especially if the outcome and predictor do not have meaningful natural units, re-estimate the model after rescaling the outcome and predictor by their standard deviations. You should also use SAS and Stata's options to calculate the standardized slope. Based on the results, is the association substantively meaningful?

Comment on whether the intercept is meaningful in your case (e.g., is zero a valid value on your predictor variable, making the *Y*-intercept within range? Is zero a valid value on your outcome variable, making the null hypothesis relevant?). If the intercept is meaningful, write a one-sentence interpretation of it.

Calculate the unconditional standard deviation of your outcome variable and compare it to the estimate of the conditional standard deviation. What does this comparison suggest about how strongly the predictor and outcome variable are related?

# Chapter 6

# BASIC CONCEPTS OF MULTIPLE REGRESSION

## CHAPTER 6: BASIC CONCEPTS OF MULTIPLE REGRESSION

Bivariate regression is often inadequate for social science theory given that we typically expect that more than one variable explains the outcome. For example, we wrote our research questions about distance to include not only the adult's earnings as a predictor of distance from the mother, but also the adult's age and number of brothers and sisters and the mother's age and years of schooling. We will add these additional variables, beyond earnings, to the model in this chapter (and in Chapter 7 we will add the adult's gender and race-ethnicity as predictors as well). In this chapter, we will focus on a basic understanding of how to interpret and test coefficients estimated from the model. In later chapters we will elaborate on this understanding, for example discussing in more detail how and why coefficient estimates change when other variables are added to the model.

## 6.1: ALGEBRAIC AND GEOMETRIC REPRESENTATIONS OF MULTIPLE REGRESSION

We will begin by examining a three-variable model, with one outcome variable and two predictors. Doing so allows us to understand the major concepts of multiple regression with simple algebra.

### 6.1.1: The Population Regression Model

The *population regression model* in multiple regression is a direct extension of the population regression model in simple regression:

$$Y_i = \beta_0 + \beta_1 X_{1i} + \beta_2 X_{2i} + \varepsilon_i$$

As in simple regression, this equation divides into a systematic and stochastic component; that is:

$$E(Y|X_{1i}, X_{2i}) = \beta_0 + \beta_1 X_{1i} + \beta_2 X_{2i}$$

is the systematic component—the conditional mean of $Y$ for given levels of $X_1$ and $X_2$. The error $\varepsilon_i$ is the random component and represents the distance that each case's outcome value, $Y_i$, falls from the conditional mean. Thus, the errors represent the additional variation in $Y$ not explained by $X_1$ and $X_2$.

We can interpret the model parameters in a similar manner as we did in bivariate regression:

- The intercept is the conditional mean of $Y$ when both predictor variables equal zero. It is the point at which the regression function (now a plane) passes through the $Y$ axis.
- The slope of $X_1$ is the amount that the conditional mean of $Y$ changes for a one-unit increase in $X_1$, controlling for $X_2$.
- The slope of $X_2$ is the amount that the conditional mean of $Y$ changes for a one-unit increase in $X_2$, controlling for $X_1$.

These interpretations can be made clear by manipulating the systematic portion of the regression function.

### Intercept
Substituting $X_1 = 0$ and $X_2 = 0$ into the equation for the conditional mean leads to all the terms dropping out of the right-hand side of the equation except for the intercept.

$$E(Y|X_1 = 0, X_2 = 0) = \beta_0 + \beta_1 * 0 + \beta_2 * 0 = \beta_0$$

### Slope
Let's consider first the slope of the $X_1$ variable. If we hold $X_2$ constant at a particular level, then that value is subsumed into the intercept and it is easy to see that the change in $Y$ when we change $X_1$ by 1 is equal to $\beta_1$.

For example, let's hold $X_2$ constant at 3 and allow $X_1$ to vary between zero and 3. Following are the conditional means when these values are substituted into the systematic portion of the regression model.

| Value of Xs are . . . | | The conditional mean of Y equals . . . | |
| --- | --- | --- | --- |
| $X_1$ | $X_2$ | | |
| 0 | 3 | $E(X_1 = 0, X_2 = 3) = \beta_0 + \beta_1 * 0 + \beta_2 * 3$ | $= (\beta_0 + 3\beta_2)$ |
| 1 | 3 | $E(X_1 = 1, X_2 = 3) = \beta_0 + \beta_1 * 1 + \beta_2 * 3$ | $= (\beta_0 + 3\beta_2) + \beta_1$ |
| 2 | 3 | $E(X_1 = 2, X_2 = 3) = \beta_0 + \beta_1 * 2 + \beta_2 * 3$ | $= (\beta_0 + 3\beta_2) + 2\beta_1$ |
| 3 | 3 | $E(X_1 = 3, X_2 = 3) = \beta_0 + \beta_1 * 3 + \beta_2 * 3$ | $= (\beta_0 + 3\beta_2) + 3\beta_1$ |

Notice that because we hold $X_2$ constant at 3, each conditional mean contains the term $\beta_0 + 3\beta_2$. The value of $X_1$, which we vary, determines how much above or below $\beta_0 + 3\beta_2$ the conditional mean falls.

As we did for bivariate regression, we can also subtract conditional means in adjacent rows to see that when we increase $X_1$ by 1 the conditional mean changes by $\beta_1$, holding $X_2$ constant. This makes very concrete our standard one sentence interpretation of the slope in multiple regression.

| Value of Xs are . . . | | Difference in conditional mean between the current row and prior row is . . . | | |
| --- | --- | --- | --- | --- |
| $X_1$ | $X_2$ | | | |
| 0 | 3 | | | |
| 1 | 3 | $[(\beta_0 + 3\beta_2) + \beta_1] - [(\beta_0 + 3\beta_2)]$ | $= \beta_0 + 3\beta_2 + \beta_1 - \beta_0 - 3\beta_2$ | |
| | | | $\beta_0 + 3\beta_2 + \beta_1 - \beta_0 - 3\beta_2 \quad = \quad \beta_1$ | |
| 2 | 3 | $[(\beta_0 + 3\beta_2) + 2\beta_1] - [(\beta_0 + 3\beta_2) + \beta_1]$ | $= \beta_0 + 3\beta_2 + 2\beta_1 - \beta_0 - 3\beta_2 - \beta_1$ | |
| | | | $\beta_0 + 3\beta_2 + 2\beta_1 - \beta_0 - 3\beta_2 - \beta_1 \quad = \quad 2\beta_1 - \beta_1 = \beta_1$ | |
| 3 | 3 | $[(\beta_0 + 3\beta_2) + 3\beta_1] - [(\beta_0 + 3\beta_2) + 2\beta_1]$ | $= \beta_0 + 3\beta_2 + 3\beta_1 - \beta_0 - 3\beta_2 - 2\beta_1$ | |
| | | | $\beta_0 + 3\beta_2 + 3\beta_1 - \beta_0 - 3\beta_2 - 2\beta_1 \quad = \quad 3\beta_1 - 2\beta_1 = \beta_1$ | |

We can also write what we will refer to as a **conditional regression equation**, in which we allow one of the variables to vary (by not holding it constant at a particular value) and fix the other variable at a particular value.

$$E(X_1, X_2 = 3) = \beta_0 + \beta_1 X_{1i} + \beta_2 * 3 = (\beta_0 + 3\beta_2) + \beta_1 X_{1i}$$

The result would be similar if we held $X_2$ constant at any other value. This would simply change the constant multiple of $\beta_2$ which we have subsumed into the intercept by holding $X_2$ constant. For example,

$$E(X_1, X_2 = 0) = \beta_0 + \beta_1 X_{1i} + \beta_2 * 0 \qquad = \beta_0 + \beta_1 X_{1i}$$
$$E(X_1, X_2 = 6) = \beta_0 + \beta_1 X_{1i} + \beta_2 * 6 \qquad = (\beta_0 + 6\beta_2) + \beta_1 X_{1i}$$

Our findings would be analogous if we held $X_1$ constant while allowing $X_2$ to vary (i.e., the value of $\beta_1$ would be subsumed in the intercept and $Y$ would change by $\beta_2$ when $X_2$ increased by 1). For example:

$$E(X_1 = 0, X_2) = \beta_0 + \beta_1 * 0 + \beta_2 X_{2i} \quad = \beta_0 + \beta_2 X_{2i}$$
$$E(X_1 = 3, X_2) = \beta_0 + \beta_1 * 3 + \beta_2 X_{2i} \quad = (\beta_0 + 3\beta_1) + \beta_2 X_{2i}$$
$$E(X_1 = 6, X_2) = \beta_0 + \beta_1 * 6 + \beta_2 X_{2i} \quad = (\beta_0 + 6\beta_1) + \beta_2 X_{2i}$$

Generally, these conditional regression equations can be expressed as follows:

$$E(X_1, X_2 = m) = \beta_0 + \beta_1 X_{1i} + \beta_2 * m \quad = (\beta_0 + m\beta_2) + \beta_1 X_{1i}$$
$$E(X_1 = n, X_2) = \beta_0 + \beta_1 * n + \beta_2 X_{2i} \quad = (\beta_0 + n\beta_1) + \beta_2 X_{2i}$$

With three variables in the equation, we could still represent the relationships graphically using a plane that occupies three-dimensional space. But, with more than three predictor variables, this process of holding other variables constant while we allow one variable to vary will allow us to display our findings graphically. These conditional regression equations reduce the multiple regression plane to straight lines. Plotting them allows us to visualize the effect of one of the $X$ variables while holding the other $X$ variable constant, making clear the interpretation of each of the slope parameters. Such graphs often facilitate interpretation (especially when the association is nonlinear, as we will see in Chapter 9).

### 6.1.2: Example 6.1

Figure 6.1 presents to the following regression plane:

$$E(Y|X_{1i}, X_{2i}) = 10 + 2X_{1i} + 3X_{2i}$$

This equation shows that the plane has an intercept of 10, a slope of 2 for the first predictor variable, and a slope of 3 for the second predictor variable. Following our general discussion above, these parameters can be interpreted as follows:

- Intercept: the conditional mean of $Y$ when both $X_1$ and $X_2$ are zero is 10.
- Slope of $X_1$: holding $X_2$ constant, $Y$ increases by 2, on average, when $X_1$ increases by 1.
- Slope of $X_2$: holding $X_1$ constant, $Y$ increases by 3, on average, when $X_2$ increases by 1.

These values are illustrated in Figure 6.1. In Figure 6.1(a), the top, left cell contains the value of 10 which is the value of $Y$ when both predictors are zero. If we look across the rows, we see increments of 3 in $Y$ corresponding to each increase of 1 in $X_2$. This is of course expected, given our interpretation of the slope for $X_2$ holding $X_1$ constant. Similarly, if we look down the columns, we see increments of 2 in $Y$ corresponding to each increase of 1 in $X_1$. Again, this is expected given our interpretation of the slope for $X_1$ holding $X_2$ constant.

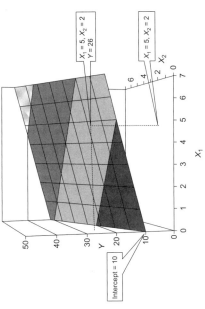

| $X_1$ | $X_2$ | | | | | | | |
|---|---|---|---|---|---|---|---|---|
| | 0 | 1 | 2 | 3 | 4 | 5 | 6 | 7 |
| 0 | 10 | 13 | 16 | 19 | 22 | 25 | 28 | 31 |
| 1 | 12 | 15 | 18 | 21 | 24 | 27 | 30 | 33 |
| 2 | 14 | 17 | 20 | 23 | 26 | 29 | 32 | 35 |
| 3 | 16 | 19 | 22 | 25 | 28 | 31 | 34 | 37 |
| 4 | 18 | 21 | 24 | 27 | 30 | 33 | 36 | 39 |
| 5 | 20 | 23 | 26 | 29 | 32 | 35 | 38 | 41 |
| 6 | 22 | 25 | 28 | 31 | 34 | 37 | 40 | 43 |
| 7 | 24 | 27 | 30 | 33 | 36 | 39 | 42 | 45 |

**(a) Conditional Means of $Y$ Given $X_1$ and $X_2$ for $E(Y|X_1,X_2) = 10 + 2X_1 + 3X_2$**

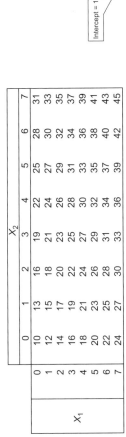

**(b) Three-Dimensional Plot of Regression Plane for $E(Y|X_1,X_2) = 10 + 2X_1 + 3X_2$**

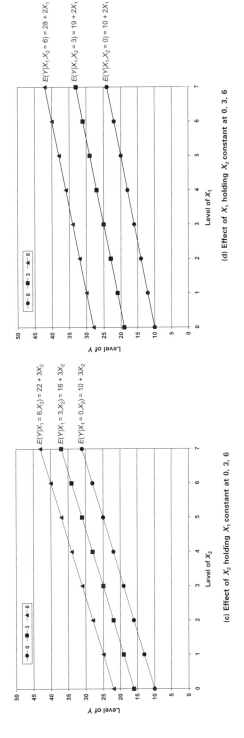

**(c) Effect of $X_2$ holding $X_1$ constant at 0, 3, 6**

**(d) Effect of $X_1$ holding $X_2$ constant at 0, 3, 6**

**Figure 6.1:** Illustration of Conditional Means and Conditional Regression Equations.

Conditional regression equations are illustrated in Figure 6.1(c) for the effect of $X_2$ when $X_1$ is held constant at zero, 3, and 6. The graph makes explicit that the rate of change of $Y$ for a given change in $X_2$ is the same at all levels of $X_1$ (equal to $\beta_2$) but these parallel lines are shifted upward (downward) by the amount of $\beta_1$ for each unit of $X_1$. In Figure 6.1(c), since we hold $X_1$ constant at values that differ by three points each between adjacent lines (i.e., $X_1 = 0, 3, 6$), the distance between the lines is $\beta_1 * 3 = 2 * 3 = 6$.

Figure 6.1(d) is analogous, giving the effect of $X_1$ holding $X_2$ constant. The lines in Figure 6.1(d) are spaced farther apart than the lines in Figure 6.1(c) owing to the higher slope for $X_2$ than $X_1$. (The distance between lines is $\beta_2 * 3 = 3 * 3 = 9$ in Figure 6.1(d).) The difference in slopes for $X_1$ and $X_2$ is also clear from the steeper lines in Figure 6.1(c) than Figure 6.1(d).

Figure 6.1(b) provides a three-dimensional plot depicting how $Y$ changes when we change both $X_1$ and $X_2$. For any $\{X_1, X_2\}$ point located at the base of the graph, the conditional mean of $Y$ is plotted in the response plane. This plot makes clear that the regression plane crosses the $Y$ axis at the value of $Y = 10$. And, the figure shows the point $\{5,2\}$ corresponding to a conditional mean of:

$$E(Y|X_1 = 5, X_2 = 2) = 10 + 2 * 5 + 3 * 2 = 26$$

The conditional regression equations we just examined make up this plane. Each line running from left to right in the plane represents the conditional effect of $X_1$, whose slope we know is 2. Each line running from front to back in the plane represents the conditional effect of $X_2$, whose slope we know is 3. At an angle (from bottom left to top right) we see the effect of simultaneously increasing $X_1$ and $X_2$. Looking back to the values in the diagonals of Figure 6.1(a) makes it clear that, as expected, $Y$ changes by 5 when both $X_1$ and $X_2$ increase by 1. $X_1$ and $X_2$ are said to have *additive* effects on $Y$ in this model (we add their two separate slopes when both change).

### 6.2: OLS ESTIMATION OF THE MULTIPLE REGRESSION MODEL

The sample **multiple regression model** with three variables is:

$$Y_i = \hat{\beta}_0 + \hat{\beta}_1 X_{1i} + \hat{\beta}_2 X_{2i} + \hat{\varepsilon}_i$$

The systematic portion of this model determines the fitted (predicted) values:

$$\hat{Y}_i = \hat{\beta}_0 + \hat{\beta}_1 X_{1i} + \hat{\beta}_2 X_{2i}$$

As in simple regression, calculus can be used to derive formulae for calculating the slopes and intercepts in multiple regression based on minimizing the sum of squared errors (Fox 2008; Kutner, Nachtsheim, and Neter 2004; Wooldridge 2009). In the model with two predictors we are now minimizing:

$$\sum \hat{\varepsilon}_i^2 = \sum (Y_i - \hat{Y})^2 = \sum (Y_i - (\hat{\beta}_0 + \hat{\beta}_1 X_{1i} + \hat{\beta}_2 X_{2i}))^2 \tag{6.1}$$

### 6.2.1: Point Estimates and Standard Errors of the Intercept and Slope

Taking the partial derivative of Equation 6.1 with respect to each unknown parameter, setting each result to zero, and solving for the unknown allows statisticians to write formulae for the intercept and slope. Writing the formulae requires matrix algebra for models with several predictors, which is not essential for our purposes (for matrix presentations, see Fox 2008; Kutner, Nachtsheim, and Neter 2004; Wooldridge 2003). But, it is informative to write the formulae for the standard errors of the slopes in the model with two predictors, and relate these back to the formulae for the standard error of the slope that we examined in Equation 5.7.

$$\sigma_{\hat{\beta}_1} = \sqrt{\frac{\sigma^2}{\sum (X_{1i} - \bar{X}_1)^2 (1 - r_{12}^2)}}$$

$$\sigma_{\hat{\beta}_2} = \sqrt{\frac{\sigma^2}{\sum (X_{2i} - \bar{X}_2)^2 (1 - r_{12}^2)}} \tag{6.2}$$

As we did in bivariate regression, we estimate $\sigma_{\hat{\beta}_1}$ and $\sigma_{\hat{\beta}_2}$ by substituting $\hat{\sigma}^2$ for $\sigma^2$, since $\sigma^2$ is unknown. In general, in multiple regression:

$$\hat{\sigma}^2 = \frac{\sum \hat{\varepsilon}_i^2}{n - k} = \frac{\sum (Y_i - \hat{Y}_i)^2}{n - k}$$

and $k$ = number of predictors in the model plus 1 for the intercept. In a model with two predictors, we have:

$$\hat{\sigma}^2 = \frac{\sum (Y_i - (\hat{\beta}_0 + \hat{\beta}_1 X_{1i} + \hat{\beta}_2 X_{2i}))^2}{n - 3}$$

As in the bivariate case, $\sigma^2$ is assumed to be homoskedastic and measures the conditional variance of the errors (and thus the conditional variance of the $Y$s).

Comparing Equation 6.2 to Equation 5.7 shows that, as in bivariate regression, the standard error of the slope in multiple regression depends directly on the conditional standard deviation of the $Y$s and depends inversely on the variation of the predictor variable and the sample size.

The new term in the denominator $(1 - r_{12}^2)$ shows that, in multiple regression, the standard error of the slope for $X_1$ also depends on the Pearson correlation between $X_1$ and $X_2$. The larger the correlation between the two predictor variables, the larger the standard error. At one extreme, if the two predictor variables are uncorrelated $(r_{12} = 0)$, then the new term in the denominator is $(1 - r_{12}^2) = 1 - 0 = 1$ and the formula reduces to the same equation as the bivariate case. At the other extreme, if the two predictor variables are perfectly correlated $(r_{12} = 1)$, then the new term in the denomination becomes $(1 - r_{12}^2) = 1 - 1 = 0$. Dividing by zero is indeterminate (the standard error approaches infinity as the correlation approaches 1). We will examine the implications of such high correlation between predictors in Chapter 11, when we consider the topic of multicollinearity.

### 6.2.2: Interpreting and Rescaling the Variables

Our interpretation for each of the slope parameters in regression models with two predictor variables is the same as in the case of bivariate regression, except for the clause "holding the other variable(s) constant."

Additional results that we used to interpret the estimates in simple regression hold as well:

■ For linear models, the change in $Y$ for a given increase in $X$ does not depend on the starting value of $X$. (We will consider nonlinear models in Chapter 9.)

■ If we want to interpret a slope parameter for an increase of $i$ units rather than an increase of 1 unit after estimation, then we multiply the relevant slope estimate by $i$.

■ We can also rescale a predictor variable prior to determining its slope so that we can read the new slope, based on the rescaled variable, directly from our output.

■ And, we can use the formulae discussed in Section 5.4.9, including the formulae for the standardized and semi-standardized coefficient estimates.

### 6.2.3: Example 6.2

Display B.6.1 provides an example of a multiple regression in which we added the variable *g1yrschl* to the model predicting *g1miles* based on *g2earn*. We will consider additional values from the output later in the chapter, but for now focus on the parameter estimates and standard errors.

### Statistical Significance of Point Estimates

First, let's consider the point estimates and hypothesis tests for each of the three rows in turn.

■ The intercept is estimated to be 78.35. This can be interpreted as follows: "Adults with no earnings whose mothers completed no schooling live nearly 80 miles away from their mothers, on average." Recall from Chapter 4 that some mothers do have no schooling and some adults have no earnings in the NSFH. As always, the default output includes a test of $\beta_0 = 0$ versus the alternative that $\beta_0 \neq 0$. We can mechanically reject this null hypothesis. The $t$-value of 2.25 exceeds 1.96 and the two-sided $p$-value of 0.025 is less than 0.05. But, recall that we have omitted adults who coreside with their mothers, therefore this null hypothesis is not strictly meaningful (although some adults report living as little as 1 mile from their mothers). Thus, in this case, the intercept is meaningful in terms of zero being a possible value for both predictors, but testing the null hypothesis that the intercept is zero is not meaningful, because zero is not a possible value for the outcome in our sample.

■ The point estimate for *g2earn* is 0.00102. The $t$-value of 4.39 has a two-sided $p$-value that is less than 0.0001. Because the $t$-value is larger than 1.96 and the $p$-value is smaller than 0.05, we can reject the null hypothesis of no linear relationship at a 5 percent alpha level. This estimate is slightly smaller than the value of 0.00113 seen in the bivariate regression in Chapter 5. In Chapter 10, we will have more to say about interpreting the change in a coefficient estimate when other variables are added to the model. The coefficient estimate can be interpreted as: "For each additional dollar earned annually, the adult lives 0.00102 miles farther away from his or her mother, on average, holding the mother's educational level constant."

■ The point estimate for *g1yrschl* is 15.92. The $t$-value of 5.36 has a two-sided $p$-value less than 0.0001. Because the $t$-value is larger than 1.96 and the $p$-value is smaller than 0.05, we can reject the null hypothesis of no linear relationship at a 5 percent alpha level. The coefficient estimate can be interpreted as: "For each additional year of schooling the mother attained, she lives nearly 16 miles farther from her child, on average, controlling for the child's earnings."

If we had directional hypotheses, we would follow the same procedures as discussed in Section 5.4.7 to test them in multiple regression as in bivariate regression.

### Substantive Size of Point Estimates

At first blush, the association between the mother's years of schooling and distance looks much larger than the association between the adult's earnings and distance. However, it is important to consider the variables' units of measurement to interpret the substantive size of these associations. As we discussed in Chapter 5, a dollar more in annual earnings is quite a small change. In multiple regression we can still translate the association into a more meaningful amount of change by multiplying the coefficient estimate by a factor $i$. Using a change in annual earnings of $i = \$10,000$, we calculate $i\beta = 10,000*0.00102 = 10.20$. Thus, the mother is expected to live about 10 miles farther away, on average, with each additional $10,000 in the adult's earnings, controlling for the mother's years of schooling.

We can also rescale variables before estimating the regression model. Display B.6.2 shows the result of rescaling *g2earn* before running the regression, using the *g2earn10000* variable that we created in Chapter 5. Notice that rescaling earnings does not affect the intercept nor the coefficient estimate for years of schooling. Even though coefficient estimates can be reinterpreted for different units of change on the predictor after estimation (e.g., with the $i\beta$ formula), rescaling the variables to a meaningful amount of change will influence how "fast readers" (who sometimes turn straight to tables) interpret the findings. Quickly scanning the parameter estimates table in Display B.6.2 versus Display B.6.1, the rescaling makes the slope for the adult's earnings seem more on a par with the slope for the mother's schooling in Display B.6.2 than it did in Display B.6.1.

Although one year of schooling seems a more meaningful amount of change than $1 of annual earnings, we also might want to rescale the mother's educational attainment to an amount of change that is a more natural benchmark for readers. For example, four years of schooling might be a relevant amount of "real world" change, representing the amount of time typically taken to complete high school or college.[1] We can multiply the coefficient estimate for *g1yrschl* in Display B.6.1 by 4 to obtain $i\beta = 4*15.92 = 63.68$. We can also rescale the variable before estimation by creating a new variable *g1yrschl4 = g1yrschl/4*. The results, shown in Display B.6.3, match within rounding error our hand calculation. And, as expected, the intercept and the coefficient estimate for *g1earn* match the values shown in Display B.6.1. We could interpret the rescaled slope for education as: "With each additional four years of schooling attained by the mother, she lives over 60 miles farther from her child, on average, controlling for the child's earnings."

## Confidence Intervals

Confidence intervals for multiple regression, like bivariate regression, are shown in the default Stata output. In SAS, we can use the /clb option to obtain them, as we did in Chapter 5. The confidence intervals are calculated in the same way in multiple regression as in bivariate regression: we add and subtract 1.96 times the estimated standard error to the point estimate for a 95 percent confidence interval.

Looking for example at Display B.6.2, the following confidence intervals are displayed in the Stata output:

| | $\hat{\beta} \pm t_{\alpha/2}\ \hat{\sigma}_{\hat{\beta}}$ | Confidence Bounds |
|---|---|---|
| G2earn10000 | 10.22048 ± 1.96 * 2.329708 | (5.65,14.79) |
| G1yrschl | 15.92156 ± 1.96 * 2.970548 | (10.10,21.75) |
| _cons | 78.34818 ± 1.96 * 34.89219 | (9.95,146.75) |

For our two predictor variables, we could interpret these as "We are 95 percent confident that the coefficient for the adult respondent's earnings, in $10,000 units, falls between 5.65 and 14.79" and "We are 95 percent confident that the coefficient for the mother's years of schooling falls between 10.10 and 21.75." These confidence bounds are helpful because they make clear the level of uncertainty in our point estimate, stating our results as a range rather than a single value (e.g., "Each additional $10,000 in earnings is associated with the respondent living between 5 and 15 miles farther from the mother, holding mother's education constant" rather than "Each additional $10,000 in earnings is associated with the respondent living about 10 miles farther from the mother, holding mother's education constant.").

## Standardized Coefficients

We can also calculate semistandardized or completely standardized coefficient estimates in multiple regression, using the same rescaling formula discussed in section 5.4.9 (i.e., completely standardized coefficient estimates are calculated by multiplying by the standard deviation of the predictor and dividing by the standard deviation of the outcome; semistandardized coefficient estimates are calculated by dividing by the standard deviation of the outcome).

The results for Display B.6.1 are as follows:[2]

| | $\hat{\beta} * (\hat{\sigma}_X/\hat{\sigma}_Y)$ | $\hat{\beta}/\hat{\sigma}_Y$ |
|---|---|---|
| G2earn | 0.001022 * (37492.04/645.0074) =0.059 | 0.001022/645.0074 =0.000 |
| G1yrschl | 15.92156 * (2.940383/645.0074) =0.073 | 15.92156/645.0074 =0.025 |

We can interpret these as we did in Chapter 5. For the completely standardized coefficient estimates, a one-standard deviation increase in earnings is associated with the adult living 0.06 standard deviations farther from the mother, on average, holding mother's education constant. And, a one-standard deviation increase in the mother's schooling is associated with a 0.07 standard deviation increase in the miles lived from the mother, on average, holding the adult's earnings constant. For the semistandardized coefficient estimates, a $1 increase in earnings is associates with zero standard deviation change in distance lived from the mother, on average, holding the mother's education constant.[3] And, when the mother has one more year of schooling, the adult lives about 0.025 of a standard deviation further away, on average, holding earnings constant.

## 6.3: CONDUCTING MULTIPLE HYPOTHESIS TESTS

When we conduct more than one hypothesis test, we are increasingly vulnerable to making a Type I error—rejecting the null hypothesis when the null is true (Bland and Altman 1995; Shaffer 1995). This concern is especially important in the context of multiple regression, where we begin to consider how a number of different predictor variables are associated with the outcome variable and thus conduct multiple tests, for example testing whether the slopes for each of the predictors differs significantly from zero. For models with dozens of predictors, this amounts to dozens of individual tests. With today's computing power, it is easy to conduct many, many tests. In this section, we will provide an intuitive example of this issue and discuss one strategy for controlling the Type I error across such numerous individual hypothesis tests. In the following sections we will discuss alternative strategies for thinking about conducting multiple hypothesis tests and selecting from among alternative regression models.

The issue of controlling the Type I error when making multiple hypothesis tests should make sense intuitively. It is a natural extension of your understanding of how sampling error relates to the probability of rejecting the null when the null is true—a Type I error—in any individual test. We know that the level of alpha we use controls the Type I error rate. The $p$-value provides us with the probability that we would obtain a $t$-statistic more extreme than the one we obtained if the null were true, given the sampling distribution of our $t$-statistic. If we use an alpha level of .05, then we reject the null hypothesis if the $p$-value is less than .05, and say that we had a 5 percent chance of making a Type I error.

However, in multiple regression models, we begin to conduct tests for numerous slope coefficients. When we conduct many tests using the same sample, then it is likely that—by chance—across all the tests we will observe some of the extreme $t$-statistics that fall in the tails of the sampling distribution. Indeed, a few such extreme $t$-statistics are expected based on the standard deviation of the sampling distributions; the problem is that it may be tempting to interpret them as real rejections of the null hypothesis in applications (and with a single sample, we cannot distinguish Type I errors from valid rejections of the null hypothesis).

More concretely, imagine that we have a data set with five measured variables that we think associate with the outcome variable. Suppose that we conduct five individual hypotheses, all at the 0.05 alpha level. Across these multiple tests, the chance of making at least one Type I error is higher than 0.05. The exact probability of making at least one Type I error across the five tests depends on how the probabilities are related among tests. We can heuristically examine the probability of making at least one Type I error by assuming that the tests are independent (note that in the regression case this assumption won't hold but it makes it easy to understand the issue).

Recall from basic statistics that, if $A$ and $B$ are independent, then $P(A \cap B) = P(A)P(B)$. In words, the probability of both $A$ and $B$ happening is the probability of $A$ times the probability of $B$, given that $A$ and $B$ are independent. Now, at alpha = .05, the probability that a test is correct with respect to Type I error (i.e., that when the null is true we fail to reject it) is $1-.05 = .95$. Under independence, we can multiply the individual probabilities for each test to get the probability that all tests are correct. This would be $0.95 * 0.95 * 0.95 * 0.95 * 0.95 = 0.95^5$ for five tests. Generally, it is $0.95^m$ where $m$ is the number of tests. The probability of the complement of this outcome, that at least one of the tests has a Type I error, is $1-0.95^m$.

| Number of Tests | Probability of at least one Type I error assuming independence of tests and alpha = .05 for each test |
|---|---|
| 1 | $1-0.95^1 = 0.05$ |
| 2 | $1-0.95^2 = 0.10$ |
| 3 | $1-0.95^3 = 0.14$ |
| 5 | $1-0.95^5 = 0.23$ |
| 10 | $1-0.95^{10} = 0.40$ |
| 30 | $1-0.95^{30} = 0.79$ |
| 50 | $1-0.95^{50} = 0.92$ |
| 100 | $1-0.95^{100} = 0.99$ |

This makes clear that as the number of tests increases, the probability of making at least one Type I error can increase rapidly.

Bland and Altman (1995) provide a number of examples in which clinical trials were simulated by randomly assigning patient records into two groups that were not actually given different treatments, and seeing whether one group appeared to have superior outcomes in any hypothesis tests. Across numerous tests, some cases were identified in which the treatment group looked better than the control group on some outcomes (even though these groups in fact did not receive different treatments). For example, one case simulated a clinical trial for two alternative treatments of coronary artery disease. The patient records were randomly assigned to one of the two fictitious treatments. When patients were subdivided into those with more and less serious coronary artery disease, one of the treatment groups appeared superior, with longer survival times. As Bland and Altman note (1995, 310): "the finding would be easy to account for by saying that the superior 'treatment' had its greatest advantage for the most severely ill patients!" even though, in fact, the two groups had been randomly created and did not receive different treatments.

One way to account for this concern would be to adjust the alpha level of the test. If we decreased the alpha level for each individual test, we would make it harder to reject each individual null hypothesis, and reduce the chances of making at least one Type I error across the tests. For example, with $\alpha = 0.05$ for five independent tests $1-0.95^5 = 0.23$. But, with $\alpha = 0.01$ for five tests $1-0.99^5 = 0.05$. Such an adjustment was developed by Carlo Emilio Bonferroni (for more detailed discussion of this approach, and other adjustments, see Bland and Altman 1995; Schochet 2008; Shaffer 1995). Under the **Bonferroni adjustment**, an overall Type I error probability is decided upon, say .05, and then this rate is divided by the number of tests to determine the alpha level for the individual tests. For example, for two tests at an overall alpha level of .05, we would use an alpha of .05/2 = .025 for each individual test. Bonferroni adjustments obviously become quite stringent in applications with many tests (e.g., .05/50 = .001).

Often, a Bonferroni-type adjustment is considered particularly important in cases in which the researcher is conducting **exploratory** rather than **confirmatory** research. In an exploratory study, the researcher may not have strong theoretical reasons to expect particular relationships between predictor variables and the outcome variable, but rather, may have collected data with numerous possible predictor variables and then looked to see which seem relevant empirically. When strong theoretical expectations are specified in advance, then the researcher can reduce the chances of overemphasizing results that are due to sampling error.

An interchange in the sociological literature provides an example of such a distinction (Schaefer 1982; House, LaRocco and French 1982). In this interchange, the critiquer, Schaefer, raises the concern that the authors of the original paper, House, LaRocco and French, conducted 225 separate analyses. With so many tests conducted, she raises the concern that surely the probability of a Type I error on at least one of the tests is nearly 100 percent. The authors responded by noting that when they went back and applied more stringent alpha levels using the Bonferroni adjustment, they still found a number of significant effects. And, most importantly, the pattern of these significant effects was consistent with their hypotheses. If the significant effects reflected pure sampling error, then they would be expected to occur as often for results that differed from their substantive expectations as for results consistent with their substantive expectations. And, they noted that their results replicated a prior study. If the results were due to sampling errors alone, it is unlikely that they would be seen across two different samples. This debate shows that a Bonferroni-type adjustment (or simply using a more conservative alpha level) may be helpful when conducting a very large number of hypothesis tests in a single study and that it is beneficial to ground hypotheses in prior research and theory whenever possible.

## 6.4: GENERAL LINEAR *F*-TEST

Multiple regression also introduces the possibility of making joint tests about more than one slope at the same time. For example, is at least one of the slopes in the model significant? Is at least one of the slopes in a subset of the predictors significant (for instance, at least one of the adult's characteristics and at least one of the mother's characteristics in our distance example)? Do two slope coefficients differ from one another? These tests are all subsumed under what is often called the **general linear *F*-test**.

The general linear *F*-test encompasses many situations in which joint hypotheses will be tested in multiple regression. Although conceptually this may seem similar to the Bonferroni-type adjustment, such approaches are distinct. Bonferroni-type adjustments deal with making a number of individual tests and controlling the Type I error across these tests. The general linear *F*-test considers hypotheses about more than one parameter simultaneously.

### 6.4.1: Basics of Decomposing the Sums of Squares

We already have considered in detail the sum of squared errors, $\Sigma(Y_i - \hat{Y}_i)^2$, which we minimize to obtain the ordinary least squares estimates. We will now abbreviate this quantitiy as SSE (for **Sum of Squared Errors**).[4] We also learned in basic statistics that the numerator in the calculation of the variance of a variable was the sum of the squared deviation of each observed value from the overall (unconditional) mean $\Sigma(Y_i - \bar{Y})^2$. We will call this quantity the **Total Sum of Squares**, or TSS. In this section, we will also examine how these two quantities are related, and define a third sum of squares, $\Sigma(\hat{Y}_i - \bar{Y})^2$ which is referred to as the **Model Sum of Squares**, or MSS.

These quantities are related to one another. $Y_i - \bar{Y} = (Y_i - \hat{Y}_i) + (\hat{Y}_i - \bar{Y})$ and TSS = SSE + MSS. Figure 6.2 shows these relationship visually.

### 6.4.2: Overview of the General Linear *F*-Test

Let's use our distance example to consider what it might mean to conduct multiple tests simultaneously. Consider the following set of models:

Model 1   $\beta_0 + \varepsilon_i$

Model 2   $\beta_0 + \beta_1 g2earn_i + \varepsilon_i$

Model 3   $\beta_0 + \beta_1 g2earn_i + \beta_2 g1yrschl_i + \varepsilon_i$

Model 4   $\beta_0 + \beta_1 g2earn_i \qquad\qquad\qquad\qquad + \beta_4 g2age_i + \beta_5 g2numbro_i + \beta_6 g2numsis_i + \varepsilon_i$

Model 5   $\beta_0 + \beta_1 g2earn_i + \beta_2 g1yrschl_i + \beta_3 g1age_i + \beta_4 g2age_i + \beta_5 g2numbro_i + \beta_6 g2numsis_i + \varepsilon_i$

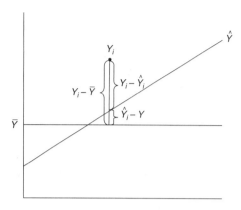

| | |
|---|---|
| Total Sum of Squares (TSS) | $\Sigma(Y_i - \bar{Y})^2$ |
| Sum of Squared Errors (SSE) | $\Sigma(Y_i - \hat{Y}_i)^2$ |
| Model Sum of Squares (MSS) | $\Sigma(\hat{Y}_i - \bar{Y})^2$ |
| | TSS = SSE + MSS |

▪ **Figure 6.2**

These models have the same outcome but differ in the set of predictor variables. We can use null hypotheses to produce one model from another, including complex hypotheses involving two or more slopes.

For example, comparing Model 3 with Model 2, we see that Model 3 contains two predictors, *g2earn* and *g1yrschl* and Model 2 contains just one predictor, *g2earn*. If we placed a restriction on Model 3 based on a null hypothesis that $\beta_2 = 0$ then Model 2 would result. Comparing Model 3 to Model 1, we see that Model 1 has no predictors. So, we would need a joint null hypothesis that $\beta_1 = 0$ and $\beta_2 = 0$ in order to produce Model 1 from Model 3.

The process of placing the constraints on Model 3 to produce Model 2 and Model 1 are shown below.

Original Model 3 $\qquad g1miles_i = \beta_0 + \beta_1 g2earn_i + \beta_2 g1yrschl_i + \varepsilon_i$

Place constraint $\beta_2 = 0$ on Model 3 $\qquad g1miles_i = \beta_0 + \beta_1 g2earn_i + 0 * g1yrshl_i + \varepsilon_i$
$\qquad\qquad\qquad\qquad\qquad\qquad\qquad\qquad \beta_0 + \beta_1 g2earn_i + \varepsilon_i$

Place constraints $\beta_1 = 0$ and $\beta_2 = 0$ on Model 3 $\qquad g1miles_i = \beta_0 + 0 * g2earn_i + 0 * g1yrschl_i + \varepsilon_i$
$\qquad\qquad\qquad\qquad\qquad\qquad\qquad\qquad \beta_0 + \varepsilon_i$

Notice that we explicitly substitute a zero for the coefficient listed in the constraint.

Model 1 is often referred to as the **intercept-only model**. Notice that it can be produced by any of the other models by constraining all of the slopes in each of the other models to zero. We will see below that the intercept-only model is used in a special case of the general linear *F*-test referred to as the **model** or **overall *F*-test**.

Whenever one model can be produced by placing constraints on the coefficients of another model, we refer to the models as **nested**. For example, we showed that Model 1 could be produced from Model 3 by constraining $\beta_1$ and $\beta_2$ to zero. And, Model 2 could be produced from Model 3 by constraining $\beta_2$ to zero. In this chapter, we will only consider constraints that a coefficient equals zero (in later chapters, we will consider other constraints, such as that two coefficients are equal). This means that, in this chapter, a nested model will contain a subset of the variables found in another model. As we just saw, Model 1 is nested in all of the other models. Model 2 is also nested in Models 3, 4, and 5. Model 3 and Model 4 are each nested in Model 5. But, Model 3 is not nested in Model 4. (Neither contains a subset of the variables in the other model. Model 3 contains one variable that is not in Model 4 and Model 4 contains three variables that are not in Model 3.)

In the terminology of the general linear $F$-test, the model with fewer variables is referred to as the **reduced (restricted** or **constrained) model** and the model with more variables is referred to as the **full (unrestricted** or **unconstrained) model**. For example, in comparing Model 3 to Model 2, Model 2 is the reduced model and Model 3 is the full model. Which is the reduced and which the full model is always relative to the particular test that you are conducting. For example, Model 3 could be a reduced model in comparison to Model 5 as a full model.

The general linear $F$-test involves four basic steps:

1. Fit the full model and obtain the error sum of squares $SSE(F)$
2. Fit the reduced model under the null hypothesis and obtain the error sum of squares $SSE(R)$
3. Calculate the $t$-statistic

$$F = \frac{SSE(R) - SSE(F)}{df_R - df_F} \div \frac{SSE(F)}{df_F} \tag{6.3}$$

   For the $F$ distribution, there are two degrees of freedom, one associated with the numerator and one associated with the denominator. Here, the **numerator degrees of freedom** is $df_R - df_F$ and the **denominator degrees of freedom** is $df_F$. The degrees of freedom for the reduced and full models ($df_R$, $df_F$) are calculated as $n-k$ where $k$ is the number of predictor variables in each model, plus one for the intercept.
4. Reject the null hypothesis if the $F$-value is greater than the critical $F$ at the given alpha level and numerator and denominator degrees of freedom; otherwise do not reject.

As noted, the full (unrestricted) model and the reduced (restricted) model differ based on the constraints (restrictions) stated in the null hypothesis. The numerator degrees of freedom (differences in degrees of freedom between the restricted and unrestricted models) will always equal the number of restrictions in the null hypothesis. For example, if $H_o$: $\beta_1 = 0$ and $\beta_2 = 0$, then the numerator degrees of freedom would be two. And, for $H_o$: $\beta_1 = 0$ the numerator degrees of freedom would be one.

Notice that the latter null hypothesis $H_o$: $\beta_1 = 0$ is the same null hypothesis that we learned to test in Chapter 5 with the $t$-value. In fact, whenever there is just one restriction in the numerator, we will see that the calculated $F$-value equals the square of a $t$-value for the same null hypothesis (and the $p$-value for the calculated $F$ statistic will equal the two-sided $p$-value for the $t$-statistic). But, the $F$-test is more general than the $t$-test because we can also use it to test more complex hypotheses about multiple parameters at once (such as $H_o$: $\beta_1 = 0$ and $\beta_2 = 0$).

It is important to emphasize that nested models evaluated by a general linear $F$-test must have the same outcome and must be estimated on the same sample size. If some predictors are missing more cases than other predictors, we need to be sure to drop cases missing values on any predictors before estimating any model to assure that all models are estimated on the same sample size. We show how to do this in Example 6.6 below.

### 6.4.3: Model *F*-test

The *model* or *overall* F test compares the estimated model to an intercept-only model, testing whether at least one of the slopes of the predictor variables in the full model is significant.[5] In this special case, we can rewrite the general linear $F$-test formula in terms of the total sum of squares and model sum of squares. We will consider the model $F$-test for the bivariate and multiple regression cases, because each provides some unique insights.

**Bivariate Regression**
We begin by examining the model $F$-test in bivariate regression.

*General Results*
The $F$-statistic for the model $F$-test in bivariate regression is identical to the square of the $t$-statistic for the slope coefficient. Both test whether the single slope coefficient is equal to zero. In the general linear $F$-test conceptualization, the null and alternative hypotheses result in the following full and reduced models for bivariate regression:

| | | |
|---|---|---|
| Reduced | $H_o$: $\beta_1 = 0$ | $Y_i = \beta_0 + \varepsilon_i$ |
| Full | $H_a$: $\beta_1 \neq 0$ | $Y_i = \beta_0 + \beta_1 X_{1i} + \varepsilon_i$ |

As discussed above, we obtain the reduced model by placing the restrictions of the null hypothesis on the full model. In this case, we take the full model $Y_i = \beta_0 + \beta_1 X_{1i} + \varepsilon_i$ and constrain the slope coefficient to zero ($\beta_1 = 0$ in the null hypothesis). Thus, the reduced model is $Y_i = \beta_0 + 0 * X_{1i} + \varepsilon_i = \beta_0 + \varepsilon_i$.

As noted above, the reduced model is often referred to as an "intercept-only model." The null hypothesis reflects a horizontal line—the conditional means are all the same, regardless of the level of the predictor variable. Thus, in the intercept-only model, the predicted value for each case is the sample mean, $\bar{Y}$. Because of this, the $SSE(R)$ is equal to the total sum of squares (TSS) for a model $F$-test when the reduced model is the intercept-only model.

Notationally, since $\hat{Y}_i = \bar{Y}$ in the reduced model, we can substitute $\bar{Y}$ for $\hat{Y}_i$ in the formula for $SSE(R)$: $\Sigma(Y_i - \hat{Y}_i)^2 = \Sigma(Y_i - \bar{Y})^2$. The degrees of freedom for each model (full and reduced) is $n\text{-}k$ where $k$ represents the number of predictors, plus one for the intercept. Since the reduced model includes only the intercept, $k$ is 1 and thus $df_R = n - 1$. Under the alternative hypothesis, the $SSE(F)$ is equal to the sum of squared errors (SSE) for the model $Y_i = \beta_0 + \beta_1 X_{1i} + \varepsilon_i$. In this case, $k = 2$ for the calculation of degrees of freedom, since the model contains one predictor variable plus the intercept, so $df_F = n - 2$.

Plugging these results into the formula for the general linear $F$-test gives a special case of the general linear $F$-test for the overall (model) $F$-test. For the bivariate case:

$$F = \frac{SSE(R) - SSE(F)}{df_R - df_F} \div \frac{SSE(F)}{df_F} = \frac{TSS - SSE}{(n-1) - (n-2)} \div \frac{SSE}{n-2} = \frac{MSS}{1} \div \frac{SSE}{n-2} \quad (6.4)$$

We will see a similar overall $F$-test for multiple regression below, with the intercept-only model as the reduced model.

Note that the numerator degrees of freedom is one in this example.[6] Thus, the $F$-statistic is equal to the square of the $t$-statistic associated with the same null hypothesis (e.g., in this case, the square of the $t$-statistic for the single slope coefficient, $\beta_1$).

*Example 6.3*
We will first apply the general linear $F$-test to our hours of chores example in the full NSFH sample, and then below we will use it with the distance example. For hours of chores predicted by number of children, the full and reduced models would be:

| | | |
|---|---|---|
| Reduced | $H_o: \beta_1 = 0$ | $hrchores_i = \beta_0 + \varepsilon_i$ |
| Full | $H_a: \beta_1 \neq 0$ | $hrchores_i = \beta_0 + \beta_1 numkid_{1i} + \varepsilon_i$ |

For instructional purposes, we will first use the general formula for the $F$-test, and explicitly estimate the reduced model and use Equation 6.3. Then, we will calculate the result with Equation 6.4. Let's start by estimating the full model, so we can see where to locate the total sum of squares (TSS), the SSE, and model sum of squares (MSS) in the ouput. Display B.6.4 shows the results. In Stata, the sums of squares are shown in the top left table. In SAS the sums of squares are in the table labeled "Analysis of Variance." The relevant column is labeled with the phrase "sums of squares" in SAS, and with the abbreviation SS in Stata. Both label the MSS row with the word Model. SAS uses Error and Stata uses Residual to label the row containing the SSE. And, Stata uses Total and SAS Corrected Total for the row containing TSS. With rounding, the values are MSS = 152,453, SSE = 2,206,415, and TSS = 2,358,868.

The intercept-only model can be run in SAS and Stata by omitting predictor variables from the regression command. In Stata `regress hrchores` and in SAS `proc reg; model hrchores=; run;`. The results are shown in Display B.6.5. Notice that, as expected, for this model the SSE equals the TSS, and as a result the MSS is zero.

The results from Display B.6.4 provide us with SSE($F$) = 2,206,415 and $df_F = n - k =$ 3,116 – 2 = 3,114. The results from Display B.6.5 provide us with SSE($R$) = 2,358,868 and $df_R$ = 3,116 – 1 = 3,115. Substituting into Equation 6.3 gives:

$$F = \frac{\text{SSE}(R) - \text{SSE}(F)}{df_R - df_F} \div \frac{\text{SSE}(F)}{df_F} = \frac{2,358,868 - 2,206,415}{(3,116-1)-(3,116-2)} \div \frac{2,206,415}{(3,116-2)} = 215.1629$$

with (3,116 – 1) – (3,116 – 2) = 1 numerator and (3,116 – 2) = 3,114 denominator degrees of freedom.

For this special case of the model or overall $F$-test, where the reduced model is an intercept-only model, we can also use Equation 6.4 to calculate the $F$-value directly from the TSS. This is convenient because we do not have to explicitly run the intercept-only model. All of the information we need is in sum of squares tables listed in Display B.6.4. Of course, this gives us the same result since TSS = SSE(R) for the intercept-only reduced model. Using just Display B.6.4 we can fill out Equation 6.4:

$$F = \frac{\text{TSS} - \text{SSE}}{(n-1)-(n-2)} \div \frac{\text{SSE}}{(n-2)} = \frac{2,358,868 - 2,206,415}{(3,116-1)-(3,116-2)} \div \frac{2,206,415}{(3,116-2)} = 215.1629$$

Because this overall or model $F$-test is commonly used, and can be calculated using the TSS for the reduced (intercept-only) model, it is always available in the SAS and Stata output. It appears in the right side of SAS's Analysis of Variance table, and is in the top right corner of the Stata output, just below the sample size (see black circles in Display B.6.4). SAS and Stata both show the associated $p$-value for this $F$-test, with one numerator and 3,114 denominator degrees of freedom (labeled `Prob > F` in Stata and `Pr > F` in SAS). In our case, the $p$-value is less than 0.0001, so we can reject the null hypothesis.

Recall that we mentioned that an $F$-test with one numerator degrees of freedom will be equivalent to a $t$-test for the same null hypothesis. In this case, our null is $H_o$: $\beta_1 = 0$, so the equivalent $t$-test should be the $t$-test for the *numkid* variable. We expected that the square of the $t$-value should equal the $F$-value. In this case, $t = 14.67$, so $t^2 = 14.67^2 = 14.67 * 14.67 = 215.21$. This matches the listed $F$-value, with rounding error.[7] Although the general linear $F$-test may seem redundant in this simple bivariate case (since we can use a $t$-statistic to test the same hypothesis), we will show many additional applications of it below and in subsequent chapters.

**Multiple Regression**
We now examine the model $F$-test for multiple regression.

*General Results*
The general linear $F$-test extends to the overall or model $F$-test for multiple regression in a straightforward fashion.

Reduced   $H_o$: $\beta_1 = \beta_2 = \ldots = \beta_p = 0$      $Y_i = \beta_0 + \varepsilon_i$

Full      $H_a$: at least one $\beta_p$ $(p = 1, \ldots, p)$   $Y_i = \beta_0 + \beta_1 X_{1i} + \beta_2 X_{2i} + \ldots + \beta_p X_{pi} + \varepsilon_i$
           is not equal to zero

Notice that the alternative is written as at least one of the slopes is not equal to zero. This is the complement to the null, which specifies that all of the slopes equal zero. If we reject the null, it could be that just one slope differs significantly from zero, that some differ from zero, or that all differ from zero. In multiple regression, if we reject the null hypothesis of the overall $F$-test, then we must inspect the individual $t$-tests for the individual slopes to see which differ significantly from zero.

In multiple regression, like bivariate regression, the reduced model of the model $F$-test is the intercept-only model. Thus, the formula for this application of the general linear $F$-test can be rewritten using the TSS.

$$F = \frac{SSE(R) - SSE(F)}{df_R - df_F} \div \frac{SSE(F)}{df_F} = \frac{TSS - SSE(X_1, \ldots, X_p)}{(n-1) - (n-k)} \div \frac{SSE(X_1, \ldots, X_p)}{(n-k)} \quad (6.5)$$

where $p$ is the number of predictors and $k$ is the number of predictors plus 1 (for the intercept). We use the notation $SSE(X_1, \ldots, p)$ for the full model because it can be helpful for clarity to list the variables in the model.

*Example 6.4*

Display B.6.6 shows the result of a multiple regression model predicting *hrchores* based on *numkid* and *hrwork*. The reduced model is the intercept-only model, shown in Display B.6.5, which is the same reduced model as we used above for the bivariate case. Substituting in the SSE($R$) for this reduced model, or equivalently the TSS from Display B.6.6, results in the following *F*-value.

$$F = \frac{\text{TSS} - \text{SSE}(numkid, hrwork)}{(n-1) - (n-3)} \div \frac{\text{SSE}(numkid, hrwork)}{(n-3)}$$

$$= \frac{2,358,868 - 2,178,881}{(3,116-1) - (3,116-3)} \div \frac{2,178,881}{3,116-3} = 128.57506$$

where k = 3 because our full model has two predictors plus the intercept. The numerator degrees of freedom works out to 2.[8] The denominator degrees of freedom is 3,116 − 3 = 3,113.

Notice that this overall *F*-test is again listed in SAS and Stata (circled in black in Display B.6.6). The *p*-value is listed as less than .0001, thus we can reject the null hypothesis that the slopes for numkid and hrwork are both zero and conclude that at least one differs from zero. Inspecting the *t*-values and two-sided *p*-values for these two variables in the SAS and Stata parameter estimates tables shows that we can reject the null hypothesis that the slope equals zero for each of them. In this case, because the numerator degrees of freedom of the *F*-test is larger than one, neither of these *t*-values squared equals the *F*-test.

### 6.4.4 Partial *F*-Test

We now examine how to use the general linear *F*-test to conduct a **partial *F*-test**.

*General Results*

The general linear *F*-test can also be used to test for the incremental or partial effect of a predictor variable, or set of predictor variables, given that one or more predictor variables are already included in the model. Logical sets of variables are common in applications, such as sets of variables measuring characteristics of the studied individual and his/her family, the sets of characteristics about each partner in a couple, or the set of characteristics about an individual's biological makeup, psychological makeup, and social network. In our distance example, we might consider the adult respondents' characteristics as one set of predictors and the mothers' characteristics as another set of predictors and specify the following reduced and full models:

Reduced  $H_0: \beta_3 = 0, \beta_4 = 0,$     $hrchores_i = \beta_0 + \beta_1 g1yrschl_i + \beta_2 g1age_i + \varepsilon_i$
$\beta_5 = 0, \beta_6 = 0$

Full     $H_a: \beta_3 \neq 0$ and/or $\beta_4 \neq 0$  $hrchores_i = \beta_0 + \beta_1 g1yrschl_i + \beta_2 g1age_i$
and/or $\beta_5 \neq 0$                       $+ \beta_3 g2earn_i + \beta_4 g2age_i$
and/or $\beta_6 \neq 0$                       $+ \beta_5 g2numbro_i + \beta_6 g2numsis_i + \varepsilon_i$

In these models, the full model includes characteristics of both the respondent and mother as predictors. The reduced model includes only the characteristics of the mother, testing the joint null hypothesis that the slopes for all of the respondents' characteristics equal zero against the alternative that the slope for at least one of the respondents' characteristics differs from zero.

## Example 6.5

We will start with a simple example of the partial $F$-test using our hours of chores example. The reduced and full models might be:

Reduced  $H_0: \beta_2 = 0$  $hrchores_i = \beta_0 + \beta_1 numkid_i + \varepsilon_i$
Full     $H_a: \beta_2 \neq 0$  $hrchores_i = \beta_0 + \beta_1 numkid_i +$
$\beta_2 hrwork_i + \varepsilon_i$

We already have the SAS and Stata results for these reduced and full models, in Display B.6.4 and Display B.6.6, respectively. Display B.6.4 shows that for this reduced model, SSE($R$) = 2,206,415. The reduced model contains one predictor in addition to the intercept, so $k = 2$ and $df_R = n - k = 3116 - 2 = 3114$. Display B.6.6 shows that for this full model, SSE($F$) = 2,178,881. The full model contains two predictors in addition to the intercept, so $k = 3$ and $df_F = n - k = 3,116 - 3 = 3,113$. Substituting into the formula for the general linear $F$-test, and using the notation to indicate which variables are in the model as shown in Equation 6.5, provides:

■ **Box 6.1**

It is sometimes confusing to students that the null hypothesis involves the coefficient for the second variable (*hrwork*) yet the reduced model contains the first variable (*numkid*). Remember that the reduced model excludes the variable whose coefficient is zero in the null hypothesis, because we write the reduced model by placing the restriction of the null hypothesis on the full model: $hrchores_i = \beta_0 + \beta_1 numkid_i + 0 * hrwork_i + \varepsilon_i = \beta_0 + \beta_1 numkid_i + \varepsilon_i$.

$$F = \frac{\text{SSE}(R) - \text{SSE}(F)}{df_R - df_F} \div \frac{\text{SSE}(F)}{df_F} = \frac{\text{SSE}(X_1) - \text{SSE}(X_1, X_2)}{(n-2) - (n-3)} \div \frac{\text{SSE}(X_1, X_2)}{(n-3)}$$

$$F = \frac{\text{SSE}(numkid) - \text{SSE}(numkid, hrwork)}{(n-2)-(n-3)} \div \frac{\text{SSE}(numkid, hrwork)}{(n-3)}$$

$$= \frac{(2,206,415 - 2,178,881)}{(3,116-2)-(3,116-3)} \div \frac{2,178,881}{(3,116-3)} = 39.34$$

The numerator degrees of freedom is 1 and the denominator degrees of freedom is $3,116 - 3 = 3,113$.[9]

Given that the reduced model is not an intercept-only model, we should not expect the calculated $F$-value to match the $F$-values listed in the default output in Display B.6.4 or Display B.6.6 (circled in black). But, given that the numerator degrees of freedom is 1, the $F$-value should match the square of a $t$-value for the null hypothesis $H_0: \beta_2 = 0$. In our example, the second variable is $hrwork$ and the null hypothesis $H_0: \beta_2 = 0$ is tested in the parameter estimates table of Display B.6.6. The square of the $t$-value listed for $hrwork$ is $t^2 = -6.27^2 = 39.31$. This matches our $F$-value with rounding error. Because the $p$-value is so small in our case ($<.0001$) we cannot tell, but the $p$-value for the $F$ test is exactly the same as the two-sided $p$-value for $hrwork$ in Display B.6.6. Since this $p$-value is less than 0.05, we conclude we can reject the null hypothesis.

### More on Decomposing the Model Sum of Squares
Before considering another application of the partial $F$-test, we will pause to visualize how the partial $F$-test is based on a decomposition of the sum of squares.

Figure 6.3 visualizes three models based on the $hrchores$ outcome. In the middle is the model with two predictors, $X_1 = numkid$ and $X_2 = hrwork$. On the left is the model with just one predictor, $X_1 = numkid$. On the right is the model with just one predictor, $X_2 = hrwork$. We have already examined the sums of squares for two of these models in Display B.6.4 (predictor: $numkid$) and Display B.6.6 (predictors: $numkid$ and $hrwork$). Display B.6.7 shows the results for a new regression with just $hrwork$ as a predictor. Note that across all of these models, the outcome is $hrchores$ and the sample size is 3,116, thus the results are appropriate for comparison with general linear $F$-tests.

At the bottom of Figure 6.3, below each of the model headings, the total sum of squares is listed. These values are taken from the last row of the sums of squares tables in the output of Display B.6.4, Display B.6.6, and Display B.6.7. In all cases, as expected, the TSS = 2,358,868. Because the total sum of squares captures variation around the unconditional mean, it is the same across all of the models (see Box 6.2).

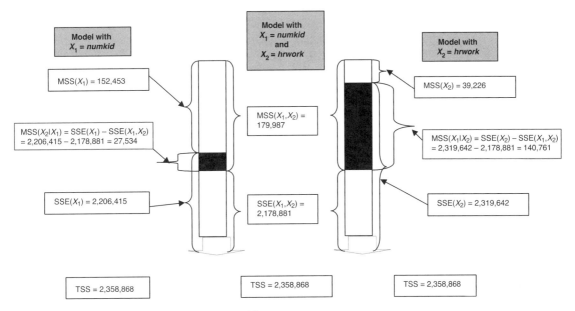

**Figure 6.3:** More on Decomposition of Sums of Squares

*Note:* The boxes are not drawn to scale.

At the top of Figure 6.3, under each model heading, is the MSS. These are taken from the top row of the sums of squares tables in the output.

From Display B.6.4,

$$\text{MSS}(X_1) = \text{MSS}(numkid) = 152{,}453.$$

---

**■ Box 6.2**

As noted above, we must be careful to be sure that the sample size remains the same across various models with different sets of predictors. Otherwise, the total sum of squares will change, because the unconditional mean will change at least slightly and we will have a different number of deviations from that mean to square and add to the sum. It is thus important to be careful to check that the sample size of your full and reduced models are the same whenever conducting general linear $F$-tests. As we remind you in the next example, this can easily be done with an *if* qualifier in SAS and Stata. Recall that SAS can also have several Model statements in a single PROC REG. Putting the Model statements together in the same PROC REG assures that the sample sizes will be the same, because SAS drops cases that are missing on any of the variables referred to in any models in the same PROC REG.

From Display B.6.6,

$$\text{MSS}(X_1, X_2) = \text{MSS}(numkid,\ hrwork) = 179{,}987.$$

From Display B.6.7,

$$\text{MSS}(X_2) = \text{MSS}(hrwork) = 39{,}226.$$

In the middle of Figure 6.3, under each model heading, is the SSE. These are taken from the middle row of the sums of squares tables in the output.

From Display B.6.4,

$$\text{SSE}(X_1) = \text{SSE}(numkid) = 2{,}206{,}415.$$

From Display B.6.6,

$$\text{SSE}(X_1, X_2) = \text{SSE}(numkid,\ hrwork) = 2{,}178{,}881.$$

From Display B.6.7,

$$\text{SSE}(X_2) = \text{SSE}(hrwork) = 2{,}319{,}642.$$

We can use algebra to relate these quantities to one another across models, drawing on the fact that TSS = SSE + MSS. One useful quantity is the additional sum of squares explained when we add one or more variables to a model. This quantity is sometimes referred to as the "extra sum of squares" (the extra amount explained by the model when the new variables are added). We denote this as $\text{MSS}(X_2|X_1)$ for a model in which $X_1$ was originally in the model and then we added $X_2$ (i.e., what amount of sum of squares are explained by $X_2$ given that $X_1$ is already in the model)? With variable names, in our example, we would write

$$\text{MSS}(X_2|X_1) = \text{MSS}(hrwork|numkid).$$

Because TSS = SSE + MSS, we can view this extra sums of squares equivalently as (a) the reduction in the sum of squares error, or (b) the increment to the model sum of squares, when we add the new variable or set of variables to the model. We will focus on the former, since this is the quantity that appears in the numerator of the general linear $F$-test: SSE($R$) – SSE($F$). For $\text{MSS}(X_2|X_1)$, the reduced model contains only $X_1$ and the full model contains both $X_1$ and $X_2$, so

$$\text{MSS}(X_2|X_1) = \text{SSE}(X_1) - \text{SSE}(X_1, X_2).$$

With variable names,

$$\text{MSS}(hrwork|numkid) = \text{SSE}(numkid) - \text{SSE}(numkid, hrwork)$$

We can use Figure 6.3 to locate the relevant values for this formula. The reduced model is on the left, and the full model is the one in the middle. Substituting the SSE from the respective models into $\text{MSS}(hrwork|numkid) = \text{SSE}(numkid) -$

SSE(*numkid,hrwork*) results in $2{,}206{,}415 - 2{,}178{,}881 = 27{,}534$. In Figure 6.3, this result is shown in the red box on the left. The solid rectangle visualizes how this value (27,534) captures the lower SSE in the full versus the reduced models (or equivalently the additional MSS in the full versus the reduced models). This is the amount of extra sums of squares explained by $X_2$ (*hrwork*) given $X_1$ (*numkid*) is already in the model.

Similarly, on the right, our reduced model contains just *hrwork* and we can also compare this model to the full model with both *numkid* and *hrwork*. Now, the extra sum of squares would be: $\text{MSS}(X_1|X_2) = \text{SSE}(X_2) - \text{SSE}(X_1,X_2)$. Written with the variable names, this would be

$$\begin{aligned} \text{MSS}(\textit{numkid}|\textit{hrwork}) &= \text{SSE}(\textit{hrwork}) - \text{SSE}(\textit{numkid,hrwork}) \\ &= 2{,}319{,}642 - 2{,}178{,}881 = 140{,}761. \end{aligned}$$

In Figure 6.3, this result is shown in the red box on the right. The solid black rectangle on the right visualizes how this value (140,761) captures the lower SSE in the full versus reduced models (with just *hrwork*) or equivalently the additional MSS in the full versus the reduced models. This is the amount of extra sums of squares explained by $X_1$ (*numkid*) given $X_2$ (*hrwork*) is already in the model.

Comparing the values of the extra sums of squares, and the size of the black boxes, clearly $X_1$ (*numkid*) explains more additional variation in hrchores than does $X_2$ (*hrwork*).

## Example 6.6

Let's now look at an example of the partial $F$-test with a larger set of predictors using our distance example. We will use as the reduced model the two-predictor model we estimated in Display B.6.1, which had *g2earn* and *g1yrschl* as predictors. We will add our other predictors to a full model (*g1age, g2age, g2numbro, g2numsis*).

| | | |
|---|---|---|
| Reduced | $H_0$: $\beta_3 = 0$, $\beta_4 = 0$,<br>$\beta_5 = 0$, $\beta_6 = 0$ | $hrchores_i = \beta_0 + \beta_1 g2earn_i + \beta_2 g1yrschl_i + \varepsilon_i$ |
| Full | $H_a$: $\beta_3 \neq 0$<br>and/or $\beta_4 \neq 0$,<br>and/or $\beta_5 \neq 0$<br>and/or $\beta_6 \neq 0$ | $hrchores_i = \beta_0 + \beta_1 g2earn_i + \beta_2 g1yrschl_i$<br>$\quad + \beta_3 g1age_i + \beta_4 g2age_i$<br>$\quad + \beta_5 g2numbro_i + \beta_6 g2numsis_i + \varepsilon_i$ |

Display B.6.8 shows the results of estimating the full model in SAS and Stata. Notice that we show explicitly how we used an *if* qualifier in SAS's DATA step and on Stata's use command to restrict the data set to cases not missing any values across the outcome and six predictor variables. We did this before running the restricted model in Display B.6.1 as well. Thus, both models are estimated on the same sample size of 5,475. Using the *if* qualifier in this case is important because there are cases that have missing values on one or more of the four new variables (*g1age, g2age, g2numbro, g2numsis*) but do not have missing values on the two original predictors (*g2earn, g1yrschl*). Thus, without this *if* qualifier, the reduced model shown in Display B.6.1 would have been estimated on a larger sample size and as a consequence its TSS would have differed from that of the full model shown in Display B.6.8.

We use the general linear *F*-test formula to conduct our hypothesis test. From the SAS output,[10] Display B.6.1 provides the $SSE(R) = 2,255,081,902$ and the degrees of freedom for the reduced model is $n - k = n - 3 = 5,475 - 3 = 5,472$ since *k* is 3 for the reduced model (two predictors plus the intercept). In Display B.6.8, the $SSE(F) = 2,238,453,227$ with degrees of freedom $n - k = n - 7 = 5,475 - 7 = 5,468$ since *k* is 7 for the full model (six predictors plus the intercept). Substituting into Equation 6.3 we get:

$$F = \frac{SSE(R) - SSE(F)}{df_R - df_F} \div \frac{SSE(F)}{df_F} = \frac{2,255,081,902 - 2,238,453,227}{(5,475 - 3) - (5,475 - 7)}$$
$$\div \frac{2,238,453,227}{(5,475 - 7)} = 10.15$$

The numerator degrees of freedom works out to 4.[11] As expected, this is the number of restrictions placed on the full model by the null hypothesis (the number of equals signs in the null hypothesis). The denominator degrees of freedom is $5,475 - 7 = 5,468$.

The reduced model in this example is not the intercept-only model, so the *F*-value we calculated does not match the *F*-value for the overall model *F*-test listed either in Display B.6.1 or Display B.6.8. The numerator degrees of freedom is also larger than 1, so the *F*-value does not match any *t*-value squared. We can use SAS and Stata's `test` commands to check our calculation and to obtain the *p*-value for the calculated *F*-value, with 4 numerator and 5,459 denominator degrees of freedom. The basic syntax is `test <varlist>` where in SAS the variables in the varlist are separated by commas. By default, when we include a variable in the variable list of the `test` command, SAS and Stata will include a constraint that its coefficient is zero in the null hypothesis.[12]

Display B.6.8 shows the commands and results of these tests. In both cases, the *F*-value matches our hand calculation of 10.15. Each lists four numerator and 5,468 denominator degrees of freedom for the test. And, both indicate that the associated *p*-value is less than .0001. Thus, at a 5 percent alpha level, we can reject the null hypothesis. We have evidence that at least one of the four new variables—*g1age, g2age, g2numbro, g2numsis*—has a coefficient that differs significantly from zero.

Examining the individual *t*-tests for these four variables in Display B.6.8 shows that we can reject the null hypothesis that the slope coefficient is zero for two variables, *g1age* and *g2numsis*. The null hypothesis for *g1age* could be written $H_0$: $\beta_3 = 0$ since it is the third variable in the model. The *t*-value for this variable is 3.17 with a two-sided *p*-value of 0.002. Since the *t*-value is larger than 1.96 and the *p*-value is smaller than 0.05, we can reject the null hypothesis at a 5 percent alpha level and conclude there is a linear relationship between *g1age* and *g1miles*. We could interpret this as: "When the mother is one year older, she lives over 4.5 miles farther from her child, on average, controlling for the adult's age, earnings, and number of brothers and sisters and for the mother's years of schooling."

The null hypothesis for *g2numsis* could be written $H_0$: $\beta_6 = 0$ since it is the sixth variable in the model. The *t*-value for this variable is 2.82 with a two-sided *p*-value of 0.005. Since the *t*-value is larger than 1.96 and the *p*-value is smaller than 0.05, we can reject the null hypothesis at a 5 percent alpha level and conclude there is a linear relationship between *g2numsis* and *g1miles*. We could interpret this as: "When adults have one more sister, they lives nearly 17 miles farther from their mothers, on average, controlling for their age, earnings, and number of brothers and for the mother's age and years of schooling."

As when we examined the associations of *g1miles* with *g2earn* and *g1yrschl* it is important to consider the substantive size of a one unit change in *g1age* and *g2numsis* when considering these statistically significant associations. At face value, the effect for *g2numsis* appears larger than the effect of *g1age* because its coefficient estimate is larger. But, we might prefer to rescale *g1age* since adding one more sister seems a greater substantive change than adding one year to the mother's age.

In fact, the standard deviation of *g1age* is 11.30 versus a standard deviation of 1.56 for number of sisters. Thus, a one unit change on *g2numsis* represents nearly two thirds of the variable's standard deviation. In contrast, a one-unit change on *g1age* represents less than a tenth of that variable's standard deviation. We could evaluate the effect of a standard deviation change on each variable by multiplying their respective coefficient estimates by their standard deviations using our formula *iβ*. This results in 11.30 * 4.56 = 51.53 for *g1age* and 1.56 * 16.94 = 26.43 for *g2numsis*.

Thus, when we put the variables on a more even footing by standardizing, the association between the mother's age and distance is larger in magnitude than the association between the number of sisters and distance. Especially if we had a conceptual rationale for another unit of change, we might also rescale *g1age* in its natural units and keep *g1miles* in its natural units (e.g., to report the coefficient estimate for a 10-year rather than one-year change on *g1age* we could create a variable *g1age10=g1age/10* before estimating the regression or multiply the coefficient estimate from Display B.6.8 by 10, 10 * 4.56 = 45.6).

### 6.4.5: Other Linear Constraints

In Chapters 7 and 8, we will consider the application of the general linear $F$-test to test other linear constraints, such as the equality of two coefficients and the sum of two coefficients.

## 6.5: $R$-SQUARED

We can also use the sums of squares to define **$R$-squared**, which is a commonly reported assessment of the goodness of fit of a regression model. For the ordinary least squares regression, there are various ways to write the $R$-squared formula (Long 1997, Chapter 4). We will use an equation that relates nicely to our discussion of the general linear $F$-test and decomposition of the sum of squares.

$$R^2 = \frac{\text{TSS} - \text{SSE}}{\text{TSS}} = 1 - \frac{\text{SSE}}{\text{TSS}}$$

Notice that the numerator of this $R$-squared formula is the numerator in the first term of the model $F$-test (Equation 6.4). Since TSS = MSS + SSE we can also write the formula equivalently as:

$$R^2 = \frac{\text{MSS}}{\text{TSS}}$$

Rewritten this way it is clear that $R$-squared measures the proportion of the total variation in $Y$ (around the unconditional mean) that is explained by the model (the variation around the conditional means). If there is a strong association between our predictor and outcome, then the variation around the conditional means should be considerable smaller than the variation around the unconditional means (see Section 5.4.3). We thus interpret $R$-squared as "The predictor variable(s) explain $R$-squared $*100$ percent of the total variation in the outcome variable." In the case of bivariate regression, $R$-squared is equal to the square of the Pearson correlation coefficient.[13]

### 6.5.1: Example 6.7

Let's first look at $R$-squared for our bivariate example predicting *g1miles* based on *g2earn*. In Display B.5.7 we saw that the correlation between these two variables was 0.0650. Thus, $r^2 = 0.0650^2 = .00423$. Looking back at Display B.5.1 we can read the MSS and TSS from the analysis of variance tables and calculated $R$-squared directly as:

$$R^2 = \frac{\text{MSS}}{\text{TSS}} = \frac{10,850,283}{2,571,327,990} = 0.00422$$

SAS and Stata also list the $R$-squared value in their default output (just to the right of the Root MSE in SAS and under the $p$-value for the model $F$-test in Stata). In Display B.5.1, both report an $R$-squared of 0.0042, again consistent with our hand calculations. We interpret the $R^2$ value as "The adult's earnings explain less than 1 percent ($0.0042 * 100 = 0.42$) of the variation in the distance his or her mother lives away."

Let's look as well at our stylized example of 50 cases, relating *hrchores* to *numkid* in Chapter 5 to see an example of a larger $R$-squared value. The Pearson correlation between these two variables in the stylized sample is 0.8364.[14] The square of this is $r^2 = 0.8364^2 = 0.8364 * 0.8364 = .6996$. This $R^2$ value would be interpreted as "The number of children in the household explains 70 percent of variation ($0.6996 * 100 = 69.96$) in the number of hours per week that employed women spend on chores." Consistent with our comparison of the size of the effects of *numkid* on *hrchores* and *g2earn* on *g1miles* in Chapter 5, comparing the $R$-squared values for these two models suggests that the association between number of children and hours of chores for employed women is much larger in our stylized example than the association between earnings and distance from mother for all adults.[15] ($R$-squared of .700 versus 0.004).

The ready interpretation of $R$-squared makes it widely used for assessing the fit of a regression model. However, $R$-squared is problematic in that it always increases if we add additional predictor variables to our model. Thus, if we chose models purely on the basis of $R$-squared, we would end up with more complicated models that have more predictor variables. In science, we usually prefer more parsimonious models (the model that explains the most variance with the fewest predictors). An **adjusted $R$-squared** is used to deal with this issue:

$$R_a^2 = 1 - \frac{\text{SSE}/(n-k)}{\text{TSS}/(n-1)}$$

This formula recognizes that each variable added to the model reduces the degrees of freedom by 1 (i.e., moving from a model with two predictors to three predictors, $k$ increases from 3 to 4). If a new variable added to a regression model does not help in explaining the variation in $Y$ (i.e., if the increase in the MSS when the variable is added to the model is small), then it may not be sufficient to overcome the loss in a degree of freedom by adding the variable to the model.

### 6.5.2: Example 6.8

We can use the results from Display B.5.1 to also calculate the adjusted $R$-squared.

$$R_a^2 = 1 - \frac{2,560,477,708 / (6350 - 2)}{2,571,327,990 / (6350 - 1)} = .00406$$

This adjusted $R$-squared value is slightly smaller than the unadjusted $R$-squared value of 0.00422. The adjusted $R$-squared is listed in the SAS and Stata output of Display B.5.1, just below $R$-squared.

### 6.5.3: Example 6.9

Finally, we will look at an example of calculating $R$-squared for our multiple regression of *g1miles* on six predictors found in Display B.6.8. Here, $R$-squared is:

$$R^2 = 1 - \frac{SSE}{TSS} = 1 - \frac{2,238,453,227}{2,277,372,983} = .01709$$

and

$$R_a^2 = 1 - \frac{SSE / (n - k)}{TSS / (n - 1)} = 1 - \frac{2,238,453,227 / (5475 - 7)}{2,277,372,983 / (5475 - 1)} = .01601$$

Again, the adjusted $R$-squared is moderately smaller than the unadjusted $R$-squared.

## 6.6: INFORMATION CRITERIA

Information criteria are increasingly used to select from alternative regression models, either in conjunction with the hypothesis testing approaches presented above or as part of an alternative approach, referred to as Bayesian (Weakliem 2004). Some scholars have critiqued the standard hypothesis testing approach, for example, because the choice of alpha level is arbitrary and because hypothesis tests are more likely to be rejected in larger sample sizes (thus leading to more complex models, with more predictors, especially in larger samples). Information criteria do not need an alpha level, are designed to penalize more complex models (with more predictors) over simpler models (with fewer predictors), and are less vulnerable to identifying more complex models in larger sample sizes.

Akaike's Information Criterion (AIC) and Bayesian Information Criterion (BIC) can be calculated in the ordinary least squares framework as follows (Fox 2008; Weakliem 2004):

$$AIC = n * \log\left(\frac{SSE}{n}\right) + 2k$$

$$BIC = n * \log\left(\frac{SSE}{n}\right) + k * log(n)$$

where $n$ is the sample size and $k$ is the number of predictors in the model plus one for the intercept (see Box 6.3).

For both AIC and BIC, models with smaller values are preferred. Guidelines for the strength of evidence favoring the model with the smaller value are also available based on the absolute difference in AICs or BICs between two models. Raftery (1995) recommended guidelines that refer to the strength of evidence in favor of the model with the smaller BIC. If the BIC difference is 0–2 then the evidence is weak, 2–6 is positive, 6–10 is strong, and greater than 10 is very strong evidence in favor of the model with the smaller BIC (Raftery 1995). Burnham and Anderson (2002, 70) recommend guidelines that refer to the strength of evidence for continuing to consider the model with the larger AIC. If the AIC difference is 0–2 then the evidence is "substantial," if it is 4–7 then the evidence is "considerably less," and if it is greater than 10 then it is "essentially none."

For the hours of chores examples, we can calculate the AIC and BIC based on the results in Displays B.6.4, B.6.5, B.6.6, and B.6.7.

> **■ Box 6.3**
>
> SAS and Stata both display the AIC and BIC based on a different formula, using what is called the log-likelihood. We do not cover models based on this approach in this book (see Fox 2008 and Long 1997). We present the formulae based on the SSE because they can be calculated based on the ordinary least squares regression results, although only one can be obtained with basic SAS and Stata commands (in SAS, by adding the option / `selection=cp aic` to the Model statement).

| Predictors | $k$ | SSE | $AIC = n * \log\left(\frac{SSE}{n}\right) + 2k$ | $BIC = n * \log\left(\frac{SSE}{n}\right) + k * \log(n)$ |
|---|---|---|---|---|
| numkid | 2 | 2,206,415 | $= 3{,}116 * \log\left(\frac{2{,}206{,}415}{3{,}116}\right) + 2 * 2$ <br> $= 20{,}452.981$ | $= 3116 * \log\left(\frac{2{,}206{,}415}{3{,}116}\right) + 2 * \log(3116)$ <br> $= 20{,}465.07$ |
| hrwork | 2 | 2,319,642 | $= 3{,}116 * \log\left(\frac{2{,}319{,}642}{3{,}116}\right) + 2 * 2$ <br> $= 20{,}608.918$ | $= 3116 * \log\left(\frac{2{,}319{,}642}{3{,}116}\right) + 2 * \log(3116)$ <br> $= 20{,}621.006$ |
| numkid, hrwork | 3 | 2,178,881 | $= 3{,}116 * \log\left(\frac{2{,}206{,}415}{3{,}116}\right) + 2 * 3$ <br> $= 20{,}415.852$ | $= 3116 * \log\left(\frac{2{,}178{,}881}{3{,}116}\right) + 3 * \log(3116)$ <br> $= 20{,}433.985$ |

Based on these results, the preferred model is the multiple regression, with both number of children and hours of work as predictors. This model has the lowest values on both AIC and BIC. The absolute value of difference in BICs between this model and the next best fitting model (the bivariate model with only number of children as a predictor) is $|20{,}433.985 - 20{,}465.07| = 31.085$, providing very strong evidence that the multiple regression is preferred. The absolute value of difference in AICs between these same models is $|20{,}415.852 - 20{,}452.981| = 37.129$, providing no evidence in favor of the bivariate model with number of children as a predictor.

## 6.7: LITERATURE EXCERPT 6.1

We present a literature excerpt here that offers an example of how to present results from multiple regression in a manuscript. We will present several additional examples in Chapter 7, and future chapters. Across chapters, we draw our examples from a range of different fields and journals to help you get used to the somewhat different styles used by each discipline and journal. When you start to format your own results for a manuscript, be sure to check the style guide in your field (e.g., American Psychological Association 2009; American Sociological Association 2007) as well as the manuscript submission guidelines and recent articles in relevant journals.

The excerpt we examine in this chapter is taken from Dooley and Prause (2005) who studied how adverse changes in maternal employment during pregnancy (such as moving from adequate employment to unemployment, having a poverty-level wage, working part-time involuntarily, or being out of the labor force) were associated with child birth weight. Literature Excerpt 6.1 shows Table 3 from their paper, with the results of four regression models that they estimated to examine these associations. Because this is the first table of results we are looking at in detail since learning some of the substance of regression analysis, we will first make some general comments about the presentation of regression results in tables.

Indeed, tables provide a succinct summary of regression results. You will become skilled at how to "read" these tables as we move forward in the book. In fact, as you become more experienced with regression analyses, you will probably find that you increasingly turn to the tables in a journal article, rather than relying on the authors' text explanations of the findings.

Although specific content and symbols will vary from subfield to subfield, most tables follow a fairly similar structure. A title summarizes the contents of the table, usually including the name of the dependent variable and key independent variables for tables of regression results. In Dooley and Prause's table, these are "Birth Weight" and

### ▨ Literature Excerpt 6.1

Table 3. Association of Change in Employment Status and Birth Weight (in grams): Ordinary Least Squares Regression (N = 1,165)

| Predictor Variable | Model 1 b | Model 1 Beta | Model 2 b | Model 2 Beta | Model 3 b | Model 3 Beta | Model 4 b | Model 4 Beta |
|---|---|---|---|---|---|---|---|---|
| Age (years) | −10.76 (−2.77)* | −.09 | −12.07 (−3.11)* | −.10 | −1.89 (−.57) | −.02 | −2.37 (−.54) | −.02 |
| Child sex (1 = male) | 92.74 (3.04)* | .09 | 93.00 (3.06)* | .09 | 90.10 (3.53)* | .08 | 85.72 (3.44)* | .08 |
| African American (1 = yes) | −158.59 (−3.76)* | −.11 | −158.50 (−3.77)* | −.11 | −141.23 (−4.00)* | −.10 | −140.24 (−4.07)* | −.10 |
| Weight prior to pregnancy (kg) | 6.46 (5.07)* | .15 | 6.40 (5.05)* | .15 | 4.66 (4.37)* | .12 | 5.23 (4.99)* | .12 |
| Alcohol use during pregnancy[a] | | | | | | | | |
|   Less than once/month | 75.86 (1.88)† | .06 | 72.07 (1.79)† | .05 | 35.12 (1.04) | .03 | 43.92 (1.33) | .03 |
|   More than once/month | 17.34 (.43) | .01 | 13.10 (.32) | .01 | 4.89 (.14) | .004 | 15.21 (.46) | .01 |
| Smoking during pregnancy (1 = yes) | −205.57 (−5.32)* | −.16 | −198.53 (−5.15)* | −.15 | −199.08 (−6.16)* | −.15 | −214.78 (−6.79)* | −.17 |
| Change from adequate employment to[b] | | | | | | | | |
|   Unemployment | −185.32 (−2.40)* | −.07 | −187.96 (−2.44)* | −.07 | −137.72 (−2.13)* | −.05 | −154.97 (−2.45)* | −.06 |
|   Poverty wage | −76.32 (−1.17) | −.03 | −75.30 (−1.16) | −.03 | −109.14 (−2.00)* | −.05 | −91.73 (−1.72)† | −.04 |
|   Involuntary part-time | −418.05 (−2.53)* | −.07 | −416.44 (−2.53)* | −.07 | −184.24 (−1.33) | −.03 | −193.36 (−1.43) | −.03 |
|   Out of the labor force | −46.95 (−1.08) | −.03 | −52.43 (−1.21) | −.04 | −85.20 (−2.34)* | −.06 | −98.05 (−2.75)* | −.07 |
| Trimester of first prenatal care visit[c] | | | | | | | | |
|   No prenatal care | | | −837.07 (−3.22)* | −.09 | −333.69 (−1.52) | −.04 | −386.33 (−1.80)† | −.04 |
|   Second or third visit | | | −50.92 (−1.10) | −.03 | −29.65 (−.76) | −.02 | −23.32 (−.61) | −.01 |
|   Missing | | | −227.32 (−1.94)* | −.06 | −200.95 (−2.04)* | −.05 | −178.93 (−1.86)† | −.04 |
| Weeks of gestation | | | | | 135.80 (22.01)* | .53 | 129.94 (21.38)* | .51 |
| Weight gain during pregnancy (kg) | | | | | | | 15.29 (7.48)* | .18 |
| Constant | 3,223.36 | | −3,273.53 | | −2,137.60 | | 1,929.75 | |
| F(df) | 8.49 (11,1153)* | | 7.97 (14,1150)* | | 42.63 (15,1149)* | | 45.37 (16,1148)* | |
| Adjusted $R^2$ | .066 | | .076 | | .349 | | .379 | |

† $p < .10$; * $p < .05$
*Note:* Numbers in parentheses are *t*-ratios.
[a] Relative to never used alcohol.
[b] Relative to remaining adequately employed.
[c] Relative to first prenatal care visit in the first trimester.

**Source:** Dooley, David and Joann Prause. 2005. "Birth Weight and Mothers' Adverse Employment Change." *Journal of Health and Social Behavior*, 46; 41–55.

"Change in Employment Status." Dooley and Prause's title also tells us the sample size ($N = 1,165$) and reminds us that the outcome is measured in grams.

The notes at the bottom of the table provide us with key details. Most often, a note tells us which symbols are used to denote which levels of $p$-values. In Dooley and Prause's table, the first note tells us that a single asterisk is used to denote a $p$-value that is less than 0.05. And, a † symbol is used to denote a $p$-value that is less than 0.10. Sometimes authors will explicitly indicate whether $p$-values are one-sided or two-sided. When unstated, we must assume that the reported $p$-values are the default two-sided values. It is important to check for such table notes about which symbols are used for which significance levels, since the symbols used vary from author to author and journal to journal (as we will see in other literature excerpts).

Notes often also provide information about which values are included in the table. The second note to Dooley and Prause's table tells us that the numbers in parentheses in the table are $t$-ratios. We should expect to see that these $t$-ratios are larger than $|1.96|$ whenever we see an asterisk (because we know that $t$-ratios of 1.96 are associated with two-sided $p$-values less than 0.05 in large samples).

The first column of a table typically lists short names of the predictor variables. Then, the following columns provide coefficient estimates (and sometimes standard errors and/or $t$-values) for these predictor variables.[16] In Dooley and Prause's table, there are four additional columns of results for four models (we will have more to say about how to compare coefficient estimates across models in Chapter 10). The labels at the top of each column sometimes denote the meaning of the tabled values (sometimes supplemented in notes, as in Dooley and Prause's note about the $t$-ratio). In Dooley and Prause's table, the lower case $b$ represents an unstandardized coefficient estimate and the word *Beta* represents a standardized coefficient estimate. Again, different professional styles and journals will have conventions for which symbols to use to represent which statistics. Tables also often present additional information from a multiple regression, such as the results of the overall $F$-test and $R$-squared. These results are provided in the final rows of Dooley and Prause's table.

Now that we understand the basic structure of the table, let's go back and look at a few specific results. First, we see that for Model 1, the overall $F$-test is significant (asterisk indicates a $p$-value less than 0.05). The numerator and denominator degrees of freedom are summarized in parentheses, as indicated in the row label F(df). For Model 1, there are 11 numerator and 1,153 denominator degrees of freedom. Counting the variables with coefficient estimates listed in Model 1, we see that there are 11 variables in the model (consistent with the 11 numerator degrees of freedom). In this model, $k = 11 + 1$, so the denominator degrees of freedom is $1,165 - 12 = 1,153$ (recall that the sample size of

1,165 is listed in the table title). We can surmise that this overall $F$-test has a null hypothesis that all 11 variables have coefficients equal to zero against an alternative hypothesis that at least one of the 11 variables has a coefficient that is different from zero. Because the calculated $F$-value is significant, we can reject the null hypothesis. We now want to examine the individual coefficient estimates to see which are significant.

We will focus for now on one of the significant predictor variables: weight prior to pregnancy (circled in red). The note next to the variable name reminds us that this variable is measured in kilograms (1 kg is about 2.2 lbs). The note in the title told us that the outcome variable is measured in grams. So, for every additional kilogram that the mother weighed prior to pregnancy, the newborn weighed about six and a half more grams, on average, controlling for the other variables in the model. Consistent with the asterisk indicating that the $p$-value is less than 0.05, the $t$-value in parentheses for this coefficient estimate is 5.07 (larger than 1.96).

Is this significant effect big or small? Given that the predictor and outcome have meaningful natural units, we can evaluate the size of the unstandardized coefficient estimate relative to real world benchmarks. In doing so, we may want to translate the coefficient estimate into a larger change in maternal prepregnancy weight. For example, multiplying the coefficient estimate by 10 reveals that an increase in maternal prepregnancy weight of 10 kilograms (or about 22 pounds) is associated with an increase in newborn birth weight of about 65 grams (which is one tenth of a pound). In the text, the authors note that babies are considered to be of low birth weight if they weigh less than 2,500 grams at birth, and the average birth weight in the sample is about 3,300 grams (pp. 144, 147). Thus, 65 grams represents less than 10 percent of the difference between being low birth weight and average weight in the sample.

The authors also provide the completely standardized coefficient estimates in their table, which allow us to interpret the results from an effect size or unitless perspective. In this case, the standardized coefficient estimate is 0.15, small in general for an effect size. Relative to the other standardized coefficient estimates in the table it appears larger (although we will have more to say about how to interpret the other coefficient estimates in the next chapter, and the completely standardized coefficient is not appropriate for many of them). One good benchmark for comparison may be weeks of gestation, a strong correlate of birth weight. The completely standardized coefficient estimate for this variable is 0.53 in the first model where it is included (Model 3, see black circle). So, the effect size of maternal prepregnancy weight is nearly 30 percent of the effect size for weeks of gestation. Thus, in its natural units and by effect size benchmarks, the effect of maternal prepregnancy weight seems modest, although it seems somewhat larger relative to the effects of other variables.

## 6.8: SUMMARY

In this chapter, we extended the concepts introduced in Chapter 5 for bivariate regression models to multiple regression models with two or more predictors. We discussed how tests of statistical significance and calculations of the substantive size of individual coefficients are quite similar to bivariate regression. We introduced the concept of conditional regression equations to aid in interpretation of multiple regression models. We also introduced several new ideas related to hypothesis testing, including being thoughtful about significance levels when conducting large numbers of individual hypothesis tests and using the general linear $F$-test to conduct hypotheses about several parameters simultaneously. We provided examples of how the general linear $F$-test is used to test whether there is evidence that any of the coefficients in the regression model differ from zero (overall $F$-test) and whether a subset of the coefficients in the regression model differ from zero (partial $F$-test). The general linear $F$-test can be used to compare nested models, in our case situations in which one model contains a subset of the variables in a larger model.

We ended by discussing measures that are often used to evaluate the performance of regression models: the $R$-squared and information criteria (AIC and BIC). $R$-squared summarizes the proportion of variation in the outcome variable explained by the set of variables in the model. Information criteria identify the regression model that best summarizes the data, allowing selection of the best model from many alternatives. Both $R$-squared and information criteria can be used to compare non-nested as well as nested models. The information criteria are also preferred by critics of classical hypothesis testing. Unlike classical hypothesis testing approaches, AIC and BIC do not rely on arbitrary significance levels (e.g., 5 percent Type I error rate in classical hypothesis testing). And, AIC and BIC are designed to identify simpler models (fewer variables in best fitting model) whereas classical hypothesis testing is more likely to identify more complex models (with more variables), especially when sample sizes are large.

## KEY TERMS

Adjusted $R$-Squared

Akaike's Information Criterion (AIC)

Bayesian Information Criterion (BIC)

Bonferroni Adjustment

Conditional Regression Equation

Confirmatory

Denominator Degrees of Freedom

Exploratory

Full Model
(also Unrestricted Model, Unconstrained Model)

General Linear $F$-Test

Intercept-Only Model

Model $F$-Test
(also Overall $F$-Test)

Model Sum of Squares (MSS)

Multiple Regression Model

Nested

Numerator Degrees Of Freedom

Partial $F$-Test

$R$-Squared

Reduced Model
(also Restricted Model, Constrained Model

Sum of Squared Errors (SSE)

Total Sum of Squares (TSS)

## REVIEW QUESTIONS

**6.1.** How do we interpret the intercept and the slope of each predictor variable in a multiple regression model with two predictors, $X_1$ and $X_2$, and the outcome of $Y$?

**6.2.** How does the formula for the standard error of the slope in multiple regression differ from the formula for the standard error of the slope in bivariate regression?

REVIEW
QUESTIONS
6

**6.3.** Write the four basic steps of the General Linear $F$-Test.

**6.4.** Write the null hypothesis for the General Linear $F$-Test that is found in standard SAS and Stata regression output. Write the alternative hypothesis and the full and reduced models associated with this test in the case of bivariate regression and multiple regression with two predictors (use $Y$ as the outcome and $X_1$ and $X_2$ as the predictors). Be sure you can show how the reduced model is obtained by placing the constraints of the null hypothesis on the full model and that you are able to read the SSE's and df's from SAS and Stata output in order to conduct the $F$-test by hand.

**6.5.** What relationship do you expect to see between the numerator degrees of freedom in the General Linear $F$-test and the null hypothesis associated with that test? Under what circumstances do you expect the $F$-test to be equivalent to a $t$-test (and the $t$-value to equal the square root of the $F$-value)?

**6.6.** How do we interpret $R$-squared in the case of simple and multiple regression? How does $R$-squared relate to the Pearson correlation in simple regression?

REVIEW
EXERCISES

6

## REVIEW EXERCISES

**6.1.** Verify the values in the conditional equations shown in Figure 6.1. What would be the expected change in $Y$ if we simultaneously increased $X_1$ by 2 and $X_2$ by 3?

**6.2.** In Example 6.2, what would the estimates of the intercept and slope be if we used the two rescaled variables, *g2earn*10000 and *g1yrschl*4 as the predictors?

**6.3.** Consider the following regression models, predicting national female labor force participation rates based on the proportion of women with a secondary education, the ratio of children to women, and the percentage of the population that is female.

$femlfp = \beta_0$
$femlfp = \beta_0 + \beta_1 femeduc$
$femlfp = \beta_0 + \beta_1 femeduc + \beta_2 kidratio$
$femlfp = \beta_0 + \beta_3 pctfem$
$femlfp = \beta_0 + \beta_1 femeduc + \beta_2 kidratio + \beta_3 pctfem$

(a) Which models are nested?

(b) For each pair of nested models, write the null hypothesis that would need to be imposed on the full model in order to produce the reduced model.

6.4. Suppose that the unadjusted $R$-squared value from a regression of wages (WAGE) on years of work experience (EXPER) is 0.25.

(a) Interpret this $R$-squared value.

(b) What is the Pearson correlation ($r$) between WAGE and EXPER?

6.5. Suppose that you know that the Pearson correlation between two variables is 0.70. What percentage of variation do these two variables share?

6.6. If you have bivariate regression results in which the $t$-value for the slope is 4, what would be the overall (model) $F$-statistic for this simple regression model?

6.7. Refer to Literature Excerpt 6.1.

(a) Write the null and alternative hypotheses for the $F$-test presented in Model 1. Make a conclusion based on the presented results.

(b) Write the null and alternative hypotheses for the $t$-ratio for Age. Make a conclusion based on the presented results.

(c) Interpret the unstandardized coefficient estimate for Age in Model 1.

(d) Interpret the standardized coefficient estimate for Age in Model 1.

6.8 Suppose that you estimate a multiple regression model using the following variables:

| | |
|---|---|
| DEPRESS | Depression scale (ranging from 0 to 105) |
| REDUC | Respondent's education level (ranging from 0 to 20) |
| REARNINC | Respondent's annual earned income (ranging from $1 to $800,000) |
| HRWORK | Respondent's hours spent at paid work/week (ranging from 1 to 95) |
| NUMKID | Number of persons <=18 in the household (ranging from 0 to 10) |

And obtain the following results:

| Source | SS | df | MS | | | |
|---|---|---|---|---|---|---|
| Model | 33076.9637 | 4 | 8269.24092 | | | |
| Residual | 1515927.29 | 5315 | 285.216799 | | | |
| Total | 1549004.25 | 5319 | 291.220954 | | | |

| | Number of obs | = 5320 |
|---|---|---|
| | F(4, 5315) | = 28.99 |
| | Prob > F | = 0.0000 |
| | R-squared | = 0.0214 |
| | Adj R-squared | = 0.0206 |
| | Root MSE | = 16.888 |

| depress | Coef. | Std. Err. | t | P>|t| | [95% Conf. Interval] | |
|---|---|---|---|---|---|---|
| reduc | -.4999201 | .0968659 | -5.16 | 0.000 | -.6898171 | -.3100231 |
| rearninc | -.0000506 | 8.40e-06 | -6.03 | 0.000 | -.0000671 | -.0000342 |
| hrwork | -.0078686 | .0180722 | -0.44 | 0.663 | -.0432975 | .0275603 |
| numkid | .6742838 | .1959539 | 3.44 | 0.001 | .2901337 | 1.058434 |
| _cons | 23.72232 | 1.517835 | 15.63 | 0.000 | 20.74674 | 26.6979 |

(a) Verify the value of $R$-squared and adjusted $R$-squared with a hand calculation.

(b) Interpret the $R$-squared value.

(c) Verify the calculation of the model $F$-value with a hand calculation. Be sure to write the null and alternative hypotheses and make a conclusion based on the displayed $p$-value.

(d) Interpret the intercept. As you do so, comment on whether the intercept is meaningful for this regression model.

(e) Interpret the point estimate for *reduc* using units of 4.

(f) Verify the confidence interval bounds for *numkid* with a hand calculation and write a one-sentence interpretation of the interval.

(g) Hand-calculate the AIC and BIC for this model.

(h) Suppose you calculated a nested alternative model with nine predictors and obtained an SSE of 1,484,564.94. Would this model be preferred over the above model based on the AIC, BIC $F$-test and adjusted $R$-squared?

**6.9.** Imagine that you conducted an exploratory regression analysis with 10 predictor variables and you wanted to control the Bonferroni alpha level across the tests of the 10 slopes at 0.05. What would be the alpha level you should use for each individual slope $t$-test?

**6.10.** How are TSS = total sum of squares, MSS = model sum of squares, and SSE = sum of squared errors mathematically related?

## CHAPTER EXERCISE

In this exercise, you will write a SAS and Stata batch program to estimate a multiple regression model, building on the batch program you wrote for Chapter 5.

Start with the batch program that you wrote for Chapter 5, using the NOS 1996 dataset with an *if expression* to only keep *the organizations that were founded after 1950* and that *do not have missing values on the mngpctfem, age, and ftsize variables.*

### 6.1  Interpretation and Hypothesis Testing in Multiple Regression

(a)  SAS/Stata tasks
   (i)  Regress *mngpctfem* on *age* and *ftsize*.

(b)  Write-up tasks
   (i)  In one sentence, interpret the intercept from the output. Is the intercept meaningful for this model and data?

   (ii)  In one sentence each, interpret the two slope estimates from the output.

   (iii)  Is each of the slope estimates statistically significant? Is each of the slope estimates substantively large in size?

   (iv)  Write the null and alternative hypotheses for the *F*-test listed in the output. Based on the output, is this test significant (use alpha = .05)?

   (v)  Interpret the *R*-squared value in one sentence.

   (vi)  Compare the conditional standard deviation of mngpctfem in this multiple regression to the simple regression and the unconditional standard deviation of mngpctfem estimated in the chapter exercise for Chapter 5. What does the comparison suggest about the fit of the multiple regression? Is this consistent with the *R*-squared value?

## COURSE EXERCISE

Re-estimate the model you developed in Chapter 5, adding one or more additional continuous predictors. Determine whether you can make directional hypotheses for any of these new predictors. Estimate the model and test your hypotheses about each slope, using two-sided or one-sided *p*-values as appropriate. In one sentence, interpret each point estimate. Request and hand calculate 95 percent confidence intervals, and interpret them, for each predictor.

Write a one-sentence interpretation of the intercept. With the new predictors added, is the intercept meaningful (i.e., is zero a valid value on all of the predictor variables?).

Write the null and alternative hypothesis for the model *F*-test listed in the output. Based on the output, is this test significant (use alpha = .05)? Hand-calculate this *F*-value.

Hand-calculate the *R*-squared value and verify your calculation using the output. Interpret the *R*-squared value in one sentence.

Determine a relevant reduced model to compare to your multiple regression model (the reduced model could be the bivariate model that you estimated in Chapter 5, or a model that includes a conceptually coherent subset of variables). Conduct the partial *F*-test for these full versus reduced models. Use the test command in SAS and Stata to confirm your hand calculation of the *F*-value and to obtain the *p*-value to make a decision based on the test. Be sure to explicitly write the null and alternative hypotheses, and the reduced and full models, for this test.

Hand-calculate AIC and BIC for the full and reduced models. Which model is preferred, based on the results?

Chapter 7

# DUMMY VARIABLES

## CHAPTER 7: USING DUMMY VARIABLES FOR NOMINAL AND ORDINAL PREDICTORS

Up until now, we have not examined the categorical predictors in our distance research questions, the respondent's gender and race-ethnicity. In this chapter, we will consider how to estimate and interpret models with such categorical predictors. We will begin by examining gender and race-ethnicity separately, and then estimate a full model with these categorical predictors and all of the continuous variables that we included in our final model in Chapter 6. Before detailing how to set up and interpret these models, however, we will take a look at a literature excerpt, to motivate the importance of models with categorical predictors, and how they relate to what we have learned already and what is still to come.

In a 1998 article in *Social Psychology Quarterly*, Jaya Sastry and Catherine Ross contemplate a possible western bias in prior studies of personal control and psychological well-being. Because Asian culture is more collectivist than western culture, they anticipate that Asians may report less personal control than nonAsians. Results of their analyses of a combination of US samples are reproduced in Literature Excerpt 7.1.

They interpret the results as follows:

> Asian Americans have significantly lower levels of personal control than do whites, the omitted group . . . Surprisingly, Asian ethnicity has a stronger relationship with the sense of control than does being black or being employed; these two variables have well-established relationships to perceived control (Sastry and Ross 1998, 108).

This excerpt illustrates the use of dummy variables to code categories of a predictor variable. Examining Sastry and Ross's table, a superscript *a* on the words *Asian* and *Black* denotes "Compared with white." Why do they interpret the results in this way? The table also shows that the unstandardized coefficient estimate (*b*) for Asian is –0.312 while the coefficient estimate for Black is –0.090. If both of these are compared with whites, as the footnote indicates, how can we tell if Asian Americans and Blacks differ from each other, on average, in personal control? In this chapter, we will answer these questions.

Note that Sastry and Ross's table includes a column labeled $\beta$ with the standardized coefficient estimates. They provide the standardized coefficient estimates for all variables. Some journals may encourage authors to do this. Or, authors may include them for all variables to be comprehensive, especially if it is easy to ask the software to calculate them. However, as we will make clear below, a one-unit change is always meaningful for dummy variables and standardized coefficient estimates do not make sense in this context. It is important, though, to evaluate the substantive size as well as the statistical significance of associations with dummy variables. We will illustrate how to do so below.

In general, we can use dummy variables in a regression model to test whether the mean of the outcome variable differs among categories of a predictor variable. Thus, we might examine research questions and hypotheses such as:

"Do men and women differ in their average incomes?"
"Among persons of Hispanic origin living in the USA, do those from Puerto Rico, Mexico, and Cuba differ in their mean length of residence?"
"When teenage girls agree versus disagree with a statement that mothers should stay home to raise children do they ultimately attain fewer years of schooling in adulthood, on average?"
"Do for-profit hospitals provide lower average quality patient care than nonprofit hospitals?"

When our original variable has only two categories (such as gender), then there is only one difference between the two groups' means. When we have a variable with more than two categories, we are sometimes primarily interested in the contrast of one category with each of the other categories. For example, we might compare whites with each other racial-ethnic group. Other times, our hypotheses might be more detailed and include contrasts among other groups (e.g., hypotheses about how African Americans may differ from Hispanics and Asians). As we will see, contrasts with one group will be easily read from our regression model. If we are interested in additional contrasts among the other groups, we will discuss three approaches for making them:

(a) re-estimating the regression model with a slightly different set of variables;
(b) using a partial $F$-test (building on what we learned in Chapter 6);
(c) testing the significance of a linear combination of coefficient estimates (something we will learn to do in this chapter).

---

■ **Literature Excerpt 7.1**

Table 2. Regression of Sense of Control on Ethnicity/Race, Socioeconomic Status, Household Status, and Sociodemographics: Domestic Survey

|  | $b$ (S.E.) | Beta |
|---|---|---|
| *Ethnicity/Race* |  |  |
| Asian[a] | −.312*** <br> (.044) | −.079 |
| Black[a] | −.090*** <br> (.020) | −.052 |
| *Socioeconomic Status* |  |  |
| Household income[b] | .001*** <br> (.000) | .083 |
| Education | .049*** <br> (.002) | .251 |
| Employed | .068*** <br> (.013) | .065 |
| *Household Status* |  |  |
| # of children | .001 <br> (.006) | .003 |
| Married | .039*** <br> (.012) | .038 |
| *Sociodemographics* |  |  |
| Age | −.004*** <br> (.000) | −.141 |
| Female | −.014 <br> (.012) | −.013 |
| Intercept | .131 |  |
| $R^2$ | .150 |  |

*Notes:*
$b$ = unstandardized coefficient; S.E. = standard error of $b$; beta = standardized coefficient
[a] Compared with white
[b] In thousands
+ $p < .10$, * $p < .05$, ** $p < .01$, *** $p < .001$ (two-tailed test)

**Source:** Sastry, Jaya and Catherine E. Ross. 1998. "Asian Ethnicity and Sense of Personal Control." *Social Psychology Quarterly*, 61(2), 110.

We will use one or more of these three approaches in the remaining chapters of the book as we interpret additional types of regression models. Although they produce the same results for a two-sided hypothesis test (thus in practice we will choose to use just one of them), looking at each in detail in this chapter offers insight into understanding dummy variable regression models. Knowing all three approaches also helps us to understand the various ways in which other researchers will present their results. And, understanding the three approaches in detail helps us to understand how to ask SAS and Stata to make the calculations for each approach.

In the remainder of this chapter, we first consider why we need a new strategy when one of the predictor variables is categorical. We will then look at the mechanics of defining dummy variables conceptually and in SAS and Stata. We will also look closely at regression equations with dummy variable predictors to help us to understand how to interpret their coefficients. We will start with a simple case with one dummy variable predictor. We will relate the results for this case to a *t*-test for the difference in means between two groups in order to solidify our understanding. We will next extend the model to cases in which we have: (a) multiple dummies representing one multicategory nominal or ordinal variable; and (b) multiple dummies representing two different nominal or ordinal variables. As we discuss the first case, we will examine in detail the three approaches to contrasting the means of all of the categories. As we discuss the second case, we will learn how to present the results in a table, including the sets of dummy variables as well as the continuous predictors we examined in Chapters 5 and 6. We end with an example from the literature to help us "put it all together."

## 7.1: WHY IS A DIFFERENT APPROACH NEEDED FOR NOMINAL AND ORDINAL PREDICTOR VARIABLES?

Thus far, we have been emphasizing our interpretation of the intercept and slope for models relating two interval-level variables to one another. But, when our predictor variable is nominal or ordinal, using a slope to relate this variable to the outcome is generally not appropriate.

Nominal variables, like a variable *race*, take on multiple categories, but the values lack both order and distance.

*race:*
1 = African American
2 = White
3 = Mexican American

4 = American Indian
5 = Other

The values could both be changed and reordered with no loss of information:

*race2:*
  4 = American Indian
 41 = Mexican American
 59 = African American
124 = White
500 = Other

Clearly, it would not make sense to enter either the *race* or *race2* variable as a single predictor in a regression model.

Similarly, education is often captured with ordinal categories that indicate progression toward degree attainment:

*hidegree:*
1 = Less than high school
2 = High school degree
3 = Some college, no degree
4 = Associate's degree
5 = Bachelor's degree
6 = Some postgraduate study
7 = Graduate degree

These values represent order, but not necessarily distance. We may be able to equivalently use other values to preserve order but imply other distances, such as:

*hidegree2:*
 9 = Less than high school
12 = High school degree
13 = Some college, no degree
14 = Associate's degree
16 = Bachelor's degree
18 = Some postgraduate study
21 = Graduate degree

Although we will focus on nominal variables in this chapter, in Chapter 9 we will consider using dummy variables to test for nonlinear effects of such ordinal predictor variables, building on the techniques learned in this chapter.

## 7.2: HOW DO WE DEFINE DUMMY VARIABLES?

To deal with these issues, nominal and ordinal variables can be analyzed in regression models using sets of **dummy variables**. Each dummy variable indicates one of the levels of the original variable (hence, these are sometimes also called **indicator variables**). Commonly, dummy variables take on two values—0 and 1—where the 1 represents the category on the original variable that the dummy variable indicates and 0 represents all other categories. Other codings are possible, although the 0/1 coding is common in the social sciences and makes it easy to interpret the results, as we will see below (see Box 7.1).

Dummy variable is the most widely used term for this approach. But, some find it confusing or misleading. One of the definitions of "dummy" in Webster's Dictionary is "An imitation or copy of something, to be used as a substitute," and the term "dummy variable" likely reflects the fact that the variable takes the place of the original category. One of the earliest article-length treatments of dummy variables notes "The dummy variable is a simple and useful method for introducing into a regression analysis information contained in variables that are not conventionally measured on a numerical scale" (Suits 1957, 548). Unfortunately, the word "dummy" also has derogatory meanings which can make it difficult for students to get used to the term. As Suits (1957, 551) noted "One occasionally encounters suspicion of dummy variables and a feeling that somehow something not quite respectable is involved in their use . . . Perhaps part of the trouble lies in the use of the term 'dummy' variable." But, the term is conventional and any and all of the categories of the original variable can be defined with "dummies" (i.e., one category is not inferior to the others, in the derogatory sense of the term dummy).

As we will see in more detail below, we represent a single categorical variable with a *set* of dummy variables. These dummy variables should be thought of together as a set of variables that capture the multiple categories of the original variable. Removing one of the dummies or recoding one of the dummies affects the interpreting of the full set of dummies. For this reason, some scientists refer to them as **dummy** or **indicator terms** rather than as dummy or indicator variables.

■ **Box 7.1**

One common alternative is effect coding. The effect variables are similar to dummies in that we create c-1 of them if a categorical variable has *c* categories and each effect variable is coded "1" to represent one of the original categories and "0" to represent all the other categories except the reference category. In the effect coding approach, the reference category is given the value "–1" on all of the effect variables. With this coding, the intercept represents the overall sample mean of the outcome (rather than the mean for the reference group). And, the coefficients of the effect variables capture the difference between the mean on the outcome for that category and the overall sample mean. In some models this coding can facilitate interpretation of the intercept, and effect coding is commonly used in Analysis of Variance models. But, dummy coding is more common for regression analysis in the social sciences. When correctly implemented, the two approaches provide equivalent results (the difference being in the interpretation).

The following table provides an example of four dummy variables used to indicate the categories of the *race* variable.

| Original Variable | Set of Dummy Variables | | | | |
|---|---|---|---|---|---|
| *race* | *aframer* | *white* | *mexamer* | *amind* | *other* |
| 1 = African American | 1 | 0 | 0 | 0 | 0 |
| 2 = White | 0 | 1 | 0 | 0 | 0 |
| 3 = Mexican American | 0 | 0 | 1 | 0 | 0 |
| 4 = American Indian | 0 | 0 | 0 | 1 | 0 |
| 5 = Other | 0 | 0 | 0 | 0 | 1 |

The table shows that a "1" on each dummy variable indicates one of the categories of the original variable. All other categories of the original variable have values of "0" on that dummy variable. For example, all cases coded "3" on *race* would be coded a "1" on the new variable named *mexamer*. And, all cases coded "1, 2, 4", and "5" on the original *race* variable would be coded "0" on the new variable named *mexamer*.

Because they are mutually exclusive (each represents one and only one of the original categories) and exhaustive (cover all of the original categories), any one of the new dummy variables is redundant with the remaining four dummy variables. For example, if we know that a case is coded a "0" on *white*, "0" on *mexamer*, "0" on *amind*, and "0" on *other*, then we can figure out that the case will be coded as "1" on *aframer*. Similarly, if we know that a case is coded a "0" on *white*, "1" on *mexamer*, "0" on *amind*, and "0" on *other*, then we can figure out that the case will be coded as "0" on *aframer*. Algebraically, any one of the dummy variables can be determined by the general equation:

$$X_1 = 1 - (X_2 + X_3 + X_4 + X_5)$$

where $X_1$ represents one of the variables and $X_2$ through $X_5$ represent the remaining four variables. In our example in this paragraph, *aframer* $= 1 - ($*white* $+$ *mexamer* $+$ *amind* $+$ *other*$)$ which works out to *aframer* $= 1 - (0 + 0 + 0 + 0) = 1 - 0 = 1$ in the first case and *aframer* $= 1 - (0 + 1 + 0 + 0) = 1 - 1 = 0$ in the second case described above.

As we will discuss further in Chapter 11, variables with this kind of redundancy cannot all be included together as predictors. Thus, we omit one of the dummy variables when we estimate a regression model. In general, to represent a variable with c categories requires c-1 dummy variables. In the *race* example, the original variable has c = 5 categories, thus we need $c - 1 = 5 - 1 = 4$ dummy variables. The group whose dummy variable is excluded is often referred to as the **reference category**. Other frequently used terms, such as **omitted category** or **excluded category**, are also strictly correct, as we have left one of the dummies for one of the categories out of the model. However, these terms can be misleading in that they imply that this *omitted* or *excluded* category is not part of the analysis, when in fact it is central to our interpretation. Which one of the five dummies we omit is arbitrary. The decision on the reference category does not affect the results of the regression analysis, although as we will see below, the coefficient estimates for the four included dummies directly show us only four contrasts (between the mean of the category indicated and the mean of the reference category). For example, if *mexamer* were excluded, then the coefficient estimate for *white* would capture the difference in means between whites and Mexican Americans. But, as we will see below, the contrasts among all the categories can be easily recovered, regardless of which one dummy from the set is excluded from the model.

Although the decision about which category to make the reference category does not affect the overall results, it does affect the ease of interpretation of results. As just noted, the difference in outcome means between that reference group and each of the other categories can be directly read from the output and directly presented in tables, thus the contrasts with the reference category will be particularly salient in the default results (see Box 7.2). In addition, as we will see below, the standard errors will be larger for contrasts involving categories that contain few cases and unless the differences in means are relatively large in size, *t*-values will be small for these contrasts. When the reference category has small cell size, all of the individual *t*-tests listed in the default output may be insignificant, even though contrasts between some of the included categories with larger sample sizes may be significant. Again, because this default output will likely stand out to you and those who read your work, and may be what you highlight in tables, it is preferable to choose a category with a relatively larger sample size as a reference.

■ **Box 7.2**

Although one reference category must be selected to estimate the model, additional contrasts can be presented by using subscripts or superscripts (see Literature Excerpt 7.2 below for an example) or by including separate rows for each contrast.

In general, it is best to avoid seemingly easy choices, such as omitting the first or last category (and to be especially careful in cases when the software package will create dummy indicators and exclude one of them for you). A better decision is to use as a reference a category that is of particular substantive interest. Prior empirical evidence and conceptual models can often be used to roughly predict the ordering of means on the outcome for some or many of the categories. Thinking through such prior evidence and rationales can help you to choose to exclude the category whose mean on the outcome you anticipate will differ from the mean on the outcome of the greatest number of other categories or with the largest magnitude of differences (for some categories, you may not be able to make predictions based on prior evidence and theory; you can include the dummy for such categories and omit as a reference a category for which you can make predictions of significant contrasts). If two or more are candidates for exclusion, then you might choose the one with the larger sample size.

### 7.2.1: How Do We Construct Dummy Variables in SAS and Stata?

Dummy variables can be easily constructed in SAS and Stata. We will begin with a straightforward approach that uses the syntax we already know for generating variables. We will then use a more compact syntax explicitly designed for dummy variables and available in both SAS and Stata.[1]

In Display B.3.1 we saw that the race variable was called M484 in the raw data file, and that the values 97 and 99 indicated missing data (refusals and no answers, respectively). Display C.7 shows the BADGIR codebook results for M484 (and for M2DP01 the variable capturing the respondent's gender, which we will use below). The codebook tells us which values correspond to which categories. Display B.7.1 shows similar results for our analytic sample. Recall that we excluded from our analytic sample respondents who were born outside the USA and who reported that their mother lived outside the USA or lived with them. These exclusion criteria may selectively affect the races. In fact, comparing Display B.7.1 to Display C.7, we see that our analytic sample has a higher percentage of white than the total sample, and most subgroups (other than American Indians) are a smaller percentage in the analytic than the total sample. Some category sizes become quite small (fewer than 30 cases) in our analytic sample, thus making it difficult to estimate conditional means in these categories with precision. We might collapse together some categories on conceptual grounds (perhaps using the empirical techniques discussed below as support in initial models). For pedagogical reasons, we separate out one of the categories with a small sample (American Indian) and group together the remaining categories with 40 or fewer cases. This allows us to demonstrate the implications of small cell size for the standard error and the conceptual challenge of interpreting a heterogeneous *other* category.

### Using the *If* Qualifier

One approach to creating the relevant dummy variables is to use the *if* qualifier. To do so, we need to have in mind the expressions that indicate either being in a category or not in a category. And, we want to pay careful attention to missing values. Table 7.1 shows how we would like each original category of the original M484 variable to be coded on the new set of dummy variables.

For nominal variables such as race, we recommend choosing names for the dummy variables that reflect which original categories are coded a "1" on the new variable.

▨ **Table 7.1: Example of Dummy Variable Coding**

| Original Variable | New Dummy Variables | | | | |
|---|---|---|---|---|---|
| M484 | aframer | white | mexamer | amind | other |
| 1 = African American | 1 | 0 | 0 | 0 | 0 |
| 2 = White | 0 | 1 | 0 | 0 | 0 |
| 3 = Mexican American | 0 | 0 | 1 | 0 | 0 |
| 4 = Puerto Rican | 0 | 0 | 0 | 0 | 1 |
| 5 = Cuban | 0 | 0 | 0 | 0 | 1 |
| 6 = Other Hispanic | 0 | 0 | 0 | 0 | 1 |
| 7 = American Indian | 0 | 0 | 0 | 1 | 0 |
| 8 = Asian | 0 | 0 | 0 | 0 | 1 |
| 9 = Other | 0 | 0 | 0 | 0 | 1 |
| 97 = Refusal | . | . | . | . | . |
| 99 = No Answer | . | . | . | . | . |

An alternative would be to use a name that contains the original variable name and category number (e.g., M484_1, M484_2, M484_3, M484_7, M484_45689). The latter can be useful in some cases (for example, in initial analyses of a variable with many categories, or ordinal variables where the numbers tie back to a meaningful rating scale). However, we find that using dummy variable names that reflect the category coded "1" usually make it easier to interpret output.

For the variable *aframer*, we can say, in words, that we would like to code the new variable a "1" if M484 is coded a "1". For all other values, except 97 and 99, we would like to code *aframer* a "0". We would like 97 and 99 to carry the value of '.' missing on *aframer*. One straightforward way to achieve this desired result in Stata is with the commands:

```
generate aframer=1 if M484==1
replace aframer=0 if M484~=1
replace aframer=. if M484>=97
```

In SAS, the comparable code would be:

```
if M484=1 then aframer=1;
if M484~=1 then aframer=0;
if M484>=97 then aframer=.;
```

We can verify that the results are what we desire by adding a `table` statement to the
`PROC FREQ` command in SAS and by using the `tabulate` command in Stata to cross-
tabulate the original variable with the new variable. Because the default results differ
between the two packages, we add some options to obtain just what we need. Stata
includes only the cell frequencies whereas SAS includes column, row, and overall
percentages by default. Because for checking purposes we only need the cell
frequencies, we can add an option in SAS to exclude the percentages. In SAS, we
would type: `PROC FREQ; tables M484*aframer /NOCOL NOROW NOPERCENT;`
`run;` We also add the option `, missing` in Stata, because by default Stata excludes
cases missing on either variable from the cross-tabulation, and we want to be able to
verify that the 97 and 99 values are coded '.' `missing` on the new dummy variable.
Thus, in Stata we would type `tabulate M484 aframer, missing`. Display B.7.2
shows the results of creating and checking this variable.

The results show that, as desired, cases coded a "1" on the original variable are a "1"
on the new *aframer* dummy. Cases that were a 99 on the original variable are missing
on the new dummy (we can see this explicitly in Stata; in SAS, the bottom of the table
notes that three cases are missing, and the 99s do not appear in the upper table). All
other categories are coded a "0". We cannot overemphasize the importance of checking
your code in this way. A little time spent verifying that the results are what you desire
will pay off immensely in the long run, when you avoid having to redo work after
fixing an early error (or worse yet, having to report that results presented or published
were in error).

The dummy variables for *white*, *mexamer*, and *amind* can be constructed similarly,
given that each indicates a single value on the original M484 variable. The *other*
dummy requires a lengthier expression to indicate its multiple categories. In words,
a "1" should be coded if M484 is between 4 and 6 or if M484 is an 8 or 9. A "0"
should be coded if M484 is between 1 and 3 or is a 7. A '.' `missing` should be
coded if M484 is a 97 or 99. The following translates these expressions from words
into Stata code:

```
generate other=1 if (M484>=4 & M484<=6) | M484==8 | M484==9
replace other=0 if (M484>=1 & M484<=3) | M484==7
replace other=. if M484>=97
```

In SAS, the comparable code would be:

```
if (M484>=4 & M484<=6) | M484=8 | M484=9 then other=1;
if (M484>=1 & M484<=3) | M484=7 then other=0;
if M484>=97 then other=.;
```

Checking that your code achieves the correct results is particularly important with such lengthier expressions. We can again use a PROC FREQ in SAS and tabulate in Stata to do so. Display B.7.3 shows the results.

### Using the True/False Evaluation

Both SAS and Stata also allow a simple syntax in which a new dummy variable is defined based on an expression that is either true or false. If the expression is true, then the new variable is coded a "1". If the expression is false, then the new variable is coded a "0". For example, for *aframer*, this expression would be stated in words as "Does M484 equal one?" For *other*, this expression would be stated in words as "Is M484 between 4 and 6 or an 8 or 9?". We can readily translate these statements into SAS or Stata syntax using the expressions we wrote above to code a "1" for each variable. In both SAS and Stata we write a statement with the new variable name on the left of the equals sign and the expression on the right. Missing data can be dealt with using an *if* qualifier.[2]

| Dummy indicator for: | SAS | Stata |
|---|---|---|
| African American | if M484<97 then aframer=(M484=1); | generate aframer=(M484==1) if M484<97 |
| White | if M484<97 then white= (M484=2); | generate white= (M484==2) if M484<97 |
| Mexican American | if M484<97 then mexamer=(M484=3); | generate mexamer=(M484==3) if M484<97 |
| American Indian | if M484<97 then amind= (M484=7); | generate amind= (M484==7) if M484<97 |
| Other | if M484<97 then other= ((M484>=4 & M484<=6) \| M484=8 \| M484=9); | generate other= ((M484>=4 & M484<=6) /// \| M484==8 \| M484==9) if M484<97 |

Let's take the dummy variable for *aframer* and see how it would be coded, based on the expression being true or false.

| Original Variable M484 | Is *if* qualifier true or false? | Is expression true or false? (Does M484 equal 1?) | | New Dummy Variable *aframer* |
|---|---|---|---|---|
| | if M484<97 | SAS M484=1 | Stata M484==1 | |
| 1 = African American | True | True | | 1 |
| 2 = White | True | False | | 0 |
| 3 = Mexican American | True | False | | 0 |
| 4 = Puerto Rican | True | False | | 0 |
| 5 = Cuban | True | False | | 0 |

| 6 = Other Hispanic | True | False | 0 |
| 7 = American Indian | True | False | 0 |
| 8 = Asian | True | False | 0 |
| 9 = Other | True | False | 0 |
| 97 = Refusal | False | n/a | . |
| 99 = No Answer | False | n/a | . |

In the first row, when the original variable is coded "1", the expression "Is M484 less than 97" is true, so the expression "Does M484 equal 1?" is evaluated. Because the latter statement is also true, the new variable *aframer* is coded "1". In contrast, in the second row, when the original variable is coded "2", the first statement is again true but the second statement "Does *M484* equal 1" is false, thus the new variable *aframer* is coded "0". The new variable is similarly coded except in the last two rows. In these cases, the first statement "Is M484 less than 97" is false, thus the second statement is not evaluated and the new variable is coded a '.' missing.

When SAS or Stata executes the code, the value on the original variable for each case is examined in relation to the expressions. The following table provides an example of what the results might look like in the datafile for the original variable M484 and the new variable *aframer*:

| id | M484 | aframer | . . . |
|----|------|---------|-------|
| 10101 | 4 | 0 | . . . |
| 10102 | 99 | . | . . . |
| 10103 | 2 | 0 | . . . |
| 20101 | 1 | 1 | . . . |
| 20102 | 1 | 1 | . . . |
| 30102 | 9 | 0 | . . . |
| 30103 | 3 | 0 | . . . |
| . . . | . . . | . . . | . . . |

## 7.3: INTERPRETING DUMMY VARIABLE REGRESSION MODELS

We will now manipulate the regression equations to help us to understand the general interpretation of dummy variable coefficients. We will also look at several examples to make this general discussion concrete. We will start with a simple case with one dummy variable predictor and then look at several extensions with multiple predictors. We will end with some conventional ways to present results for dummy variable predictors.

### 7.3.1: Single Dummy Variable Predictor

We will begin by manipulating the equation for a simple model with a single dummy variable predictor to help us to understand why we can interpret the coefficient of the dummy variable as the difference in means between the group coded "1" on the dummy variable and the reference group coded "0" on the dummy variable.

We will represent the dummy variable with the letter $D$ to distinguish it from integer-level predictor variables which we have been denoting $X$ in earlier chapters. Then, a model with a single dummy predictor can be represented as follows:

$$Y_i = \beta_0 + \beta_1 D_i + \varepsilon_i$$

Suppose that $Y$ represents the variable *g1miles* from our prior examples (the distance in miles that an adult lives from his or her mother) and that $D$ represents a dummy variable *female* which equals "1" to indicate women and "0" for men. It is conventional still to refer to $\beta_0$ as the intercept, but we will call $\beta_1$ the dummy variable coefficient (rather than the slope).

Recall that our original variable (*M2DP01*) has two levels (coded 1 = men and 2 = women). Thus, the number of categories, $c$, is two. We can create two dummy variables based on this original two-category variable: (a) dummy *male* which indicates the *M2DP01* category of 1 (men); and, (b) a dummy *female* which indicates the *M2DP01* category of 2 (women). Based on our rule, we need $c - 1 = 2 - 1 = 1$ of these dummy variables in the regression model. We'll start by including *female*, although we will see in Display B.7.4 below that the significance tests do not depend on which dummy variable we exclude (although interpretation of the default output does).

We can use our familiar process of writing expected values (conditional means) for particular levels of our predictor variable to understand how we interpret this model with the female dummy. Because we include just one dummy variable that takes on only two values in this simple model, there are only two possible expected values. When $D = 0$, then we have the expected value for men. When $D = 1$, then we have the expected value for women. Substituting in the values of $D$ gives the following results:

$$E(Y|D_i = 0) = \beta_0 + \beta_1 \times 0 = \beta_0 \qquad \text{the average distance from the mother for men} \qquad (7.1)$$

$$E(Y|D_i = 1) = \beta_0 + \beta_1 \times 1 = \beta_0 + \beta_1 \qquad \text{the average distance from the mother for women} \qquad (7.2)$$

We can take the difference between these two results to help us to interpret the dummy variable coefficient $\beta_1$.[3]

In words:   (the average distance from the mother for women) minus
            (the average distance from the mother for men)

In symbols:   $(\beta_0 + \beta_1) - (\beta_0) = \beta_1$                    (7.3)

So, how do we interpret $\beta_0$ and $\beta_1$ in this model with a single dummy variable predictor? As we just showed in Equation 7.1, the intercept ($\beta_0$) is the average distance from the mother for men. And, as we saw in Equation 7.3, the dummy variable coefficient ($\beta_1$) is the difference in average distance from the mother between women and men. As seen in Equation 7.2, in order to calculate the actual average for women, we need to add together the intercept ($\beta_0$) and the dummy variable coefficient ($\beta_1$).

Our typical null hypothesis about the slope in earlier chapters has been that it equals zero, which represented a horizontal line or no linear relationship between $Y$ and $X$. How would we interpret the case in which the dummy variable coefficient, $\beta_1$, is zero? Looking back at Equation 7.2, if $\beta_1 = 0$, then the expected value for women would be $\beta_0 + \beta_1 = \beta_0 + 0 = \beta_0$. This is the same expected value as calculated for men in Equation 7.1. Thus, when the dummy variable coefficient is zero, the mean for women is the same as the mean for men. In general, our familiar process of calculating the $t$-statistic for the hypothesis that $\beta_1 = 0$ is still a meaningful hypothesis (although now it represents no difference in the mean on $Y$ between the two groups, rather than no linear relationship between $Y$ and $X$). We can use the same formulae we learned in Chapters 5 and 6 to calculate the coefficients, standard errors, and $t$-values for this test. The standard $t$-value (ratio of the coefficient to its standard error) and $p$-value that SAS and Stata will show will be a test of this null hypothesis that the dummy variable coefficient is zero ($H_0$: $\beta_1 = 0$; the means are the same in the two groups) against an alternative hypothesis that the dummy variable's coefficient is not zero ($H_0$: $\beta_1 \neq 0$; the means differ between the two groups). It is also possible to test directional hypotheses using the techniques discussed in Section 5.4.7.

In sum, in a simple model with a single dummy variable, we interpret the intercept as the average on the outcome for the reference category (in this case, men). And, we interpret the coefficient of the dummy variable as the difference in outcome means between the **included category** (in this case, women) and the reference category (in this case, men). If the dummy variable's coefficient is not significantly different from zero, then the outcome means for the two groups are statistically equivalent. If the dummy variable's coefficient is significant, then this means that we have statistical evidence that the included category (in this case, women) has a different average value on the outcome than does the reference category (in this case, men).

Let's now look at an example to make this concrete. Display B.7.4 contains the results of actually regressing *g1miles* on *female* in the NSFH data. We use the syntax explained above to create a dummy variable *female* based on evaluating the expression "Is *M2DP01* equal to two?" (Recall that *M2DP01* has no missing values; see again Display C.7). Notice that our command for the regression is the same as we have been using already for interval predictors, but we now use our dummy variable predictor (*female*).

The results give us the following estimated regression equation:

$$\hat{Y} = 323.76 - 59.39 \times female$$

Both the intercept and dummy coefficient differ significantly from zero. Their $t$-values (23.56 and –3.35) are both greater than our cutoff for a 5 percent two-sided test for our large sample (critical $t$-value of 1.96). And, both of the $p$-values are less than 0.05.

Using the general interpretation for the intercept and dummy coefficient that we just worked out, we would say:

> On average, adult men live about 324 miles from their mothers.
> Adult women live about 60 miles closer to their mothers than do men, on average.
> Adult women, on average, live $323.76 - 59.39 = 264.37$ miles from their mothers.

What happens if we switch the reference category from men to women? Display B.7.4 also contains the results of creating a new dummy variable called *male* and regressing *g1miles* on *male* in the NSFH data. Notice that our code is very similar to the code for the top model in Display B.7.4, although the expression used to define the dummy *male* now asks "Is M2DP01 equal to 1?" to indicate men, and the predictor variable in the regression command is *male* rather than *female*.

The results give us the following estimated regression equation:

$$\hat{Y} = 264.37 + 59.39 \times male$$

Several important points are made clear by comparing these results to those in the top panel of Display B.7.4:

■ The dummy coefficient for *male* is identical in magnitude to the dummy coefficient for *female* but opposite in sign (59.39 and –59.39, respectively). Just as the mean for females is about 60 miles less than the mean for men, the mean for men is about 60 miles more than the mean for women.

■ The intercept is now 264.37, which we had calculated based on the top panel of Display B.7.4 to be the average distance for women. So, as expected, with the dummy variable switched to *male*, the intercept is the average distance for women, who are now the reference category.

■ The mean distance for men can be calculated as the intercept plus the dummy coefficient in the bottom panel of Display B.7.4. That is, $264.37 + 59.39 = 323.76$. As expected, within rounding, this is the same mean distance for men as we read from the intercept in the top model of Display B.7.4, when men were the reference category.

### Correspondence with a *t*-Test for the Difference between Two Means

The simple regression model we just examined, with one dummy variable predictor, is directly analogous to the *t*-test for a difference in means between two groups taught in introductory statistics courses.[4] Let's conduct the same hypothesis test that we did in Display B.7.4 with a *t*-test to help relate our new concepts to these familiar concepts. We just saw that in the regression context, one way to write our null hypothesis was that the dummy variable coefficient (difference in means between the two groups) was zero. We can write our hypotheses for the *t*-test of the difference between two means equivalently as:

$$H_o: \mu_{male} - \mu_{female} = 0$$

$$H_a: \mu_{male} - \mu_{female} \neq 0$$

Display B.7.5 shows the results of conducting a *t*-test of these hypotheses with *M2DP*01 as our grouping variable. Recall that the *t*-test command calculates the mean on the interval-variable (here, *g1miles*) within each of the groups and the difference between these means. These match up to the values we saw in Display B.7.4. The mean for men (*M2DP*01 = 1, labeled MALE in Display B.7.5) is 323.76. The mean for women (*M2DP*01 = 2, labeled FEMALE in Display B.7.5) is 264.37.

The difference between the two means in Display B.7.5 is 59.39 and the *t*-value is 3.35. This is consistent with the results we saw in the dummy variable regression. The sign (positive) matches the bottom model in Display B.7.4 because the difference in means for the *t*-test is calculated by subtracting the higher-coded value (*M2DP*01 = 2, FEMALE) from the lower coded value (*M2DP*01 = 1, MALE).

If we can test the same hypothesis with the *t*-test for group means as with dummy-variable regression, why learn them both? As we have seen in earlier chapters, it is rare in the social sciences that we have data (such as experimental data) that make the two-group *t*-test (or simple dummy variable model) informative.[5] As we will see in the next series of sections, in the regression context we can extend the model by adding additional predictor variables to test more complex hypotheses and control for confounding variables. Social scientists often use *t*-tests for group means when initially describing data, but for testing central research hypotheses, they generally rely on the regression approach.

### 7.3.2: Nominal/Ordinal Variable with More than One Category

In this section, we will extend our concepts to models with multiple dummy variables representing an original categorical variable with more than two levels.

For example, a model that includes three dummy variables to represent four categories of a categorical variable can be written as follows:

$$Y_i = \beta_0 + \beta_1 D_{1i} + \beta_2 D_{2i} + \beta_3 D_{3i} + \varepsilon_i \qquad (7.4)$$

Again, the outcome variable might be *g1miles* and the dummy variables might represent three of the race-ethnicity categories in the NSFH. For simplicity in this example, we will define the reference category as African Americans and exclude the relatively small number (100) of NSFH cases in our analytic sample who identified themselves as Puerto Rican, Cuban, other Hispanic, Asian, or Other. Thus, our example will take $D_1 = amind$, $D_2 = mexamer$, and $D_3 = white$. We already defined the relevant variables above, and reproduce the table below, shading the categories that we will drop from the data set before estimating the regression.

| Original Variable | New Dummy Variables | | | | |
|---|---|---|---|---|---|
| M484 | aframer | white | mexamer | amind | other |
| 1 = African American | 1 | 0 | 0 | 0 | 0 |
| 2 = White | 0 | 1 | 0 | 0 | 0 |
| 3 = Mexican American | 0 | 0 | 1 | 0 | 0 |
| 4 = Puerto Rican | 0 | 0 | 0 | 0 | 1 |
| 5 = Cuban | 0 | 0 | 0 | 0 | 1 |
| 6 = Other Hispanic | 0 | 0 | 0 | 0 | 1 |
| 7 = American Indian | 0 | 0 | 0 | 1 | 0 |
| 8 = Asian | 0 | 0 | 0 | 0 | 1 |
| 9 = Other | 0 | 0 | 0 | 0 | 1 |
| 97 = Refusal | . | . | . | . | . |
| 99 = No Answer | | | | | |

Notice that excluding these categories from the data set is different from excluding the *other* category to make it the reference category. When we drop cases coded 4 to 6, 8, 9, or missing, our data set will now be comprised of the subset of the cases who report being African American, white, Mexican American, or American Indian, and any three of the four dummies presented below can be used in a regression model.

| Original Variable | New Dummy Variables | | | |
| --- | --- | --- | --- | --- |
| *M484* | *aframer* | *white* | *mexamer* | *amind* |
| 1 = African American | 1 | 0 | 0 | 0 |
| 2 = White | 0 | 1 | 0 | 0 |
| 3 = Mexican American | 0 | 0 | 1 | 0 |
| 7 = American Indian | 0 | 0 | 0 | 1 |

In Equation 7.4, the intercept represents the mean for the reference category (in our first example, African Americans). The dummy variable coefficients in Equation 7.4 assess the difference between the mean of this reference category and the mean of each of the categories indicated by the three included dummy variables. We can see this by writing the expected values.

$$E(Y|D_{1i} = 0, D_{2i} = 0, D_{3i} = 0)$$
$$= \beta_0 + \beta_1 \times 0 + \beta_2 \times 0 + \beta_3 \times 0 = \beta_0$$
$$= \beta_0$$

average distance from mother for African Americans (7.5)

$$E(Y|D_{1i} = 1, D_{2i} = 0, D_{3i} = 0)$$
$$= \beta_0 + \beta_1 \times 1 + \beta_2 \times 0 + \beta_3 \times 0$$
$$= \beta_0 + \beta_1$$

average distance from mother for American Indians (7.6)

$$E(Y|D_{1i} = 0, D_{2i} = 1, D_{3i} = 0)$$
$$= \beta_0 + \beta_1 \times 0 + \beta_2 \times 1 + \beta_3 \times 0$$
$$= \beta_0 + \beta_2$$

average distance from mother for Mexican Americans (7.7)

$$E(Y|D_{1i} = 0, D_{2i} = 0, D_{3i} = 1)$$
$$= \beta_0 + \beta_1 \times 0 + \beta_2 \times 0 + \beta_3 \times 1$$
$$= \beta_0 + \beta_3$$

average distance from mother for whites (7.8)

Let's subtract the expected value for the reference category (African American) from the expected value for each of the included categories to see why this is the correct interpretation for the dummy variable coefficients, similarly to what we did above when we interpreted the single dummy for *female* in Equation 7.3.

| Difference in expected value in words: | Difference in expected value in symbols: |
| --- | --- |
| American Indian minus African Americans | $(\beta_0 + \beta_1) - (\beta_0) = \beta_1$ |
| Mexican Americans minus African Americans | $(\beta_0 + \beta_2) - (\beta_0) = \beta_2$ |
| Whites minus African Americans | $(\beta_0 + \beta_3) - (\beta_0) = \beta_3$ |

These results make it explicit that each dummy variable coefficient captures how much higher or lower the outcome mean is for the included category than the outcome mean for the reference category.[6] A common mistake for students when interpreting models like this one is to interpret each of the dummies as the contrast of the group indicated by the dummy versus all other groups. The results above make clear that this interpretation is not correct. Rather, the comparison group is the reference category (in this case African Americans).[7]

Display B.7.6 shows the SAS and Stata code for creating these new dummy variables and the regression command which lists the names of the three dummy variables following the format that we used for equations with multiple predictors in Chapter 6. We list them in the order we associated them with our general dummies ($D_1$ = *amind*, $D_2$ = *mexamer*, and $D_3$ = *white*) so that it is easy to relate them to our general formulae.

Looking at the parameter estimates, we see that just one of the dummy variable coefficients is significantly different from zero (*t*-value greater than 1.96 and *p*-value less than 0.05): the coefficient for white, which is positive in sign. So, American Indians and Mexican American adults do not differ significantly from African American adults in the mean distance they live from their mothers. But, white adults live about 60 miles further away from their mothers than do African American adults, on average.

The actual predicted values are:

| | | |
|---|---|---|
| African Americans | 240.97 | |
| American Indians | 240.97 + 157.03 = 398.00 | **(7.9)** |
| Mexican Americans | 240.97 – 8.25 = 232.72 | **(7.10)** |
| Whites | 240.97 + 56.72 = 297.69 | |

So, Mexican Americans live an average of 233 miles from their mothers. African Americans live just slightly further from their mothers, averaging 241 miles. Whites live further, nearly 300 miles on average from their mothers. The mean for American Indians is substantially larger than any other group, at nearly 400 miles.

But, American Indians have the smallest sample size and their estimate has the largest standard error (in Display B.7.6, $\hat{\sigma}_{\hat{\beta}_1}$ = 129.49 whereas $\hat{\sigma}_{\hat{\beta}_2}$ = 54.48 and $\hat{\sigma}_{\hat{\beta}_3}$ = 24.77). Thus, although the average distance they live from their mothers is estimated to be substantially larger than the three other groups, the estimate has considerable error. This reflects the different sample sizes among the groups. Recall that we excluded cases missing any variables (including the interval variables we will add to the model below), so the sample sizes are even smaller than what was seen in Display B.7.1.

Among the 5,401 cases included in the regression model, there are just 25 American Indians as opposed to 166 Mexican Americans and 4,432 whites. The reference category, African American, has 778 cases. Seeing the implications of these sample sizes for the standard errors reinforces our comment above of not selecting a category with few cases as the reference category. And, it reinforces the general importance of selecting a data set or designing new data collection to contain sufficient sample sizes across the groups of interest (the NSFH data are not well suited to regression models that focus on American Indians).

## Testing Differences in Means among the Included Groups

As mentioned above, we can recover the differences in means among all groups, based on the default output with any of the categories used as a reference. We do so by calculating the difference between the estimated coefficients for the two included dummy variables.

To start, we can use the predicted values (means within groups) that we figured out above (Equations 7.9 and 7.10) to calculate directly the difference in means between Mexican Americans and American Indians.

$$(\bar{X}_{mexamer} - \bar{X}_{aframer}) - (\bar{X}_{amind} - \bar{X}_{aframer})$$

$$(\bar{X}_{mexamer} - \bar{X}_{amind})$$

$$232.72 - 398.00$$

$$= -165.28 \qquad\qquad (7.11)$$

This result (−165.28) should match the difference in coefficients between Mexican Americans and American Indians. Indeed, recall that our interpretations of the coefficient estimates mirror the difference in means written above. When African Americans are the reference category, the coefficient estimate for Mexican Americans $(\hat{\beta}_2)$ reflects the difference in means between Mexican Americans and African Americans; and, the coefficient estimate for American Indians $(\hat{\beta}_1)$ reflects the difference in means between American Indians and African Americans. Thus, we can calculate the difference in means between Mexican Americans and American Indians based on their respective coefficient estimates from the regression with African Americans as the reference category.

$$\hat{\beta}_2 - \hat{\beta}_1$$

$$(-8.25) - (157.03) = -165.28$$

This difference of −165.28 is larger than the difference between either group and African Americans. But, is this difference between the two included groups (Mexican

Americans and American Indians) itself significantly different from zero? That is, does the average difference that Mexican Americans live from their mothers differ significantly from the average difference that American Indians live from their mothers?

In general, what if we also had hypothesized about the differences in means between the pairs of included groups as well as differences from the reference category mean? We might, for example, expect that white adults live further from their mothers than do American Indian and Mexican American adults, due to emphases on familial ties in American Indian and Mexican cultures.[8] We will now look at three strategies for testing for differences among the included groups (including whether the $-165.28$ calculated in Equation 7.11 is significant):

(a) re-estimating the regression model with a slightly different set of variables;
(b) using a partial $F$-test;
(c) testing the significance of a linear combination of coefficients.

We'll conclude by talking about the relative strengths and weaknesses of each approach.

### Re-estimating the Model with a Different Reference Category

One way to test the differences in means among the included groups is to re-estimate the model with a different reference category, similar to the strategy we took above when we had a single dummy predictor. When we have more than one dummy predictor, this approach can become cumbersome because we may need to re-estimate the model multiple times, depending on the number of categories on our original variable. In this case, we have three included dummies, so there are three pairwise comparisons among them (American Indian versus Mexican American, American Indian versus white, and Mexican American versus white). We will capture two of these three contrasts the first time that we exclude a different dummy variable from the set. But, we'll need to re-estimate the model a third time to pick up the final contrast.

Importantly, as we will see, there is no need to re-estimate the model, since all of the information we need to contrast the included categories is available upon our first estimation (this is evident, for example, in the fact that the sum of squares and overall $F$-test will be the same across models which use different reference categories). But, it is pedagogically helpful for some students to see this concretely.

Display B.7.7 provides the results of re-estimating the regression model first with American Indians as the reference category, then with Mexican Americans as the reference category, and finally with whites as the reference category. Note that each time, we included three of the four dummy variables in this set. We must do so in order to get equivalent results as our initial model, from Display B.7.6.

First, notice that each regression model in Display B.7.7 repeats one of the coefficient estimates from Display B.7.6, but reversed in sign. This is because we are calculating the difference in means between the same pair of groups, but flipping the reference category.

For the contrast between American Indians and African Americans:

■ The coefficient estimate for amind was 157.03 in Display B.7.6 (with African Americans excluded). The coefficient estimate for aframer in the top model of Display B.7.7 (with American Indians excluded) is –157.03.

For the contrast between Mexican Americans and African Americans:

■ The coefficient estimate for *mexamer* was –8.25 in Display B.7.6 (with African Americans excluded). The coefficient estimate for *aframer* in the middle model of Display B.7.7 (with Mexican Americans excluded) is 8.25.

For the contrast between whites and African Americans:

■ The coefficient estimate for *white* was 56.72 in Display B.7.6 (with African Americans excluded). The coefficient estimate for *aframer* in the bottom model of Display B.7.7 (with whites excluded) is –56.72.

This is directly analogous to what we saw in the simple regression model for gender: when we switched the dummy variable that we excluded, the coefficient estimate was the same in magnitude, but opposite in sign. The same is true in these models with a set of dummies for a multicategory variable, but only one contrast remains the same when we switch the reference category. The other contrasts offer new comparisons among some of the originally included variables.

In the top model of Display B.7.7, because the group "American Indians" is now the reference category, we now have estimates of the difference in means between Mexican Americans and American Indians (–165.28) and between whites and American Indians (–100.31). Neither is significant under a two-sided test (both $t$-values less than 1.96 and $p$-values greater than 0.05). Notice that the former value of –165.28 is exactly the same amount as the difference in coefficient estimates we calculated in Equation 7.11. This re-emphasizes the fact that all of the information about the contrasts is captured in a single estimation of the model. What we could not read off of the original regression output in Display B.7.6, however, was whether this difference in average distance between Mexican Americans and American Indians was significant.

Re-estimating the model with American Indians excluded in Display B.7.7 shows us that it is not (even though the difference is large in absolute terms, its standard error is also large, due to the small sample size of American Indians, making the *t*-value just –1.21).

The two values we just examined are also repeated in the middle and bottom models of Display B.7.7. The contrast of American Indians and Mexican Americans is found in the middle model, the same magnitude but reversed in sign from what we saw in the top model (the coefficient estimate for *amind* with Mexican Americans excluded is 165.28 in the middle model). Similarly, the contrast of whites and American Indians is found in the bottom model, the same magnitude but reverse sign from what we saw in the top model (the coefficient estimate for *amind* with whites excluded is 100.31 in the bottom model).

The middle and bottom panels also contain one new contrast that was not found in the top panel of Display B.7.7 nor in Display B.7.6. The difference between whites and Mexican Americans is just over 60 miles (64.97 in the middle model for *white* with Mexican Americans excluded; –64.97 in the bottom panel for *mexamer* with whites excluded). This difference is not significant for a two-sided test ($p = .197$).[9]

As noted, although this approach—re-estimating the regression model with a different excluded dummy variable—gives us all of the contrasts that we desire and is quite simple and concrete, it can become cumbersome. Depending on the number of dummy variables in the set, we must re-restimate the model several times and we obtain some redundant information in each re-estimation. We will now look at two different approaches, each of which can be implemented in SAS and Stata with commands that follow a single estimation of the model.

### Using the Partial F-test

A second approach to testing hypotheses about differences between the included variables is with the partial *F*-test which we learned in Chapter 6. We can use the SAS and Stata test commands that we learned in that chapter, but first we will examine the logic of the partial *F*-test for our new application in this chapter and we will calculate one test "by hand" to make the logic concrete.

Recall that we used the partial *F*-test to compare a full and reduced model, where we wrote the reduced model by placing the restrictions of the null hypothesis on the full model. What are the restrictions when we are comparing the two included coefficients? Let's use the coefficients in Equation 7.4 to figure out what would need to be true for the means for American Indians and whites to differ. To do so, we will write the differences between the expected values for the included variables found in Equations 7.6 to 7.8 in words and with symbols:[10]

| Difference in expected value in words | Difference in expected value in symbols | |
|---|---|---|
| American Indians minus Mexican Americans | $(\beta_0 + \beta_1) - (\beta_0 + \beta_2)$ <br> $= \beta_0 + \beta_1 - \beta_0 - \beta_2 = \beta_1 - \beta_2$ | **(7.12)** |
| American Indian minus whites | $(\beta_0 + \beta_1) - (\beta_0 + \beta_3)$ <br> $= \beta_0 + \beta_1 - \beta_0 - \beta_3 = \beta_1 - \beta_3$ | **(7.13)** |
| Mexican Americans minus whites | $(\beta_0 + \beta_2) - (\beta_0 + \beta_3)$ <br> $= \beta_0 + \beta_2 - \beta_0 - \beta_3 = \beta_2 - \beta_3$ | **(7.14)** |

Thus, for each pair of included groups, the difference in their means is captured by the difference between their respective dummy variables' coefficients. What if we wanted to test the null hypothesis that the difference in means between a pair of included groups is zero? For example, what if $H_o$: $\beta_1 = \beta_2$ (or equivalently $H_o$: $\beta_1 - \beta_2 = 0$)? Looking back at the expected values in Equations 7.6 and 7.7, we can see that if $\beta_1$ and $\beta_2$ are equivalent to one another, then the two groups' means must be the same, because we add that same value $\beta_1 = \beta_2$ to the intercept ($\beta_0$) to calculate each group's expected value. More concretely, suppose $\beta_0 = 30$ and $\beta_1 = \beta_2 = 10$. Then the expected value for American Indians would be: $\beta_0 + \beta_1 = 30 + 10 = 40$. And, the expected value for Mexican Americans would be: $\beta_0 + \beta_2 = 30 + 10 = 40$.

How do we use this null hypothesis ($H_o$: $\beta_1 = \beta_2$) to construct the partial $F$-test? Our full model is Equation 7.4. Our reduced model results from applying the restriction of the null hypothesis (that the coefficients for the first two dummy variables are equal) to this full model. If $\beta_1 = \beta_2$, then we can substitute a new symbol to represent this single coefficient in the model. Let's call it $\beta_{12}$.

$$Y_i = \beta_0 + \beta_{12}D_{1i} + \beta_{12}D_{2i} + \beta_3 D_{3i} + \varepsilon_i$$

We can now simplify this equation algebraically, because of the common term $\beta_{12}$:

$$Y_i = \beta_0 + \beta_{12}(D_{1i} + D_{2i}) + \beta_3 D_{3i} + \varepsilon_i$$

What does the new term $D_{1i} + D_{2i}$ imply? To make this clear, let's look back at the values of our dummy variables (reordered to match the order of the dummy variables in our regression model).

|  | Dummy Indicator for: | | | |
| Original Variable | American Indian $D_{1i}$ | Mexican American $D_{2i}$ | White $D_{3i}$ | American Indian or Mexican American $D_{1i} + D_{2i}$ |
| M484 | | | | |
| --- | --- | --- | --- | --- |
| 7 = American Indian | 1 | 0 | 0 | 1 |
| 3 = Mexican American | 0 | 1 | 0 | 1 |
| 2 = White | 0 | 0 | 1 | 0 |
| 1 = African American | 0 | 0 | 0 | 0 |

So, we could think of $D_{1i} + D_{2i}$ as a new dummy indicator of being either American Indian or Mexican American. Suppose we explicitly defined a new variable $D_{12i} = D_{1i} + D_{2i}$, then we could estimate the following as our reduced model:

$$Y_i = \beta_0 + \beta_{12}D_{12i} + \beta_3 D_{3i} + \varepsilon_i$$

The sum of squared errors and degrees of freedom from this reduced model, along with the sum of squared errors and degrees of freedom from the full model (Equation 7.4), would be used to calculate a partial F test of the null hypothesis that $H_0$: $\beta_1 - \beta_2 = 0$ (or equivalently $H_0$: $\beta_1 = \beta_2$) against the alternative that $H_a$: $\beta_1 - \beta_2 \neq 0$ (or equivalently $H_a$: $\beta_1 \neq \beta_2$).

Display B.7.8 shows the results of estimating the reduced model. The sums of squares error for the reduced model is 2,192,402,062 with degrees of freedom 5,398. The sum of squared errors for the full model is shown in Display B.7.6 as 2,191,808,493 with degrees of freedom 5,397. The difference in degrees of freedom between the two models is one, as expected (5,398 − 5,397 = 1) because there is one restriction in the null hypothesis (that the two coefficients are equivalent).

Substituting into our standard formula for the partial $F$-test provides the following result:

$$F = \frac{SSE(R) - SSE(F)}{df_R - df_F} \div \frac{SSE(F)}{df_F} = \frac{(2,192,402,062 - 2,191,808,493)}{5,398 - 5,397}$$

$$\div \frac{2,191,808,493}{5,397} = 1.4616$$

Estimating the reduced model and calculating the $F$-value by hand like this is not time-saving, relative to our first approach (re-estimating the model with a different reference category). But, we can use the `test` command which we learned in Chapter 6, using the variable names to tell SAS and Stata what restriction to place on the full model (e.g., in the example we just calculated, where the null hypothesis is $H_0: \beta_1 = \beta_2$ we would write `test amind=mexamer`). Recall that SAS and Stata use the variable names on the test commands to refer to their associated coefficients, so `test amind=mexamer` is interpreted by SAS and Stata as "Test the null hypothesis that the coefficient for American Indians is equal to the coefficient for Mexican Americans." Because we ordered our predictor variables so that the first was American Indian and the second Mexican American, this null hypothesis can be written symbolically as $H_0: \beta_1 = \beta_2$.

Display B.7.9 shows the results of these test commands. The results confirm our calculation for the contrast of American Indians to Mexican Americans, providing a calculated $F$-value of 1.46. The $p$-value of 0.2267 is larger than 0.05. Thus, we fail to reject the null hypothesis that the means for American Indians and Mexican Americans are equivalent. The additional two tests also have $p$-values that are larger than 0.05, thus giving statistical evidence that the means do not differ significantly between American Indians and whites, $F(1,5397) = 0.62$, $p = 0.4326$, and between Mexican Americans and whites, $F(1,5397) = 1.66$, $p = 0.1972$.

### Testing the Linear Combination of Coefficients

We will now consider a final approach for testing the difference in means among the included categories. To do so, we will introduce a new statistical concept: how to test the significance of a linear combination of coefficients. Generally, in our models, a **linear combination** will involve the addition of two coefficients or the subtraction of two coefficients.

We saw above that one way to write the null hypothesis when we tested the difference in means of our first two included variables was $H_0: \beta_1 - \beta_2 = 0$. This difference is a linear combination of coefficients. Statistical theory tells us how to calculate the standard error of a linear combination, and we can divide the difference in estimated coefficients by this standard error. The resulting value is a $t$-statistic, and thus we can use our standard procedures for calculating a $p$-value or comparing the calculated $t$-value to a critical $t$-value in order to make a conclusion about the null hypothesis.

The new statistical concept we need in order to implement this approach is the **variance-covariance matrix of the estimated regression coefficients**. We emphasized in Chapter 5 that the estimated regression coefficients have sampling distributions; that is, the estimate of the population parameter (the population regression coefficient) will differ somewhat when calculated on each successive sample that we might draw from the population. The variances of the regression coefficients capture that variation. We use the square root of these variances—the standard errors—whenever we calculate a $t$-statistic for an individual regression coefficient. As in a variance–covariance matrix for variables, the variances of the regression coefficients fall "on the diagonal" of the matrix. The parameter estimates also covary (in some applications, when the intercept is large, the slope is also large; in other applications, when the intercept is large, the slope is small; in yet other applications, the intercept and slope are hardly associated). The covariances—the values that fall "off the diagonal" of the matrix—capture this covariation among the coefficients. We use these covariances in the formula for the standard error of a linear combination of coefficients.

Display B.7.10 shows how to ask SAS and Stata for the variance–covariance matrix of the estimates. To see that a portion of this new variance–covariance matrix contains some familiar quantities, let's first compare the square root of the diagonal elements of this matrix with the standard errors in the model, which are displayed in Display B.7.6.

■ For *amind* the variance on the diagonal of the matrix (found by looking down the column labeled *amind* and across the row labeled *amind*) is 16,766.643. The square root of 16,766.643 is 129.49, which matches the standard error shown in Display B.7.6 for *amind*.

■ For *mexamer* the variance on the diagonal of the matrix is 2,968.4825. The square root of 2,968.4825 is 54.48, which matches the standard error shown in Display B.7.6 for *mexamer*.

■ For *white* the variance on the diagonal of the matrix is 613.63279. The square root of 613.63279 is 24.77, which matches the standard error shown in Display B.7.6 for *white*.

We also need the off-diagonal elements to calculate the standard error for the linear combination of coefficients. We find the covariance between the coefficients for *amind* and *mexamer* by reading down the column labeled *amind* and across the row labeled *mexamer*. This value is 522.0001. Notice that in our current model, all of these off-diagonal elements (covariances) are identical because we do not have any control variables in the model. In models with additional predictor variables (such as the model shown in Display B.7.12), the covariances would differ.

The formula for the **standard error of the difference** between two coefficients is (see Box 7.3):

$$\sqrt{\hat{\sigma}^2_{\hat{\beta}_1} + \hat{\sigma}^2_{\hat{\beta}_2} - 2\widehat{\text{cov}}_{\hat{\beta}_1\hat{\beta}_2}}$$

where $\hat{\sigma}^2_{\hat{\beta}_1}$ is the estimated variance of the first estimated coefficient (in our case the variance of *amind* which we saw was 16,766.643), $\hat{\sigma}^2_{\hat{\beta}_2}$ is the estimated variance of the second estimated coefficient (in our case the variance of *mexamer* which we saw was 2,968.4825), and $\widehat{\text{cov}}_{\hat{\beta}_1\hat{\beta}_2}$ is the estimated covariance between the first and second

---

**■ Box 7.3**

---

This formula is based on the general formula for the variance of a linear combination of two variables:

  Linear Combination:  $aX_1 + bX_2$

  Variance of Linear Combination: $a^2\text{Var}(X_1) + b^2\text{Var}(X_2) + 2ab\text{Cov}(X_1, X_2)$

Note that $a$ and $b$ can be 1 and they can be negative in sign. So, if we had $a = 1$ and $b = -1$, the resulting formula would be:

  $\text{Var}(X_1) + \text{Var}(X_2) - 2\text{Var}(X_1, X_2)$

which is a general version of the formula we used for the difference between two coefficients. Similarly, if $a = 1$ and $b = 1$ then the formula would be:

  $\text{Var}(X_1) + \text{Var}(X_2) + 2\text{Cov}(X_1, X_2)$

which is a formula we can use if we need to test the sum of two coefficients (e.g., $\beta_1 + \beta_2$). For example, we could use this formula to calculate the standard error of the predicted value for women based on estimating Equation 7.2 (i.e., $\beta_0 + \beta_1$).

---

estimated coefficients (in our case the covariance of *amind* and *mexamer* which we saw was 522.0001). Thus, the standard error of the difference in this case is:

$$\sqrt{16,766.643 + 2,968.4825 - 2 \times 522.0001} = 136.71549$$

To calculate the *t*-value, we divide the difference in coefficient estimates by this standard error:

$$t = \frac{(\hat{\beta}_1 - \hat{\beta}_2) - 0}{\sqrt{\sigma_{\hat{\beta}_1}^2 + \sigma_{\hat{\beta}_2}^2 - 2 \times \text{cov}_{\hat{\beta}_1 \hat{\beta}_2}}} = \frac{[(157.0283) - (-8.254855)] - 0}{136.71549} = \frac{165.28315}{136.71549} = 1.208957$$

This *t*-value is smaller than the critical *t*-value of 1.96 for a 5 percent significance level in our large sample, thus we cannot reject our null hypothesis. Notice that the standard error and *t*-value match those seen in the middle panel of Display B.7.7 for the *amind* coefficient estimate (with Mexican American as the excluded category).

For testing linear combinations, Stata also offers a specific command called `lincom`. This command is useful because it calculates the difference in coefficient estimates, the standard error, the *t*-value, and the *p*-value, thus reducing the chances that we will make a calculation error and providing all of the details we would need for reporting results in a paper. Display B.7.11 shows the results of using the lincom command for all three contrasts among the included variables shown in Equations 7.12 to 7.14.

Again, these results match the corresponding contrasts found in Display B.7.7 when we re-estimated the model with a different reference category, emphasizing that all of the relevant contrasts can be tested after a single model estimation. This final linear combination approach is thus the preferred approach because it gives the complete information about all contrasts (coefficient estimates, standard errors, $t$-values, and $p$-values) available based on a single estimation of the model.

### Summary of Approaches to Testing Contrasts Among Included Categories

We'll now take a closer look at the similarity in results across the three approaches, and then discuss their relative strengths. Let's start by summarizing the results for each approach, drawn from Display B.7.7, Display B.7.9, and Display B.7.11.

| In words | In symbols | Approach 1: Re-estimating the Model (Display B.7.7) | | | | Approach 2: Partial F test (Display B.7.9) | | Approach 3: Linear Combination (Display B.7.11) | | | |
|---|---|---|---|---|---|---|---|---|---|---|---|
| | | Diff | se | $t$-value | $p$-value | $F$-value | $p$-value | Diff | se | $t$-value | $p$-value |
| Mexican American minus American Indian | $\beta_2 - \beta_1$ | −165.28 | 136.72 | −1.21 | 0.23 | 1.46 | 0.23 | −165.28 | 136.72 | −1.21 | 0.23 |
| White minus American Indian | $\beta_3 - \beta_1$ | −100.31 | 127.81 | −0.78 | 0.43 | 0.62 | 0.43 | −100.31 | 127.81 | −0.78 | 0.43 |
| White minus Mexican American | $\beta_3 - \beta_2$ | 64.97 | 50.38 | 1.29 | 0.20 | 1.66 | 0.20 | 64.97 | 50.38 | 1.29 | 0.20 |

The table makes clear that the three approaches provide equivalent results, although the first and third approaches provide more information:

▪ All approaches provide the same calculated test statistics. (Recall, for a test with one numerator degree of freedom, the square root of the $F$-value equals the $t$-value in magnitude. In the first row, $\sqrt{1.46} = 1.21$. In the second row, $\sqrt{0.62} = 0.79$. In the third row, $\sqrt{1.66} = 1.29$.)
▪ All approaches provide the same $p$-values for two-sided tests.
▪ The first and third approaches also calculate the difference in coefficient estimates and the standard error of this difference.
▪ The first and third approaches provide the direction of the difference and can be used to test directional hypotheses (one-sided tests) by adjusting the $p$-value based on whether its sign is consistent with the alternative hypothesis.

For most models, today's computing power is quick enough that re-estimating a model with a different reference category takes only a few minutes. But, for some data sets and for some of the more complicated types of model we will preview in Chapter 12, it can take days to re-estimate a model. In these situations, the second and third approaches are time-saving. The second and third approaches also provide compact code, which is helpful for organizing and proofing our batch programs. Each approach is also useful in the broader context of learning statistical methods for the social sciences. Approach 2 reinforces our understanding of the partial $F$-test. Approach 3 introduces the new concepts of the linear combination and variance–covariance matrix of the estimates which we will see again in future chapters and in reading the literature.

What approach should we use? If we only want to report the significance of contrasts among included variables, then the `test` command will give us the information and can be implemented in either SAS or Stata. If we want to report the exact difference between the included variables' coefficient estimates and the standard error of these differences, then Stata's `lincom` command is convenient. Alternatively, to obtain the difference in coefficient estimates and their standard errors, we can also re-estimate the model with a different reference category. This approach will not be time-consuming as long as our sample is not quite large and our model is fairly simple.

We will now end our discussion of dummy variable predictors with an example of including more than one multicategory predictor in the model and adding interval-level variables to the model with sets of dummy indicators. Because most of the interpretation follows what we have already discussed, we will focus on the presentation and interpretation of the results, using the `test` command to contrast the included categories.

## 7.3.3: More than One Nominal/Ordinal Predictor

Putting together our models from Section 7.3.1 and Section 7.3.2, we now estimate a model that predicts how far adults live from their mothers based both on their gender and their race-ethnicity.

$$Y_i = \beta_0 + \beta_1 D_{1i} + \beta_2 D_{2i} + \beta_3 D_{3i} + \beta_4 D_{4i} + \varepsilon_i \tag{7.15}$$

To combine our two earlier models, we now represent *female* with $D_4$ and we maintain the use of $D_1$ for American Indians, $D_2$ for Mexican Americans, and $D_3$ for whites. As we will emphasize in the interpretation below, this model is additive. It assumes that the association between race and distance is the same for men and women. And, it assumes that the association between gender and distance is the same across races. We will discuss in Chapter 8 how to test and loosen this assumption.

Before we look at the results of estimating this model, let's look at the general expected values to help us understand the interpretations. We make these calculations by using the dummy variables to indicate each group, then adding together the effects that are relevant for each group.

| MEN | Average distance for: |
|---|---|
| 1 $\quad E(Y|D_1 = 0, D_2 = 0, D_3 = 0, D_4 = 0) = \beta_0 + \beta_1 \times 0 + \beta_2 \times 0 + \beta_3 \times 0 + \beta_4 \times 0$ <br> $\quad = \beta_0$ | African-American men |
| 2 $\quad E(Y|D_1 = 1, D_2 = 0, D_3 = 0, D_4 = 0) = \beta_0 + \beta_1 \times 1 + \beta_2 \times 0 + \beta_3 \times 0 + \beta_4 \times 0$ <br> $\quad = \beta_0 + \beta_1$ | American Indian men |
| 3 $\quad E(Y|D_1 = 0, D_2 = 1, D_3 = 0, D_4 = 0) = \beta_0 + \beta_1 \times 0 + \beta_2 \times 1 + \beta_3 \times 0 + \beta_4 \times 0$ <br> $\quad = \beta_0 + \beta_2$ | Mexican-American men |
| 4 $\quad E(Y|D_1 = 0, D_2 = 0, D_3 = 1, D_4 = 0) = \beta_0 + \beta_1 \times 0 + \beta_2 \times 0 + \beta_3 \times 1 + \beta_4 \times 0$ <br> $\quad = \beta_0 + \beta_3$ | White men |

| WOMEN | |
|---|---|
| 5 $\quad E(Y|D_1 = 0, D_2 = 0, D_3 = 0, D_4 = 1) = \beta_0 + \beta_1 \times 0 + \beta_2 \times 0 + \beta_3 \times 0 + \beta_4 \times 1$ <br> $\quad = \beta_0 + \beta_4$ | African-American women |
| 6 $\quad E(Y|D_1 = 1, D_2 = 0, D_3 = 0, D_4 = 1) = \beta_0 + \beta_1 \times 1 + \beta_2 \times 0 + \beta_3 \times 0 + \beta_4 \times 1$ <br> $\quad = \beta_0 + \beta_1 + \beta_4$ | American Indian women |
| 7 $\quad E(Y|D_1 = 0, D_2 = 1, D_3 = 0, D_4 = 1) = \beta_0 + \beta_1 \times 0 + \beta_2 \times 1 + \beta_3 \times 0 + \beta_4 \times 1$ <br> $\quad = \beta_0 + \beta_2 + \beta_4$ | Mexican-American women |
| 8 $\quad E(Y|D_1 = 0, D_2 = 0, D_3 = 1, D_4 = 1) = \beta_0 + \beta_1 \times 0 + \beta_2 \times 0 + \beta_3 \times 1 + \beta_4 \times 1$ <br> $\quad = \beta_0 + \beta_3 + \beta_4$ | White women |

In examining these expected values, we see that the interpretations are similar to our earlier examples with one categorical variable, although they now control for the other variable(s) in the model:

■ The intercept is the average on the outcome for the group coded zero on all of the included dummy variables, in this case African-American men (*amind* = 0, *mexamer* = 0, *white* = 0, *female* = 0). Equivalently, when all of the predictor variables are dummies, then, the intercept is the mean for the group that is indicated by the reference category of each of the original categorical variables (in the example, for race-ethnicity: African Americans; for gender: men).

■ Looking within race-ethnicity and between genders, we see that the coefficient estimate for $D_4$ (which represents the *female* dummy variable) still captures the difference in means between women and men, but now holding constant race-ethnicity. So, holding race constant as African American, the difference between the expected value in row 5 (African-American women) and row 1 (African-American men) is $\beta_4$. The difference between the expected values in the remaining rows within race-ethnicity is similarly $\beta_4$ (row 6 minus row 2; row 7 minus row 3; row 8 minus row 4).

■ Looking within gender and across race-ethnicity reveals that the coefficient estimates for each of the included dummy variables for race-ethnicity captures the difference between the indicated category and the excluded race-ethnicity (African Americans), but now controlling for gender. For example, the difference in expected values between row 6 (American Indian women) and row 5 (African-American women) is $\beta_1$; and, the difference in expected values between row 2 (American Indian men) and row 1 (African-American men) is $\beta_1$. So, controlling for gender, $\beta_1$ represents the average difference distance between American Indians and African Americans. Similarly, the difference between row 7 and row 5 and between row 3 and row 1 is $\beta_2$. And, the difference between row 8 and row 5 and between row 4 and row 1 is $\beta_3$.

■ Finally, for contrasts among the included categories of race-ethnicity within gender, we would still test the equality of their coefficients (or equivalently that the difference between their coefficients is zero). For example, the difference in expected values between row 6 (American Indian women) and row 7 (Mexican-American women) is $\beta_1 - \beta_2$. Likewise, the difference between row 2 (American Indian men) and row 3 (Mexican-American men) is $\beta_1 - \beta_2$. Similarly, the difference between row 6 and row 8 and between row 2 and row 4 is $\beta_1 - \beta_3$. And, the difference between row 7 and row 8 and between row 3 and row 4 is $\beta_2 - \beta_3$.

### 7.3.4: Adding Interval Variables and Presenting the Results

We will now further extend the model by adding interval predictors, and interpret the results in a table and figure format similar to that we might use in a publication. The purpose of adding more variables to the model and putting these results in presentation format is to examine a more realistic model and practice presentation style, not to suggest the substantive relevance of these findings. In a publication, we would flesh out the conceptual rationale for our hypothesis tests, and add additional controls. We will look at a substantively interpretable example from the published literature in the final section of this chapter.

We can extend Equation 7.15 by adding the full list of interval variables we examined at the end of Chapter 6.

$$Y_i = \beta_0 + \beta_1 D_{1i} + \beta_2 D_{2i} + \beta_3 D_{3i} + \beta_4 D_{4i} + \beta_5 X_{5i} + \beta_6 X_{6i} + \beta_7 X_{7i}$$
$$+ \beta_8 X_{8i} + \beta_9 X_{9i} + \beta_{10} X_{10i} + \varepsilon_i$$

where $X_5 = g2earn$, $X_6 = g2age$, $X_7 = g2numbro$, $X_8 = g2numsis$, $X_9 = g1yrschl$, $X_{10} = g1age$.

This model is still additive, and the interpretation of the dummy variable coefficients remains as discussed above; however, we now will remind the reader in our interpretation that we have controlled for this fuller set of variables.

Display B.7.12 provides the results of estimating this equation in SAS and Stata. And, Table 7.2 shows how these results might be presented in a publication table.

■ **Table 7.2: Regression Model of Distance Adults Live from their Mothers (in miles)**

|  | Coefficient (s.e.) |
| --- | --- |
| Respondent's Characteristics | |
| Race-ethnicity[a] | |
| American Indian | 185.14 |
| | (128.54) |
| Mexican American | 61.85 |
| | (55.32) |
| White | 35.07 |
| | (25.22) |
| Female | −33.79+ |
| | (18.80) |
| Annual Earnings[b] | 6.68* |
| | (2.51) |
| Age (years) | −0.04 |
| | (1.65) |
| Number of Brothers | 5.20 |
| | (5.96) |
| Number of Sisters | 18.56* |
| | (6.03) |
| Mother's Characteristics | |
| Years of Schooling | 21.12* |
| | (3.28) |
| Age (years) | 4.61* |
| | (1.44) |
| Intercept | −293.51 |
| | (71.86) |

*Source:* National Survey of Families and Households, Wave I.
Notes: Standard errors in parentheses. $n = 5,401$.
[a] Reference category is African American.
[b] In $10,000 increments, adjusted to 2007 dollars.
* $p < .05.$ + $p < .10$ (two-sided tests).

Examining the results for our dummy variables, we see that, controlling for the respondent's race-ethnicity, earnings, age, number of bothers and number of sisters, and the mother's age and years of schooling, adult women live 34 miles closer to their mothers than do adult men, on average. Under a two-sided test, this effect is marginally significant, with a *p*-value less than 0.10.[11] Controlling for gender, the racial-ethnic groups do not differ significantly from one another in distance lived from the mother.[12]

Notice that the intercept in this model is –293.51, a value that is indicated to differ significantly from zero. This nonsensical negative value on distance reflects the mechanical interpretation of the intercept as the predicted value when all variables in the model equal zero. Although zero is a possible value on many variables, it is not possible in this sample of adults for *g2age* and *g1age*.[13] Thus, the estimated intercept is not interpretable and the hypothesis test that its value equals zero is not meaningful. It is useful to list the estimate in the table, however, so that more meaningful predictions can be made from the model, as we will illustrate below.

The table also shows that the significant associations seen in Chapter 6 for the respondent's earnings and number of sisters and mother's age and years of schooling remain significant once we also control for the respondent's race-ethnicity and gender.

We will use predictions from the model to illustrate the association of distance to one of these—mother's years of schooling—along with the differences of conditional means by gender. There are various strategies for making such predictions in multiple regression models with many predictors. We will discuss these various techniques in Chapter 9, as we use graphs more intensively to illustrate nonlinear effects. For now, we will use an approach in which we hold all variables constant at their sample means except one or two variables which we will vary systematically. For example, we could make predictors for men and women by substituting in values of 0 and 1 on the female variable and we could make predictions for different levels of the mother's years of schooling, for example, by substituting in values of 10 and 14 on *g1yrschl* and we could use the sample means for the remaining variables (obtained from the Stata summarize and SAS proc means commands introduced in Chapter 4).

For example, the full prediction equation is:

$$g1miles = -293.51 + 185.14 * amind + 61.85 * mexamer + 35.07 * white$$
$$- 33.79 * female + 6.68 * g2earn10000 - 0.04 * g2age + 5.20 * g2numbro$$
$$+ 18.56 * g2numsis + 21.12 * g1yrschl + 4.61 * g1age$$

Substituting in the mean values for the interval variables other than *g1yrschl* (see Display B.7.13), and indicating the most common race-ethnicity in the data (white) results in:

$$glmiles = -293.51 + 185.14 * 0 + 61.85 * 0 + 35.07 * 1 - 33.79 * female$$
$$+ 6.68 * 3.075228 - 0.04 * 34.35271 + 5.20 * 1.432883 + 18.56 * 1.381781$$
$$+ 21.12 * glyrschl + 4.61 * 60.13498$$

Now, all of the values are fixed, except for *female* and *glyrschl*. We can work through the algebra to rewrite the equation as:

$$glmiles = 71.047519 - 33.79 * female + 21.12 * glyrschl$$

Since *female* can only take on two values, we can further write this as two conditional regression equations:

| | |
|---|---|
| female = 0 | $glmiles = 71.047519 - 33.79 * 1 + 21.12 * glyrschl$ |
| | $= 71.05 + 21.12 * glyrschl$ |
| female = 1 | $glmiles = 71.047519 - 33.79 * 1 + 21.12 * glyrschl$ |
| | $= 37.26 + 21.12 * glyrschl$ |

These results are two regression lines, with the same slope (the coefficient estimate for *glyrschl*) but different intercepts. We need at least two points to plot each of these lines. One point is given by the intercept, but we will use two more typical values for the mother's years of schooling in the data set: 10 and 14.

| | | |
|---|---|---|
| female = 0 | | $glmiles = 71.05 + 21.12 * glyrschl$ |
| | glyrschl = 10 | $glmiles = 71.05 + 21.12 * 10 = 282.25$ |
| | glyrschl = 14 | $glmiles = 71.05 + 21.12 * 14 = 366.73$ |
| female = 1 | | $glmiles = 37.26 + 21.12 * glyrschl$ |
| | glyrschl = 10 | $glmiles = 37.26 + 21.12 * 10 = 248.46$ |
| | glyrschl = 14 | $glmiles = 37.26 + 21.12 * 14 = 332.94$ |

We asked Stata to make additional predictions for us, for the levels of schooling from 8 to 16, and show the plot of the results in Figure 7.1.

This figure reinforces the fact that we are using an additive model. We have not allowed the effect of gender to differ by mother's years of schooling (nor allowed the effect of mother's years of schooling to differ by gender). The slope of each regression line is 21.12, the coefficient estimate for *glyrschl*. And, the distance between the regression lines is –33.79, the coefficient estimate for *female*. We will examine how to allow for interactions in Chapter 8, and we will see that figures are often particularly useful in displaying results when interactions are significant.

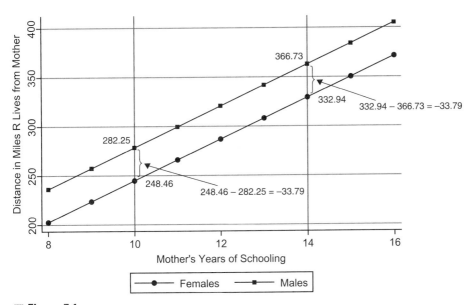

■ **Figure 7.1**

*Source:* National Survey of Families and Households, Wave 1.

We can use techniques similar to those used in Chapters 5 and 6 to evaluate the substantive magnitude of the marginally significant difference in average distances by gender. Because the predictor is a dummy indicator, a one-unit increase signifies a change from the category coded "0" to the category coded "1" (males to females using the *female* variable from the model in Display B.7.12). Rescaling the dummy variable for interpretation thus does not make sense (including dividing by the standard deviation to produce a completely standardized coefficient). But, it is possible to rescale the outcome variable for interpretation purposes. For example, we might divide the coefficient estimate of *female* by the standard deviation of *g1miles* to evaluate the effect size using a semistandardized coefficient estimate. This is directly analogous to Equation 5.9 where the numerator is the difference in group means on the outcome and the denominator is the outcome standard deviation.

Based on `Proc Means` in SAS and `summarize` in Stata we find that for the subsample of 5,401 in our current model, the standard deviation of *g1miles* was 637.5 (nearly the same as the value for the subsample used in Chapters 5 and 6). So, $\hat{\beta}_4/\hat{\sigma}_Y = -33.79/637.5 = -0.053$. Thus, the difference in average miles from the mother by gender is small using Cohen's criteria. And, in the outcome's natural units, an average difference of about 33 miles for men versus women may seem substantively modest as well, at least for private transportation (although perhaps larger than the semistandardized result).

## 7.4: PUTTING IT ALL TOGETHER

We use two literature excerpts to reinforce the concepts learned in this chapter, and connect them back to those learned in Chapter 6.

### 7.4.1: Literature Excerpt 7.2

A 2005 article published in *Sociology of Education* by Kelly Raley, Michelle Frisco, and Elizabeth Wildsmith provides a nice example of testing substantively based hypotheses about a multicategory predictor variable, including the choice of reference category and presentation of tests among included variables. Examining their results helps us to verify our understanding of the interpretation of dummy variable models in a substantive context.

The authors used the NSFH to examine how children's living arrangements influence their educational outcomes. They took advantage of the rich information that parents reported about family transitions since the child's birth, gathered retrospectively in the NSFH, to classify children into five categories of *Family Experiences*:

- those who were always living with two biological or adoptive parents (*two parent*);
- those who were always living with one parent (*always single*);
- those who at some point lived with a step-parent who was married to their biological parent (*married stepfamily*);
- those who at some point lived with a cohabiting partner of their biological parent (*cohabiting stepfamily*);
- those whose parents divorced but never repartnered or remarried (*divorced*).

As they point out, separating out cohabiting stepfamilies is important, given the rapid rise in this family form, but has been relatively neglected in the literature.

The authors anticipate that two mechanisms predict how family type, including cohabiting, will relate to educational success: family resources and family instability. Based on prior literature, they expect that children in cohabiting stepfamilies will have more resources (parental time and money) than children in single-parent families, but less than those in two-parent families. In contrast, they expect that children in cohabiting families will experience more family instability than single-parent and other families (more transitions, include dissolution of the partnership, and more conflict). If the latter mechanism (family instability) outweighs the former mechanism (family resources), then they expect that children in cohabiting families will have poorer educational outcomes than children in single-parent families.

Literature Excerpt 7.2 shows the results of the authors' OLS regression models using these prior family experiences to predict the mothers' reports of the child's current grades (*Grades*) and mothers' expectations for how much schooling the child would complete (*Mother's Educational Expectations*).

---

### ▥ Literature Excerpt 7.2

**Table 3.** OLS Estimates for a Model Predicting Parent's Educational Expectations and Grades at the First Interview

| | Mother's Educational Expectations | | | | Grades | | | |
| | Model 0 | | Model 1 | | Model 0 | | Model 1 | |
| Variables | B | p-value | B | p-value | B | p-value | B | p-value |
|---|---|---|---|---|---|---|---|---|
| *Family Experiences (two parent)* | | | | | | | | |
| Always single | −.85 | .00 | −.17 | .53 | −.55 | .02 | −.33 | .19 |
| Divorced | −.31 | .07 | .02 | .90[a] | −.37 | .02[b] | −.23 | .13[a] |
| Married stepfamily | −.47 | .02 | −.32 | .09 | −.22 | .21 | −.24 | .17[a] |
| Cohabiting stepfamily | −.64 | < .01 | −.40 | .01 | −.72 | < .01 | −.68 | < .01 |
| *Race-Ethnicity (non-Hispanic white)* | | | | | | | | |
| Black | | | .37 | .02 | | | .29 | .05 |
| Hispanic | | | .31 | .11 | | | .39 | .03 |
| Other race | | | .34 | .37 | | | .68 | .05 |
| Female | | | .19 | .05 | | | .50 | < .01 |
| Grade level | | | −.01 | .60 | | | −.06 | .01 |
| Not in school | | | −1.77 | < .01 | | | | |
| *Mother's Education (less than high school)* | | | | | | | | |
| High school | | | .77 | < .01 | | | .33 | .01 |
| Some college | | | 1.47 | < .01 | | | .62 | < .01 |
| College graduate | | | 2.10 | < .01 | | | 1.18 | < .01 |
| Missing | | | 1.48 | .02 | | | .10 | .86 |
| *Income (lowest quintile)* | | | | | | | | |
| 20–39% | | | .27 | .14 | | | .17 | .30 |
| 40–59% | | | .35 | .07 | | | .36 | .05 |
| 60–79% | | | .66 | < .01 | | | .37 | .04 |
| 80%+ | | | 1.18 | < .01 | | | .55 | .00 |
| Missing | | | .42 | < .01 | | | .03 | .85 |
| *Unweighted Sample Size* | 1,161 | | 1,161 | | 1,100 | | 1,100 | |
| $R^2$ | | | .02 | .22 | | | .03 | .13 |

*Note:* Analyses are weighted.
[a] Significantly different from cohabiting stepfamilies, $p < .05$.
[b] Significantly different from cohabiting stepfamilies, $p < .10$.

**Source:** Raley, R. Kelly, Michelle L. Frisco, and Elizabeth Wildsmith. 2005. "Maternal Cohabitation and Educational Success." *Sociology of Education*, 78: 155.

Before examining the results, we will first compare the basic layout and contents of this table to the example we examined in Literature Excerpt 6.1. Raley, Frisco, and Wildsmith use the notation $B$ (rather than $b$) for unstandardized coefficient estimates, and do not present standardized coefficients. They also provide the level of the $p$-values (rather than using asterisks to designate significance), and do not provide the standard error or $t$-value.[14] The sample sizes are also listed for each model within the table (rather than in the title). (In Chapter 12, we briefly introduce the concept of weighting, which the authors refer to by the sample size and in their first table note). The contents of the table in Literature Excerpt 7.2 also shows the $R$-squared, but not the overall $F$-test.

For some details needed in substantive interpretation, we looked within the text of the article. There we saw that the child's *Grades* is reported by mothers on a 9-point scale, from 1 = mostly *F*s to 9 = mostly As. The variable averages between 7 and 7.3 across the five family experiences' categories. *Mother's Educational Expectations* for the child ranges from 1 to 7, with 1 indicating not finishing high school and 7 indicating completing a master's or doctorate degree. This variable has an average of 4 to 4.9 across the five family experiences' categories (pp. 149–50, 152).[15]

As just described, the authors have five categories of family experience. They explicitly defined these categories to be "mutually exhaustive and mutually exclusive" (p. 150). Thus, there are $c = 5$ categories on their original family experiences variable and they can include $c = 5 - 1 = 4$ dummy variables to indicate these categories. They tell us how they chose their reference category:

> Following common practice, we present models with two-parent families as the reference group. However, because our hypotheses are predictions about how children who experience maternal cohabitation compare to others, we note when cohabiting stepfamilies are significantly different from the other groups (p. 153).

In their Table 3 (see again Literature Excerpt 7.2), the reference category is indicated by the word "*two parent*" next to *Family Experiences*. Looking at the other variables listed in their table, we can see that they similarly indicate the reference category for *Race-Ethnicity* (*non-Hispanic white*), *Mother's Education* (*less than high school*), and *Income* (*lowest quintile*). The indentation of their row headings helps us to distinguish the levels of these multicategory variables from other variables. They also include single dummy variables to indicate gender (*female*) and school enrollment status (*not in school*). As is often done in articles, they do not explicitly note the reference category for these single dummy variables, but by aligning them on the left, we can easily recognize that these represent original variables that have two categories.

Because the authors are particularly interested in contrasts with children living in cohabiting families, they use a superscript letter to indicate significant differences between the coefficients for cohabiting stepfamilies and the coefficients for the other included categories (*Always single, Divorced*, and *Married stepfamily*). As the notes to the authors' table describe, the superscript *a* indicates *p*-values less than .05 for these contrasts. The superscript *b* indicates *p*-values less than .10 for these contrasts.[16]

For each of their outcome variables, the authors present two models. Model 0 includes only the set of four *Family Experiences* dummies. Model 1 adds variables that they treat as controls (race-ethnicity, gender, grade level, school enrollment status, maternal education, and income). Because our interest in this chapter is in understanding the interpretation of the dummy variables, we will examine just one of the models (Model 1) in order to simplify the discussion. The estimates for the set of dummy coefficients are circled in red for these models in Literature Excerpt 7.2. We will discuss further in Chapter 10 how we could compare the results of the two models to see how the coefficient estimates for the dummy variables change when the control variables are added.

For *Mother's Educational Expectations* in Model 1, the *p*-values indicate that only one of the contrasts with the reference category (two-parent families) is significant: mothers who had cohabited with a partner expected their child to complete less education (.40 of a point less), on average, controlling for race-ethnicity, gender, grade level, school enrollment status, maternal education, and income. We conclude that this difference is significant because the *p*-value for this dummy variable (0.01) is less than an alpha of 0.05.

The superscript *a* for the *Divorced* variable indicates that mothers also reported lower educational expectations for children when they had cohabited with a partner than when they had divorced but not repartnered, on average with controls. The coefficient estimate for the contrast indicated by the superscript *a* is not explicitly shown in the table, but we can figure it out based on the values that are shown. The coefficient estimate for *Divorced* of 0.02 is the difference in means between the *Divorced* category and the reference category, *Two-Parent* families; that is, $\hat{Y}_{\text{Divorced}} - \hat{Y}_{\text{TwoParent}} = 0.02$. Similarly the *B* coefficient estimate for *Cohabiting Stepfamily* of $-0.40$ is the difference in means between the *Cohabiting Stepfamily* category and the reference category, *Two Parent* families; that is, $\hat{Y}_{\text{CohabitingStep}} - \hat{Y}_{\text{TwoParent}} = -0.40$. Thus, the difference between Cohabiting Stepfamilies and Divorced families is:

$$(\hat{Y}_{\text{CohabitingStep}} - \hat{Y}_{\text{TwoParent}}) - (\hat{Y}_{\text{Divorced}} - \hat{Y}_{\text{TwoParent}}) = -0.40 - (0.02) = -0.42$$

This difference between cohabiting and divorced families of $-0.42$ is slightly larger than the difference between cohabiting and two-parent families of $-0.40$.

Turning to the results for *Grades* in Model 1, we again see that only one of contrasts with the reference category (two-parent families) is significant: mothers reported lower grades for children when they had cohabited (the magnitude of the coefficient estimate is nearly 0.70 of a point on the scale). In addition, the superscript $a$'s indicate that children in cohabiting families have significantly lower mother-reported grades than children in divorced families and married stepfamilies, on average with controls (differences of $[-0.68] - [-0.24] = -0.44$ and $[-0.68] - [-0.23] = -0.45$).

As the authors cover in detail in their discussion, there are many cautions to interpreting these results causally, even with their set of controls. Nonetheless, their results have important implications and appropriately using dummy variable models helped them to reveal these findings. Many studies have not separated cohabiting families from other families (sometimes because information was not collected to identify such families). The authors took advantage of the NSFH design (oversampling cohabiting families) in order to do so. The results indicate that this distinction is important, because children in cohabiting families often have lower education-related outcomes than children in other family types.[17] As the authors conclude: "Research that does not distinguish among different forms of unmarried-mother families is likely to assign the negative effects of cohabitation to experience in a single-parent family" (p. 158). The authors also conclude that the results "suggest that family instability has negative effects on educational outcomes over and above the negative effects that are due to the lower resources available to children in cohabiting stepfamilies" (p. 158), although they could not examine these mechanisms directly. (We will discuss how to do so in Chapter 10.) These results point to directions for future research (quantitative, qualitative, and mixed methods), including the importance of using designs that allow for separating out cohabiting families and for exploring the mechanisms through which children's educational outcomes may be limited by the composition of parents in their family.

### 7.4.2: Literature Excerpt 7.3

Our second literature excerpt, published in the journal *Public Health Reports*, provides an example of presenting confidence intervals for regression coefficients, a practice that is standard in public health but less common in social science journals. The authors, Leo Morales, Peter Guitierrez, and Jose Escarce were interested in extending the established finding that Hispanic children have higher blood lead levels than nonHispanic children by exploring variation in blood lead levels among Hispanic children.

Literature Excerpt 7.3 shows Table 3 from the article which presents point estimates and confidence intervals from two multiple regression models predicting children's blood lead levels. Let's first examine the similarities and differences between this table's layout and contents and those of Literature Excerpt 7.2 and Literature Excerpt 6.1. Most salient is that the authors present the point estimate and confidence interval, but not the standard error nor $t$-ratio, within the table. The notes tell us that all significance levels are indicated

▨ **Literature Excerpt 7.3**

Table 3. Continuous measure of lead among Mexican-American youth 1 to 17 years of age

| | Blood lead levels μg/dl (β [95% CII]) | | | |
|---|---|---|---|---|
| | Model 1 | | Model 2 | |
| Sex | | | | |
| Male | 0.63 | [0.41, 0.85][a] | 0.58 | [0.37, 0.80][a] |
| Female | | — | | — |
| Age | | | | |
| 1 to 4 | 1.82 | [1.43, 2.20][a] | 1.76 | [1.38, 2.13][a] |
| 5 to 11 | 0.78 | [0.53, 1.03][a] | 0.81 | [0.52, 1.09][a] |
| 12 to 17 | | — | | — |
| Generational status | | | | |
| First | 0.84 | [0.30, 1.38][a] | 0.66 | [0.99, 1.22][b] |
| Second | 0.08 | [−0.31, 0.47] | 0.00 | [−0.42, 0.42] |
| Third or higher | | — | | — |
| Home language | | | | |
| Spanish | 0.79 | [0.28, 1.30][a] | 0.83 | [0.28, 1.37][a] |
| Spanish and English | 0.46 | [−0.17, 1.10] | 0.47 | [−0.16, 1.09] |
| English | | — | | |
| Family income (poverty-income ratio) | | | | |
| <50% | 1.18 | [0.60, 1.77][a] | 1.19 | [0.58, 1.81][a] |
| 50% to 100% | 0.96 | [0.39, 1.52][a] | 0.85 | [0.29, 1.41][a] |
| 100% to 200% | 0.50 | [0.13, 0.87][b] | 0.34 | [0.02, 0.67][b] |
| 200% or more | | — | | — |
| Educational attainment of household head (years) | | | | |
| 6 or less | 0.62 | [0.14, 1.09][b] | 0.57 | [0.47, 1.10][b] |
| 7 to 12 | 0.39 | [0.04, 0.74][b] | 0.28 | [−0.10, 0.66] |
| 13 or more | | — | | — |
| Age of housing | | | | |
| Built before 1946 | | | 0.77 | [0.17, 1.36][b] |
| Built 1946 to 1973 | | | 0.51 | [0.08, 0.95][b] |
| Built 1974 or later | | | | — |
| Drinking water | | | | |
| Tap | | | 0.48 | [0.13, 0.82][a] |
| Well | | | 0.50 | [−0.69, 1.68] |
| Bottled | | | | — |

*Note:* Regressions also include indicators for region.
[a] *p* < 0.01
[b] *p* < 0.05
CI − confidence interval

**Source:** Morales, Leo S., Peter Guitierrez, and Jose J. Escarce. 2005. "Demographic and Socioeconomic Factors Associated with Blood Lead Levels among American Children and Adolescents in the United States." *Public Health Reports*, 120(4): 448–454.

with superscript letters (*a* for $p < .01$ and *b* for $p < .05$). Scanning the results, we can see that the only cases which do not have a superscript letter are confidence intervals that include zero, as expected (i.e., second-generational status; Spanish and English home language; educational attainment of 7 to 12 in Model 2; and well drinking water).

All of the predictor variables are coded with dummy indicators of two or more categories. The authors explicitly list each reference category in the table. For example, the first variable is the child's sex. *Male* is the included category and *Female* the reference category. In Model 2, the point estimate tells us that males have average blood levels that are 0.58 ug/dL higher than females, controlling for the other variables in the model. The confidence interval tells us that males have blood levels that are between 0.37 and 0.80 ug/dL higher than females, controlling for the other variables in the model.

An example of a predictor variable with more than two categories is *Age of housing* which indicates structures *Built before 1946* and structures *Built 1946 to 1973* versus those *Built 1974 or later*, the reference category. The first point estimate tells us that children who live in homes built before 1946 have blood levels that are 0.77 ug/dL higher, on average, than children living in homes built after 1973, controlling for the other variables in the model. The first confidence interval tells us that children who live in homes built before 1946 have blood levels that are between 0.17 and 1.36 ug/dL higher than children living in homes built after 1973, controlling for the other variables in the model. The second point estimate tells us that children who live in homes built between 1946 and 1973 have blood levels that are 0.51 ug/dL higher, on average, than children living in homes built after 1973, controlling for the other variables in the model. The second confidence interval tells us that children who live in homes built between 1946 and 1973 have blood levels that are between 0.08 and 0.95 ug/dL higher than children living in homes built after 1973, controlling for the other variables in the model.

The results for the age of housing are estimated with less precision than the effect of gender, something made clear by the confidence intervals. Thus, although the point estimate for *Built before 1946* is larger than the point estimate for gender, the range of the confidence intervals is wider.

How large are these results, substantively? Looking to the Method section of the article tells us that blood levels are measured in micrograms per deciliter. Although the mean and standard deviation are not indicated, the authors report the range is 0.07 to 71.8 and that 10 was the CDC threshold for intervention at the time of publication (p. 449). Thus, the difference between boys and girls is about one twentieth of the intervention threshold.[18] The confidence interval bounds range about one third less than and larger than this point estimate. For age of housing, whereas the point estimates are close to (or slightly larger than) the point estimate for gender, they range from much smaller (as small as one one-hundredth of the CDC threshold) to much larger (over one tenth of the CDC threshold).[19]

## 7.5: SUMMARY

Dummy variables allow us to use nominal and ordinal variables as predictors in regression models. Each dummy variable indicates one category of the original variable. If the original variable has $c$ categories, then we can include $c - 1$ of these dummies in the regression model. The coefficient for each dummy variable captures the difference in means on the outcome variable between the indicated category and this reference category, adjusting for any other variables in the model.

When an original variable has more than two categories, we may be interested in testing hypotheses about the differences in means between pairs of the included categories. We can use one of three approaches to do so:

(a) re-estimating the regression model with a different reference category;
(b) using a partial $F$-test;
(c) testing the significance of a linear combination of coefficients.

These approaches provide equivalent results for two-sided hypothesis tests. Understanding each of them helps us to learn concepts and understand the literature, but we will use only one in practice. The second approach is easy to implement in SAS and Stata, although it does not calculate the difference in means between the included categories, nor the standard error of these differences. The first and third approaches calculate these values, in addition to the test statistics and $p$-values, and allow for directional (one-sided) tests. The first approach can be cumbersome, but is not unwieldy for small samples and relatively simple models. But, we recommend using the third approach (if you have access to Stata) because it produces the simplest and most complete results while reinforcing that all information is contained in a single estimation of the model.

## KEY TERMS

Dummy Term (also Indicator Term)

Dummy Variable (also Indicator Variable)

Included Category

Linear Combination

Reference Category (also Omitted Category, Excluded Category)

Standard Error of the Difference

Variance–Covariance Matrix of the Estimated Regression Coefficients

## REVIEW QUESTIONS

**7.1.** Conceptually, how do you define a set of dummy variables based on an original nominal or ordinal variable?

**7.2.** Why is one dummy variable excluded when you estimate a regression model? How do you choose this reference category?

**7.3.** In the case of an original variable with two categories and an original variable with three categories, how would the intercept and dummy variable coefficient(s) change if you changed the reference category?

**7.4.** What are the three approaches to testing the differences in means among the included groups?

**7.5.** Explain how the interpretation of the dummy coefficients relates to the difference between the predictions (the Y-hats) in a model that contains only dummy variables.

## REVIEW EXERCISES

**7.1.** Suppose you estimated the following regression equation:

$$\widehat{EARNINGS} = 32{,}000 - 12{,}000 * FEMALE$$

where EARNINGS is a respondent's annual earnings and FEMALE is a dummy variable coded 1 for women and 0 for men.

(a) How can you interpret the intercept and dummy coefficient from this model?

(b) What would the new prediction equation be if you regressed earnings on a new dummy variable, *MALE*, coded 1 for men and 0 for women?

**7.2.** Suppose you have sample data in which average household income was $50,000 per year for nonHispanic white adults, $34,000 per year for nonHispanic Black adults, and $40,000 for Hispanic adults. Write the prediction equations (including the intercept and slope) that you would see if you estimated separately each of the following regression models:

(a) Regress household income on two dummy variables indicating nonHispanic white and nonHispanic Black adults, with Hispanic adults as the reference category.

(b) Regress household income on two dummy variables indicating nonHispanic white and Hispanic adults, with nonHispanic Black adults as the reference category.

(c) Regress household income on two dummy variables indicating nonHispanic Black and Hispanic adults, with nonHispanic white adults as the reference category.

---

**7.3.** Suppose that you obtain a new software package and want to verify that you are correctly using its regression command. You know that in your data set the average number of days that low birthweight newborns stay in the hospital (NUMDAYS) is 15 for white babies and 30 for Latino babies. Write the prediction equations (including the intercept and slope) that you should obtain from the software package if you estimated separately each of the two following bivariate regression models:

(a) regress *NUMDAYS* on a dummy variable *WHITE*, coded "1" for white babies and "0" for Latino babies.

(b) regress *NUMDAYS* on a dummy variable *LATINO*, coded "1" for Latino babies and "0" for white babies.

---

**7.4.** Refer to Literature Excerpt 7.1. The authors present the items for the outcome variable, Sense of Control, in the Appendix (p. 117). Sense of Control is the average of eight items in which the respondent claims or denies control over good and bad outcomes. The average ranges from −2 to 2. The mean (standard deviation) of the outcome are 0.52 (0.51) for Asians and 0.68 (0.50) for nonAsians.

(a) Interpret the *R*-squared value.

(b) The article indicates that "marital status is coded (1) for respondents who are currently married or living together as married, and (0) for the divorced separated, widowed, or never married" (p. 107). Write the null and alternative hypotheses being tested for this Married variable.

(c) Interpret the unstandardized coefficient estimate for the Married variable.

(d) Calculate a semistandardized estimate for the Married variable, using the standard deviation of the outcome variable. (Because the authors only provided the standard deviation within the Asian and nonAsian subgroups, you can choose one of the group's standard deviations or calculate the semistandardized estimate once using each standard deviation.)

(e) Calculate and interpret a 95 percent confidence interval for the Married Variable using the unstandardized coefficients.

**7.5.** Refer to Model 2 of Literature Excerpt 7.3.

  (a) Write the null and alternative hypotheses for each of the "Educational attainment of household head" dummy variables. Make a conclusion based on the presented results.

  (b) Interpret the point estimates for each of the "Educational attainment of household head" dummy variables.

  (c) Interpret the confidence intervals for each of the "Educational attainment of household head" dummy variables.

---

**7.6.** Below are SAS results from a hypothetical data set using the following variables:

|          |                                                            |
|----------|------------------------------------------------------------|
| WAGE     | Outcome variable ranging from $2/hour to $15/hour          |
| EDUC     | Education level ranging from 8 to 13                       |
| UNION    | Dummy variable (1 = in a union job; 0 = not in a union job) |
| JOBCLUB  | Dummy variable (1 = got job from job club; 0 = didn't get job from job club) |

| Analysis of Variance | | | | | |
|------------------|-----|----------------|-------------|---------|--------|
| Source           | DF  | Sum of Squares | Mean Square | F Value | Pr > F |
| Model            | 3   | 1881.98291     | 627.32764   | 145.71  | <.0001 |
| Error            | 996 | 4287.98767     | 4.30521     |         |        |
| Corrected Total  | 999 | 6169.97058     |             |         |        |

| | | | | |
|----------------|---------|----------|--------|
| Root MSE       | 2.07490 | R-Square | 0.3050 |
| Dependent Mean | 8.20074 | Adj R-Sq | 0.3029 |

| Parameter Estimates | | | | | |
|-----------|----|-----------------------|----------------|---------|----------|
| Variable  | DF | Parameter Estimate    | Standard Error | t Value | Pr > \|t\| |
| Intercept | 1  | −0.97396              | 0.47246        | −2.06   | 0.0395   |
| educ      | 1  | 0.83490               | 0.04462        | 18.71   | <.0001   |
| jobclub   | 1  | 0.65761               | 0.15841        | 4.15    | <.0001   |
| union     | 1  | 1.19360               | 0.15553        | 7.67    | <.0001   |

a) Write the prediction equation based on this model.

b) Interpret the point estimate for the *jobclub* variable.

c) Interpret the point estimate for the *union* variable.

d) Calculate and interpret the confidence interval for the *jobclub* variable.

e) Calculate and interpret the confidence interval for the *union* variable.

## CHAPTER EXERCISE

CHAPTER
EXERCISE
**7**

In this exercise, you will write a SAS and Stata batch program to estimate a multiple regression model, with dummy variables, building on the batch programs you wrote for Chapters 5 and 6.

Again, use the NOS 1996 data set with an *if expression* to only keep *the organizations that were founded after 1950* and that *do not have missing values on the mngpctfem, age, and ftsize variables*.

Before you begin the SAS and Stata tasks, use the `codebook` and `proc freq` commands to get the distribution of *a*5 and *a*16.

**Note:** In all of the questions you can use *either* explicit coding of each level of the dummy variable using multiple lines of commands or the succinct dummy variable syntax. And, be careful to ensure that cases with missing values on the original variables *a*5 and *a*16 variables are '.' missing on your dummy variables. (Note that you will need to look back to your answers to the Chapter Exercise in Chapter 3 to identify the original missing variable codes for these variables.)

### 7.1: Two-Category Original Variable

(a) SAS/Stata tasks

  (i) Create a dummy variable to indicate organizations that *are* required to report the demographic composition of their workforce to the government, based on the original *a*5 variable. (Be careful to account for missing value codes.)

  (ii) Regress *mngpctfem* on this dummy variable (we will refer to this as Regression #1 in the Write-up tasks).

  (iii) Create another dummy variable to indicate organizations that are not required to report the demographic composition of their workforce to the government, based on the original *a*5 variable. (Be careful to account for missing value codes.)

(iv) Regress *mngpctfem* on this dummy variable (we will refer to this as Regression #2 in the Write-up tasks).

(b) Write-up tasks

(i) Interpret the point estimates of the intercept and slope in Regression #1.

(ii) Interpret the point estimates of the intercept and slope in Regression #2.

(iii) Discuss how Regression #1 and Regression #2 are related to each other (e.g., show how you could have figured out the intercept and slope for Regression #2 based only on Regression #1).

## 7.2: MultiCategory Original Variable

(a) SAS/Stata tasks

(i) Create four dummy variables to represent the four levels of the *a*16 variable. (Be careful to account for missing value codes.)

(ii) Regress *mngpctfem* on all four dummy variables (we will refer to this as Regression #3 in the Write-up tasks).

(iii) Regress *mngpctfem* on three of the four dummy variables, using organizations whose type of work is only making a product (a16 = 1) as the reference category (we will refer to this as Regression #4).

(1) Use the test command to test whether there are any differences in the mean percentage of managers who are females among the four types of work category (we will refer to this as Test #1 in the Write-up tasks).

(2) Use the test command to test whether the means differ between each pair of the three categories of a16 included in the regression model (three tests all together; we will refer to this as Test #2 in the Write-up tasks).

(iv) Regress *mngpctfem* on three of the four dummy variables, using organizations whose type of work is only making a service (a16=2) as the reference category (refer to this as Regression #5 in the Write-up tasks).

(1) Use the Stata lincom command to test whether the means differ between each pair of the three categories of a16 included in the regression model (three tests all together; we will refer to this as Test #3 in the Write-up tasks).

(b) Write-up tasks

(i) Describe the results in Regression #3. Are they what you expected? How should you pick the reference category in a model with dummy variables?

(ii) Write the null and alternative hypotheses and a conclusion based on the results of the Test #1 command. Is this test shown in the output of any of the regressions (Regression #3, Regression #4, Regression #5)? If it is, why? If not, why not?

(iii) Interpret the point estimates of the intercept and dummy variable coefficients in the default output for Regression #4.

(iv) Interpret the point estimates of the intercept and dummy variable coefficients in the default output for Regression #5.

(v) Indicate which of the tests in Test #2 and Test #3 correspond to which of the tests in the default output of Regression #4 and Regression #5. Be sure to note whether any of the tests in Test #2 and Test #3 are not shown in either set of default output.

## COURSE EXERCISE

COURSE
EXERCISE
7

Using the data set you created in the course exercise to Chapter 4, create a set of dummy variables to indicate one of your categorical predictor variables.

Choose one of the categories as a reference, and regress your continuous outcome variable on the remaining dummy variables. Use the `test` command in SAS and Stata and the `lincom` command in Stata to test the contrasts among the included categories. Before estimating the code, think about whether you could make directional hypotheses, in advance, about any of the pairwise contrasts (i.e., do you have enough theoretical basis or prior empirical evidence to expect that the mean for one group is larger than the mean for another group?). In cases where you can specify a directional hypothesis, use a one-sided hypothesis test.

Write the prediction equation and calculate the predicted value for each category on the original variable by substituting "0"s and "1"s for the dummy variables.

In one sentence, interpret the point estimates of the intercept and each dummy variable coefficient.

Calculate and interpret a 95 percent confidence interval for each dummy variable.

# Chapter 8

# INTERACTIONS

## CHAPTER 8: INTERACTIONS

The models we have considered thus far have all involved additive relationships. The coefficient for one variable is the same, regardless of the level of another variable. In this chapter, we will discuss **interactions** between variables: situations in which the relationship between a predictor variable and the outcome variable differs depending on the level of another predictor variable.

We will begin by looking at interactions between dummy variables, an approach that might address questions such as "Is the gender gap in wages the same for African Americans and whites?" "Are the health benefits that distinguish married from single adults the same for women and men?" "Does the gap in educational attainment between teenage and adult mothers differ depending on their living arrangements (i.e., alone, extended family, married, cohabiting)?"

We will then consider interactions between a dummy variable and an interval variable, an approach that might address questions such as "Are the returns to education (the link between educational attainment and earnings) greater for whites than other race/ethnicity groups?" "Is the association between job satisfaction and complexity of job tasks stronger for people with a college degree than those with a high school degree?" "Is the link between the level of crime in a community and the level of social disorganization in a community weaker in suburban or rural areas than in urban areas?"

We will also present the **Chow Test** to measure the degree to which an entire regression model differs between two or more groups, extending these approaches to the multiple regression context. For example, if we build a model of educational attainment that includes parents' educational attainment, educational aspirations of peers in high school, and a youth's score on a standard achievement test in high school as predictors, are all of the coefficients in the model the same for boys and girls, or do some differ by gender?

We will end by considering interactions between two interval variables. This approach might be used for hypotheses such as "Job stress is associated with harsher parenting, but this association weakens with each additional increment of social support received from family and friends" or "Living in a community where adults have higher average educational attainment relates to the educational aspirations of youth, but more strongly when a higher proportion of neighborhood adults are the same race/ethnicity as the youth."

Interactions require more steps for interpretation than do additive models, and graphs are often useful in presenting the results succinctly. For each type of interaction, we will begin by using algebra to understand the interpretation of conditional regression equations. We separately examine three types of interaction (dummy by dummy variable, dummy by interval variable, and interval by interval variable) to facilitate understanding, although we will see that there are many similarities in specification and interpretation across these types. We will discuss how to estimate the models in SAS and Stata and how to graph the prediction equations and present the results using Excel.[1] We will use our hours of chores example, but we will include men as well as women in the sample so that we can consider gender differences in the overall number of hours spent on chores and in the effects of marital status, hours of paid work, and number of children.[2]

We will see that each type of interaction uses a product term between variables to capture the interaction, and that we can build on what we have learned about the general linear $F$-test and linear combinations to calculate conditional effects when interpreting the models. We introduce the concept of centering variables to aid in interpretation of models involving one or more interval predictors. We will also look at two literature excerpts to help us to understand why and how interactions are used in research, and how graphing the results can facilitate interpretation (see also Jaccard and Turrisi 2003 for a nice, detailed treatment of interactions in multiple regression).

## 8.1: LITERATURE EXCERPT 8.1

Peter Marsden, Arne Kalleberg, and Cynthia Cook published a study in the journal *Work and Occupations* in 1993 that examined gender differences in organizational commitment. The authors hypothesized not only that there might be an overall average difference in organizational commitment between men and women, captured by a dummy variable, but also why such differences might exist. They further speculated that the constructs that explain organizational commitment might differ for men and women. As they note:

> The final analyses that we report here examine the possibility that there may be gender differences in the processes leading to organizational commitment. If family roles compete more strongly with work roles among women than among men, for example, then we should expect some interactions of such variables with gender in their effects on commitment (p. 382).

They also expect gender differences in effects because prior research has found "the way in which age, autonomy, and occupational status are associated with job involvement differs between men and women" (p. 382).

The authors used the 1991 General Social Survey to examine their hypotheses. This nationally representative survey was an improvement over prior studies of gender differences in organizational commitment which focused on a single or small number of organizations, making it difficult to know the generalizability of the results. The outcome of organizational commitment was captured with six questions such as "I am willing to work harder than I have to in order to help this organization succeed" and "I find that my values and the organization's are similar" rated from 1 = *strongly disagree* to 4 = *strongly agree*. The sum of the organizational commitment items averaged 2.87 with a standard deviation of 0.54.

The authors measure 15 aspects of jobs and families that might predict organizational commitment, as well as two controls (race-ethnicity and education). They estimate regression models that allow them to examine how each of these 17 variables predicts organizational commitment for men and for women, and whether the effects of these 17 variables differ between men and women. Literature Excerpt 8.1 shows the results from their Table 5. They use a general linear $F$-test and find "at most, weak evidence of differences between men and women" across the 17 variables (The $F$-statistic reported in the table notes of 1.218 with 17 and 699 degrees of freedom and $p > 0.10$; see value circled in red in the Literature Excerpt).

Although most of the 17 variables have statistically equivalent effects for men and women, the authors use asterisks in the middle column of the table to indicate that the coefficient estimates for these three variables do differ significantly for men and women (see black circles in the Literature Excerpt). In each of these three cases, the predictor is associated with the outcome only for men and not for women (see red circles in Literature Excerpt).

▨ **Literature Excerpt 8.1**

Table 5:  Gender-Specific Regressions for Organizational Commitment (All Employed Respondents)

| Explanatory Variables | Regression Coefficients | | |
|---|---|---|---|
| | Women | | Men |
| Work position | | | |
| Position in authority hierarchy | .015 | | .075** |
| Autonomy | .159** | | .135** |
| Perceived quality workplace relations | .191** | | .142** |
| Promotion procedures (dummy) | .072 | | .035 |
| Nonmerit reward criteria | −.010 | *** | −.135** |
| Workplace size (log) | .002 | | −.012 |
| Self-employed (dummy) | .314** | | .273** |
| Career experiences | | | |
| Years with employer | .018 | | .012 |
| Advances with this employer | .042 | | .053 |
| Hours worked last week (or typical) | .001 | | .000 |
| Compensation | | | |
| Annual earnings (log) | .023 | | −.040 |
| Number of fringe benefits | .011 | | .026* |
| Family affiliations | | | |
| Currently married (dummy) | −.011 | *** | .145** |
| Number of persons aged 12 or less in household | .027 | *** | −.050† |
| Frequency of job-home conflict | −.056* | | −.009 |
| Sociodemographic controls | | | |
| White (dummy) | .044 | | −.120† |
| Years education | −.009 | | .004 |
| Constant | 1.450** | | 1.704** |
| $R^2$ | .355 | | .385 |
| $N$ | 369 | | 366 |

*Note:* The gender-specific equations presented here are derived from an equation that includes interaction terms between gender and all other variables; $R^2$ for that equation is .364. $F$ statistic for test of the hypothesis that there are no gender differences between equations is 1.218 on 17 and 699 degrees of freedom, $p > .10$.
† $p < .10$; * $p < .05$; ** $p < .01$; *** $t$ statistic for gender difference in coefficients exceeds 2.0.

**Source:** Marsden, Peter V., Arne L. Kalleberg, and Cynthia R. Cook. 1993. "Gender Differences in Organizational Commitment: Influences of Work Positions and Family Roles." *Work and Occupations,* 20, 368–90.

■  For *Nonmerit reward criteria,* the association is significantly negative for men (−.135) but insignificant for women (−.010). Thus, if men perceive that nonmerit criteria (race, gender, and "favored relationship with the boss") are used in decisions about raises and promotions, then they report less organizational commitment (p. 385). Among women, perceptions of nonmerit criteria are not associated with organizational commitment.

- For the dummy indicator of current marital status, men report higher organizational commitment when they are married (.145) but marital status is not associated with organizational commitment for women (–.011).
- For school-age children, aged 12 or less, in the household, the association is marginally negative for men (–0.050) but not significant for women (0.027). Thus, men tend to report less organizational commitment when there are more children in the household, but number of children is not associated with organizational commitment for women.

We will learn below how the authors estimated this model, and how they are able to interpret the results as they do, as well as additional strategies for presenting results from models that include interactions.

## 8.2: INTERACTIONS BETWEEN TWO DUMMY VARIABLES

In Chapter 7, we considered additive models that included two sets of dummy variables constructed based on the levels of two different categorical variables. We will begin examining interactions by considering interactions between such dummy variables.

### 8.2.1: Review of Additive Model

We start with the additive model, so we can easily relate its results to the interaction model. We will predict hours spent on chores by indicators of being female and being married. We will begin with a simple conceptualization, and then further complicate it below. For gender, we might initially consider various social processes, including socialization practices, through which women learn how to accomplish housekeeping tasks and take responsibility for keeping an orderly home, resulting in adult women reporting spending more time each week on household chores than adult men. For marital status, we might initially expect that people who are married versus single reside in households with more need for chores (for example, the laundry to be done, amount of food to prepare, amount of shopping to be done, etc., all increase with one more person in the household).

We will specify this model as follows:

$$Y_i = \beta_0 + \beta_1 D_{1i} + \beta_2 D_{2i} + \varepsilon_i$$

where $D_1$ is a dummy variable *married* coded 1 = married and 0 = unmarried and $D_2$ is a dummy variable *female* coded 1 = female and 0 = male and $Y_i$ is *hrchores*, the hours per week spent on chores. As we learned in Chapter 7, the intercept in this model provides the average number of hours per week spent on chores for men who do not have a spouse (the reference category on each dummy). The coefficient estimate for the first dummy indicates how many more or less hours married persons spend on chores, controlling for gender. And, the coefficient estimate for the second dummy indicates how many more or less hours women spend on chores, controlling for having a spouse.

Recall that in this additive model the difference in means between married and single persons is the same regardless of gender and the difference in means between men and women is the same regardless of marital status. In this simple model, with two dummy predictors of two separate variables, there are four possible expected values (conditional means) from the model, shown in Table 8.1.

■ **Table 8.1: Conditional Means with Two Dummy Variables (Additive Model)**

| | | | |
|---|---|---|---|
| Unmarried Men | $E(Y\|D_1 = 0, D_2 = 0)$ | $= \beta_0 + \beta_1 * 0 + \beta_2 * 0$ | $= \beta_0$ |
| Married Men | $E(Y\|D_1 = 1, D_2 = 0)$ | $= \beta_0 + \beta_1 * 1 + \beta_2 * 0$ | $= \beta_0 + \beta_1$ |
| Unmarried Women | $E(Y\|D_1 = 0, D_2 = 1)$ | $= \beta_0 + \beta_1 * 0 + \beta_2 * 1$ | $= \beta_0 + \beta_2$ |
| Married Women | $E(Y\|D_1 = 1, D_2 = 1)$ | $= \beta_0 + \beta_1 * 1 + \beta_2 * 1$ | $= \beta_0 + \beta_1 + \beta_2$ |

Thus, the effect of being married for men is $\beta_1$ (difference between row 2 and row 1). And, the effect of being married is also $\beta_1$ for women (difference between row 4 and row 3). If we compare row 3 to row 1 and row 4 to row 2, we similarly see that the effect of being female is the same among single persons as it is among married persons.

We can also see the constant effect of one variable within levels of the other variable by writing the conditional regression equations where we hold gender constant (at either $0 = male$ or $1 = female$) and allow marital status to vary (between $0 =$ unmarried and $1 =$ married).

■ **Table 8.2: Conditional Regression Equation for Effect of Being Married Within Gender (Additive Model)**

| | | | |
|---|---|---|---|
| Men | $E(Y\|D_1, D_2 = 0)$ | $= \beta_0 + \beta_1 * D_{1i} + \beta_2 * 0$ | $= \beta_0 + \beta_1 D_{1i}$ |
| Women | $E(Y\|D_1, D_2 = 1)$ | $= \beta_0 + \beta_1 * D_{1i} + \beta_2 * 1$ | $= (\beta_0 + \beta_2) + \beta_1 D_{1i}$ |

These results show that the effect of being married ($D_1$) is the same for men and for women ($\beta_1$ in both cases). And, the interpretations of the conditional intercepts match the conditional means in Table 8.1. In each conditional regression equation, the intercept is the mean for people coded "0" on $D_{1i}$, that is unmarried people. So, the intercept for men in Table 8.2 matches the conditional mean in row 1 of Table 8.1 (unmarried men). And, the intercept for women in Table 8.2 matches the conditional mean in row 3 of Table 8.1 (unmarried women).

We can similarly write the conditional regression equations where we hold marital status constant (at 0 = unmarried or 1 = married) and allow gender to vary (between 0 = *male* and 1 = *female*).

**■ Table 8.3: Conditional Regression Equation for Effect of being Female within Marital Status (Additive Model)**

| Unmarried | $E(Y|D_1 = 0, D_2)$ | $= \beta_0 + \beta_1 * 0 + \beta_2 * D_{2i}$ | $= \beta_0 + \beta_2 D_{2i}$ |
|-----------|---------------------|-----------------------------------------------|------------------------------|
| Married   | $E(Y|D_1 = 1, D_2)$ | $= \beta_0 + \beta_1 * 1 + \beta_2 * D_{2i}$ | $= (\beta_0 + \beta_1) + \beta_2 D_{2i}$ |

These results show that the effect of being female ($D_2$) is the same for unmarried and married persons ($\beta_2$ in both cases). And, the interpretations of the conditional intercepts match the conditional means in Table 8.1. In Table 8.3, the intercept is the mean for people coded "0" on $D_{2i}$, that is, men. So, the intercept for unmarried persons in Table 8.3 matches the conditional mean in row 1 of Table 8.1 (unmarried men). And, the intercept for married persons in Table 8.3 matches the conditional mean in row 2 of Table 8.1 (married men).

## 8.2.2: Introduction of Interaction Model

Restricting the effect of each dummy to be the same within levels of the other variable is frequently inconsistent with theory and prior research. In our example, we might hypothesize that there is an interaction between marriage and gender in predicting hours spent on chores. As we noted above, a household with a married couple has greater chore demands (e.g., more laundry, cooking, etc.) than would a household comprised by either partner in that couple alone. But, there are also two people to accomplish those chores, and the total time spent on chores across the two people may be less than each would spend in their own households owing to shared tasks. For example, one larger meal can be cooked and shared for less than the time it takes to cook two separate meals. Various conceptual models and prior empirical studies suggest that women and men will not evenly divide such chores in the shared household. Rather, women will do more than men (West and Zimmerman 1987; Grossbard-Shechtman 1993). This suggests that married men should spend less time on chores than unmarried men, but married women should spend more time on chores than unmarried women. And, we should also see that the gender difference in time spent on chores is larger for married than unmarried persons.[3]

In regression models, we test for the presence of such interactions by introducing a **product term** into the regression model. This product term is a variable created by multiplying together the two predictor variables of interest. This product term is added to the regression model along with the two predictor variables of interest.

$$Y_i = \beta_0 + \beta_1 D_{1i} + \beta_2 D_{2i} + \beta_3 D_{1i} D_{2i} + \varepsilon_i$$

We can write the conditional means as we did before, by substituting in the four possible combinations of zeros and ones across the two separate dummy variables. We substitute in the appropriate value (0 or 1) within the product term as well as the individual variables.

■ **Table 8.4: Conditional Means with Two Dummy Variables (Interaction Model)**

| | | | |
|---|---|---|---|
| Unmarried Men | $E(Y \mid D_1 = 0, D_2 = 0)$ | $= \beta_0 + \beta_1 * 0 + \beta_2 * 0 + \beta_3 * 0 * 0$ | $= \beta_0$ |
| Married Men | $E(Y \mid D_1 = 1, D_2 = 0)$ | $= \beta_0 + \beta_1 * 1 + \beta_2 * 0 + \beta_3 * 1 * 0$ | $= \beta_0 + \beta_1$ |
| Unmarried Women | $E(Y \mid D_1 = 0, D_2 = 1)$ | $= \beta_0 + \beta_1 * 0 + \beta_2 * 1 + \beta_3 * 0 * 1$ | $= \beta_0 + \beta_2$ |
| Married Women | $E(Y \mid D_1 = 1, D_2 = 1)$ | $= \beta_0 + \beta_1 * 1 + \beta_2 * 1 + \beta_3 * 1 * 1$ | $= \beta_0 + \beta_1 + \beta_2 + \beta_3$ |

Comparing rows within this table makes clear that the effect of being married is no longer required to be the same for men and women and the effect of being female is no longer required to be the same for persons who are and are not married. For example, the difference in average hours of chores between married and unmarried men is $\beta_1$ but the difference in average hours of chores between married and unmarried women is $\beta_1 + \beta_3$. Similarly, the difference in average hours of chores between unmarried men and unmarried women is $\beta_2$ whereas the difference in average hours of chores between married men and married women is $\beta_2 + \beta_3$.

When we write these results in terms of conditional regression equations, we similarly see that now both the intercept and the effect of one dummy differs, depending on the level of the other dummy. First, holding gender constant at its two levels, and allowing marital status to vary results in the following two conditional regression equations. Notice that we again substitute the appropriate value into the product term for the variable that is held constant.

■ **Table 8.5: Conditional Regression Equation for Effect of being Married within Gender (Interaction Model)**

| | | | |
|---|---|---|---|
| Men | $E(Y \mid D_1, D_2 = 0)$ | $= \beta_0 + \beta_1 * D_{1i} + \beta_2 * 0 + \beta_3 * D_{1i} * 0$ | $= \beta_0 + \beta_1 * D_{1i}$ |
| Women | $E(Y \mid D_1, D_2 = 1)$ | $= \beta_0 + \beta_1 * D_{1i} + \beta_2 * 1 + \beta_3 * D_{1i} * 1$ | $= (\beta_0 + \beta_2) + (\beta_1 + \beta_3) * D_{1i}$ |

These equations let us directly calculate the result we saw, based on comparing the conditional means in Table 8.4. The effect of being married for men is $\beta_1$. The effect of being married for women is $\beta_1 + \beta_3$. The coefficient on the product term, $\beta_3$, captures the difference in the effect of being married between men and women. If $\beta_3 = 0$ then the effect of being married would be the same for men and women.

As in the additive model, the intercepts differ in the two conditional regression equations. In the first equation, $\beta_0$ is the average hours of chores for men who are coded a zero on $D_1$ (are unmarried). In the second equation, $\beta_0 + \beta_2$ is the average hours of chores for women who are coded "0" on $D_1$ (are unmarried). These match the predicted means we calculated above for rows 1 and 3 of Table 8.4.

We can similarly write the conditional regression equations where we hold marital status constant (at 0 = unmarried or 1 = married) and allow gender to vary (between 0 = *male* and 1 = *female*).

**Table 8.6: Conditional Regression Equation for Effect of being Female within Marital Status (Interaction Model)**

| | | | |
|---|---|---|---|
| Unmarried | $E(Y \mid D_1 = 0, D_2)$ | $= \beta_0 + \beta_1 * 0 + \beta_2 * D_{2i} + \beta_3 * 0 * D_{2i}$ | $= \beta_0 + \beta_2 * D_{2i}$ |
| Married | $E(Y \mid D_1 = 1, D_2)$ | $= \beta_0 + \beta_1 * 1 + \beta_2 * D_{2i} + \beta_3 * 1 * D_{2i}$ | $= (\beta_0 + \beta_1) + (\beta_2 + \beta_3) * D_{2i}$ |

These equations also directly calculate the results we saw above by comparing the predicted values from Table 8.4. The effect of being female for unmarried adults is $\beta_2$. The effect of being female for married adults is $\beta_2 + \beta_3$. The coefficient on the product term, $\beta_3$, captures the difference in the effect of being female between unmarried and married adults. If $\beta_3 = 0$, then the effect of being female would be the same for married and unmarried adults.

In terms of the intercept, in the first equation, $\beta_0$ is the average hours of chores for unmarried adults who are coded "0" on $D_2$ (are male). In the second equation, $\beta_0 + \beta_1$ is the average hours of chores for married adults who are coded "0" on $D_2$ (are male). These match the predicted means we calculated above for rows 1 and 2 of Table 8.4.

### 8.2.3: Creating the Product Term in SAS and Stata

We can easily create the product term needed to estimate these models in SAS and Stata using the general syntax for creating a new variable and the multiplication operator. We suggest choosing a name for the product term that signals the variables in the interaction. We recommend still restricting to 12 characters, to improve readability of output, and perhaps using symbols (like the underscore; in Stata, capitalization might also be used). It may be necessary to abbreviate the original variable names to restrict the name of the product term to 12 characters. For example, we could use the name *fem_marr* for the product term between *female* and *married*.

In SAS, we would create the variable with the command `fem_marr=female*married;` in Stata, we would similarly type `generate fem_marr=female*married` (see Box 8.1). The new variable will contain the result of multiplying the values of the two original variables together. Any variable coded '.' `missing` on either of the variables will be coded missing on the product term. For example, we might have in the data file:

| MCASEID | female | married | fem_marr | . . . |
|---------|--------|---------|----------|-------|
| 54638 | 0 | 1 | 0 | . . . |
| 33865 | 0 | 1 | 0 | . . . |
| 76453 | 1 | 0 | 0 | . . . |
| 99857 | 0 | . | . | . . . |
| 26374 | 0 | 0 | 0 | . . . |
| 11948 | 1 | 1 | 1 | . . . |
| 95443 | 1 | 1 | 1 | . . . |
| 66730 | 1 | 0 | 0 | . . . |
| 22647 | . | 1 | . | . . . |
| 88740 | 0 | 0 | 0 | . . . |
| . | . | . | . | . |
| . | . | . | . | . |
| . | . | . | . | . |

■ **Box 8.1**

Remember that in SAS the new variable creation must happen within the Data Step. All variables needed for any regression models estimated anywhere in the batch program, including product terms, must be created in the Data Step. In contrast, in Stata, the new variable creation can occur anywhere in the batch program, including just before the model using that variable. For example, product terms could be created in the line immediately preceding the interaction model. We usually put our variable creation in Stata together at the beginning of the batch program, because we find that it helps to organize our batch program for proofing.

## 8.2.4: Further Understanding the Product Term

It can be helpful for students to think of the product term between two dummy variables as an indicator for something unique about having a particular characteristic on both variables. We allow the mean for people coded "1" on *both* dummies to differ from the mean for people coded "1" on only *one* of the dummies. We can see that by multiplying the two original dummies, the results are exactly equivalent to what we would have created if we had set out to make a dummy to indicate cases coded "1" on both dummies.

|  | Married? | Female? | Married* Female | Married and Female? |
|---|---|---|---|---|
|  | $D_1$ | $D_2$ | $D_1{}^*D_2$ | $D_3$ |
| Not Married/Male | 0 | 0 | 0 | 0 |
| Married/Male | 1 | 0 | 0 | 0 |
| Not Married/Female | 0 | 1 | 0 | 0 |
| Married/Female | 1 | 1 | 1 | 1 |

Now, to calculate the mean we expect for people who are married and female, we cannot simply sum the basic effects of being married and being female captured by $D_1$ and $D_2$. Instead, we need to add in something unique about having both of the characteristics—to being both married and being female. This uniqueness is captured by $D_3$.

The following table replicates the results we saw above, but now using the dummy indicator of being married and female. Although equivalent to what we have already seen, this table may be helpful to students who find it useful to think in terms of the indicator of something unique about people with both of the characteristics (here married and female) as they become familiar with the product term used in interactions.

▒ **Table 8.7: Conditional Means in Model with Indicator of being Married and Female**

| | | | |
|---|---|---|---|
| Unmarried Men | $E(Y \mid D_1 = 0, D_2 = 0, D_3 = 0)$ | $= \beta_0 + \beta_1 * 0 + \beta_2 * 0 + \beta_3 * 0$ | $= \beta_0$ |
| Married Men | $E(Y \mid D_1 = 1, D_2 = 0, D_3 = 0)$ | $= \beta_0 + \beta_1 * 1 + \beta_2 * 0 + \beta_3 * 0$ | $= \beta_0 + \beta_1$ |
| Unmarried Women | $E(Y \mid D_1 = 0, D_2 = 1, D_3 = 0)$ | $= \beta_0 + \beta_1 * 0 + \beta_2 * 1 + \beta_3 * 0$ | $= \beta_0 + \beta_2$ |
| Married Women | $E(Y \mid D_1 = 1, D_2 = 1, D_3 = 1)$ | $= \beta_0 + \beta_1 * 1 + \beta_2 * 1 + \beta_3 * 1$ | $= \beta_0 + \beta_1 + \beta_2 + \beta_3$ |

This coding allows for something different about having both characteristics indicated by the two dummies for the original variable, as opposed to having just one of the characteristics.[4]

### 8.2.5: Example 8.1

Display B.8.1 shows the commands for creating the product term and estimating the interaction model in SAS and Stata. The estimated regression equation is:

$$\hat{Y}_i = 22.31 - 2.60D_{1i} + 10.20D_{2i} + 6.64D_{1i}D_{2i} \tag{8.1}$$

Or, written with our variable names:

$$\widehat{hrchores}_i = 22.31 - 2.60married_i + 10.20female_i + 6.64fem\_marr_i$$

■ **Box 8.2**

Our conceptual models might allow us to specify the direction of the interaction. For example, in our case, we hypothesized that the effect of being female on doing more chores should be accentuated among married persons. This implies that the product term should be positive. In our case, the product term is in fact positive, consistent with our hypothesis, so we could conduct a one-sided hypothesis test by dividing the p-value in half. If the sign on the product term was inconsistent with our hypothesis, then we would subtract half the p-value from one.

The default hypothesis test for *fem_marr* tests whether a significant interaction is present. As we discussed above, if the coefficient on the product term is zero ($\beta_3 = 0$), then the effect of each variable is the same within levels of the other variable. If the coefficient on the product term is not zero ($\beta_3 \neq 0$), then the effect of each variable differs within levels of the other variable (see Box 8.2). In our case, the t-value of 5.29 is greater than the critical t of 1.96 for our large sample, and the p-value is smaller than 0.05, so we can reject the null hypothesis.

Let's use the prediction equation (Equation 8.1) to calculated predicted values (Table 8.8) and the conditional regression equations, and then we will revisit the interpretation of the values in the prediction equation.

Table 8.8 shows that married men are estimated to spend the least amount of time on chores, on average, at about 20 hours per week. Unmarried men spend slightly more time on chores, at about 22 hours per week. Unmarried women spend nearly ten more hours per week on chores, at about 33 hours per week. And, married women spend the most time on chores, at over 36 hours per week.

▨ **Table 8.8: Predicted Values (Y-hats) from Estimated Interaction Model with Two Dummy Variables**

| | | | |
|---|---|---|---|
| Unmarried Men | $(\hat{Y}\,|\,D_1 = 0, D_2 = 0)$ | $= 22.31 - 2.60 * 0 + 10.20 * 0 + 6.64 * 0 * 0$ | 22.31 |
| Married Men | $(\hat{Y}\,|\,D_1 = 1, D_2 = 0)$ | $= 22.31 - 2.60 * 1 + 10.20 * 0 + 6.64 * 1 * 0$ | 19.71 |
| Unmarried Women | $(\hat{Y}\,|\,D_1 = 0, D_2 = 1)$ | $= 22.31 - 2.60 * 0 + 10.20 * 1 + 6.64 * 0 * 1$ | 32.51 |
| Married Women | $(\hat{Y}\,|\,D_1 = 1, D_2 = 1)$ | $= 22.31 - 2.60 * 1 + 10.20 * 1 + 6.64 * 1 * 1$ | 36.55 |

The conditional regression equations allow us to readily see the differences in effects within levels of the other variable, and will allow us to assess the significance of these **conditional effects**. We can again use the prediction equation, Equation 8.1, to hold one variable constant while allowing the other to vary.

**Table 8.9: Estimated Conditional Regression Equation for Effect of being Married, within Gender**

| Men | $(\hat{Y}\mid D_1, D_2 = 0)$ | $= 22.31 - 2.60 * D_{1i} + 10.20 * 0 + 6.64 * D_{1i} * 0$ | $= 22.31 - 2.60 D_{1i}$ |
|---|---|---|---|
| Women | $(\hat{Y}\mid D_1, D_2 = 1)$ | $= 22.31 - 2.60 * D_{1i} + 10.20 * 1 + 6.64 * D_{1i} * 1$ | $= (22.31 + 10.20) + (-2.60 + 6.64) D_{1i}$ |
| | | | $= 32.51 + 4.04 D_{1i}$ |

**Table 8.10: Estimated Conditional Regression Equation for Effect of being Female, within Marital Status**

| Unmarried | $(\hat{Y}\mid D_1 = 0, D_2)$ | $= 22.31 - 2.60 * 0 + 10.20 * D_{2i} + 6.64 * 0 * D_{2i}$ | $= 22.31 + 10.20 D_{2i}$ |
|---|---|---|---|
| Married | $(\hat{Y}\mid D_1 = 1, D_2)$ | $= 22.31 - 2.60 * 1 + 10.20 * D_{2i} + 6.64 * 1 * D_{2i}$ | $= (22.31 - 2.60) + (10.20 + 6.64) * D_{2i}$ |
| | | | $= 19.71 + 16.84 D_{2i}$ |

Notice that we can read the coefficient estimates for one of the conditional regression equations—the equation for the category coded "0" on the other variable—directly from the SAS and Stata output. The other requires us to sum two coefficient estimates.

For the conditional equation that we can read directly off the output (men in Table 8.9; unmarried in Table 8.10), we can immediately see the results of the test of whether that conditional effect differs from zero. For men, the $t$-value for *married* ($D_{1i}$) from Display B.8.1 is $-2.88$, which is larger in magnitude than 1.96, and the associated $p$-value is smaller than 0.05. Thus, for men, we can reject the null hypothesis that the effect of being married is zero. Likewise, for the unmarried the $t$-value for *female* ($D_{2i}$) from Display B.8.1 is 11.15, which is larger than 1.96, and the associated $p$-value is smaller than 0.05. Thus, for unmarried persons, we can reject the null hypothesis that the effect of being female is zero.

### Estimating the Conditional Regression Equation for Included Category
What about the conditional effect of being married for women? And, the conditional effect of being female for married persons? We can use the same three approaches we introduced in Chapter 7 to calculate the significance of these conditional effects:

(a) re-estimating the regression model using different reference categories;
(b) using a general linear $F$-test;
(c) testing the significance of a linear combination of coefficients.

*Changing the Reference Category*

If we re-estimate the model reversing the reference category for the conditioning variable, we will be able to read the other conditional equations from the default results (conditional regression equation for women in Table 8.9; conditional regression equation for married in Table 8.10). Display B.8.2 shows the results for doing this.

In the top panel, we created a new dummy indicator *male* and a new product term based on this variable called *male_marr*. When we use these new variables in the regression, substituting *male* for *female* and *male_marr* for *fem_marr*, we can read the conditional effect of being married for women directly from the output. The coefficient estimate for *married* is now 4.03, matching our hand calculation for the effect of being married among women in Table 8.9. The associated $t$-value of 4.63 is greater than 1.96 and the $p$-value is smaller than 0.05, indicating that the effect of being married is significantly different from zero for women, as it was for men. But, for women, the coefficient estimate is positive in sign, whereas for men it was negative in sign.

In the bottom panel of Display B.8.2, we create a new dummy indicator *unmarried* and a new product term based on this variable called *fem_unmarr*. When we use these new variables in the regression, substituting *unmarried* for *married* and *fem_unmarr* for *fem_marr*, we can read the conditional effect of being female for married persons directly from the output. The coefficient estimate for *female* is now 16.84, matching our hand calculation for the effect of being female among married adults in Table 8.10. The associated $t$-value of 19.58 is greater than 1.96 and the $p$-value is smaller than 0.05, indicating that the effect of being female is also significantly different from zero for married adults, as it was for unmarried adults. Both conditional effects of being female are positive in sign, but the **magnitude of the effect** is larger for married adults than unmarried adults.

*General Linear F-test*

We can also use a general linear $F$-test to test the conditional effects. Although the results will not provide us with the point estimate of the conditional effect, they will tell us whether our hand-calculated conditional effects are significantly different from zero. To request these tests in SAS and Stata, we mimic the way the conditional effects are written in Tables 8.5 and 8.6.

| Conditional Effect of . . . | In symbols | In SAS and Stata |
| --- | --- | --- |
| Married for Women (from Table 8.5) | $\beta_1 + \beta_3$ | test married+fem_marr=0 |
| Female for Married Persons (from Table 8.6) | $\beta_2 + \beta_3$ | test female+fem_marr=0 |

The SAS and Stata commands will test the null hypothesis that the sum of the coefficients for the listed variables (*married* and *fem_marr* in the top row; *female* and *fem_marr* in the bottom row) equal zero, against the alternative that they are not equal to zero.

The results are shown in Display B.8.3. Notice that because these test the same conditional effects as shown in Display B.8.2, when we reverse the reference categories, the *F*-values should match the square of the respective *t*-value in Display B.8.2.

■ For the conditional effect of being married for women, we had $t = 4.63$ in Display B.8.2 and we have $F = 21.44$ in Display B.8.3. As expected, $t^2 = 4.63 * 4.63 = 21.44$ which matches the *F*-value.

■ For the conditional effect of being female for married adults, we had $t = 19.58$ in Display B.8.2 and we have $F = 383.30$ in Display B.8.3. As expected, $t^2 = 19.58 * 19.58 = 383.38$ which matches the *F*-value, within rounding error.

■ The *p*-values for the *F*-tests in Display B.8.3 also match the respective two-sided *p*-values for the *t*-tests shown in Display B.8.2 (all $p < 0.0001$).

## Linear Combination of Coefficients

We can also use the linear combination of coefficients approach that we learned in Chapter 7 to calculate the conditional effects. In this case, our linear combination involves the sum rather than the difference between two coefficients, and the formula for estimating the standard error of the sum of two coefficients is:

$$\sqrt{\hat{\sigma}^2_{\hat{\beta}_1} + \hat{\sigma}^2_{\hat{\beta}_2} + 2\widehat{cov}_{\hat{\beta}_1\hat{\beta}_2}}$$

Notice that for the sum of two coefficients, we add rather than subtract twice the covariance between the coefficients in the formula (see Box 7.3 in Chapter 7 for additional details). We obtain the covariances using the /covb option in SAS and the estat vce option in Stata, as we did in Chapter 7. The results are shown in Display B.8.4.

For the hypothesis that $H_0: \beta_1 + \beta_3 = 0$ we calculate the relevant *t*-value using the following formula and results from Display B.8.4:

$$t = \frac{(\hat{\beta}_1 + \hat{\beta}_3) - 0}{\sqrt{\hat{\sigma}^2_{\hat{\beta}_1} + \hat{\sigma}^2_{\hat{\beta}_3} + 2\widehat{cov}_{\hat{\beta}_1\hat{\beta}_3}}} = \frac{(-2.60 + 6.64) - 0}{\sqrt{0.81770038 + 1.5770922 + (2 * -0.81770038)}}$$

$$= \frac{4.03}{.8714309} = 4.62$$

Notice that the standard error in the denominator, and the final *t*-value, match the results shown in the top panel of Display B.8.2 where we reversed the dummy variable to estimate the conditional effect of being married for women directly.

Similarly, for the hypothesis that $H_0: \beta_2 + \beta_3 = 0$ we calculate the relevant $t$-value using the following formula and results from Display B.8.4:

$$t = \frac{(\hat{\beta}_2 + \hat{\beta}_3) - 0}{\sqrt{\hat{\sigma}^2_{\hat{\beta}_1} + \hat{\sigma}^2_{\hat{\beta}_3} + 2\widehat{\text{cov}}_{\hat{\beta}_1 \hat{\beta}_3}}} = \frac{(10.20 + 6.64) - 0}{\sqrt{0.83731317 + 1.5770922 + (2 * -0.83731317)}}$$

$$= \frac{16.84}{.86010408} = 19.58$$

Notice that the standard error in the denominator, and the final $t$-value, match the results shown in the bottom panel of Display B.8.2 where we reversed the dummy variable to estimate the conditional effect of being female for married persons directly.

In Stata, we can also use the `lincom` command to ask Stata to calculate the point estimate and standard error for a linear combination of coefficients for us. Display B.8.5 shows the results of doing so, which match our hand calculations above and the results in Display B.8.2. As we noted in Chapter 7, this approach is the preferable among the three, because it provides the fullest information (point estimate, standard error, $t$-value, and $p$-value) for the conditional effects after a single estimation, although the direct command is available only in Stata. In the remainder of the chapter, when we use the linear combination approach, we will use the Stata command.

### 8.2.6: Summary of Interpretation

In a model with an interaction between two dummy variables:

$$Y_i = \beta_0 + \beta_1 D_{1i} + \beta_2 D_{2i} + \beta_3 D_{1i} D_{2i} + \varepsilon_i$$

we interpret the results as follows.

■ Intercept ($\beta_0$): average of the outcome variable for cases coded "0" on both dummies.
■ Coefficient of First Dummy ($\beta_1$): conditional effect of first dummy when the second dummy is zero.
■ Coefficient of Second Dummy ($\beta_2$): conditional effect of second dummy when the first dummy is zero.
■ Coefficient of Product Term ($\beta_3$): difference in conditional effects (amount larger or smaller than the conditional effect of one dummy is for cases coded "1" versus "0" on the other dummy).

As we see below, if we add additional control variables to the model, then our interpretation of the coefficients for the dummy variables and product term remain the same, although we add the phrase "controlling for the other variables in the model."

When other variables are in the model, the intercept may no longer be meaningful, if zero is not a valid value on some of those variables. We can also use the same three techniques presented above to test the conditional effect of the variable coded one on the other dummy, when controls are in the model.

### 8.2.7: Presenting Results and Interpreting their Substantive Size

We will end this section by showing how to calculate predicted values based on the coefficient estimates and graph the results. In Appendix H.8.1, we show how to use Excel to do this. Excel is simple to use, widely accessible, and can be used with coefficient estimates produced from any software package, including SAS and Stata (or coefficient estimates from a published article). We will use this technique in Chapter 9 as well, when we interpret nonlinear models estimated in OLS. SAS and Stata both also have dedicated commands for predicting values and graphing them which we will preview in Chapter 12.

Figure 8.1 shows the chart which we created in Excel (again, see Appendix H.8.1 for the detailed steps for creating such a chart). The values calculated by Excel and plotted in the chart match those that we calculated "by hand" in Table 8.8. But, graphing the values can help the reader to visualize the interaction results. Figure 8.1 makes clear that the gender gap is wider for married than unmarried adults. Likewise, men do fewer chores when they are married than unmarried. Women do more chores when they are

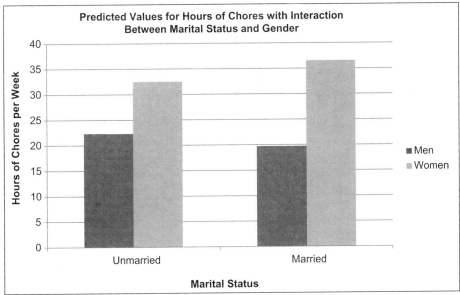

**Figure 8.1:** Example of Chart of Predicted Values, Created in Excel.

married than unmarried. The graph also helps us to visualize the substantive significance of the contrasts which we found were all statistically significant.[5] For example, among married adults, women spend an average of over 16 more hours on chores per week than do men. If this additional time on chores were equally distributed throughout the week, it would amount to more than two more hours spent on chores per day for wives than husbands. This difference between wives and husbands is also over three quarters of the standard deviation of hours of chores ($\hat{\sigma}_y = 25.16$) in the sample of men and women combined. Thus, this effect size is between medium and large using Cohen's rules of thumb ($16.84/25.16 = 0.67$, between 0.50 and 0.80; see again Section 5.4.9).

## 8.3: INTERACTION BETWEEN A DUMMY AND AN INTERVAL VARIABLE

**▨ Box 8.3**

Note that the words weaker and stronger refer to the magnitude of the effect. In both cases the effect may be negative, but a weaker effect for women would mean that it was less negative for women than men and a stronger effect for women would mean that it was more negative for women than men.

We will continue with our example of hours of chores per week as the outcome variable, but now we will focus on hours of paid work per week as a predictor and conceptualize ways in which we might think of gender moderating the relationship between hours of employment and hours of chores. One hypothesis might come from the "second shift" concept, expecting that women bear most of the burden of household chores, even if they are employed. We might hypothesize that, on average, men tend to spend a fairly limited and fixed amount of time on chores per week, and that they decide on this level of chores pretty independently from their hours of employment. Women, on the other hand, may take on as much of the burden of chores as they can, cutting back only when their paid work hours force them to do so. Under this hypothesis, the association between hours of employment and hours of chores should be negative for both genders, but stronger (have a steeper slope) for women than for men (see Box 8.3).

### 8.3.1: Conditional Regression Equations

The mechanics of estimating a regression model with an interaction between a dummy and interval variable are the same as the mechanics of estimating an interaction between two dummy variables: we introduce a product term that is constructed by multiplying together the dummy and the interval predictor variable.

$$Y_i = \beta_0 + \beta_1 X_{1i} + \beta_2 D_{2i} + \beta_3 X_{1i} D_{2i} + \varepsilon_i \tag{8.2}$$

where $X_1$ is an interval measure of hours of paid work per week, *hrwork*, $D_2$ is the dummy indicator of being *female*, and $Y$ is our interval measure of hours spent on household chores per week, *hrchores*.

Unlike the interaction between two dummy variables that we examined in the previous section, where the model had just four predicted values, there are now many possible predictions across the hours of paid work per week. In the NSFH, the variable *hrwork* ranges from zero to 93 for men and from zero to 88 for women, the variable takes on nearly every value in between, although most values fall between 10 and 50 hours of work per week.[6]

### Conditional Regression Equations for Interval Variable

We will begin by considering the conditional regression equations holding gender constant at its two levels and allowing *hrwork* to vary.

**■ Table 8.11: Conditional Regression Equation for Hours of Paid Work within Gender (Interaction Model)**

| | | | |
|---|---|---|---|
| Male | $E(Y \mid X_1, D_2 = 0)$ | $= \beta_0 + \beta_1 * X_{1i} + \beta_2 * 0 + \beta_3 * X_{1i} * 0$ | $= \beta_0 + \beta_1 * X_{1i}$ |
| Female | $E(Y \mid X_1, D_2 = 1)$ | $= \beta_0 + \beta_1 * X_{1i} + \beta_2 * 1 + \beta_3 * X_{1i} * 1$ | $= (\beta_0 + \beta_2) + (\beta_1 + \beta_3) * X_{1i}$ |

This is very similar to what we saw above in Table 8.5 for the conditional effect of being married within gender. But now, with an interval predictor, $\beta_3$ captures any difference in the slope relating $X$ (hours of paid work) to $Y$ (hours of chores) for women versus men. And, $\beta_2$ captures any difference in the mean hours of chores where these lines cross the $Y$ axis (when $X$ is zero).

It is also instructive to reinforce these interpretations by comparing the above result to the additive model that included a dummy variable and an interval variable. Recall from Chapter 7 that such a model results in two conditional regression equations that differed only in the intercept. In that case, the coefficient for the dummy variable measured the constant difference between the two parallel regression lines (see Section 7.3.4).

$$Y_i = \beta_0 + \beta_1 X_{1i} + \beta_2 D_{2i} + \varepsilon_i$$

**■ Table 8.12: Conditional Regression Equation for Hours of Paid Work within Gender (Additive Model)**

| | | | |
|---|---|---|---|
| Male | $E(Y \mid X_1, D_2 = 0)$ | $= \beta_0 + \beta_1 * X_{1i} + \beta_2 * 0$ | $= \beta_0 + \beta_1 * X_{1i}$ |
| Female | $E(Y \mid X_1, D_2 = 1)$ | $= \beta_0 + \beta_1 * X_{1i} + \beta_2 * 1$ | $= (\beta_0 + \beta_2) + \beta_1 * X_{1i}$ |

This model allows the intercepts to differ but constrains the association between $X$ (hours of paid work) and $Y$ (hours of chores) to be the same for men and women, resulting in parallel lines. Thus, adding the interaction term to this model allows us to examine not only whether being in one group versus the other (e.g., male versus female) shifts the regression line up or down but also whether the slope of the regression line is steeper or shallower for one group versus the other (e.g., males versus females).

If we also omitted the dummy variable for female, by restricting $\beta_2 = 0$, we would force men and women to have exactly the same regression line, $(Y|X_1) = \beta_0 + \beta_1 X_{1i}$. It is helpful to keep this in mind because these are clearly nested models, and we will see below how we can use the general linear $F$-test to compare these models and statistically test whether allowing for different intercepts and different slopes improves the fit of the models.

### Conditional Effects for Dummy Variable

As noted above, in the interaction model, if the lines are not parallel, then the difference between the groups captured by the dummy (in our case, men and women) is not constant. The difference will be larger at some levels of the interval variable than at other levels of the interval variable. The coefficient on the dummy variable captures the average difference between groups when the interval variable is zero (see again Table 8.11). Especially if zero is not a valid or meaningful value on the interval variable, we may want to examine the conditional effects of the dummy variable at other values of the interval variable.

■ **Table 8.13: Conditional Effect of Being Female within Hours of Paid Work Per Week**

| Hours | | | |
|---|---|---|---|
| 10 | $E(Y|X_1 = 10, D_2)$ | $= \beta_0 + \beta_1 * 10 + \beta_2 * D_2 + \beta_3 * 10 * D_2$ | $= (\beta_0 + 10\beta_1) + (\beta_2 + 10\beta_3)D_2$ |
| 20 | $E(Y|X_1 = 20, D_2)$ | $= \beta_0 + \beta_1 * 20 + \beta_2 * D_2 + \beta_3 * 20 * D_2$ | $= (\beta_0 + 20\beta_1) + (\beta_2 + 20\beta_3)D_2$ |
| 30 | $E(Y|X_1 = 30, D_2)$ | $= \beta_0 + \beta_1 * 30 + \beta_2 * D_2 + \beta_3 * 30 * D_2$ | $= (\beta_0 + 30\beta_1) + (\beta_2 + 30\beta_3)D_2$ |
| 40 | $E(Y|X_1 = 40, D_2)$ | $= \beta_0 + \beta_1 * 40 + \beta_2 * D_2 + \beta_3 * 40 * D_2$ | $= (\beta_0 + 40\beta_1) + (\beta_2 + 40\beta_3)D_2$ |
| 50 | $E(Y|X_1 = 50, D_2)$ | $= \beta_0 + \beta_1 * 50 + \beta_2 * D_2 + \beta_3 * 50 * D_2$ | $= (\beta_0 + 50\beta_1) + (\beta_2 + 50\beta_3)D_2$ |

So, in the first row, the effect of being female $(D_2)$ is $\beta_2 + 10\beta_3$. This captures the difference in average hours of chores between women and men who are employed 10 hours per week.

Similarly, in the last row, the effect of being female $(D_2)$ is $\beta_2 + 50\beta_3$. This captures the difference in average hours of chores between women and men who are employed for 50 hours per week.

### 8.3.2: Creating the Product Term in SAS and Stata

We can again create the product term easily in SAS and Stata using the general syntax for creating a new variable and the multiplication operator. Using a consistent convention for creating interactions within a project will help you to keep track of the interactions and read the output. The convention is up to you. We will again use the *fem_* at the beginning of the name of our product term, as we did for the interaction between gender and marital status. In SAS, we will create the variable with the command `fem_hrwork=female*hrwork;` In Stata, we will similarly type `generate fem_hrwork=female*hrwork`.

Similarly to the interaction between two dummy variables, the new variable will contain the result of multiplying the values of the two original variables together. Any variable coded '.' missing on either of the variables will be coded missing on the product term. For example, we might have in the data file:

| MCASEID | female | hrwork | fem_hrwork | . . . |
|---------|--------|--------|------------|-------|
| 54638 | 0 | 25 | 0 | . . . |
| 33865 | 0 | 40 | 0 | . . . |
| 76453 | 1 | 0 | 0 | . . . |
| 99857 | 0 | . | . | . . . |
| 26374 | 0 | 0 | 0 | . . . |
| 11948 | 1 | 45 | 45 | . . . |
| 95443 | 1 | 30 | 30 | . . . |
| 66730 | 1 | 0 | 0 | . . . |
| 22647 | . | 50 | . | . . . |
| 88740 | 0 | 0 | 0 | . . . |
| . | . | . | . | . |
| . | . | . | . | . |
| . | . | . | . | . |

### 8.3.3: Calculating Conditional Regression Equations in SAS and Stata

In Section 8.2 we saw that the conditional effects for the category coded "0" on the dummy (e.g., men and unmarried adults in our example) could be read right off the regression output. And, there were three ways to obtain the significance of the other conditional effects:

(a) re-estimating the model with a different reference category;
(b) using the test commands to conduct general linear $F$-tests;
(c) calculating the standard error and $t$-value for a linear combination of coefficients.

We can use similar approaches for the conditional regression equations when we have an interaction between a dummy and interval variable. We will again present these approaches in some detail, to see the comparable results produced by these approaches in this new context. In the remaining sections, we will use the third approach, with the Stata `lincom` command, because it succinctly gives us all the results we desire (point estimates, standard errors, *t*-values, *p*-values) after a single estimation.

### Re-estimating the Regression Equation

For the conditional effect of the interval variable, we can re-estimate the model with the other category on the dummy variable as the reference. In our case, we could use the *male* dummy variable and a new product term `male_hrwork=male*hrwork`. This will allow us to read directly from the default output the conditional regression equation for women shown in Table 8.11.

For the conditional effects of being female within levels of hours of paid work, we can also re-estimate the model with a different level of hours of work indicated by the value of zero. In particular, we will *center* the conditioning variable so that zero represents the conditioning value.

Let's look at an explicit example to make this concrete. If we were interested in the condition of "Hours of Paid Work Per Week = 10," then we would center the hours of paid work variable by creating a new variable that subtracts the value of 10 from *hrwork* (i.e., `hrworkC10=hrwork-10`).

| hrwork | hrworkC10 = hrwork − 10 |
|--------|-------------------------|
| 0 | −10 |
| 5 | −5 |
| 10 | 0 |
| 15 | 5 |
| 20 | 10 |
| 25 | 15 |
| 30 | 20 |
| 35 | 25 |
| 40 | 30 |
| 45 | 35 |
| 50 | 40 |

Notice that a zero on the new variable, *hrworkC10*, represents a 10 on the original variable, *hrwork*. If we create a new product term with this centered variable and re-estimate the regression, then all the estimates will remain unchanged except the intercept and the coefficient estimate for the dummy variable in the interaction. The intercept will now be the average hours of chores for men who work 10 hours per week.

And, the coefficient estimate for female will now represent the difference in average chores between women and men who work 10 hours per week.

To implement this approach, we need to follow three steps in SAS and Stata for each conditioning value:

1. Calculate the centered variable.
2. Create a product term using the new centered variable.
3. Estimate the regression model using the new centered variable and the new product term.

For example, for the level of 10 on hours of work, we would do the following.

|  | SAS | Stata |
|---|---|---|
| Variable Creation | ```data interact;``` | |
| | ```  set "c:\hrchores\interact";``` | ```generate hrworkC10=hrwork-10``` |
| | ```  hrworkC10=hrwork-10;``` | ```generate fem_hrwC10=female*hrworkC10``` |
| | ```  fem_hrwC10=female*hrworkC10;``` | |
| | ```run;``` | |
| Regression Model | ```proc reg;``` | |
| | ```  model hrchores=hrworkC10``` | ```regress hrchores hrworkC10 ///``` |
| | ```    female fem_hrwC10;``` | ```  female fem_hrwC10``` |
| | ```run;``` | |

We purposefully showed the Data Step in SAS to emphasize that the new variable creation must happen within the Data Step (not interspersed with procedures, such as immediately before the relevant Proc Reg).

**Using the Test Command**

To use the `test` command to obtain the significance of the coefficients in the conditional regression equations, we mirror the ways in which we represented each coefficient in Tables 8.11 and 8.13 above.

For the effect of hours of work among women, the conditional effect shown in Table 8.11 was $\beta_1 + \beta_3$. The sum of the coefficient for the interval variable and the product term. The corresponding test command would be `test hrwork+fem_hrwork=0`. Based on this command, SAS and Stata would test the null hypothesis that the sum of the coefficients for these two variables was zero against the two-sided alternative that their sum differed from zero.

For the effect of being female within hours of paid work, we can use the following test statements, mirroring the conditional effects shown in Table 8.13.

| Hours | Symbols | SAS/Stata Syntax |
|-------|---------|------------------|
| 10 | $\beta_2 + 10\beta_3$ | test female+10*fem_hrwork=0 |
| 20 | $\beta_2 + 20\beta_3$ | test female+20*fem_hrwork=0 |
| 30 | $\beta_2 + 30\beta_3$ | test female+30*fem_hrwork=0 |
| 40 | $\beta_2 + 40\beta_3$ | test female+40*fem_hrwork=0 |
| 50 | $\beta_2 + 50\beta_3$ | test female+50*fem_hrwork=0 |

Note that the syntax for these test commands is exactly the same for SAS and Stata test. Each tests whether the sum of the coefficient for female and the indicated multiple of the coefficient for the product term equals zero against the alternative that the sum is not equal to zero.

### Linear Combination of Coefficients

We can similarly mimic the conditional effects with the Stata lincom command. For the conditional effect of hours of work among women this would be lincom hrwork+fem_hrwork.

For the effect of being female within hours of paid work this would be:

| Hours | Stata Syntax |
|-------|--------------|
| 10 | lincom female+10*fem_hrwork |
| 20 | lincom female+20*fem_hrwork |
| 30 | lincom female+30*fem_hrwork |
| 40 | lincom female+40*fem_hrwork |
| 50 | lincom female+50*fem_hrwork |

### 8.3.4: Example 8.2

We begin by estimating Equation 8.2 in SAS and Stata. Display B.8.6 shows the commands used to create the product term and estimate the regression, and the results. The estimated regression equation is:

$$\hat{Y}_i = 26.12 - 0.12X_{1i} + 16.99D_{2i} - 0.13X_{1i}D_{2i} \tag{8.3}$$

Or, written with our variable names:

$$\widehat{hrchores}_i = 26.12 - 0.12hrwork_i + 16.99female_i - 0.13fem\_hrwork_i$$

The default hypothesis test for `fem_hrwork` shown in SAS and Stata tests whether the interaction is significant. If we cannot reject the null hypothesis that the product term is zero ($H_0$: $\beta_3 = 0$), then the effect of each variable is statistically equivalent within levels of the other variable. If we can reject the null hypothesis that the product term is zero, then we have evidence that the effect of each variable differs within levels of the other variable ($H_a$: $\beta_3 \neq 0$). In our case, the *t*-value for *fem_hrwork* of –2.87 is greater in magnitude than the cutoff of 1.96 for our large sample, and the *p*-value is smaller than 0.05, so we can reject the null hypothesis.[7]

## Conditional Regression Equations for Effect of Hours of Paid Work

We can use the prediction equation (Equation 8.3) to calculate predicted values and the conditional regression equations. Let's begin with the conditional regression equations for our interval variable, *hrwork*.

■ **Table 8.14: Estimated Conditional Regression Equation for Effect of Hours of Paid Work, Within Gender**

| | | | |
|---|---|---|---|
| Men | $(\hat{Y}\|X_1, D_2 = 0)$ | $= 26.12 - 0.12X_{1i} + 16.99 * 0 - 0.13 * X_{1i} * 0$ | $= 26.12 - 0.12X_{1i}$ |
| Women | $(\hat{Y}\|X_1, D_2 = 1)$ | $= 26.12 - 0.12X_{1i} + 16.99 * 1 - 0.13 * X_{1i} * 1$ | $= (26.12 + 16.99) - (0.12 + 0.13)X_{1i}$ |
| | | | $= 43.11 - 0.25X_{1i}$ |

Notice that similar to the interaction of two dummy variables, we can read the coefficient estimates for one of the conditional regression equations—the equation for the category coded zero on the other variable, *men*—directly from the SAS and Stata output (see Box 8.4). The other (conditional regression equation for *women*) requires us to sum two coefficients to calculate the intercept and to sum two coefficients to calculate the slope.

For men, we can refer back to Display B.8.6 to see the results of the test of whether the conditional effect of hours of work differs from zero. The *t*-value for *hrwork* ($X_{1i}$) is –3.96, which is larger in absolute magnitude than 1.96, and the associated *p*-value is smaller than 0.05. Thus, for men, we can reject the null hypothesis that the effect of hours of work on hours of chores is zero.

For women, we hand-calculated the conditional effect in Table 8.14. To test whether this association between hours of work and hours of chores is significantly different from zero for women requires us to use one of the three strategies discussed above. Display B.8.7 presents the results of each

■ **Box 8.4**

The conditional effect that we can read off the default output is sometimes referred to as the *main effect*. We find the term *conditional effect* preferable, however, because as we emphasize throughout the chapter, whenever an interaction is present, we must interpret the effect of one variable conditional on levels of the other variable in the interaction. Depending on the number of levels of the other variable, there may be numerous conditional effects.

of the three approaches, re-estimating the regression in the top panel, $F$-test in the middle panel, and linear combination in the bottom panel.

The first and third approaches provide the point estimate and standard errors, as well as the hypothesis test. The values are the same, with the estimate of the conditional effect of *hrwork* for women being estimated to be –0.25 in both cases. This matches, within rounding error, our hand calculation above.[8] As expected, the $F$-value in the second approach equals the square of the $t$-value from the first and third approaches ($t^2 = -8.18 * -8.18 = 66.91$), within rounding error. In all cases, the $p$-value is less than 0.05, indicating that we can reject the null hypothesis that the conditional effect of hours of work is zero among women.

### Conditional Effects of Gender

We can similarly use Equation 8.3 to calculate the conditional effects of being female within hours of paid work.

▨ **Table 8.15: Estimated Effect of Being Female, Within Hours of Paid Work**

| Hours | | Conditional Regression Equations | |
|---|---|---|---|
| 10 | $(\hat{Y}|X_1 = 10, D_2)$ | $= 26.12 - 0.12 * 10 + 16.99D_{2i} - 0.13 * 10 * D_{2i}$ | $= (26.12 - 1.20) + (16.99 - 1.30)D_{2i}$ <br> $= (24.92) + (15.69)D_{2i}$ |
| 20 | $(\hat{Y}|X_1 = 20, D_2)$ | $= 26.12 - 0.12 * 20 + 16.99D_{2i} - 0.13 * 20 * D_{2i}$ | $= (26.12 - 2.40) + (16.99 - 2.60)D_{2i}$ <br> $= (23.72) + (14.39)D_{2i}$ |
| 30 | $(\hat{Y}|X_1 = 30, D_2)$ | $= 26.12 - 0.12 * 30 + 16.99D_{2i} - 0.13 * 30 * D_{2i}$ | $= (26.12 - 3.60) + (16.99 - 3.90)D_{2i}$ <br> $= (22.52) + (13.09)D_{2i}$ |
| 40 | $(\hat{Y}|X_1 = 40, D_2)$ | $= 26.12 - 0.12 * 40 + 16.99D_{2i} - 0.13 * 40 * D_{2i}$ | $= (26.12 - 4.80) + (16.99 - 5.20)D_{2i}$ <br> $= (21.32) + (11.79)D_{2i}$ |
| 50 | $(\hat{Y}|X_1 = 50, D_2)$ | $= 26.12 - 0.12 * 50 + 16.99D_{2i} - 0.13 * 50 * D_{2i}$ | $= (26.12 - 6.00) + (16.99 - 6.50)D_{2i}$ <br> $= (20.12) + (10.49)D_{2i}$ |

The results show that the gender gap in hours of chores narrows as work hours increase. Among adults who work 10 hours per week, women spend nearly 16 more hours a week on chores than do men. Among adults who work 50 hours per week, the gender difference is just over 10 hours per week.

Do these conditional gender gaps differ significantly from zero? We will use the three approaches outlined above to ask SAS and Stata to conduct hypothesis tests for the conditional effects of gender. The results are shown in Displays B.8.8 to B.8.10. Display B.8.8 shows the results of re-estimating the regression model after centering hours of work. Display B.8.9 shows the results of the `test` command. Display B.8.10 shows the results of the `lincom` command. The first and third approaches provide the point estimates of the effect of gender within each level of hours of paid work. The results match our hand calculations in Table 8.15 with rounding error.[9] The $F$-values from the second approach equal the square of the $t$-values from the first and third approaches.

| Hours of paid work per week | $F$-value (Approach #2) | Square of $t$-value (Approach #1 and #3) |
|---|---|---|
| 10 | 121.13 | $t^2 = 11.01 * 11.01 = 121.22$ |
| 20 | 187.62 | $t^2 = 13.70 * 13.70 = 187.69$ |
| 30 | 304.23 | $t^2 = 17.44 * 17.44 = 304.15$ |
| 40 | 344.12 | $t^2 = 18.55 * 18.55 = 344.10$ |
| 50 | 178.79 | $t^2 = 13.37 * 13.37 = 178.76$ |

In all cases, the $p$-values are less than 0.05, so we can reject the null hypothesis that the gender difference is zero within each levels of hours of paid work.

### 8.3.5: Summary of Interpretation

In a model with an interaction between a dummy variable and an interval variable:

$$Y_i = \beta_0 + \beta_1 X_{1i} + \beta_2 D_{2i} + \beta_3 X_{1i} D_{2i} + \varepsilon_i$$

we interpret the results as follows.

■ Intercept ($\beta_0$): average of the outcome variable for cases coded "0" on the dummy and zero on the interval variable. If zero is not a valid value on the interval variable, then the intercept is not meaningful.
■ Coefficient of Interval Variable ($\beta_1$): conditional slope for interval variable for the category coded "0" on the dummy.
■ Coefficient of Dummy Variable ($\beta_2$): conditional effect of dummy variable when the interval variable is zero. If zero is not a valid value on the interval variable, then this conditional effect is not meaningful.
■ Coefficient of Product Term ($\beta_3$): difference in conditional effects for a one-unit change on the other variable (amount larger or smaller that the conditional slope is for cases coded "1" on the dummy; amount conditional effect of dummy changes as the interval variable increases by one).

As we see below, if we add additional control variables to the model, then our interpretation of these coefficient estimates remain the same, although we add the phrase "controlling for the other variables in the model." When other variables are in the model, the intercept may no longer be meaningful, if zero is not a valid value on some of those variables. We can use the same three techniques presented above to test the conditional effects, when controls are in the model.

### 8.3.6: Presenting Results and Interpreting their Substantive Size

We can use similar techniques as discussed in Appendix H.8.1 to present the results of an interaction between a dummy and interval variable in Excel. Figure 8.2 shows the results (see Appendix H.8.2 for details on creating this type of graph).

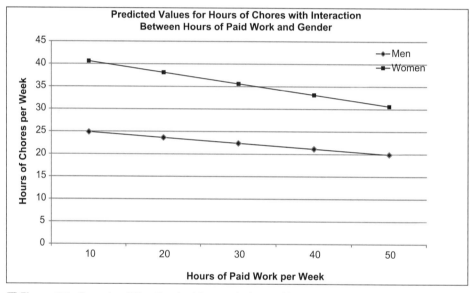

■ **Figure 8.2:** Example of Line Graph of Dummy by Interval Interaction, Created in Excel.

As with the interaction of two dummy variables, the graph helps us to visualize the conditional regression effects calculated above. The negative association between hours of paid work and hours of chores looks steeper for women than for men, as expected. Changing from 10 to 50 hours of paid work per week, women's weekly hours of chores decreases by 10 hours, whereas men's weekly hours of chores decrease by less than 5 hours. And, the gap between women and men narrows as hours of paid work increases, dropping from a gap of over 15 hours at 10 hours of weekly employment to a gap of about 10 hours at 50 hours of weekly employment.

These differences seem large in a real world sense. It is also straightforward to calculate their size relative to the standard deviation of hours of chores of 25.16. Appendix H.8.2

shows how we used Excel to calculate these differences in predicted values relative to the outcome standard deviation, to create the semi-standardized effects or approximate effect size. As shown in Display H.8.8, the gender gap drops from over 60 percent to just 40 percent of the standard deviation of hours of chores when we move from 10 to 50 hours of work per week. Using Cohen's rule of thumb, the approximate effect size drops from above to below the cutoff for medium size (see again Section 5.4.9). In terms of the conditional effects of hours of work, for a large increase in employment of 40 hours per week (from 10 to 50 hours), the effect size is under the medium cutoff, but twice as large for women (0.40) as for men (0.20).

In addition to using changes in hours of employment that have real world meaning in its natural units, we might also calculate fully standardized conditional effects for hours of paid work. The standard deviation of *hrwork* is 14.70.[10] We can calculate standardized conditional effects of hours of work for men and for women by multiplying the conditional effects by the ratio of the standard deviations of the predictor and outcome variables. The standardized effects are small for both genders. Among women, a one-standard deviation increase in weekly hours of work is associated with a 0.15 standard deviation decrease in weekly hours of chores [$= (-0.12432 - 0.1256) *$ $14.70/25.16 = -0.15$]. Among men, a one-standard deviation increase in weekly hours of work is associated with a 0.07 standard deviation decrease in weekly hours of chores ([$= (-0.12432) * 14.70/25.16 = -0.07$; see also Display H.8.9 regarding calculating these standardized effects in Excel).

## 8.4: CHOW TEST

We can build on the concepts just introduced to examine whether a full multiple regression model differs between two groups. In our example, suppose that we theorized that hours spent on chores per week was determined jointly by the two variables considered above—marital status and hours of paid work per week—and also by race-ethnicity and the number of children in the household; and, that the effects of all of these variables differed by gender.[11]

An interaction between gender and number of children in predicting chores can be motivated conceptually, similar to the interactions between gender and marital status or hours of paid work. Children are a net addition to chores, especially when young (they require cooking, laundry, cleaning, driving to and from activities, etc., but can contribute little to these chores). We might anticipate that women complete more of the chores around the house associated with having children. Thus, we might expect that any increase in chores associated with having children is larger for women than for men.

■ Box 8.5

For this example, we will exclude from the sample the relatively small number of families in the NSFH of other race-ethnicities. An alternative to excluding them from the sample would be to group them together in an *other* category, but it is difficult to develop hypotheses and interpret results for such a heterogeneous group. And, the example is simplified by using three homogeneous groups.

Similarly, we might look to literature on egalitarianism in African American families, suggesting less of a gender difference in chores within these families, and traditionalism in Mexican American families, suggesting more of a gender difference in chores within these families (see Box 8.5). We will use whites as the reference group since they are the largest category (with nearly three-quarters of the NSFH sample) and we expect them to differ from both African Americans and Mexican Americans, although in opposite ways. Because of these familial differences, we also expect greater differences by race-ethnicity in hours of chores among women than among men.

So, generally, we have a multiple regression model with five predictors.

$$Y_i = \beta_0 + \beta_1 D_{1i} + \beta_2 X_{1i} + \beta_3 X_{2i} + \beta_4 D_{3i} + \beta_5 D_{4i} + \varepsilon_i \tag{8.4}$$

where $D_1$ is a dummy indicator of marital status (*married*), $X_1$ is hours of paid work per week (*hrwork*), $X_2$ is number of children in the household (*numkid*), and $D_3$ and $D_4$ are a pair of dummy variables indicating African American (*aframer*) and Mexican American (*mexamer*) race-ethnicities, respectively, with whites being the reference category.

Not shown is $D_2$, the dummy indicator of gender (*female*). Because we expect all of the coefficients for all five predictors shown in Equation 8.4 to differ by gender, an intuitive approach to estimating the model would be to run the regression twice, once on the subsample of men and once on the subsample of women. This approach is the intuitive basis for what is referred to as a *Chow test* (Chow 1960; see Ghilagaber 2004 for a more recent treatment of the test and Giordano, Longmore, and Manning 2006; Jarrell and Stanley 2004; and Fairweather 2005 for recent applications). Alternatively, we could use our new understanding of product terms to specify a single model with the dummy for gender and interactions between gender and each of the five other predictors. Gujarati (1970a, 1970b) showed that this approach provides an equivalent *F*-test for any group differences in the full regression model as the Chow test. We will confirm this in our example below and see that these two approaches provide the same coefficient estimates although somewhat different standard errors. Both allow us to test whether any coefficients differ between the groups (in our case between men and women). But, the second approach also allows us to determine readily whether individual coefficients differ by gender.

### 8.4.1: Separate Regression Models Within Groups

We will write the separate regression equations for men and women using subscripts to differentiate the coefficients for men (M) and for women (W).

Men $\quad Y_i = \beta_{M0} + \beta_{M1}D_{1i} + \beta_{M2}X_{1i} + \beta_{M3}X_{2i} + \beta_{M4}D_{3i} + \beta_{M5}D_{4i} + \varepsilon_i$ **(8.5)**

Women $\quad Y_i = \beta_{W0} + \beta_{W1}D_{1i} + \beta_{W2}X_{1i} + \beta_{W3}X_{2i} + \beta_{W4}D_{3i} + \beta_{W5}D_{4i} + \varepsilon_i$ **(8.6)**

We will similarly use subscripts to designate the model run on the total sample (T) as:

Total Sample $\quad Y_i = \beta_{T0} + \beta_{T1}D_{1i} + \beta_{T2}X_{1i} + \beta_{T3}X_{2i} + \beta_{T4}D_{3i} + \beta_{T5}D_{4i} + \varepsilon_i$ **(8.7)**

We will estimate the gender-specific models in SAS or Stata by using a *by* option to estimate the regression model within the levels of gender (once for men and once for women). The data first need to be sorted by gender (so that cases coded "0" on the group variable are in the initial rows of the data file and cases coded "1" on the group variable are in the remaining rows of the data file). In SAS, the sorting takes place with the procedure `proc sort` and the `by` statement is used in both the `sort` and `reg` procedures. In Stata, the sorting can be combined with a `by` request using the `bysort` prefix preceding the `regress` command.

---

SAS

---

```
proc sort;
  by female;
run;

proc reg;
  model hrchores=married hrwork numkid aframer mexamer;
  by female;
run;
```

---

Stata

---

```
bysort female: regress hrchores married hrwork numkid ///
aframer mexamer
```

---

### 8.4.2: The Fully Interacted Model and General Linear *F*-Test

We can also write a single equation that captures Equations 8.5 and 8.6 (for men and women) using a straightforward extension of the concepts we have already considered for interactions. This equation is called a **fully interacted model** and includes one product term between every variable in the model and the grouping variable (gender in our example).

Our hypothesis is that the entire model shown in Equation 8.4 is moderated by gender. This means that we have to allow all of the parameters in the model—the intercept, the coefficients for the three dummies and the slope coefficients for the two interval variables—to vary by gender. We can do so by adding a dummy variable indicating females to the model (to allow the intercept to vary by gender) and by adding five product terms, each created by multiplying the variable from the basic model by the female dummy (to allow the coefficients for the three dummies and the slopes for the two interval variables to vary by gender). In our example, we would have:

$$Y_i = \beta_0 + \beta_1 D_{1i} + \beta_2 X_{1i} + \beta_3 X_{2i} + \beta_4 D_{3i} + \beta_5 D_{4i}$$
$$+ \beta_6 D_{2i} + \beta_7 D_{2i} D_{1i} + \beta_8 D_{2i} X_{1i} + \beta_9 D_{2i} X_{2i} + \beta_{10} D_{2i} D_{3i} + \beta_{11} D_{2i} D_{4i} + \varepsilon_i \quad \text{(8.8)}$$

Recall that $D_2$ is the dummy indicator for females.

Now, we can see that this specification produces the two desired conditional regression equations:

▪ **Table 8.16: Conditional Regression Equations within Gender**

| Male | $E(Y\|D_1,X_1,X_2,D_3,D_4,D_2 = 0)$ | $= \beta_0 + \beta_1 D_{1i} + \beta_2 X_{1i} + \beta_3 X_{2i} + \beta_4 D_{3i} + \beta_5 D_{4i}$ |
|---|---|---|
| | | $+ \beta_6 * 0 + \beta_7 * 0 * D_{1i} + \beta_8 * 0 * X_{1i} + \beta_9 * 0 * X_{2i} + \beta_{10} * 0 * D_{3i} + \beta_{11} * 0 * D_{4i}$ |
| | | $= \beta_0 + \beta_1 D_{1i} + \beta_2 X_{1i} + \beta_3 X_{2i} + \beta_4 D_{3i} + \beta_5 D_{4i}$ |
| Female | $E(Y\|D_1,X_1,X_2,D_3,D_4,D_2 = 1)$ | $= \beta_0 + \beta_1 D_{1i} + \beta_2 X_{1i} + \beta_3 X_{2i} + \beta_4 D_{3i} + \beta_5 D_{4i}$ |
| | | $+ \beta_6 * 1 + \beta_7 * 1 * D_{1i} + \beta_8 * 1 * X_{1i} + \beta_9 * 1 * X_{2i} + \beta_{10} * 1 * D_{3i} + \beta_{11} * 1 * D_{4i}$ |
| | | $= (\beta_0 + \beta_6) + (\beta_1 + \beta_7)D_{1i} + (\beta_2 + \beta_8)X_{1i} + (\beta_3 + \beta_9)X_{2i} + (\beta_4 + \beta_{10})D_{3i} + (\beta_5 + \beta_{11})D_{4i}$ |

The top row provides the conditional regression equation for men and the bottom row the conditional regression equation for women. Note that it is clear from this table that each of the product terms captures the difference between men and women on each of the model parameters: $\beta_6$ the difference in the intercept, $\beta_7$ the difference in the effect of being married ($D_1$), $\beta_8$ the difference in the effect of hours of paid work per week ($X_1$), $\beta_9$ the difference in the effect of number of children ($X_2$), $\beta_{10}$ the difference in the effect of being African American versus white ($D_3$), and $\beta_{11}$ the difference in the effect of being Mexican American versus white ($D_4$).

The conditional regression equations make clear that for the entire regression model to be the same for men and women, all of these coefficients, from $\beta_6$ through $\beta_{11}$, must equal zero. We can use the general linear $F$-test to test such a joint null hypothesis against the alternative that at least one of the coefficients, from $\beta_6$ through $\beta_{11}$, does not equal zero. If we reject the joint null hypothesis, then we can look at the individual coefficients for the *female* dummy and the product terms to see which individual coefficients differ significantly from zero.

Table 8.17 summarizes the null and alternative hypotheses and the full and restricted models of this general linear $F$-test.

■ **Table 8.17: General linear $F$-test of gender differences based on fully interacted model**

| | |
|---|---|
| $H_0$ | $\beta_6 = 0$, $\beta_7 = 0$, $\beta_8 = 0$, $\beta_9 = 0$, $\beta_{10} = 0$, $\beta_{11} = 0$ |
| $H_a$ | $\beta_6 \neq 0$ and/or $\beta_7 \neq 0$ and/or $\beta_8 \neq 0$ and/or $\beta_9 \neq 0$ and/or $\beta_{10} \neq 0$ and/or $\beta_{11} \neq 0$ |
| | (that is, at least one of the coefficients from $\beta_6$ to $\beta_{11}$ is not equal to zero) |
| Full | $Y_i = \beta_0 + \beta_1 D_{1i} + \beta_2 X_{1i} + \beta_3 X_{2i} + \beta_4 D_{3i} + \beta_5 D_{4i}$ |
| | $\quad + \beta_6 D_{2i} + \beta_7 D_{2i} D_{1i} + \beta_8 D_{2i} X_{1i} + \beta_9 D_{2i} X_{2i} + \beta_{10} D_{2i} D_{3i} + \beta_{11} D_{2i} D_{4i} + \varepsilon_i$ |
| Reduced | $Y_i = \beta_0 + \beta_1 D_{1i} + \beta_2 X_{1i} + \beta_3 X_{2i} + \beta_4 D_{3i} + \beta_5 D_{4i}$ |
| | $\quad + 0 * D_{2i} + 0 * D_{2i} D_{1i} + 0 * D_{2i} X_{1i} + 0 * D_{2i} X_{2i} + 0 * D_{2i} D_{3i} + 0 * D_{2i} D_{4i} + \varepsilon_i$ |
| | $\quad = \beta_0 + \beta_1 D_{1i} + \beta_2 X_{1i} + \beta_3 X_{2i} + \beta_4 D_{3i} + \beta_5 D_{4i} + \varepsilon_i$ |

The reduced model is the same as Equation 8.7. And, as we shall see, the full model (fully interacted model) has the same residual sums of squares as Equations 8.5 and 8.6 added together (see Box 8.6).

---

■ **Box 8.6**

It is worth examining how the conditional regression equation for men in Table 8.16 differs from the reduced model in Table 8.17, since in the abstract they look equivalent. First, one is based on constraining the values of a predictor (conditioning) and the other is based on constraining coefficients to zero (based on the null hypothesis). That is, in Table 8.16, we obtain the conditional regression equation for men by substituting in zeros for $D_2$ in the fully interacted model. In contrast, in Table 8.17, we obtain the reduced model by substituting in zeros for $\beta_6$ to $\beta_{11}$ in the fully interacted model. Second, they are based on different subsamples. The full model in Table 8.17 is estimated using the total sample, with the dummy *female* and the product terms omitted. It assumes (based on the null hypothesis) that the coefficients do not differ by gender. Thus, if the null hypothesis is correct, the best estimates of the effects of our five predictors in Equation 8.4 are based on men and women combined. In contrast, the conditional regression equation for men in Table 8.16 based on the fully interacted model is equivalent to Equation 8.5, the model estimated just on the male subsample. It assumes (based on the alternative hypothesis) that the coefficients do differ by gender. Thus, if the alternative hypothesis is correct, the best estimates of our five predictors in Equation 8.4 are based on men and women separately.

## 8.4.3: Example 8.3

Display B.8.11 shows the results of estimating the reduced model (Equation 8.7). Display B.8.12 shows the results of estimating Equations 8.5 and 8.6. Display B.8.13 shows the results of estimating the fully interacted model. Display B.8.13 also shows the results of asking SAS and Stata to calculate the general linear $F$-test shown in Table 8.17 which we will use to verify our hand calculations. Display B.8.14 uses Stata's `lincom` command to calculate the conditional effects for women, shown in Table 8.16.

We will begin by comparing the models, including verifying that the models run within gender are equivalent to the fully interacted model in terms of of the point estimates and residual sums of squares.

- The sum of the sample size of the regression model estimated on the male subsample and the sample size of the regression model estimated on the female subsample ($n_M + n_W = 2,849 + 2,971 = 5,820$ from Display B.8.12) is equal to the sample size of the total sample ($n_T = 5,820$ in Display B.8.11) and the fully interacted model ($n = 5,820$ in Display B.8.13). For Display B.8.12 versus Display B.8.11, this is expected, since the subsample models in Display B.8.12 simply selected two groups that comprise the full sample. For Display B.8.13 versus Display B.8.11, this verifies that the full and reduced models are estimated on the same sample, which as we emphasized in Chapter 6 is required for a general linear $F$-test.
- The sum of the residual sum of squares for the models run on the male and female subsamples ($SSE_M + SSE_W = 1,142,398.38 + 2,081,280.21 = 3,223,678.6$ in Display B.8.12) is also the same as the sums of squares error for the fully interacted model ($SSE = 3,223,678.59$ in Display B.8.13).
- And, the sum of the error degrees of freedom for the models run on the male and female subsamples (*error* $df_M$ + error $df_W = 2,843 + 2,965 = 5,808$ in Display B.8.12) is the same as error degrees of freedom for the fully interacted model (error df = 5,808 in Display B.8.13).

Thus, both approaches (running models within gender or running a fully interacted model) give us equivalent inputs for the full model in the general linear $F$-test.

$$F = \frac{SSE(R) - SSE(F)}{df_R - df_F} \div \frac{SSE(F)}{df_F} = \frac{3,456,551.62 - 3,223,678.59}{5,814 - 5,808}$$

$$\div \frac{3,223,678.59}{5,808} = 69.93$$

This matches the result shown in Display B.8.13 for the $F$-test of the null hypothesis laid out in Table 8.17 (that the *female* dummy and the product terms all have coefficients equal to zero). The numerator degrees of freedom is $5,814 - 5,808 = 6$ which equals the number of constraints in the null hypothesis (six coefficients equal zero in the null hypothesis).

The test command in SAS and Stata provides the $p$-value for the $F$-test, which is less than 0.05. Thus, we reject the null hypothesis and conclude that at least one of the coefficients differs between men and women. An advantage of the fully interacted model over the models run within gender is that we can then look at the individual product terms and the *female* dummy to see which individual coefficients differ from zero.

The point estimates from the models run within gender match the point estimates for the fully interacted model (including the subsequent `lincom` commands for the conditional effects among women). The standard errors for the individual coefficients (and thus the $t$-values and $p$-values) differ somewhat. But, in our case, the conclusions for two-sided hypothesis tests are the same in both approaches (see Box 8.7).

■ **Box 8.7**

We recommend reporting all results from the fully interacted model, as it is used to test the difference in coefficients by gender. Note that in our example, there is not a uniform advantage for one approach over the other in terms of standard errors. The standard errors for men are larger in the fully interacted model (Display B.8.13), whereas the standard errors for women are larger in the subsample model (Display B.8.12).

▨ **Table 8.18: Point Estimates and Significance Tests from Models Run within Gender and Fully Interacted Model.**

| | Within Gender | | Fully Interacted Model | | |
|---|---|---|---|---|---|
| | Men (Display B.8.12) | Women (Display B.8.12) | Men (Display B.8.13) | Difference (Display B.8.13) | Women (Display B.8.14) |
| married | −3.18* | 1.30 | −3.18* | 4.48* | 1.30 |
| hrwork | −0.11* | −0.22* | −0.11* | −0.11* | −0.22* |
| numkid | 1.45* | 6.24* | 1.45* | 4.79* | 6.24* |
| aframer | 6.04* | 4.06* | 6.04* | −1.98 | 4.06* |
| mexamer | 5.77* | 14.26* | 5.77* | 8.49* | 14.26* |
| intercept | 25.18* | 34.69* | 25.18* | 9.51* | 34.69* |

* $p < 0.05$ (two-tailed test).

## 8.4.4: Presenting Results

Appendix H.8.3 shows how to create a graph of the interaction between gender and hours of paid work, similar to the graph we showed in Figure 8.2, but using the fully interacted model. This illustrates how to make predictions with additional variables in the model, beyond those in the interaction.

Indeed, typically, in a manuscript, our hypotheses focus on a subset of the variables in a multiple regression model. Or, if we have numerous hypotheses, we focus on one at a time for interpretation. Graphs can be useful to aid in interpretation in these multiple regression contexts, similar to what we saw above in Section 8.2 and 8.3 with models that just included the two variables in the interaction. But, when we have additional variables in a multiple regression model, we are immediately confronted with the question: how should we deal with the other variables when we want to make predictions while varying just one or two variables? It makes sense to hold these variables constant, so that we can isolate the effect of changing the predictors we are focusing on interpreting (in our case, hours of work and gender). But, what values should we use? A number of alternative choices are available (Long 1997; Long and Freese 2006), but we will use one standard approach, in which we hold all other variables constant at their means (see Box 8.8).

Figure 8.3 shows the results of using this approach. Comparing the graphs in Figures 8.2 and 8.3 and the partially and fully standardized effects in Displays H.8.9 and H.8.11 with those in Figure 8.3 shows generally similar results, although the association between hours of work and hours of chores is a bit less steep in Figure 8.3 than Figure 8.2 (slope of –0.22 versus –0.25 for women).

### 8.4.5: Summary of Interpretation

Following are general interpretations for a fully interacted model:

■ Intercept ($\beta_0$): average of the outcome variable for cases coded "0" on all variables in the model. If zero is not a valid value on one or more predictors, then the intercept is not meaningful.

■ Coefficient of individual variables: conditional effect of variable for the category coded "0" on the grouping variable.

■ Coefficient of grouping variable: difference in intercept between category coded "1" and category coded "0" on the grouping variable. If zero is not a valid value on one or more predictors, then this coefficient is not meaningful.

■ Coefficient of product terms: amount larger or smaller that the conditional effect for each variable is for cases coded "1" on the grouping variable.

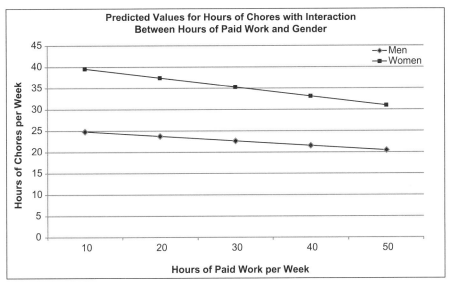

**Figure 8.3:** Predicted Values from Fully Interacted Model.

We can use any of three techniques to test the conditional effects for the group coded "1" on the grouping variable (re-estimate the model with the other group as a reference on the grouping variable or use general linear $F$-tests or tests of linear combinations).

The standard calculation of the Chow test includes the dummy variable for the two groups in the null hypothesis. This captures whether the intercept differs between the groups. However, it will often make sense to exclude this dummy from the general linear $F$-test in the fully interacted model. As we have seen, if one or more of the product terms are significant, then the gender gap will vary within levels of the other variables. And, if zero is not a valid value on all predictors, then testing whether the $Y$ intercept differs by gender is not meaningful.

It is also possible to specify a partially interacted model, rather than a fully interacted model, if we expect that some but not all effects will vary by the grouping variable. For example, we might expect some variables to have similar effects on hours of chores for men and women (such as the size of the house, which may affect the overall volume of chores).

And, a fully or partially interacted model can be set up for grouping variables with more than two groups. In this case, one group will be the reference and sets of product terms will be created using the dummy indicator of each of the remaining groups. These product terms, and the dummy indicators of each of the included groups, will be included in the full regression model (see Gujarati 1970b for an example).

### 8.4.6: Revisiting Literature Excerpt 8.1

Looking back at Literature Excerpt 8.1, we now understand how and why the authors estimated a fully interacted model to estimate the presented coefficient estimates that they describe in the note to their table (see sentence underlined in red). They indicate that an $F$-test for gender differences was insignificant (see sentence underlined in black). Because they report 17 numerator degrees of freedom for this test, and the gender-specific equations contain 17 predictors, it appears that they excluded the dummy variable for gender from this test. Excluding the gender dummy is consistent with their goal in this section of the paper to answer the question "Are correlates of commitment gender-specific?" (p. 382).

The fact that the authors do not use graphics to help to understand the interactions is consistent with the limited evidence of interactions by gender in terms of statistical and substantive significance.

As noted previously, just three of the 17 individual covariates had coefficient estimates that differed significantly by gender. We can use the reported descriptive statistics in the paper to fully standardize the two scale predictors and partially standardize the dummy predictor. Doing so shows that the significant differences are all small in size.

| Predictor Variable | Unstandardized Interaction Coefficient | Standard Deviation of Predictor | Standardized Interaction Coefficient[a] |
|---|---|---|---|
| Nonmerit reward criteria | $-0.010 - (-0.135) = 0.125$ | 0.72 | 0.17 |
| Currently married | $-0.011 - 0.145 = -0.156$ | n/a | -0.29 |
| Number of persons aged 12 or less in household | $0.027 - (-0.050) = 0.077$ | 0.91 | 0.13 |

*Note:* Standard deviation of outcome (organizational commitment) is 0.54 (p. 376). Standard deviation of predictor variables are taken from the paper's appendix.

[a] Completely standardized coefficient for 1st and 3rd predictor variables. Semistandardized coefficient for 2nd predictor variable (because it is a dummy variable).

And, as the authors note in the text, they had limited a priori hypotheses for how the 17 predictors might have different effects for men versus women. One interaction is consistent with prior research (association of marital status significant and positive for men and insignificant for women), but another is inconsistent with expectations (e.g., effect of number of children being significant and negative in sign for men and insignificant for women) and one has no "ready interpretation" (p. 383) from the authors (effect of nonmerit reward criteria significant and negative for men but insignificant for women).

## 8.5: INTERACTION BETWEEN TWO INTERVAL VARIABLES

Interactions between two interval variables often seem more complicated to interpret than interactions involving at least one dummy variable. But, as we will see, the concepts we've considered for interactions between two sets of dummy variables and between a dummy and interval variable extend directly to interactions between two interval variables.

### 8.5.1: Conditional Regression Equations

Setting up an interaction model for two interval variables proceeds similarly to interactions involving at least one dummy variable: we introduce a product term between the two interval variables into the model to test for an interaction.

We will continue with our example of hours of chores per week as the outcome, but now we will introduce an interaction between number of children and hours of paid work per week.

To begin, let's look at the additive version of the model:

$$Y_i = \beta_0 + \beta_1 X_{1i} + \beta_2 X_{2i} + \varepsilon_i$$

where $X_1$ measures hours of paid work per week (*hrwork*) and $X_2$ measures number of children in the household (*numkid*; see Box 8.9).

In this additive model, the association between hours of chores and hours of paid work does not depend on the number of children in the household. Likewise, the association between hours of chores and the number of children in the household does not depend on the hours of paid work. We can see this clearly by looking at the conditional regression equations (see Table 8.19).

> **■ Box 8.9**
>
> To simplify the presentation, we excluded families with more than four children from the sample throughout this chapter.

It is clear from these tables that the condition we place on one of the variables is absorbed in the intercept and the slope for the other variable is the same regardless of the condition. Thus, the conditional regression equations for an additive model with two interval variables produce many parallel lines. This result is similar to what we saw with an additive model with a dummy and interval variable in Section 7.3.4, but with only two parallel lines.

The interaction model introduces a product term between hours of paid work and number of children:

$$Y_i = \beta_0 + \beta_1 X_{1i} + \beta_2 X_{2i} + \beta_3 X_{1i} X_{2i} + \varepsilon_i \tag{8.10}$$

▨ **Table 8.19: Conditional Regression Equations (Additive Model)**

Effect of Hours of Paid Work Per Week ($X_1$), Conditional on Number of Children ($X_2$)

| | | | |
|---|---|---|---|
| No Children | $E(Y\|X_1, X_2 = 0)$ | $= \beta_0 + \beta_1 * X_{1i} + \beta_2 * 0$ | $= \beta_0 + \beta_1 * X_{1i}$ |
| One Child | $E(Y\|X_1, X_2 = 1)$ | $= \beta_0 + \beta_1 * X_{1i} + \beta_2 * 1$ | $= (\beta_0 + \beta_2) + \beta_1 * X_{1i}$ |
| Two Children | $E(Y\|X_1, X_2 = 2)$ | $= \beta_0 + \beta_1 * X_{1i} + \beta_2 * 2$ | $= (\beta_0 + 2\beta_2) + \beta_1 * X_{1i}$ |
| Three Children | $E(Y\|X_1, X_2 = 3)$ | $= \beta_0 + \beta_1 * X_{1i} + \beta_2 * 3$ | $= (\beta_0 + 3\beta_2) + \beta_1 * X_{1i}$ |
| Four Children | $E(Y\|X_1, X_2 = 4)$ | $= \beta_0 + \beta_1 * X_{1i} + \beta_2 * 4$ | $= (\beta_0 + 4\beta_2) + \beta_1 * X_{1i}$ |

Effect of Number of Children ($X_2$), Conditional on Hours of Paid Work Per Week ($X_1$)

Hours
paid work/week

| | | | |
|---|---|---|---|
| 10 | $E(Y\|X_1 = 10, X_2)$ | $= \beta_0 + \beta_1 * 10 + \beta_2 * X_{2i}$ | $= (\beta_0 + 10\beta_1) + \beta_2 * X_{2i}$ |
| 20 | $E(Y\|X_1 = 20, X_2)$ | $= \beta_0 + \beta_1 * 20 + \beta_2 * X_{2i}$ | $= (\beta_0 + 20\beta_1) + \beta_2 * X_{2i}$ |
| 30 | $E(Y\|X_1 = 30, X_2)$ | $= \beta_0 + \beta_1 * 30 + \beta_2 * X_{2i}$ | $= (\beta_0 + 30\beta_1) + \beta_2 * X_{2i}$ |
| 40 | $E(Y\|X_1 = 40, X_2)$ | $= \beta_0 + \beta_1 * 40 + \beta_2 * X_{2i}$ | $= (\beta_0 + 40\beta_1) + \beta_2 * X_{2i}$ |
| 50 | $E(Y\|X_1 = 50, X_2)$ | $= \beta_0 + \beta_1 * 50 + \beta_2 * X_{2i}$ | $= (\beta_0 + 50\beta_1) + \beta_2 * X_{2i}$ |

Now it is clear from the conditional regression equations that if the interaction term is significant, then the effect of number of children depends on the hours of paid work per week and the effect of hours of paid work depends on the number of children (see Table 8.20).

Clearly, the slope of each variable can now differ, depending on the level of the other variable. In the top panel, each time we raise the number of children by one, we add $\beta_3$ to the effect of hours of paid work. Similarly, in the bottom panel, each time we raise the hours of paid work by 10, we add $10 * \beta_3$ to the effect of number of children.

### 8.5.2: Example 8.3

Display B.8.15 presents the estimates of Equation 8.10. Display B.8.16 presents the conditional effects of Table 8.20, calculated using the Stata `lincom` command.[12]

Based on Display B.8.15, we see that the interaction term is significantly different from zero. The $t$-value for *numk_hrw* is –6.63, greater in magnitude than 1.96, and the $p$-value is smaller than 0.05. Thus, the effect of each variable depends on the level of the other variable, meaning that the conditional regression equations are needed in order to interpret the results.

### ■ Table 8.20: Conditional Regression Equations (Interaction Model)

Effect of Hours of Paid Work Per Week ($X_1$), Conditional on Number of Children ($X_2$)

| No Children | $E(Y|X_1, X_2 = 0)$ | $= \beta_0 + \beta_1 * X_{1i} + \beta_2 * 0 + \beta_3 * X_{1i} * 0$ | $= \beta_0 + \beta_1 * X_{1i}$ |
|---|---|---|---|
| One Child | $E(Y|X_1, X_2 = 1)$ | $= \beta_0 + \beta_1 * X_{1i} + \beta_2 * 1 + \beta_3 * X_{1i} * 1$ | $= (\beta_0 + \beta_2) + (\beta_1 + \beta_3) * X_{1i}$ |
| Two Children | $E(Y|X_1, X_2 = 2)$ | $= \beta_0 + \beta_1 * X_{1i} + \beta_2 * 2 + \beta_3 * X_{1i} * 2$ | $= (\beta_0 + 2\beta_2) + (\beta_1 + 2\beta_3) * X_{1i}$ |
| Three Children | $E(Y|X_1, X_2 = 3)$ | $= \beta_0 + \beta_1 * X_{1i} + \beta_2 * 3 + \beta_3 * X_{1i} * 3$ | $= (\beta_0 + 3\beta_2) + (\beta_1 + 3\beta_3) * X_{1i}$ |
| Four Children | $E(Y|X_1, X_2 = 4)$ | $= \beta_0 + \beta_1 * X_{1i} + \beta_2 * 4 + \beta_3 * X_{1i} * 4$ | $= (\beta_0 + 4\beta_2) + (\beta_1 + 4\beta_3) * X_{1i}$ |

Effect of Number of Children ($X_2$), Conditional on Hours of Paid Work Per Week ($X_1$)

Hours
paid work/week

| 10 | $E(Y|X_1 = 10, X_2)$ | $= \beta_0 + \beta_1 * 10 + \beta_2 * X_{2i} + \beta_3 * 10 * X_{2i}$ | $= (\beta_0 + 10\beta_1) + (\beta_2 + 10\beta_3) * X_{2i}$ |
|---|---|---|---|
| 20 | $E(Y|X_1 = 20, X_2)$ | $= \beta_0 + \beta_1 * 20 + \beta_2 * X_{2i} + \beta_3 * 20 * X_{2i}$ | $= (\beta_0 + 20\beta_1) + (\beta_2 + 20\beta_3) * X_{2i}$ |
| 30 | $E(Y|X_1 = 30, X_2)$ | $= \beta_0 + \beta_1 * 30 + \beta_2 * X_{2i} + \beta_3 * 30 * X_{2i}$ | $= (\beta_0 + 30\beta_1) + (\beta_2 + 30\beta_3) * X_{2i}$ |
| 40 | $E(Y|X_1 = 40, X_2)$ | $= \beta_0 + \beta_1 * 40 + \beta_2 * X_{2i} + \beta_3 * 40 * X_{2i}$ | $= (\beta_0 + 40\beta_1) + (\beta_2 + 40\beta_3) * X_{2i}$ |
| 50 | $E(Y|X_1 = 50, X_2)$ | $= \beta_0 + \beta_1 * 50 + \beta_2 * X_{2i} + \beta_3 * 50 * X_{2i}$ | $= (\beta_0 + 50\beta_1) + (\beta_2 + 50\beta_3) * X_{2i}$ |

Display B.8.16 presents the conditional effects of hours of work, within number of children, on the left and the conditional effects of number of children, within hours of work, on the right. All of the conditional effects differ significantly from zero, with *p*-values less than 0.05. Thus, the association of one variable with the outcome is significant across levels of the other variable, but the steepness of the conditional slopes vary. For employment, we see that the effect of an additional hour of paid work ranges from –0.19 when no children are in the household to –0.70 when four children are in the household. For children, the effect of one more child ranges from 7.79 when the adult works 10 hours per week to 2.69 when the adult works 50 hours per week. We will examine the substantive magnitude of these associations further in the next section.

## 8.5.3: Presenting Results

Appendix H.8.4 shows how to adapt the steps for calculating predictions and graphing the results in Excel for this interaction between two interval variables.

Figure 8.4 shows how we used this approach to graph the effects of hours of work within levels of number of children. We graphed the conditional slopes for three rather than all five levels of number of children to make it easier for the eye to pick up the difference in steepness across the lines.

In Display H.8.12, we also calculated the partially (dividing by the standard deviation of *hrchores* of 25.16) and fully standardized (multiplying by the standard deviation of

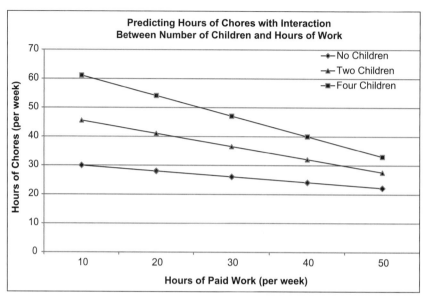

**Figure 8.4:** Presenting Interaction Between Two Interval Variables, Graph Created in Excel.

the predictor, *hrwork*, of 14.70 and dividing by the standard deviation of *hrchores* of 25.16) conditional effects. For hours of work, we show the unstandardized and partially standardized effects of a more substantively meaningful 10 hour rather than 1 hour change (columns I and J of Display H.8.12).

Most of the effects are fairly modest in size, although the effect of hours of work for adults with four children is approaching a medium size (−0.41). Ten more hours of paid work among these families is associated with an unstandardized 7 hour decrease in time spent on chores per week (about 1 hour less per day, if spread equally throughout the week).

### 8.5.4: Summary of Interpretation

Following are general interpretations for interactions between two interval variables:

■ Intercept ($\beta_0$): average of the outcome variable for cases coded zero on both interval variables. If zero is not a valid value on at least one of the predictors, then the intercept is not meaningful.
■ Coefficient of individual variables: conditional effect of one variable when the other variable is zero. These conditional effects are not meaningful if zero is not a valid value on the other variable.
■ Coefficient of product terms: amount larger or smaller that the conditional effect for one variable when the other variable in the interaction increase by one.

It is common to conduct interactions between interval variables one by one (as we will see in the next literature excerpt). We could also place the interval interaction between hours of paid work and number of children within our fully interacted model. This would lead to a three-way interaction. This three-way interaction makes conceptual sense in our example, since we might expect that the interaction between hours of work and number of children depends on gender.

$$
\begin{aligned}
Y_i = {} & \beta_0 + \beta_1 D_{1i} + \beta_2 X_{1i} + \beta_3 X_{2i} + \beta_4 D_{3i} + \beta_5 D_{4i} \\
& + \beta_6 D_{2i} + \beta_7 D_{2i} D_{1i} + \beta_8 D_{2i} X_{1i} + \beta_9 D_{2i} X_{2i} \\
& + \beta_{10} D_{2i} D_{3i} + \beta_{11} D_{2i} D_{4i} + \varepsilon_i \\
& + \beta_{12} X_{1i} X_{2i} + \beta_{13} D_{2i} X_{1i} X_{2i} + \varepsilon_i
\end{aligned}
\tag{8.9}
$$

The conditional regression equations based on this model are as follows.

■ **Table 8.21: Conditional Regression Equations within Gender**

| | | |
|---|---|---|
| Male | $E(Y \mid D_1, X_1, X_2, D_3, D_4, D_2 = 0)$ | $= \beta_0 + \beta_1 D_{1i} + \beta_2 X_{1i} + \beta_3 X_{2i} + \beta_4 D_{3i} + \beta_5 D_{4i}$ |
| | | $\quad + \beta_6 * 0 + \beta_7 * 0 * D_{1i} + \beta_8 * 0 * X_{1i} + \beta_9 * 0 * X_{2i} + \beta_{10} * 0 * D_{3i} + \beta_{11} * 0 * D_{4i}$ |
| | | $\quad + \beta_{12} X_{1i} X_{2i} + \beta_{13} * 0 * X_{1i} X_{2i}$ |
| | | $= \beta_0 + \beta_1 D_{1i} + \beta_2 X_{1i} + \beta_3 X_{2i} + \beta_4 D_{3i} + \beta_5 D_{4i} + \beta_{12} X_{1i} X_{2i}$ |
| Female | $E(Y \mid D_1, X_1, X_2, D_3, D_4, D_2 = 1)$ | $= \beta_0 + \beta_1 D_{1i} + \beta_2 X_{1i} + \beta_3 X_{2i} + \beta_4 D_{3i} + \beta_5 D_{4i}$ |
| | | $\quad + \beta_6 * 1 + \beta_7 * 1 * D_{1i} + \beta_8 * 1 * X_{1i} + \beta_9 * 1 * X_{2i} + \beta_{10} * 1 * D_{3i} + \beta_{11} * 1 * D_{4i}$ |
| | | $\quad + \beta_{12} X_{1i} X_{2i} + \beta_{13} * 1 * X_{1i} X_{2i}$ |
| | | $= (\beta_0 + \beta_6) + (\beta_1 + \beta_7) D_{1i} + (\beta_2 + \beta_8) X_{1i} + (\beta_3 + \beta_9) X_{2i} + (\beta_4 + \beta_{10}) D_{3i} + (\beta_5 + \beta_{11}) D_{4i}$ |
| | | $\quad + (\beta_{12} + \beta_{13}) X_{1i} X_{2i}$ |

Now, each gender-specific conditional regression equation includes the interaction between the two interval variables.

The three-way interaction term is created in SAS and Stata using a three-variable product term (e.g., `fem_nk_hw=female*numkid*hrwork` in SAS; `generate fem_nk_hw=female*numkid*hrwork` in Stata).

If the coefficient on this three-way interaction term ($\beta_{13}$) is significantly different from zero, then the interaction between hours of paid work and number of children differs significantly between men and women. The conditional effect of this interaction can be read off the default output for men ($\beta_{12}$). For women, we could use Stata's lincom command to easily calculate ($\beta_{12} + \beta_{13}$) and its significance (or we could re-estimate the model with the *male* dummy and *male* product terms, or use the test command in SAS or Stata).

## 8.6: LITERATURE EXCERPT 8.2

We end with an example from the literature of an interaction between two interval variables.[13]

In 2007, Rory McVeigh and Julianna Sobolewski published a study in the *American Journal of Sociology* that aims to examine how socioeconomic characteristics relate to voting, especially how occupation and wealth relate to voting Democratic or Republican. As the authors state (p. 448):

> Social scientists have given substantial attention to relationships between voting and social class . . . Most of this research assumes that, were it not for other factors, individuals possessing limited economic resources would naturally prefer candidates who promise to do more than their opponents to promote vertical redistribution of wealth. More prosperous voters, on the other hand, should prefer candidates who promote policies that will help them to preserve their accumulated wealth.

The authors add to this literature by considering not only income inequality within a single group (all Americans) but also between groups (between men and women and whites and nonwhites) and especially how these inequalities are linked to occupational segregation. In addition, whereas much of the literature has focused on individual voters, the authors look at local contexts (counties). They hypothesize that (p. 449):

> Republican candidates should receive the most electoral support in locations where large proportions of the community benefit from a conservative political agenda that preserve inequalities based on categorical distinctions. Support for Republican candidates should be especially strong in locations where there is a high degree of occupational segregation based on gender and racial categories, and where these categorical boundaries are most vulnerable to penetration.

**■ Box 8.10**

This can be a useful strategy for making the tables interpretable when zero is not a meaningful value on one or more of the variables in the interaction(s), although it is important to remember to adjust for this centering when calculating other conditional effects by hand or in SAS or Stata.

The authors note in the article that "To facilitate interpretation of our interaction effects, we center all of our independent variables on their mean values" (p. 472). Thus, in the tables, the coefficient estimate of each variable involved in the interaction is its conditional effect when the other variable is at its mean (see Box 8.10). They also use graphs to help the reader to visualize significant interactions.

Although the authors test a number of interactions, we will focus on one presented in the first column of their Table 3 (see Literature Excerpt 8.2a). Notice that in this table, six models were run, each with one interaction between sex segregation and another variable. As we noted above, unless one of the variables in the interaction is a grouping variable, it is common to reduce complexity and facilitate interpretation by running separate models, each with a single interaction.

The interaction we focus on is circled in red in Literature Excerpt 8.2a. The conditional effects of the two variables in the interaction are circled in black. One is a measure of horizontal inequality (occupational sex segregation) which the authors measure by the *index of dissimilarity* in the county. The variable can range from zero to 1 and

---

■ **Literature Excerpt 8.2a**

Table 3.  Percentage Voting Republican in 2004 Presidential Election: Occupational Sex Segregation Interacting with Threats to Segregation, U.S. Counties

| | Model 1 | Model 2 | Model 3 | Model 4 | Model 5 | Model 6 |
|---|---|---|---|---|---|---|
| Theoretical variables: | | | | | | |
| Occupational sex segregation | 13.927*** | 12.060*** | 13.104*** | 11.984*** | 10.862*** | 10.020*** |
| | (2.742) | (2.729) | (2.793) | (2.737) | (2.746) | (2.769) |
| Occupational race segregation | .841 | 1.506 | 1.401 | 1.160 | .846 | 1.327 |
| | (.933) | (.956) | (.940) | (.950) | (.950) | (.949) |
| %women in labor force | −.001 | −.007 | .020 | .014 | .030 | .008 |
| | (.033) | (.034) | (.033) | (.033) | (.032) | (.033) |
| %nonwhite | −.119*** | −.113*** | −.116*** | −.112*** | −.118*** | −.110*** |
| | (.016) | (.016) | (.016) | (.016) | (.016) | (.016) |
| %nonwhite, squared | −.003*** | −.003*** | −.003*** | −.003*** | −.003*** | −.003*** |
| | (.0004) | (.0004) | (.0004) | (.0004) | (.0004) | (.0004) |
| %management | .241*** | .256*** | .291*** | .256*** | .225*** | .198*** |
| | (.049) | (.050) | (.049) | (.050) | (.048) | (.050) |
| %professional | .156** | .180*** | .148** | .169** | .145** | .168** |
| | (.055) | (.055) | (.055) | (.056) | (.055) | (.055) |
| %construction and extraction | −.188*** | −.177** | −.196*** | −.206*** | −.248*** | −.194*** |
| | (.058) | (.059) | (.058) | (.059) | (.061) | (.059) |
| %production workers | −.038 | −.027 | −.039 | −.037 | −.056 | −.045 |
| | (.032) | (.032) | (.031) | (.032) | (.031) | (.031) |
| %women with bachelor's degree | −.593*** | −.561*** | −.575*** | −.609*** | −.653*** | −.603*** |
| | (.041) | (.043) | (.041) | (.043) | (.041) | (.041) |
| %nonwhite with bachelor's degree | −.039** | −.034* | −.035* | −.036** | −.043** | −.039** |
| | (.014) | (.014) | (.014) | (.014) | (.014) | (.014) |
| Residential mobility | .160*** | .157*** | .156*** | .161*** | .166*** | .162*** |
| | (.027) | (.027) | (.027) | (.027) | (.027) | (.027) |
| Sex segregation × women labor force | .967*** | | | | | |
| | (.249) | | | | | |
| Sex segregation × women with degree | | .967*** | | | | |
| | | (.224) | | | | |
| Sex segregation × residential mobility | | | 1.132*** | | | |
| | | | (.194) | | | |

*(Continued overleaf)*

Table 3.  (continued)

|  | Model 1 | Model 2 | Model 3 | Model 4 | Model 5 | Model 6 |
|---|---|---|---|---|---|---|
| Sex segregation × professional |  |  |  | .892** (.339) |  |  |
| Sex segregation × construction |  |  |  |  | .733 (.456) |  |
| Sex segregation × production |  |  |  |  |  | −.853*** (.211) |
| Control variables: |  |  |  |  |  |  |
| Republican partisanship | 3.844*** (.084) | 3.805*** (.084) | 3.844*** (.084) | 3.842*** (.085) | 3.885*** (.084) | 3.859*** (.085) |
| Unemployment, 2004 | −.775*** (.097) | −.779*** (.096) | −.786*** (.099) | −.812*** (.097) | −.804*** (.097) | −.800*** (.097) |
| Log of population density | −1.250*** (.127) | −1.222*** (.129) | −1.129*** (.127) | −1.232*** (.128) | −1.310*** (.128) | −1.204*** (.130) |
| Median family income | .182*** (.027) | .184*** (.028) | .147*** (.027) | .170*** (.027) | .170*** (.027) | .174*** (.027) |
| Income inequality | .016 (.049) | .000 (.050) | .017 (.050) | .002 (.050) | .014 (.050) | .011 (.050) |
| %Evangelical | .068*** (.009) | .067*** (.009) | .064*** (.009) | .066*** (.009) | .065*** (.009) | .065*** (.009) |
| %Catholic | −.049*** (.010) | −.049*** (.009) | −.049*** (.009) | −.049*** (.009) | −.049*** (.010) | −.050*** (.010) |
| %married | .214*** (.032) | .198*** (.032) | .217*** (.031) | .214*** (.032) | .229*** (.032) | .224*** (.032) |
| %part-time | .023 (.028) | .029 (.028) | .054 (.028) | .043 (.028) | .040 (.028) | .032 (.028) |
| Median age | −.338*** (.044) | −.326*** (.043) | −.351*** (.044) | −.317*** (.044) | −.301*** (.043) | −.303*** (.044) |
| Exposure CBSA sex segregation | 11.075*** (2.515) | 10.623*** (2.488) | 10.396*** (2.457) | 9.564*** (2.474) | 8.626*** (2.484) | 8.817*** (2.541) |
| $R^2$ | .868 | .869 | .870 | .868 | .868 | .868 |

*Note.*—Fixed effects estimates with controls for state-level effects (robust SE's in parentheses).
* $P < .05$.
** $P < 01$.
*** $P < .001$.

**Source:** McVeigh, Rory and Julianna M. Sobolewski. 2007. "Red Counties, Blue Counties, and Occupational Segregation by Sex and Race." *American Journal of Sociology*, 113, 446–506.

represents the "proportion of either men or women who would have to change occupations to produce a distribution where sex is completely uncorrelated with occupational categories" (p. 465). The other variable in the interaction is a variable that they view as a potential threat to occupational sex segregation: the percentage of the county's women who are in the labor force. The outcome is the percentage voting Republican among those who voted in the 2004 presidential election in the county.

The results reveal a positive and significant interaction (indicated by asterisks on 0.967). As noted, the authors centered the variables at their means before creating the product terms. So, when sex segregation is at its mean level, the percentage of women in the labor force is unrelated to voting Republican (–0.001). In contrast, the conditional effect of occupational sex segregation on voting when female labor force participation is at its mean level is significant (13.927 with asterisks). Scanning the table, this coefficient estimate appears quite large relative to the other values, but recall that occupational sex segregation ranges from zero to 1. Thus, this estimate indicates that when sex segregation changes from its minimum to maximum value in counties with average female labor force participation, the percentage voting Republican increases by 14 points, controlling for numerous other characteristics of the counties.

The authors' Figure 2, shown in Literature Excerpt 8.2b, presents conditional effects of occupational sex segregation within three levels of female labor for participation: 45 percent, 55 percent and 65 percent. Among counties with the greatest threat (65 percent women in the labor force) the figure shows that moving from the minimal to

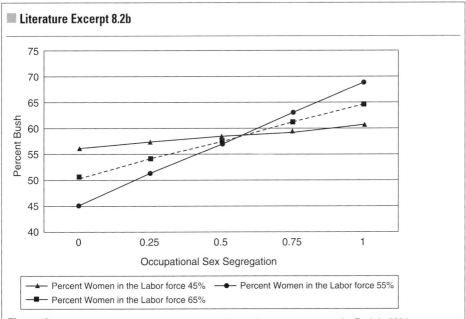

**Literature Excerpt 8.2b**

**Figure 2:** The effect of occupational sex segregation on the percentage vote for Bush in 2004 at varying levels of percentage women in the labor force.

**Source:** McVeigh, Rory and Julianna M. Sobolewski. 2007. "Red Counties, Blue Counties, and Occupational Segregation by Sex and Race." *American Journal of Sociology*, 113, 446–506.

the maximal level of sex segregation is associated with nearly a 25 point increase in the percentage of the counties voters who voted Republican. In contrast, among counties with the lowest threat (45 percent women in the labor force) the increase in Republican voting associated with maximal change in sex segregation is less than 5 points.[14]

## 8.7: SUMMARY

In this chapter, we showed how to estimate and interpret interactions. Adding interactions to regression models allows us to test for moderation effects anticipated in our conceptual models. Estimating interactions are important to correctly specifying regression models. Depending on the form of the interaction, effects estimated on the full sample may understate, overstate, or fully mask the conditional effects within levels of the other variable.

Although we looked separately in the chapter at interactions between two dummy, a dummy and interval, and two interval variables, all follow a similar process of adding a product term to the regression model to capture the interaction. A significant coefficient estimate on the product term indicates statistical evidence for moderation. Calculating conditional effects (the association between one variable and the outcome within levels of the other variable) facilitates interpretation of the significant interaction. Three approaches can be used to test the significance of these conditional effects: (a) re-estimating the model after excluding a different reference category on a dummy variable or after recentering an interval variable, (b) using the general linear $F$-test, or (c) testing the linear combination of coefficients. Stata's `lincom` command provides an easy way to implement the third approach, and obtain all the information needed after a single estimation (the estimate, standard error, and significance of each conditional effect). We also used Excel to calculate predicted values and graph the conditional effects, and to calculate partially and fully standardized conditional effects. Such graphs and calculations are vital to the substantive interpretation of statistically significant effects. Finally, we showed how to estimate a fully interacted model and use the Chow test to examine whether an entire regression model differs within groups.

## KEY TERMS

Chow Test

Conditional Effect(s)

Fully Interacted Model

Interaction(s)

**SUMMARY 8**

**KEY TERMS 8**

Magnitude of (the) Effect (also weaker and stronger)

Product Term

## REVIEW QUESTIONS

**8.1.** For dummy*dummy interaction models:

(a) What is the difference between the conditional means and the conditional effects?

(b) How do you write each conditional effect based on the general regression equation and/or the prediction equation?

(c) How do you interpret each coefficient in the model (the values that you would see in the default SAS and Stata output), including the intercept?

(d) How would your results change if you used a different reference category on one or both dummy variables?

**8.2.** For dummy*interval interaction models:

(a) In a basic model, with one interval predictor (X), one dummy predictor (D), and the product term of the dummy times the interval predictor (D*X), how would you interpret the intercept and three variables' coefficients?

(b) How would you write the conditional regression equations for this model within the two levels of the dummy variable?

(c) How would you write the conditional regression equations for this model within levels of the interval predictor variable?

(d) What would be the null and alternative hypotheses, and the corresponding reduced and full models, for the Chow test based on this model?

(e) What would you expect the results to be if you reverse coded the dummy variable, omitting the category that had previously been coded "1", created a new interaction term based on this new reverse coded dummy, and re-estimated the model with the new dummy and new interaction term?

**8.3.** For interval*interval interaction models:

(a) How do you interpret the coefficients in a model with an interaction between two interval variables (the intercept, the coefficient for each of the predictor variables and the coefficient on the product term)?

(b) How would you write conditional regression equations (holding one variable constant at a particular value) based on the interval by interval interaction model?

## REVIEW EXERCISES

**8.1.** Consider the research question: "Greater job stress is associated with harsher parenting, but this association weakens with each additional increment of social support received from family and friends." How would you set up a regression model to examine this research question?

**8.2.** Imagine that you hypothesize that the racial gap in earnings is smaller for women than for men (that is, the difference in average earnings between African Americans and whites is smaller for women than for men). You estimate the following prediction equation:

EARNINGS = 35,000 – 15,000 * FEMALE – 10,000 * AFRAMER + 10,000 * FEMALE * AFRAMER

in which FEMALE is coded "0" for men and "1" for women and AFRAMER is coded "0" for whites and "1" for African Americans.

(a) What are the estimates of the racial earnings gap for men and women based on this model (i.e., hint: write the conditional regression equations, holding gender constant at zero and one, and indicate the coefficient for AFRAMER in these equations).

(b) Calculate the four conditional means based on this prediction equation.

**8.3.** Suppose that you hypothesize that earnings is explained by a person's education level (EDUC, years of schooling) and experience (EXPER, years in the occupation) but that this regression model differs for men and women (FEMALE, 0 = men, 1 = women).

(a) Write the general equation for a fully interacted model that would test this hypothesis.

(b) Write the null and alternative hypotheses for the Chow test based on this fully interacted model.

(c) Write the reduced model that results from placing the constraints of the null hypothesis on the full model.

**8.4.** Refer to Model 2 in Literature Excerpt 8.2a. Note that the author's measure *% women with bachelor's degree* is: "a measure of the percentage of women ages 25 or older who have earned a bachelor's degree (including those who have also earned a graduate or professional degree" (p. 467).

  (a) Interpret the coefficient estimate for *Occupational sex segregation* in Model 2.

  (b) Interpret the coefficient estimate for *% women with bachelor's degree* in Model 2.

  (c) Interpret the coefficient estimate for Sex segregation x women with degree in Model 2.

  (d) Calculate the conditional effect of Occupational *sex segregation when % women with bachelor's degree* is 10 and 30 (recall that the variables in the interaction are centered; note that the mean of % women with bachelor's degree is 16.06).

**8.5.** Based on the two regression models shown below, do the following:

  (a) Statistically compare Model 1 and Model 2 using the Chow test (use a cutoff of F = 2.22 for a significant test at alpha = 0.05). *Be sure to list the null and alternative hypotheses for the test.*

  (b) Write the prediction equation based on Model 2. You may round all coefficients to two nonzero decimal places (e.g., round 0.00034561 to 0.00035).

  (c) Based on the prediction equation that you wrote out in Question 8.5b, write the conditional regression equations for married and unmarried persons based on Model 2. You may round all coefficients to two nonzero decimal places (e.g., round 0.00034561 to 0.00035).

  (d) Based on the results of Model 2, which coefficients in the conditional equations you wrote in Question 8.5c differ significantly for married versus unmarried persons? Justify your response by writing the relevant t-values and/or p-values (use a cutoff alpha of 0.05). Be sure to consider the intercept as well as the four predictor variables.

| DEPRESS | Depression scale (ranging from 0 to 105) |
| REDUC | Respondent's education level (ranging from 0 to 20) |
| REARNINC | Respondent's annual earned income (ranging from $1 to $800,000) |
| HRWORK | Respondent's hours spent at paid work/week (ranging from 1 to 95) |

| NUMKID | Number of persons <=18 in the household (ranging from 0 to 10) |
| MARRY | Dummy indicator of respondent's marital status (1 = married, 0 = not married) |
| MRY_EDUC | Product of MARRY*REDUC |
| MRY_EARN | Product of MARRY*REARNINC |
| MRY_HRWK | Product of MARRY*HRWORK |
| MRY_NUMK | Product of MARRY*NUMKID |

### Model 1

`. regress depress reduc rearninc hrwork numkid`

| Source | SS | df | MS | | |
|---|---|---|---|---|---|
| Model | 33076.9637 | 4 | 8269.24092 | | |
| Residual | 1515927.29 | 5315 | 285.216799 | | |
| Total | 1549004.25 | 5319 | 291.220954 | | |

```
Number of obs  =   5320
F(4, 5315)     =  28.99
Prob > F       = 0.0000
R-squared      = 0.0214
Adj R-squared  = 0.0206
Root MSE       = 16.888
```

| depress | Coef. | Std. Err. | t | P>|t| | [95% Conf. Interval] |
|---|---|---|---|---|---|
| reduc | -.4999201 | .0968659 | -5.16 | 0.000 | -.6898171 -.3100231 |
| rearninc | -.0000506 | 8.40e-06 | -6.03 | 0.000 | -.0000671 -.0000342 |
| hrwork | -.0078686 | .0180722 | -0.44 | 0.663 | -.0432975 .0275603 |
| numkid | .6742838 | .1959539 | 3.44 | 0.001 | .2901337 1.058434 |
| _cons | 23.72232 | 1.517835 | 15.63 | 0.000 | 20.74674 26.6979 |

### Model 2

`. regress depress reduc rearninc hrwork numkid marry mry_reduc mry_earn`
`mry_hrwk mry_numk`

| Source | SS | df | MS | | |
|---|---|---|---|---|---|
| Model | 64439.3136 | 9 | 7159.92374 | | |
| Residual | 1484564.94 | 5310 | 279.579084 | | |
| Total | 1549004.25 | 5319 | 291.220954 | | |

```
Number of obs  =   5320
F(9, 5310)     =  25.61
Prob > F       = 0.0000
R-squared      = 0.0416
Adj R-squared  = 0.0400
Root MSE       = 16.721
```

| depress | Coef. | Std. Err. | t | P>|t| | [95% Conf. Interval] |
|---|---|---|---|---|---|
| reduc | -.2655434 | .160663 | -1.65 | 0.098 | -.5805089 .0494222 |
| rearninc | -.0000641 | .0000134 | -4.77 | 0.000 | -.0000904 -.0000378 |
| hrwork | -.012559 | .0293056 | -0.43 | 0.668 | -.0700101 .0448921 |
| numkid | 1.862119 | .3300898 | 5.64 | 0.000 | 1.215007 2.509231 |
| marry | -.5102804 | 3.130079 | -0.16 | 0.871 | -6.64652 5.62596 |
| mry_reduc | -.2853036 | .2009015 | -1.42 | 0.156 | -.679153 .1085458 |
| mry_earn | .0000308 | .0000172 | 1.80 | 0.073 | -2.83e-06 .0000644 |
| mry_hrwk | .0043745 | .0370308 | 0.12 | 0.906 | -.0682211 .0769701 |
| mry_numk | -1.46112 | .4100886 | -3.56 | 0.000 | -2.265062 -.6571779 |
| _cons | 23.00074 | 2.479346 | 9.28 | 0.000 | 18.1402 27.86128 |

# CHAPTER EXERCISE

In this exercise, you will write a SAS and a Stata batch program to estimate a multiple regression model, with interactions.

**To begin**, prepare the shell of your batch program, including the commands to save your output and initial commands (e.g., SAS library, turning *more* off and closing open logs in Stata).

Start with the NOS 1996 dataset that you created in Chapter 4 and use the following command on the use command in Stata or in the data step in SAS to exclude cases with missing data:

```
if mngpctfem~=. & age~=. & ftsize~=. & (a5>-999 & a5<3) & a3<3
```

*Notice that we are not restricting the analyses to organizations founded since 1950 for this exercise.*

In all cases, conduct two-sided hypothesis tests. Use a 5 percent alpha unless otherwise indicated.

## 8.1. Interaction Between Two Dummy Variables

(a) SAS/Stata tasks:

(i) Create a dummy variable to indicate organizations that are independent based on the original variable a3. (Note that the if statement on the use command in Stata or in the data step in SAS already excluded cases with missing data on a3).

(ii) Re-create the dummy variable that you created in Chapter 7 to indicate organizations that are *required to report* the demographic composition of their workforce to the government, based on the original a5 variable. (Note that the if statement on the use command in Stata or in the data step in SAS already excluded cases with missing data on *a5*.)

(iii) Create a product term interacting the dummy variables for *independent* organizations and organizations that are *required to report* their demographic composition to the government.

(iv) Calculate the mean of *mngpctfem* for each of the four subgroups created by the two dummies (i.e., independent/required to report; independent/not required to report; not independent/required to report; not independent/not required to report). Use the proc sort and by commands in SAS and the bysort command in Stata, similar to Section 8.4.1, but in combination with the `proc means` command in SAS and the `summarize` command in Stata. We will refer to these results as Summarize #1 in the write-up.

(v) Regress *mngpctfem* on the two dummy variables (we will refer to this additive model as Regression #1 in the Write-up tasks).

(vi) Regress *mngpctfem* on the two dummy variables and the product term (we will refer to this interaction model as Regression #2 in the Write-up tasks).

 (1) Use the Stata `lincom` command to test the conditional effect of being independent for organizations that are required to report their demographic composition to the government (we will refer to this as Test #1 in the Write-up tasks).

 (2) Use the Stata `lincom` command to test the conditional effect of being required to report their demographic composition to the government among independent organizations (we will refer to this as Test #2 in the Write-up tasks).

(b) Write-up tasks:

(i) Write the prediction equation for Regression #2. Calculate the four possible conditional means for Regression #2 by hand. Discuss how these means relate to the results from Summarize #1.

(ii) Write the prediction equation for Regression #1. Calculate the four possible conditional means for Regression #1 by hand. Discuss how these means relate to the results from Summarize #1.

(iii) Interpret the intercept, each dummy variable coefficient, and the product term in Regression #2. Which differ significantly from zero at an alpha of 0.05? What about an alpha of 0.10?

(iv) Interpret the coefficients estimated in Test #1 and Test #2. Which differ significantly from zero at an alpha of 0.05? What about an alpha of 0.10?

(v) Show how to calculate the coefficients in Regression #2 and the coefficients estimated in Test #1 and Test #2 by hand using the results from Summarize #1.

## 8.2. Interaction Between a Dummy and Interval Variable

(a) SAS/Stata tasks:

(i) Create a new rescaled variable for *ftsize* in units of 100.

(ii) Create a product term interacting the dummy variable indicating *independent* organizations with the rescaled *ftsize* variable that captures the number of full-time employees.

(iii) Regress *mngpctfem* on the independent dummy, the rescaled ftsize variable, and their interaction (we will refer to this as Regression #3 in the Write-up tasks).

  (1) Use the SAS and Stata `test` commands to conduct a Chow test that the intercept and slope for the association of rescaled *ftsize* with *mngpctfem* differ for organizations that are independent and those that are part of a larger organization (we will refer to this as Test #3 in the Write-up tasks).

  (2) Use the Stata `lincom` command to calculate the conditional slope of rescaled *ftsize* for organizations that are not *independent* (we will refer to this as Test #4 in the Write-up tasks).

  (3) Use the Stata `lincom` command to calculate the conditional effect of *independent* for organizations that have 5, 50, and 500 full-time employees (we will refer to this as Test #5 in the Write-up tasks). *Be careful to rescale the values in the* `lincom` *command (e.g., 5/100 = 0.05).*

(iv) Regress *mngpctfem* on the rescaled *ftsize* variable (we will refer to this as Regression #4 in the Write-up tasks).

(b) Write-up tasks:

(i) Use Regression #3 and Regression #4 to calculate the F-value by hand for the Chow test that you asked SAS and Stata to conduct in Test #3. Verify that the value matches the value you obtained from SAS and Stata. Write the null and alternative hypotheses for this test and make a conclusion using the SAS or Stata output (use a 5 percent alpha level).

(ii) Calculate by hand the coefficient for the test that you asked Stata to conduct in Test #4.

(iii) Calculate by hand the coefficients for the tests that you asked Stata to conduct in Test #5. *Be careful to rescale the values in your hand calculations (e.g., 5/100 = 0.05).*

### 8.3. Interaction Between Two Interval Variables

(a) SAS/Stata tasks:

(i) Create a new rescaled variable for *age* in units of 10.

(ii) Create a product term interacting the *rescaled age* variable that captures the years since the organization was founded and the rescaled *ftsize* variable that captures the number of full-time employees.

(iii) Regress *mngpctfem* on the *rescaled age* variable, the rescaled *ftsize* variable, and their interaction (we will refer to this as Regression #5 in the Write-up tasks).

    (1) Use the Stata `lincom` command to calculate the conditional effects of rescaled age for organizations that have 5, 50, and 500 full-time employees (we will refer to this as Test #6 in the Write-up tasks). Be careful to rescale the values in the `lincom` command (*e.g., 5/100 = 0.05*).

    (2) Use the Stata `lincom` command to calculate the conditional effects of rescaled ftsize for organizations that are 1, 10, and 100 years old (we will refer to this as Test #7 in the Write-up tasks). *Be careful to rescale the values in the* `lincom` *command (e.g., 1/10 = 0.10).*

(b) Write-up tasks:

(i) Based on the parameter estimates in Regression #5, is there evidence of an interaction between rescaled age and rescaled *ftsize* in predicting *mngpctfem*? Justify your response by stating the null and alternative hypotheses and the *t*-value and/or *p*-value.

(ii) Calculate by hand the coefficients for the tests that you asked Stata to conduct in Test #6. *Be careful to rescale the values in the* `lincom` *command (e.g., 5/100 = 0.05).*

(iii) Calculate by hand the coefficients for the tests that you asked Stata to conduct in Test #7. *Be careful to rescale the values in the* `lincom` *command (e.g., 1/10 = 0.10).*

## COURSE EXERCISE

Identify at least one interaction model that you might estimate, based on the data set that you created in the course exercise to Chapter 4. Ideally, think about conceptual reasons why you might expect interactions between some of the dummy and/or interval variables available in the data set. Alternatively, choose a few variables and explore whether interactions are identified empirically in the data set.

Create the product term between the two variables in your interaction. Estimate the interaction model, including both the variables and their interaction. Refer to the sections of the chapter summarizing how to interpret the coefficients based on each type of model, and write a one-sentence interpretation of each coefficient. Use the strategies outlined in the chapter to calculate and test the significance of conditional effects, depending on the types of variable in your model (dummy and/or interval). Follow the steps outlined in Appendices H.B.1 to H.8.4 to graph the results. Discuss whether the conditional effects that you graph are statistically significant and substantively meaningful.

# NONLINEAR RELATIONSHIPS

## CHAPTER 9: NONLINEAR RELATIONSHIPS

To date, we have assumed that the association between an interval predictor variable and the outcome is linear. A one-unit increase in the predictor results in the same change in the outcome across all levels of the predictor. A nonlinear relationship, in contrast, allows the effect of the predictor to differ at different starting values. For example, the difference in school achievement may be greater between families with $11,000 and $10,000 annual incomes than families with $101,000 and $100,000 annual incomes. In this chapter we present techniques for modeling nonlinear relationships between the predictor and outcome within the context of OLS estimation and demonstrate how to use various nonlinear shapes to match our conceptual models. In Chapter 12, we will provide a roadmap to additional strategies for modeling nonlinear relationships, outside of OLS.

## 9.1: NONLINEAR RELATIONSHIPS

### 9.1.1: Some Possible Shapes of Nonlinear Relationships

We will discuss three of the most common approaches to modeling nonlinear relationships within OLS in the social sciences (see Box 9.1):

(a)  transforming $X$ or $Y$ using natural logs;[1]
(b)  using a quadratic form of $X$;
(c)  using dummy variables for $X$.

Figure 9.1 illustrates some possible nonlinear shapes associating $X$ and $Y$. Notice that in the middle and bottom figures, the steepness of the slope varies, but it never changes sign, across the levels of $X$. In the middle two figures, the slope is always positive, and

---

■ **Box 9.1**

We focus on these three because they are commonly used, but other transformations are possible. Some allow for an exploratory search for the best-fitting nonlinear curve associating a predictor with an outcome (e.g., developed by Box and Cox 1964 and Tukey 1977; see Fox 2008, for an introduction to transformations in general and their application in regression analysis). Other *spline* or *piecewise linear* models allow a linear relationship between the predictor and outcome that varies in steepness in different ranges of the predictor (Fox 2008; Gujarati 2003). These models can be quite useful in some contexts, but it is rare in the social sciences for theoretical models to suggest where the slopes would change, and the quadratic and logarithmic approaches are more commonly used.

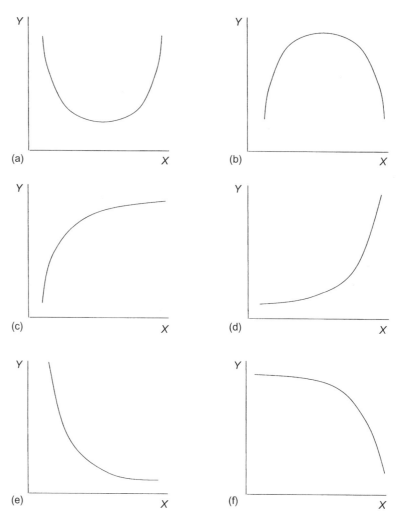

■ **Figure 9.1:** Some Possible Nonlinear Shapes Associating *X* with *Y*

in the bottom two figures the slope is always negative. In contrast, in the top two figures, the slope reverses sign. In Figure 9.1(a), the slope starts out negative and then becomes positive. In Figure 9.1(b), the slope starts out positive and then becomes negative.

Figure 9.2 adds dashed lines drawn to just touch the curves at various points. These lines are referred to as **tangent lines**. They help us to visualize the slope at different levels of *X*. Starting in the bottom two figures, we see in Figure 9.2(e) that the slope begins steeply negative at lower values of *X* and then approaches zero at higher values of *X*. In contrast, in Figure 9.2(f), the slope begins near zero, and then becomes steeply negative. We refer to the association in Figure 9.2(e) as **decreasing at a decreasing rate** and the association in Figure 9.2(f) as **decreasing at an increasing rate**. In Figure 9.2(c), the slope begins steeply positive and then approaches zero. In Figure 9.2(d), the slope begins close to zero and then becomes steeply positive. We refer to Figure 9.2(c) as **increasing at a decreasing rate** and Figure 9.2(d) as **increasing at an increasing rate**. The top two figures put together two of these types of association. In Figure 9.2(a), the association is decreasing at a decreasing rate at lower values of

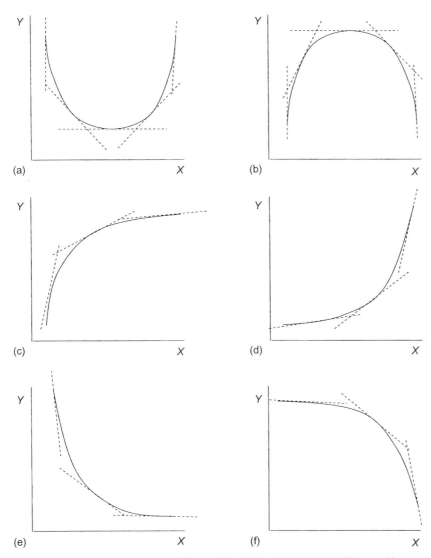

**Figure 9.2:** Some Possible Nonlinear Shapes Associating *X* with *Y*: With Tangent Lines

$X$ and then increasing at an increasing rate at higher values of $X$. At some point in the middle, the slope becomes zero. In Figure 9.2(b), the association is increasing at a decreasing rate at lower values of $X$ and decreasing at an increasing rate at higher values of $X$. At some point in the middle, the slope becomes zero.

We can imagine some substantive relationships that might produce shapes like those in Figure 9.2.

■ In Figure 9.2(a), we might imagine associating financial or physical dependency as an outcome and age as a predictor. Children are quite dependent on adults for their physical needs, especially when very young. With the transition to adulthood, they become able to provide for their own needs. Then, with old age, they become increasingly dependent on others.

■ In Figure 9.2(b), we could envision an association between income and age. Children have little personal income, especially during preschool and school age. As they move through adolescence and into adulthood, they begin to produce some earnings. At some point in middle age, their incomes peak and then begin to decline as they retire and enter old age.

■ In Figure 9.2(c), we might think of predicting a positive developmental outcome (such as reading comprehension) based on family income. We might imagine that each increment to income associates more strongly with increased reading comprehension among lower than higher income families. At higher income levels, additional income may have very little association with reading comprehension.

■ In Figure 9.2(e), we might similarly imagine associating a negative developmental outcome (such as problem behaviors) with family income. We might imagine that each increment to income associates more strongly with decreased problem behaviors among lower-income families. At higher-income levels, additional income may have very little association with behavior problems.

■ In Figure 9.2(d), we can imagine associating earnings as the outcome with education as a predictor. We might expect that returns to education are larger at higher education levels (e.g., that the increment to earnings for adults with one additional year of graduate school is larger than the increment to earnings for adults with one additional year of high school).

■ In Figure 9.2(f), we might imagine a negative association of health behavior, such as smoking or eating processed foods, with education, expecting that these behaviors might diminish more rapidly at higher than at lower levels of education.

Each of the graphs in Figure 9.2 shows a smooth relationship between $X$ and $Y$. We might also have conceptual models that suggest nonlinear associations that are not smooth. For example, consider the example of returns to education that we posed for Figure 9.2(d). Rather than a smoothly increasing association between education and

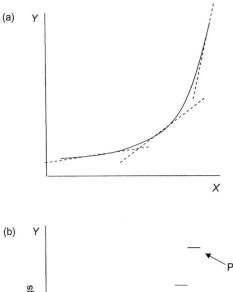

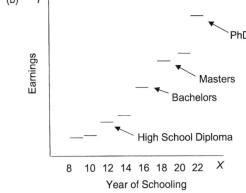

■ **Figure 9.3:** Stair Step Nonlinear Relationship Between $X$ and $Y$

earnings, as shown in Figure 9.3(a), we might expect more of a stair-step type of association, as shown in Figure 9.3(b). Earnings may increase more with years of schooling associated with degree completion (e.g., from 11 to 12 and from 15 to 16) than those associated with one more year toward a degree (e.g., 10 to 11 and 14 to 15). Indeed, we might treat different categories of education as ordinal categories rather than interval numbers, like years of schooling.

## 9.1.2: Literature Excerpts

Four literature excerpts help to illuminate further how nonlinear associations might be used conceptually.[2]

### Literature Excerpt 9.1: Age and Household Income

Andrea Willson (2003) examined the kind of age–income trajectory we conceptualized above, but focused particularly on women. She was interested in how women's income trajectories were influenced not only by their own employment but also by their marital status (which could give access to a husband's earnings and retirement benefits) and race. Her Figure 3, reproduced in Literature Excerpt 9.1, plots predicted values from a regression model that allows for a **curvilinear** association between age and household income. The plots show the predicted peak in income during middle age, with the association being especially curved for white married women who were always employed (top curve). In contrast, the association is slightly but less sharply curved for black women who were never married and not always employed (bottom curve). One of the conclusions from the figure is that income inequality is greatest in midlife and diminishes in older age (although as Willson points out, her outcome does not capture potential asset differentials in later life, which might reveal additional inequality in older age).

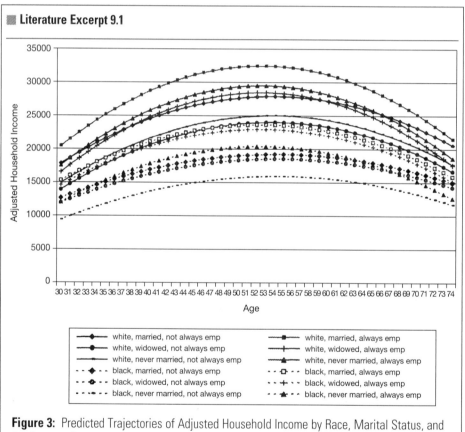

**Figure 3:** Predicted Trajectories of Adjusted Household Income by Race, Marital Status, and Employment Pattern

**Source:** Willson, Andrea E. 2003. "Race and Women's Income Trajectories: Employment, Marriage, and Income Security over the Life Course." *Social Problems*, 50, 87–110.

## Literature Excerpt 9.2:  Age and Social Contact

Cornwell, Laumann, and Schumm (2008) take up the often held notion that the elderly are socially isolated. As they note: "Contrary to the image of older adults as either helpless victims of modernization or authors of their own isolation, this line of research portrays older individuals as resilient to potentially isolating events like retirement and bereavement" (p. 186). Although they find that many associations with social connectedness are linear, they find that the association of age with the volume of contact with network members is curvilinear (see Literature Excerpt 9.2). They speculate about the reasons for the U-shaped pattern.

> Contact volume may decrease through the 50s and 60s because social roles begin to dissipate around this time. Contact volume is lowest for those in their late 60s and early 70s, but it may increase as respondents grow older and they adapt to the loss of social roles, friends, and family members (p. 193).

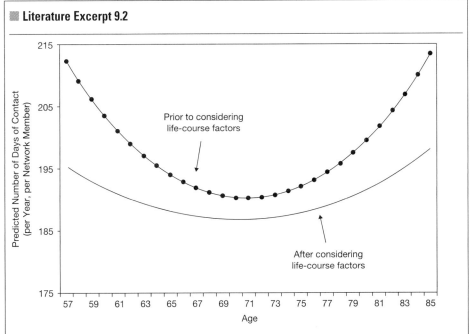

■ **Literature Excerpt 9.2**

**Figure 2:** Older Adult's Predicted Volume of Interaction with Network Members, per Year and by Age

**Source:** Cornwell, Benjamin, Edward O. Laumann, and L. Philip Schumm. 2008. "The Social Connectedness of Older Adults: A National Profile." *American Sociological Review*, 73, 185–203.

The authors note that their interpretation of this curvilinear association is speculative, since little prior conceptual and empirical work has anticipated or found such nonlinear relationships.

### Literature Excerpt 9.3: Per Capita Income and Health

Pritchett and Summers (1996) consider the nonlinear way in which "wealthier is healthier" at the country level. Their Figure 1, reproduced in Literature Excerpt 9.3, shows that as per capita income increases, infant mortality declines and life expectancy increases, but these associations are particularly rapid at lower income levels. They interpret the association using techniques we will discuss below, noting that "The

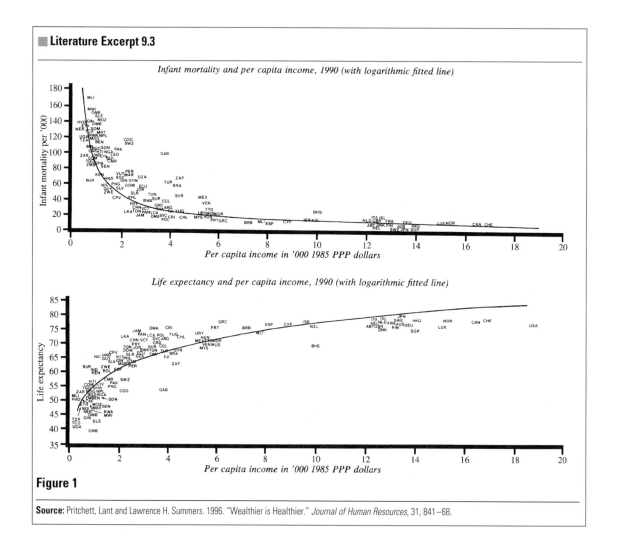

■ **Literature Excerpt 9.3**

*Infant mortality and per capita income, 1990 (with logarithmic fitted line)*

*Life expectancy and per capita income, 1990 (with logarithmic fitted line)*

**Figure 1**

**Source:** Pritchett, Lant and Lawrence H. Summers. 1996. "Wealthier is Healthier." *Journal of Human Resources*, 31, 841–68.

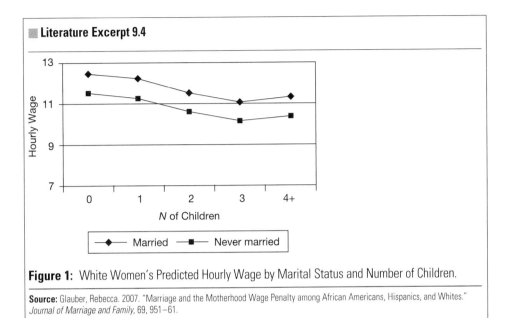

**Figure 1:** White Women's Predicted Hourly Wage by Marital Status and Number of Children.

**Source:** Glauber, Rebecca. 2007. "Marriage and the Motherhood Wage Penalty among African Americans, Hispanics, and Whites." *Journal of Marriage and Family*, 69, 951–61.

estimates imply that if income were 1 percent higher in the developing countries, as many as 33,000 infant and 53,000 child deaths would be averted annually" (p. 844).

### Literature Excerpt 9.4: Number of Children and Women's Hourly Wage

Glauber (2007) examines the wage penalty associated with motherhood. She dummy-coded the number of children into five categories (grouping together women with four or more children), allowing for a relationship between children and hourly wage that could take any form, including nonlinear. Her Figure 1, shown in Literature Excerpt 9.4, shows the association separately for married and never married women. This flexible dummy variable specification reveals that wages generally decline with more children, but the decrease is sharpest between one and two children.

### 9.1.3: Specifying Nonlinear Models

We will cover the typical statistical techniques for modeling the shapes illustrated in Figures 9.2 and 9.3. Mathematically, the figures in the top of Figure 9.2 are **quadratic** functions, and can be modeled in regression by adding a squared term for the predictor. The shapes shown in Figures 9.2(c)–(e) are typically modeled with a natural log transformation of the outcome and/or the predictor. The shape shown in Figure 9.2(f) is often modeled using a quadratic function in which only a portion of the curve is observable in the range of $X$s in the data. And, the stair-step type of relationship shown in Figure 9.3 is modeled with a set of dummy variables for the predictor.

## Quadratic Form

How do we capture curves like those shown in Figures 9.2(a) and (b) in regression models? In order to specify a quadratic functional form, we add a term to the regression model for the square of the predictor variable.

$$Y_i = \beta_0 + \beta_1 X_i + \beta_2 X_i^2 + \varepsilon_i \tag{9.1}$$

The U-shape shown in Figure 9.2(a) occurs when $\beta_2 > 0$ and the inverted U-shape shown in Figure 9.2(b) occurs when $\beta_2 < 0$. If $\beta_2 = 0$, then the model reduces to the linear functional form:

$$Y_i = \beta_0 + \beta_1 X_i + 0 * X_i^2 + \varepsilon_i$$
$$= \beta_0 + \beta_1 X_i + \varepsilon_i$$

This means that the linear model is nested within the quadratic model, and we can thus use a general linear $F$-test to test whether a quadratic form fits the data better than a linear form.

The value of $X$ at which the slope becomes zero—the horizontal tangent lines in Figure 9.2—can be calculated using the formula $-\beta_1/2\beta_2$.[3] Calculating this point can be useful, to discern whether this point at which the slope changes direction is within the range of $X$ values in the data set, or alternatively is to the left or to the right of the observed $X$ values. Indeed, within applications, only a portion of the U or inverted-U may actually fit the data, such as in Figure 9.2(f).

As illustrated in Figure 9.2 and the examples, the slope is no longer constant in a regression model with a quadratic functional form. Thus, we can no longer make a general statement that a one-unit change in $X$ is associated with a given expected change in $Y$. Indeed, clearly, it is possible for increases in $X$ to be associated with decreases in $Y$ over some range of the $X$ values, to be associated with no change in $Y$ at some point, and to be associated with increases in $Y$ over another range of the $X$ values. To illustrate these nonlinear relationships, it is often useful to calculate predicted values over the range of $X$ in the sample and plot them, or to calculate predicted values at key values of $X$ to display in a table or paper text. The slope at particular values of $X$ can also be calculated using the equation $\beta_1 + 2\beta_2 X_i$.[4] Note that if $X_i = 0$, then this equation works to $\beta_1 + 2\beta_2 * 0 = \beta_1$. So, the coefficient for $\beta_1$ is the slope associating the predictor with the outcome when the predictor is zero. If zero is not a valid value on the predictor, then $\beta_1$ is not meaningful.

## Logarithmic Transformation

The shapes in Figures 9.2(c)–(e) can be specified by taking the **natural log** of the outcome (9.2(c), 9.2(e)), predictor (9.2(d), 9.2(e)), or both (all three shapes).

With the **logarithmic transformations,** we model the linear association between the log of $X$ and/or log of $Y$. We can then transform the coefficient estimates and predicted values back into the natural units of $X$ and $Y$ for interpretation, and reveal the nonlinear relationship between $X$ and $Y$.

Models in which $Y$ but not $X$ is transformed are sometimes referred to as log-lin models (see Box 9.2). Similarly, models in which both $Y$ and $X$ are transformed are log-log models and models in which $X$ but not $Y$ is transformed are called lin-log models.

### Percentage Change

To begin, let's firmly define four types of change: **absolute change, factor change, proportionate (or relative) change,** and **percentage change.** Factor change and percentage change are frequently used when interpreting models based on the logarithmic transformation in OLS. Recall that in the Literature Excerpt 9.3 we quoted the authors' interpretation of the effect, which referred to a 1 percent change rather than a one-unit change in the predictor. The following table provides general formulae and two examples for four types of change.

| Type of Change | General Formula | Examples | |
|---|---|---|---|
| | | $X_{new} = 102, X_{old} = 101$ | $X_{new} = 2, X_{old} = 1$ |
| Absolute | $X_{new} - X_{old}$ | $102 - 101 = 1$ | $2 - 1 = 1$ |
| Factor | $\dfrac{X_{new}}{X_{old}}$ | $\dfrac{102}{101} = 1.009901$ | $\dfrac{2}{1} = 2$ |
| Relative | $\dfrac{\left( X_{new} - X_{old} \right)}{X_{old}}$ | $\dfrac{(102 - 101)}{101} = 0.009901$ | $\dfrac{(2-1)}{1} = 1$ |
| Percentage | $\left[ \dfrac{\left( X_{new} - X_{old} \right)}{X_{old}} \right] * 100$ | $\left[ \dfrac{(102 - 101)}{101} \right] * 100 = 0.99\%$ | $\left[ \dfrac{(2-1)}{1} \right] * 100 = 100\%$ |

Note that we wrote the table with $X$s (change on predictor variables) but the results are the same for $Y$s (change on outcome variables).

The examples make clear that even when absolute change is the same for two different starting values (one unit of absolute change in our examples), factor, relative, and percentage change depend upon the starting value. In our example, an increase in 1 is a larger proportional change at smaller values of the variable (starting at 1) than at larger values of the variable (starting at 101).

Notice that relative and percentage change can be rewritten in terms of factor change.

Relative Change
$$\frac{(X_{new} - X_{old})}{X_{old}} = \frac{X_{new}}{X_{old}} - \frac{X_{old}}{X_{old}} = \frac{X_{new}}{X_{old}} - 1 \qquad = \text{Factor Change} - 1 \qquad (9.2)$$

Percentage Change
$$\frac{X_{new} - X_{old}}{X_{old}} * 100 = \left[\frac{X_{new}}{X_{old}} - \frac{X_{old}}{X_{old}}\right] * 100 = \left[\frac{X_{new}}{X_{old}} - 1\right] * 100 = (\text{Factor Change} - 1) * 100 \quad (9.3)$$

### Log-Lin Model

We write the log-lin regression model as follows:

$$\ln(Y_i) = \beta_0 + \beta_1 X_{1i} + \varepsilon_i \qquad (9.4)$$

For interpretation, it is useful to exponentiate both sides of the equation resulting in:[5]

$$Y_i = \exp(\beta_0 + \beta_1 X_{1i} + \varepsilon_i) = e^{\beta_0 + \beta_1 X_{1i} + \varepsilon_i} \qquad (9.5)$$

We write the prediction equation based on Equation 9.4 as $\widehat{\ln(Y_i)} = \hat{\beta}_0 + \hat{\beta}_1 X_{1i}$. Thus, a prediction based on the estimated coefficients is in log $Y$ units.

We need a few technical details to accurately transform these predictions back into the natural units of $Y$. It is intuitive to transform the predictions back to $Y$'s natural units by taking the exponential of both sides of this prediction equation:

$$\hat{Y}_i = \exp(\hat{\beta}_0 + \hat{\beta}_1 X_{1i}) = e^{\hat{\beta}_0 + \hat{\beta}_1 X_{1i}} \qquad (9.6)$$

But, if $Y$ really follows a lognormal distribution, then it can be shown that the predicted values should in fact be adjusted for the estimate of the conditional variance (i.e., $\hat{\sigma}^2$ the mean square error; Greene 2008).

$$\hat{Y}_i = \exp(\hat{\sigma}^2/2)\exp(\hat{\beta}_0 + \hat{\beta}_1 X_{1i}) = e^{\hat{\sigma}^2/2} e^{\hat{\beta}_0 + \hat{\beta}_1 X_{1i}} \qquad (9.7)$$

The larger the mean square error, the more this adjustment will affect the predictions.

Wooldridge (2009, 210–13) also recommends another adjustment which does not assume normality of the error terms:

$$\hat{Y}_i = \widehat{\exp(\varepsilon)}\exp(\hat{\beta}_0 + \hat{\beta}_1 X_{1i}) = \widehat{e^{(\varepsilon)}}e^{\hat{\beta}_0 + \hat{\beta}_1 X_{1i}} \tag{9.8}$$

where $\widehat{\exp(\varepsilon)}$ is estimated by calculating the residuals from Equation 9.4 and then exponentiating them and taking their average. This will produce a single coefficient estimate which is the estimate of $\widehat{\exp(\varepsilon)}$ that is used in adjusting the predictions for interpretation. We illustrate how to use this approach below.

### Exact Interpretation of Coefficient

We can also examine change in predicted values as we did in earlier chapters to help us to interpret the coefficient estimates from the log-lin regression, although rather than considering the absolute change in the outcome, we will consider its factor changes (the ratio of predicted values rather than the difference in predicted values). The results are the same using either Equation 9.6, 9.7, or 9.8, so we will use the simpler 9.6. We substitute into the equation the generic notation for $X_1 = X$ and $X_1 = X + 1$ to denote a one-unit increase in $X$ and rewrite the results using the laws of exponents.[6]

▨ **Table 9.1: Expected Values at Two Levels of *X***

| | | | |
|---|---|---|---|
| $E(Y\|X_1 = X+1)$ | $\exp(\beta_0 + \beta_1(X+1))$ | $\exp(\beta_0 + \beta_1 X + \beta_1)$ | $\exp(\beta_0)\exp(\beta_1 X)\exp(\beta_1)$ |
| $E(Y\|X_1 = X)$ | $\exp(\beta_0 + \beta_1 X)$ | $\exp(\beta_0 + \beta_1 X)$ | $\exp(\beta_0)\exp(\beta_1 X)$ |

We can now calculate the factor change in the expected value: the ratio of the predicted value when $X$ is incremented by 1 over the predicted value when $X$ is at its original value.

$$\frac{E(Y|X_1 = X+1)}{E(Y|X_1 = X)} = \frac{\exp(\beta_0)\exp(\beta_1 X)\exp(\beta_1)}{\exp(\beta_0)\exp(\beta_1 X)} = \exp(\beta_1)$$

We can rewrite the final expression as follows:

$$E(Y|X_1 = X + 1) = \exp(\beta_1) * E(Y|X_1 = X)$$

This makes it clear that taking the exponential of the slope coefficients gives us the factor by which $Y$ is expected to change, when $X$ increases by 1. For example, if $\exp(\beta_1) = 1.25$, then we would say "$Y$ is expected to be 1.25 times larger when $X$ increases by 1." Thus, for the log-lin model, we can concisely interpret the nonlinear relationship between $X$ and $Y$ in terms of the factor change. It is sometimes convenient also to convert this factor change to relative or percentage change. Using Equations 9.2 and 9.3, relative change would be $\exp(\beta_1) - 1$ and percentage change would be

$[\exp(\beta_1) - 1] * 100$. So, our factor change of 1.25 would be a relative change of 0.25 and a percentage change of 25 percent. We could say "$Y$ is expected to be 25 percent larger when $X$ increases by 1."

It is straightforward to show that the interpretation of the effect of a dummy variable is similar. For the model $\ln (Y_i) = \beta_0 + \beta_1 X_{1i} + \beta_2 D_{1i} + \varepsilon_i$, taking the exponential of $\beta_2$ gives the factor by which $Y$ is expected to differ for cases indicated by $D = 1$ versus cases indicated by $D = 0$, holding $X_1$ constant. Subtracting 1 from this exponentiated value, and multiplying by 100, gives us the percentage by which cases indicated by $D = 1$ differ on $Y$ from cases coded $D = 0$, holding $X_1$ constant.

In log-lin models, absolute change does not reduce to such a concise interpretation.

$$E(Y|X_1 = X + 1) - E(Y|X_1 = X) = [\exp(\hat{\sigma}^2/2)\exp(\beta_0)\exp(\beta_1 X)\exp(\beta_1)]$$
$$- [\exp(\hat{\sigma}^2/2)\exp(\beta_0)\exp(\beta_1 X)]$$

To interpret the absolute change requires calculating predicted values for specific values of $X$ and summarizing these various values in the text, a table, or a graph.

### Approximate Interpretation of Coefficient

In publications, you may sometimes see authors interpret the coefficient estimate from a log-lin regression model as relative change, without exponentiating the value and subtracting one. This approximation holds when the coefficient estimate is small and the change in $X$ is small. However, as is seen in the table below, clearly this approximation is adequate only for small coefficients and with today's computing power it is easy to make the exact calculations.

| Value of Coefficient | Percentage Change Interpretation (%) | |
| --- | --- | --- |
| $\beta_1$ | Exact<br>$[(\exp(\beta_1) - 1)]*100$ | Approximate<br>$\beta_1 *100$ |
| .05 | 5.13 | 5 |
| .10 | 10.52 | 10 |
| .25 | 29.40 | 25 |
| .50 | 64.87 | 50 |
| 1.00 | 171.83 | 100 |

### Log-Log and Lin-Log Models

In some models, we may also transform an interval *predictor* variable by taking its log.

lin-log     $Y_i = \beta_0 + \beta_1 \ln (X_{1i}) + \varepsilon_i$

log-log     $\ln (Y_i) = \beta_0 + \beta_1 \ln (X_{1i}) + \varepsilon_i$

One approach to interpretation for these models uses the same kind of approximation that we discussed for the log-lin model.

1.  In log-log models, the coefficient for the logged predictor $\beta_1$ can be interpreted *approximately* as: a 1 percent increase in $X$ is associated with a $\beta_1$ percent change in $Y$, all else constant.
2.  In lin-log models, the coefficient for the logged predictor $\beta_1$ can be interpreted *approximately* as: a 1 percent increase in $X$ is associated with a $\beta_1/100$ change in $Y$, all else constant.

The first interpretation above is used frequently in economics, and in subfields of other disciplines that intersect with economics. The effect of a 1 percent change in $X$ on the percent change in $Y$ captures elasticity, a central concept in economics. In addition, wages and incomes are typically positive values with a right skew and often become closer to the normal distribution with a log transformation. A recent paper published in the *American Journal of Sociology* provides an example. Sharkey (2008, 950–1) writes:

> Table 2 contains results from the basic model of intergenerational mobility, where the log of the average neighborhood income of parents predicts the log of average neighborhood income of children as adults. Under ordinary least squares (OLS), the intergenerational elasticity is .64 meaning a 1 percent change in the parent's neighborhood income is associated with a .64 percent change in the child's neighborhood income as an adult.

Alternatively, the results of log-log and lin-log models can be interpreted using predicted values. In log-log models, the values of $X$ must be transformed into $ln (X)$ before making the predictions and the predicted values must be transformed back into the original units by taking the exponential of the predicted values (and adjusted, as shown in Equations 9.7 or 9.8). In lin-log models, the values of $X$ must be transformed into $ln (X)$ before making the predictions, but the predicted values will be in the natural units of $Y$.

## Dummy Variables

Dummy variables can be used to model an association between an interval or ordinal predictor and the outcome, allowing for a flexible relationship between the predictor and outcome. This approach is especially useful for ordinal variables, where there are a relatively small number of categories and substantial sample sizes in each category, such as the Literature Excerpt 9.4 which examined the number of children as a predictor. When categories are numerous and/or when substantive rationales apply, original categories can also be grouped to form larger categories. For example, in Literature Excerpt 9.4, mothers with four or more children were grouped together.

$$wage = \beta_0 + \beta_1 onekid + \beta_2 twokid + \beta_3 threekid + \beta_4 fourpkid + \varepsilon_i$$

Tests among the included categories can be conducted as discussed in Chapter 7 using a different reference category, test commands, or linear combinations (e.g., tests of $\beta_1 = \beta_2$ or equivalently $\beta_1 - \beta_2 = 0$).

When we use dummy variables to indicate the categories of ordinal or interval variables, we can also test the linearity of the relationship between the predictor and outcome (see Box 9.3). If the association is in fact linear, then the change in $Y$ between adjacent categories of $X$ should be equivalent across levels of $X$. For the model in Literature Excerpt 9.4 if the change is linear, we expect

$$(\beta_1) = (\beta_2 - \beta_1) = (\beta_3 - \beta_2) = (\beta_4 - \beta_3)$$

■ **Box 9.3**

For interval variables, we might group several adjacent categories together to produce a large enough sample size for a dummy variable specification. For example, for age as a predictor, we might group age in 5- or 10-year intervals. Then, the dummies no longer capture a one-unit change on the predictor (rather, for example, a 5- or 10-unit change). But, using this approach can be helpful for visualizing the shape of the relationship between a predictor and outcome.

These constraints can be tested with a general linear $F$-test. (The last might be excluded in this example, since *fourpkid* groups together women with five and more children.)

### 9.1.4: Examples

We will now look at an example based on our NSFH distance outcome. We will illustrate the quadratic and logarithmic models by predicting distance based on the mother's years of schooling. We might expect that distance would associate with years of schooling with an "increasing at an increasing rate" shape. As noted previously, education may lead to greater distance. For example, education may associate with more job opportunities in distant areas and greater income for moves, especially at the highest education levels.[7] We will illustrate the dummy variable approach based on the adult's number of sisters. This variable is a count with a small number of values. Thus, its association with the outcome can be flexibly modeled with a small number of dummy variables. We might expect that an adult with more sisters can rely on these sisters to remain close to an aging parent, allowing the adult to live further away.[8] We will begin with a simple model where just the outcome and predictor of interest (mother's years of schooling; number of sisters) are in the model, but we will add the rest of the control variables from our data set when we discuss outliers and heteroskedasticity in Chapter 11.

## Quadratic and Logarithmic Models

We can create the needed squared and logged variables using operators in SAS and Stata. The squared term can be calculated using the multiplication operator, which we used for interactions. But, now the variable is multiplied by itself. In both SAS and Stata the `log` command takes the natural log of the variable. When using the log transformation, it is important to remember that the log of a negative value and the log of zero are undefined. Because of these properties, logarithmic transformations are best applied to variables that naturally take on only positive values. For variables in which possible values also include zero, the transformation is sometimes taken after adding some constant, typically one (e.g., $\log X = \log(X + 1)$). This adjustment works well and conforms with the interpretations described above as long as there are not many zeros and the maximum value in the natural units is not small. Both *g1miles* and *g1yrschl* are non-negative; in our analytic data set *g1miles* does not contain zeros, but *g1yrschl* does contain zeros. So, our commands would be as follows.

|  | **SAS** | **Stata** |
| --- | --- | --- |
| squared predictor | sqg1yrschl=g1yrschl*g1yrschl; | generate sqg1yrschl=g1yrschl*g1yrschl |
| logged predictor | logg1yrschl=log(g1yrschl+1); | generate logg1yrschl=log(g1yrschl+1) |
| logged outcome | logg1miles=log(g1miles); | generate logg1miles=log(g1miles) |

### Techniques for Choosing among Models

We estimate six models:

(a) a linear model, in which neither the outcome nor predictor are logged (which we'll refer to as lin-lin);
(b) a lin-log model;
(c) a quadratic model with a linear outcome (which we'll refer to as lin-sq);
(d) a log-lin model;
(e) a log-log model;
(f) a quadratic model with a logged outcome (which we'll refer to as log-sq).

We will estimate each model and then make predictions from the models to examine the shape of the estimated relationship.

Four of the models, the log-log, log-lin, lin-sq, or log-sq, might capture the expected "increasing at an increasing rate" association between the mother's years of schooling and distance. If each shows this association, which should we choose as the best? For nested models, we can use the general linear $F$-test to compare models. Only two sets of models are nested, however. The lin-sq model is nested within the lin-lin model and the log-sq model is nested within the log-lin model. In these cases, the $t$-test for the squared term conducts the general linear $F$-test of whether the quadratic form fits better than the linear form for the relationship.

As long as the two models have the same outcome variable, we can also compare the size of the adjusted $R$-squared values to compare them, even when not nested. We can also use AIC and BIC to compare non-nested models. In our case, we can use the adjusted $R$-squared to compare among each set of three models with the same outcome: the linear outcome (lin-lin, lin-log, lin-sq) and the log outcome (log-lin, log-log, log-sq).

It is not appropriate to compare the $R^2$ or adjusted $R^2$ values of models with different outcomes, including our non-logged versus logged outcomes. As a critique of early applications of models with logged outcomes emphasized:

> The authors make [the choice between the models with non-logged and logged outcomes] by comparing the $R^2$s. But the $R^2$ yielded by estimation of [the logged model] is not the proportion of the variance of Y explained by the regression. It is rather the proportion of the variance of the logarithm of $Y$ explained by the regression: logy, not $Y$, is the dependent variable. The two regression models are explaining different sources of variation, and the $R^2$s are therefore in principle incomparable (Seidman 1976).

It is possible to calculate a version of $R^2$ for the model with a logged outcome, by converting predictions based on the logged model back into their natural units and then regressing the outcome in its natural units on these predictions (for example, this is the approach suggested by Seidman 1976). The rescaling adjustments to the predicted values discussed above for log outcomes will not affect this $R$-squared value, so any of the predictions (Equations 9.6, 9.7, and 9.8) can be used to make the calculation.

### Estimated Models

Display B.9.1 provides the SAS and Stata results for the three models with the nonlogged outcome. Display B.9.2 provides similar results for the three models with the logged outcome. Display B.9.3 shows how to calculate $\widehat{\exp(\varepsilon)}$ to adjust the predictions in logged units and to calculate the approximate $R$-squareds to compare the logged to the linear models. Finally, Display H.9.1 summarizes the results of the models in Excel, and calculates and graphs predicted values for specific levels of $X$.

More specifically, in Display B.9.1, we create the logged and squared terms for the predictor, and then regress *g1miles* on the three forms of the predictor: natural units (lin-lin), logged (lin-log), and quadratic (lin-sq). Because all of these models have the same outcome, we can compare the adjusted $R$-squared values. Although all of the values are quite small, the value for the quadratic model is largest (0.0082). In addition, we can test whether the quadratic model is preferred over the linear model using the significance of the squared variable, *sqg1yrschl*, in Display B.9.1.[9] It has a $t$-value of 3.55, which is larger than 1.96, giving us evidence that the model with the squared term explains more variation in distance than the linear model.

In Display B.9.2, we create the logged outcome, and regress it on the three forms of the predictor. The three models in this box have the same outcome, *logg1miles*, and thus may be compared on their adjusted $R$-squared values. Again, the adjusted $R$-squared values are all small in magnitude, but largest for the quadratic model (0.0223). In addition, the $t$-value for the squared term is significantly different from zero in the log-sq model ($t$-value of 3.98).

We can also compare the models with AIC and BIC, using the formulae from Chapter 6 (see next page). The results again indicate the quadratic models are preferred (have the smallest AIC and BIC).

So, the quadratic model is preferred for both the logged outcome and the outcome in its natural units. But, which of these two has the best fit? And, how different is the shape of these different models? In Display B.9.3 we show how to calculate the adjustment to the predicted values of the logged outcome and how to calculate its approximate $R$-squared value in the natural units of $Y$. We show the procedure for the *log-sq* model as an example. The process can be similarly implemented for the *log-lin* and *log-log* models.

There are several steps in Display B.9.3, including some new commands:

■ Stata's `predict` command (alone and with the `,` `residuals` option) and SAS's `output` statement with the `predicted=` and `residual=` options. The predictions are calculated by SAS and Stata substituting into the prediction equation the value(s) of $X$ for each case. For example, a case with *g1yrschl* = 12 and *sqg1yrschl* = 12 ∗ 12 = 144 would have the following prediction based on the third model in Display B.9.2: $\widehat{\log{(Y)}} = 2.809548 - 0.0635735 * 12 + 0.0087078 * 144 = 3.3005892$. Using these commands, rather than making hand calculations, is quite useful for large data sets and large models. In our data set, Stata and SAS calculate the predicted value for all 5,472 cases. SAS and Stata's commands work a bit differently.

 ■ In Display B.9.3, in Stata syntax, we type `predict logYhat` to ask Stata to calculate the predicted values and store them in the new variable that we call *logYhat*. We also type `predict logResid, residuals` to ask Stata to calculate the residuals and store them in the new variable that we call *logResid*. Stata adds

| Model | k | SSE | $AIC = n*\log\left(\frac{SSE}{n}\right) + 2k$ | $BIC = n*\log\left(\frac{SSE}{n}\right) + k*\log(n)$ |
|---|---|---|---|---|
| lin-lin | 2 | 2,262,599,142 | $= 5472 * \log\left(\frac{2,262,599,142}{5,472}\right) + 2*2$ <br> $= 70,769.987$ | $= 5472 * \log\left(\frac{2,262,599,142}{5,472}\right) + 2*\log(5472)$ <br> $= 70,783.201$ |
| lin-log | 2 | 2,269,835,869 | $= 5472 * \log\left(\frac{2,269,835,869}{5,472}\right) + 2*2$ <br> $= 70,787.46$ | $= 5472 * \log\left(\frac{2,269,835,869}{5,472}\right) + 2*\log(5472)$ <br> $= 70,800.675$ |
| lin-sq | 3 | 2,257,385,363 | $= 5472 * \log\left(\frac{2,257,385,363}{5,472}\right) + 2*3$ <br> $= 70,759.363$ | $= 5472 * \log\left(\frac{2,257,385,363}{5,472}\right) + 3*\log(5472)$ <br> $= 70,779.185$ |
| log-lin | 2 | 31,239 | $= 5472 * \log\left(\frac{31,239}{5,472}\right) + 2*2$ <br> $= 9,536.3506$ | $= 5472 * \log\left(\frac{31,239}{5,472}\right) + 2*\log(5472)$ <br> $= 9,549.5654$ |
| log-log | 2 | 31,467 | $= 5472 * \log\left(\frac{31,467}{5,472}\right) + 2*2$ <br> $= 9,576.1433$ | $= 5472 * \log\left(\frac{31,467}{5,472}\right) + 2*\log(5472)$ <br> $= 9,589.3581$ |
| log-sq | 3 | 31,149 | $= 5472 * \log\left(\frac{31,149}{5,472}\right) + 2*3$ <br> $= 9,522.5629$ | $= 5472 * \log\left(\frac{31,149}{5,472}\right) + 3*\log(5472)$ <br> $= 9,542.3851$ |

the new variables to the existing data file, so it is easy to use them in further analyses.

■ In SAS, we need to save the predictions and residuals in a new data file. We can keep other variables we need in that data file as well (such as *g1miles* in our example) and then refer to that data file in future analyses. To do so, we type:

```
output out=predict(keep=g1miles logYhat logResid)
        predicted=logYhat residual=logResid;
```

where `out=` tells SAS to create a new data file called *predict*, `keep=` tells SAS which variables to keep (in our example, *g1miles logYhat*, and *logResid*) `predicted=` asks SAS to calculate predicted values and store them in the new variable named *logYhat*, and `residual=` asks SAS to calculate residuals and store them in the new variable named *logResid*.

■ We then use the exponential function in SAS and Stata—exp—to convert the predicted values and residuals into their nonlogged natural units. We type `explogYhat= exp(logYhat);` and `explogResid=exp (logResid);` in SAS and `generate  explogYhat=exp(logYhat)` and `generate explogResid=exp (logResid)` in Stata. Note that in SAS we must first set the new data set *predict*. We named the new destination data set, where we save the exponentiated values, *predict2*.

■ To calculate the scaling value to adjust for the downward bias on the predictions we use the `proc means` command in SAS and the `summarize` command in Stata. The resulting coefficient estimate of 10.75774 is the value we will use momentarily to adjust our prediction equation for this *log-sq* model in Display H.9.1.

■ The final regression model in Display B.9.3, which includes a constant term for the regression of *g1miles* on *explogYhat* provides the approximate $R$-squared value for the log-sq model in the natural units of Y: 0.0073.

Display H.9.1 uses Excel to summarize the results across the models, building on the techniques introduced in Appendix H.8. The $R$-squared values and coefficient estimates from each model are shown in cells A1 to G8. In cells A10 to G21, predictions are calculated for each model for values of *g1yrschl* ranging from 6 to 16.[10]

The prediction equation shown in the Excel formula bar shows the prediction for the values of *g1yrschl*=12 which we hand-calculated in the first bullet above. But, in Excel, we take the exponential of this prediction and multiply by the scale factor from Display B.9.3 of 10.75774. We can confirm with a calculator that 10.75774 ∗ exp(3.3005892) = 291.84 which is the result from the Excel formula.

We will walk through an example of each prediction equation in the text, to emphasize the importance of appropriately accounting for the transformations when making predictions. We must be sure to use the appropriate transformation on the right-hand side as well as the left-hand side of the equations. As we show below, it is easy to get the predictions right by thinking back to how each variable is created in SAS and Stata. We must also take care to make any adjustments to the prediction to translate back to its natural units. To get this right, it is helpful to write the outcome variable name, to emphasize if it is in its logged or natural units.[11]

Examples of prediction equations for *g1yrschl* = 12

| Prediction Equation | $\log \widehat{g1miles}$ | $\widehat{g1miles}$ |
|---|---|---|
| **Lin-lin** | | |
| $\widehat{g1miles} = 92.86 + 17.39 * \text{g1yrschl}$ | — | 301.54 |
| $\widehat{g1miles} = 92.86 + 17.39 * 12$ | | |
| **Lin-log** | | |
| $\widehat{g1miles} = 24.65 + 107.62 * \text{logg1yrschl}$ | | |
| $\widehat{g1miles} = 24.65 + 107.62 * \log(\text{g1yrschl} + 1)$ | — | 300.69 |
| $\widehat{g1miles} = 24.65 + 107.62 * \log(12 + 1)$ | | |
| **Lin-sq** | | |
| $\widehat{g1miles} = 293.78 - 25.70 * \text{g1yrschl} + 2.09 * \text{sqg1yrschl}$ | | |
| $\widehat{g1miles} = 293.78 - 25.70 * \text{g1yrschl} + 2.09 * \text{g1yrschl} * \text{g1yrschl}$ | — | 286.34 |
| $\widehat{g1miles} = 293.78 - 25.70 * 12 + 2.09 * 12 * 12$ | | |
| **Log-lin** | | |
| $\widehat{\log g1miles} = 1.97 + 0.116 * \text{g1yrschl}$ | 3.36 | 10.8991*exp(3.36) |
| $\widehat{\log g1miles} = 1.97 + 0.116 * 12$ | | 313.78 |
| **Log-log** | | |
| $\widehat{\log g1miles} = 1.28 + 0.814 * \text{logg1yrschl}$ | | |
| $\widehat{\log g1miles} = 1.28 + 0.814 * \log(\text{g1yrschl} + 1)$ | 3.37 | 11.102*exp(3.37) |
| $\widehat{\log g1miles} = 1.28 + 0.814 * \log(12 + 1)$ | | 322.83 |
| **Log-sq** | | |
| $\widehat{\log g1miles} = 2.81 - 0.0636 * \text{g1yrschl} + 0.00871 * \text{sqg1yrschl}$ | | |
| $\widehat{\log g1miles} = 2.81 - 0.0636 * \text{g1yrschl} + 0.00871 * \text{g1yrschl} * \text{g1yrschl}$ | 3.30 | 10.75774*exp(3.30) |
| $\widehat{\log g1miles} = 2.81 - 0.0636 * 12 + 0.00871 * 12 * 12$ | | 291.84 |

These results match the values in the row for *g1yrschl* = 12 in Display H.9.1 (Row 17), within rounding error.

The predicted values for all levels of *g1yrschl* are plotted in the six graphs in Display H.9.1. Three of the six plots look curvilinear to the eye: log-lin, lin-sq, and log-sq. These are three of the specifications that we anticipated might capture the expected "increasing at an increasing rate" relationship. These three models also show the largest adjusted $R$-squared values, although there is not a single best choice based on our model comparisons. Based on the within-outcome adjusted $R$-squares, and general linear $F$-test for the squared term, the *log-sq* model outperforms the *log-lin* model. But, the approximate *R*-squared in the natural *Y* units is slightly higher for the *log-lin* than for the *log-sq* model, and both are slightly lower than the value for the *lin-sq* model (0.0077 and 0.0075 in the fourth row in the Excel worksheet; 0.0082 in the third row of the Excel worksheet). Even so, we will see in Chapter 11 some other reasons to prefer logging the outcome.

## Dummy Variable Flexible Forms

We now turn to our example of predicting distance based on number of sisters, using dummy variables to allow a flexible form of the relationship. The results are similar for the outcome in its natural units and log form and we will focus on the logged outcome.

Display B.9.4 shows the commands we used to create the seven dummy variables and the results of estimating the regression model using these dummies. We collapsed together adults with seven or more sisters, because fewer than 30 cases had each observed values above six, and use adults with no sisters as the reference category. We graph the predicted values based on this model in Display H.9.2. We use the technique shown in Display B.9.3 to calculate the scale factor for the exponentiated predicted values. The shape of the graph suggests that the flexible dummy specification is useful because the shape is not linear but is also not like any of the shapes shown in Figure 9.2. Distance is fairly constant, hovering around 300 miles, for adults with zero to three sisters, then it begins to fall for those with four to six sisters, and finally jumps to nearly 600 miles for those with seven or more sisters.

The figure suggests that we might further simplify the model by collapsing together some of the categories. We used the test command for each pair of included categories to see whether there was empirical support for such collapsing (e.g., test g2numsis1= g2numsis2; test g2numsis1=g2numsis3; test g2numsis1=g2numsis4, etc). The following table summarizes the *p*-values for those tests.

| | g2numsis0 | g2numsis1 | g2numsis2 | g2numsis3 | g2numsis4 | g2numsis5 | g2numsis6 |
|---|---|---|---|---|---|---|---|
| g2numsis1 | 0.18 | | | | | | |
| g2numsis2 | 0.62 | 0.10 | | | | | |
| g2numsis3 | 0.76 | 0.56 | 0.52 | | | | |
| g2numsis4 | 0.25 | 0.07 | 0.42 | 0.23 | | | |
| g2numsis5 | 0.04 | 0.01 | 0.07 | 0.04 | 0.30 | | |
| g2numsis6 | 0.05 | 0.02 | 0.07 | 0.04 | 0.22 | 0.72 | |
| g2numsis7p | 0.03 | 0.07 | 0.02 | 0.05 | 0.01 | 0.00 | 0.00 |

The pattern of results suggests that adults with zero to three sisters could be grouped together. The means do not differ among the groups with zero to three sisters (solid red rule box), but each of these small sister sizes differs from those with five or six sisters (solid black box), at the alpha level of 0.07 or smaller. Likewise, adults with five and six sisters have statistically equivalent means (dashed black rule box). Those with seven or more sisters differ from all other groups (red dashed rule box), suggesting they should be a separate group. But, those with four sisters do not differ from any of groups, except those with seven or more sisters (unboxed values), making it unclear whether to group those with four sisters with the smaller or larger sister sizes. In cases like this, a judgment will have to be made, ideally with some conceptual rationale, and it is useful to examine how sensitive the results are to different choices about the four-sister category (e.g., grouping them with those with three or fewer sisters, grouping them with those with five or six sisters, leaving them as their own category). We will present the model which leaves adults with four sisters as their own category, which we found slightly better fitting and the most substantively revealing (in a paper, it might make sense to summarize in a footnote or paragraph the results of the alternative specifications).

The results of the collapsed model are presented in Display B.9.5 and the predicted values are calculated and plotted in Display H.9.3. In the collapsed model, like the original model, we still fail to reject the null hypothesis that those with four sisters differ from those with zero to three or those with five or six, and the predicted values shows that the group with four sisters live nearly 60 miles closer to their mothers than those with zero to three sisters and about 60 miles further from their mothers than those with five or six sisters, on average.

### Box 9.4

The $p$-value can be looked up in an $F$-table in any reference book or reliable online reference (e.g., http://stattrek.com/Tables/F.aspx). It is also easy to use Stata to calculate such probabilities. The syntax for density functions can be found by typing `help functions` on the command line. For the $F$ distribution, the syntax is `F(n1,n2,f)` where $n1$ is the numerator and $n2$ is the denominator degrees of freedom and $f$ is the calculated $F$-value. Stata's display command can be used to ask for this value to be calculated and displayed. In our case `display F(4,5464,0.816)` returns .48528348.

We can formally test that the model in Display B.9.5 is a better fit to the data than the model in Display B.9.4 using the general linear $F$-test. The null hypothesis is that $\beta_1 = \beta_2 = \beta_3 = 0$ (adults with zero to three sisters live the same average distance from their mothers) and $\beta_5 = \beta_6$ (adults with five or six sisters live the same average distance from their mothers). Imposing these constraints on the model in Display B.9.4 results in the model in Display B.9.5. We can use their respective sums of square errors and error degrees of freedom to calculate the $F$-value.

$$F = \frac{31,780.1391 - 31,761.1625}{5,468 - 5,464} \div \frac{31,761.1625}{5,464} = 0.816$$

This $F$-value, with 4 numerator and 5,464 denominator degrees of freedom, is not significant, with a $p$-value of 0.51

(see Box 9.4). This indicates that we cannot reject the null hypothesis that those with zero to three sisters live the same average distance from their mothers and that those with five or six sisters live the same average distance from their mothers. So, the model shown in Display B.9.5 is preferred over the model shown in Display B.9.4.

Overall, these results suggest that the linear effect of number of sisters that we reported previously (in Chapter 6) is not a good approximation of the actual relationship between number of sisters and distance from the mother. Adults who have three or fewer sisters live about the same average distance of nearly 300 miles from their mothers. Adults with four sisters live somewhat closer, at less than 240 miles. And, adults with five or six sisters live somewhat closer still, at about 170 miles. In contrast, those with seven or more sisters live substantially further away: close to 600 miles from their mothers on average. The vast majority of adults have zero to three sisters, and only 54 adults in the sample have seven or more sisters. It is possible that these very large families differ from other families in some way that explains this much greater distance. On the other hand, the graphs in Displays H.9.2 and H.9.3 should also give an analyst concerns about an influential observation (possibly a data error) that pulls the average distance upward among this small set of very large families. We will consider this possibility in Chapter 11.

## 9.2: SUMMARY

In this chapter, we showed how to specify several common forms of nonlinear relationships between an interval predictor and outcome variable using the quadratic function and Logarithmic Transformation. We discussed how these various forms might be expected by conceptual models and how to compare them empirically. We also learned how to calculate predictions to show the estimated forms of the relationships, and used Excel to facilitate these calculations and to graph the results. We also showed how to use dummy variables to estimate a flexible relationship between an ordinal or interval predictor and the outcome.

## KEY TERMS

Absolute Change

Curvilinear

Decreasing at a Decreasing Rate

Decreasing at an Increasing Rate

Factor Change

Increasing at a Decreasing Rate

Increasing at an Increasing Rate

Logarithmic Transformation

Natural Log

Percentage Change

Proportionate (or Relative) Change

Quadratic

Tangent Lines

## REVIEW QUESTIONS

**9.1.** How would you choose among three models that predicted an outcome variable with: (a) a linear predictor variable, (b) a logged predictor variable, and (c) a quadratic predictor variable?

**9.2.** How would you create a logged predictor variable when the predictor includes zero (but no negative values)?

    (a) How would you write the prediction equation when a linear outcome variable is regressed on this variable?

    (b) Suppose 5 was a valid value on the predictor. How would you make a prediction for this value of the predictor?

**9.3.** What are the approximate and exact interpretations of the coefficient estimate for a logged outcome variable when the predictor is in its natural units (log-lin model)?

**9.4.** What are the approximate interpretations for a model with a logged predictor when the outcome is also logged (log-log model) and when the outcome is not logged (lin-log model)?

**9.5.** How would you write a general prediction equation for a quadratic functional form?

    (a) Suppose 5 was a valid value on the predictor. How would you make a prediction for this value of the predictor?

    (b) What do the significance and sign of the coefficient estimate for the quadratic term tell us?

(c) What are the formulae for calculating the slope at each level of $X$ and for calculating the point at which the slope becomes zero?

**9.6.** How can you calculate the various types of change (absolute, factor, relative, and percentage) for two values?

REVIEW
EXERCISES
9

## REVIEW EXERCISES

**9.1.** Imagine that your collaborator regressed the log of respondents' earnings on their years of schooling, and obtained the following results:

$$\text{LN}\hat{\text{E}}\text{ARN} = 5.31 + 0.22 * \text{YRSCHL}$$

(a) Help your collaborator interpret the coefficient for YRSCHL with a factor and percentage change approach (i.e., *calculate the factor and percentage change and interpret them in words*; use the *exact* not the approximate approaches).

**9.2.** Suppose that your collaborator regressed the log of income on a dummy indicator of being African American versus white and found that the coefficient for the African American dummy variable was –0.511.

(a) Help your collaborator interpret this coefficient with a factor and percentage change approach (i.e., *calculate the factor and percentage change and interpret them in words;* use the *exact* not the approximate approaches).

**9.3.** Consider the following regression equation:

$$\text{EAR}\hat{\text{N}}\text{INGS} = 1,000 + 1,600 * \text{AGE} - 20 * \text{SQAGE}$$

where SQAGE is the square of AGE, AGE is measured in years, and EARNINGS is annual earnings.

(a) If you plotted predicted values from this prediction equation, would you expect to see a U- or inverted U-shape? Why?

(b) At what value of AGE does the slope switch directions?

(c) What are predicted earnings when AGE is 20, 40, and 60?

CHAPTER
EXERCISE
9

## CHAPTER EXERCISE

In this exercise, you will write SAS and Stata batch programs to estimate nonlinear relationships.

To begin, prepare the shell of your batch program, including the commands to save your output and initial commands (e.g., SAS library, turning *more* off and closing open logs in Stata).

Start with the NOS 1996 dataset that you created in Chapter 4 and use the following command on the `use` command in Stata or in the `data step` in SAS to exclude cases with missing data:

```
if mngpctfem~=. & mngpctfem~=0 & ftsize~=. & ftsize~=0
```

*Notice that we are not restricting the analyses to organizations founded since 1950 for this exercise.*

*In all cases, conduct two-sided hypothesis tests. Use a 5 percent alpha unless otherwise indicated.*

---

### 9.1 Non-linearity

(a)  SAS/Stata tasks:

   (i)    Summarize the *mngpctfem* and *ftsize* variables to verify that neither variable has zero values.

   (ii)   Create a logged version of each variable, adding one before logging is not needed because the original variables do not contain zeros. (*Do not rescale the variables before taking the log.*)

   (iii)  Create a rescaled version of *ftsize*, rescaling to units of 100.

   (iv)  Create a quadratic term for the *rescaled ftsize* variable.

   (v)   Regress the original nonlogged *mngpctfem* variable on the rescaled *ftsize*, on the logged version of *ftsize*, and the linear and quadratic terms for rescaled *ftsize*.

   (vi)  Regress the logged version of the *mngpctfem* variable on the rescaled *ftsize*, on the logged version of *ftsize*, and the linear and quadratic terms for rescaled *ftsize*.

   (vii) Calculate the adjustment factor for making predictions for the log-log model.

   (viii) For the log-log model, calculate the $R$-squared values discussed in the chapter that approximate the $R$-squared value in $Y$ units.

(b)  Write-up tasks:

   (i)    Calculate AIC and BIC for each model.

(ii)   Based on the significance of individual coefficients, the AIC/BIC values, and the *R*-squared values for the models, which of the six models is preferred?

(iii)  Interpret the coefficient in the log-lin model using the *approximate* and *exact* approaches to interpretation.

(iv)   Interpret the coefficient in the log-log model using the *approximate* interpretation.

(v)    Using only the sign of the coefficient on the quadratic term, does this quadratic form have a U-shape or inverted U-shape? Calculate the value of *ftsize* at which the slope becomes zero and calculate the slope in organizations with 10, 100, and 1,000 employees. *Be sure to account for the rescaling of ftsize in your calculations.*

## COURSE EXERCISE

COURSE
EXERCISE

9

Identify at least one nonlinear relationship that you might estimate based on the data set that you created in the course exercise to Chapter 4. Ideally, think about conceptual reasons why you might expect a nonlinear relationship between one of the predictors and outcomes. Alternatively, choose a predictor and explore whether a nonlinear relationship is identified empirically in the data set. Ideally, choose a continuous predictor to practice quadratic and logged models and choose a categorical predictor to practice examining stair-step-type models with dummy variables.

For your continuous predictor, estimate the six models discussed in the chapter (lin-lin, lin-log, lin-sq, log-lin, log-log, and log-sq models). Compare the performance of the models using individual *t*-tests (where appropriate), *R*-squared values, and AIC/BIC values. Plot predicted values from the preferred model, transforming the predictor and/or outcome as needed when making predictions.

For your categorical predictor, use dummy variables to flexibly estimate the relationships among the categories and outcome. Use tests among the dummy variables to potentially collapse groups. Calculate predicted values and graph the results.

# INDIRECT EFFECTS AND OMITTED VARIABLE BIAS

## CHAPTER 10: INDIRECT EFFECTS AND OMITTED VARIABLE BIAS

Multiple regression is ideally suited to social science research because it allows researchers to adjust for confounding variables when experimental designs are not possible. But, such confounders are not always measured. When they are not, researchers can use their understanding of multiple regression to anticipate the direction of bias due to their omission. Multiple regression can also be used to determine whether the effect of a variable on the outcome operates through the mechanisms laid out in theories, by adding measures for those mechanisms to the model. But, care must be taken to interpret these mechanisms, because study designs often do not allow the direction of effects to be unambiguously determined.

## 10.1: LITERATURE EXCERPT 10.1

We begin with an excerpt from the literature that illustrates the use of multiple regression to adjust for confounders and to examine mediators. Stacey Brumbaugh and colleagues (2008) examined the predictors of attitudes toward gay marriage in three states that considered but ultimately rejected (Minnesota) or had recently passed (Arizona and Louisiana) changes to marriage laws. Called covenant marriage, these changes increase the bar for entering and exiting marriage, requiring counseling before marriage and extended waiting periods before fault-based divorce. The authors hypothesize that individuals are more likely to oppose gay marriage when they perceive a cultural weakening of heterosexual marriage as a threat. In their model, the people most threatened would be those who have a stake in marriage, such as married persons. They further expect and test for a mechanism for this effect: "Based on our threat model, we hypothesize that religious, political and social attitudes, especially attitudes regarding the deinstitutionalization of marriage and marriage promotion efforts, mediate the effects of marital, parenthood, and cohabitation histories on attitudes toward gay marriage" (p. 349). The authors also control for a number of sociodemographic and socioeconomic variables, which adjust for their potential effect on both the predictors of interest and the outcome. For example, they note that prior results associating race-ethnicity and attitudes about gay rights depend on whether and how researchers had adjusted for socioeconomic status.

Literature Excerpt 10.1 reproduces Table 3 from the paper. The organization of the results follows a standard strategy for presenting the effects of adding confounders and mediators. Model 1 begins with a set of sociodemographic controls (female, black, other race, and age) and dummy indicators for two of the three states included in the study. Model 2 adds controls for socioeconomic status (dummy indicators of education and employment and log family income). Model 3 adds dummy indicators of marital,

■ **Literature Excerpt 10.1**

Table 3.  Regression Models Predicting Attitudes Against Gay Marriage (N = 976)

| Variable | Model 1 | | Model 2 | | Model 3 | | Model 4 | |
|---|---|---|---|---|---|---|---|---|
| | $B$ | $pr^2$ | $B$ | $pr^2$ | $B$ | $pr^2$ | $B$ | $pr^2$ |
| Intercept | 2.92*** | | 1.73*** | | 2.88*** | | −.02 | |
| Sociodemographic controls | | | | | | | | |
| Female | −.42*** | (2.7%) | −.42*** | (2.5%) | −.44*** | (2.9%) | −.42*** | (3.2%) |
| Black | .24* | (0.3%) | .32** | (0.6%) | .32** | (0.6%) | .22* | (0.4%) |
| Other race | −.18 | (0.1%) | −.12 | (0.0%) | −.07 | (0.0%) | −.12 | (0.0%) |
| Age | .02*** | (5.6%) | .02*** | (5.7%) | .01*** | (0.8%) | .01** | (0.5%) |
| Stare context | | | | | | | | |
| Arizona | −.06 | (0.0%) | −.06 | (0.0%) | −.04 | (0.0%) | −.06 | (0.0%) |
| Louisiana | .24** | (0.6%) | .23** | (0.5%) | .22** | (0.6%) | −.04 | (0.0%) |
| Socioeconomic status | | | | | | | | |
| High school | | | .52*** | (1.1%) | .50*** | (1.0%) | .44*** | (1.0%) |
| Some college | | | .36** | (0.5%) | .36** | (0.5%) | .33** | (0.6%) |
| College | | | .33* | (0.4%) | .30* | (0.3%) | .21 | (0.2%) |
| Postcollege | | | .07 | (0.0%) | .11 | (0.0%) | .17 | (0.1%) |
| Employed full time | | | −.09 | (0.1%) | −.05 | (0.0%) | −.00 | (0.0%) |
| Log family income | | | .08 | (0.3%) | .02 | (0.0%) | .00 | (0.0%) |
| Income missing | | | −.32** | (0.5%) | −.21 | (0.2%) | −.17 | (0.2%) |
| Marital, cohabitation, and parenthood histories | | | | | | | | |
| Divorced/separated | | | | | −.33*** | (1.0%) | −.08 | (0.0%) |
| Never married | | | | | −.36*** | (0.7%) | −.15 | (0.1%) |
| Child or children | | | | | .34*** | (0.9%) | .24** | (0.6%) |
| Recently cohabited | | | | | −.52*** | (3.4%) | −.20** | (0.6%) |
| Religious, political, and social attitudes | | | | | | | | |
| Religiosity | | | | | | | .10* | (2.6%) |
| Political conservatism | | | | | | | .27*** | (4.5%) |
| Attitudes toward divorce | | | | | | | .04*** | (1.2%) |
| Perceived blameworthiness for family breakdown | | | | | | | .07*** | (3.0%) |
| Attitudes toward covenant marriage | | | | | | | .02*** | (0.7%) |
| $F$ Statistic | 15.80*** | | 9.11*** | | 11.05*** | | 22.85*** | |
| Nested $F$ | N/A | | 3.18*** | | 15.55*** | | 52.81*** | |
| $R^2$ | .09 | | .11 | | .16 | | .35 | |
| Adjusted $R^2$ | .08 | | .10 | | .15 | | .33 | |

*Note.* In parentheses are the effect sizes, measured by squared partial correlation coefficients ($pr^2$), indicating the percentage of remaining variance explained.
* $p < .05.$ ** $p < .01.$
*** $p < .001$, one-tailed tests.

**Source:** Brumbaugh, Stacey M., Laura A. Sanchez, Steven L. Nock and James D. Wright. 2008. "Attitudes Toward Gay Marriage in States Undergoing Marriage Law Transformation." *Journal of Marriage and Family,* 70, 345–59.

cohabitation, and parenthood histories. Model 4 adds religious, political and social attitudes.

The table setup makes it easy to look across a row and see how a coefficient estimate changes when other variables enter the model. For example, we see that the coefficient estimate for the dummy indicator of female changes little across the models (being −0.42 in three of the models, and −0.44 in one model). In contrast, the coefficient estimate for black race-ethnicity (versus the white reference category) increases to 0.32 from 0.24 in Model 2 versus Model 1, remains at 0.32 in Model 3, and then falls to 0.22 in Model 4. Turning to the coefficient estimates for the marital, cohabitational, and parenthood histories, these change considerably between Models 3 and 4. The coefficient estimates all fall in magnitude between Model 3 and Model 4, and become nonsignificant for the contrasts of *divorced/separated* and of *never married* with the reference category (*married or widowed*).

We will discuss below how to examine in more detail how and why a coefficient changes when another variable enters the model. In this literature excerpt, the pattern of coefficient estimates for black race-ethnicity is difficult to interpret because the difference between blacks and whites unexpectedly widens, rather than falls, with adjustments for socioeconomic status.[1] The pattern of marital, cohabitation, and parenthood histories is easier to interpret, and in line with the authors' expectations. Those with a higher stake in marriage, including those who are currently married, have not recently cohabited, and are not parents have more negative views toward gay marriage. These differences are reduced substantially when attitudes toward divorce and covenant marriage, the breakdown of the family, political conservatism, and religiosity are added to the model. This result can be interpreted as demonstrating that attitudes are a reason for the differences among those with different marital, parenthood, and cohabitation experiences, although as the authors note, the direction of association is difficult to determine with cross-sectional data. It is possible that attitudes predict family structure rather than family structure predicting attitudes. Justification for interpreting a variable as a confounder or mediator, and placing a variable as a predictor, mediator, or outcome, cannot be provided empirically by cross-sectional data. The researcher uses theory and prior empirical research to guide placement of variables in the model. More advanced techniques and longitudinal designs build on the basic concepts of OLS better to identify the ordering and causal nature of associations (see roadmap in Chapter 12).

## 10.2: DEFINING CONFOUNDERS, MEDIATORS, AND SUPRESSOR VARIABLES

Frequently, a coefficient estimate decreases in size when other variables are added to the model. Such a change is interpreted as reflecting spuriousness or mediation, depending on the conceptual model and study design. It is also possible for a coefficient estimate to increase when another variable is added to the model. Such change is typically interpreted as reflecting suppression.

## 10.2.1: Confounders

As we discussed in Chapter 5, regression models are asymmetrical. One variable is selected as the outcome and other variables are predictors of that outcome. This setup implies a causal association between the predictor and outcome. But, with observational data, it is difficult to justify a causal interpretation. We also saw in Chapter 5 that a bivariate regression coefficient is nothing more than a rescaled correlation coefficient. Simply selecting one variable as the outcome and the other as the predictor in a regression model does not elevate the association from correlational to directional.

Similarly, the well-known phrase *Correlation does not imply causation* holds for bivariate and multiple regression. If we use a dummy variable that indicates teenage motherhood to predict years of schooling in a bivariate regression with correlational data, and find a significant negative coefficient estimate, we cannot conclude that teenage parenthood causes adolescent girls to limit their schooling. Because we did not randomly assign young mothers to be teenage mothers or not, they may differ on many other characteristics that predict both their status as teenage mothers and the years of schooling they complete (such as growing up in a low-income neighborhood or a low-income family). If the association is really due to these other variables, then social programs and policies aimed at reducing teenage motherhood would not increase the years of schooling these girls complete. Rather, in this case, we would need to address the root cause—family and community background—to simultaneously reduce teenage motherhood and increase educational attainment.

Statistically, this situation is referred to as a **spurious relationship**. The other characteristics that predict both our predictor of interest and the outcome are referred to as **confounders** or **common causes** or **extraneous variables**. A spurious relationship exists when an apparent association between one variable and another is really due to a third variable. It is helpful to display this and other associations in this chapter visually.

Figures 10.1(a) and (b) help us visualize a spurious relationship using conventions we will follow throughout this chapter. The arrows point away from predictor variables toward outcomes. Throughout our examples, we will use the subscript 2 to denote our predictor of interest and the subscript 3 to depict the third variable, leaving the subscript 1 for our ultimate outcome of interest.

Figure 10.1(a) illustrates a directional relationship from $X_2$ (predictor) to $Y$ (outcome). The lack of arrow between $X_2$ and $Y$ in Figure 10.1(b) indicates that $X_2$ is no longer

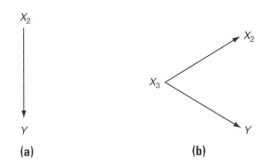

**Figure 10.1**

associated with $Y$ once we account for $X_3$. Instead, the arrows from $X_3$ to $X_2$ and $Y$ show that when $X_3$ changes, it produces change in both $X_2$ and $Y$. Ignoring $X_3$ makes it appear that $X_2$ leads to $Y$ because both $X_2$ and $Y$ covary with $X_3$; that is, $X_2$ and $Y$ seem to "move together" because of their shared association with $X_3$.

Table 10.1 provides a hypothetical example of what the regression results would look like in the case of a completely spurious association. In Model 1, *Years of Schooling* is predicted by *Age at First Birth* in a bivariate regression. The coefficient estimate is positive and significant. In Model 2, when we add a measure of *Parents' SES* to the model, the coefficient estimate for *Age at First Birth* falls to zero and is insignificant. Only *Parents' SES* significantly predicts *Years of Schooling* in Model 2.

**Table 10.1: Completely Spurious Association between Age at First Birth and Years of Schooling**

|  | Dependent Variable: Years of Schooling | |
| --- | --- | --- |
|  | Model 1 | Model 2 |
| Age at First Birth | 0.16* | 0.00 |
|  | (0.02) | (0.01) |
| Parents' SES | — | 0.99* |
|  |  | (0.01) |
| Intercept | 7.64 | 10.48 |
|  | (0.31) | (0.10) |

* p < .05 —indicates excluded from model.
*Source:* Hypothetical data created for illustration purposes.

Although the word "spurious" is typically used to denote a situation in which a third variable completely explains away an association, as is illustrated in Table 10.1, it is also possible for an association to be reduced but not eliminated by adjusting for a confounder. In Table 10.2 we use hypothetical data to illustrate a partially spurious association between the predictor and outcome. In Model 2 of Table 10.2, when we control for *Parents' SES*, the coefficient estimate for *Age At First Birth* decreases but remains positive and significant. And, the association between *Parents' SES* and *Years of Schooling* is smaller in magnitude in Table 10.2 versus Table 10.1, reflecting the fact that in the second hypothetical example, we created the hypothetical data set to have a weaker association between *Parents' SES* and *Years of Schooling*.

■ **Table 10.2: Partially Spurious Association between Age at First Birth and Years of Schooling**

|  | Dependent Variable: Years of Schooling | |
| --- | --- | --- |
|  | Model 1 | Model 2 |
| Age at First Birth | 0.16* | 0.09* |
|  | (0.02) | (0.01) |
| Parents' SES | — | 0.59* |
|  |  | (0.02) |
| Intercept | 7.64 | 8.95 |
|  | (0.31) | (0.25) |

\* p < .05 —indicates excluded from model.
*Source:* Hypothetical data created for illustration purposes.

### 10.2.2: Example 10.1

We will also use our NSFH hours of work variable to illustrate various types of association throughout the chapter. We will begin by estimating how men's ratings of their own health predict their hours of work, and examine to what extent the men's ages confound this association. We anticipate that older age predicts both poorer health and reduced work hours (or retirement). Hours of work and age are variables we considered in prior chapters. The health rating comes from a standard question, "Compared with other people your age, how would you describe your health?" which is rated on a scale from 1 = *very poor* to 5 = *excellent*.[2]

Display B.10.1 presents the results of estimating two models in SAS and Stata. The first model predicts hours of work in a bivariate regression based on the health rating. The second adds the respondent's age in a multiple regression model. For ease of comparison, and to demonstrate how we might table the results for a manuscript, we summarize the results below.

▨ **Table 10.3: Predicting Hours of Work by Self-Reported Health in the NSFH, with and without Controls for Age**

|  | Dependent Variable: Hours of Work Per Week | |
|---|---|---|
|  | Model 1 | Model 2 |
| Self-Reported Health | 1.14* | 1.02* |
|  | (0.31) | (0.31) |
| Age (Years) | — | −0.10* |
|  |  | (0.02) |
| Intercept | 37.86 | 42.27 |
|  | (1.29) | (1.51) |

\* $p < .05$ —indicates excluded from model.
*Source:* NSFH, Wave 1, Sample of Men, $n = 3{,}742$.

Model 1 shows that, before controlling for age, each additional point higher that the respondent rated his health is associated with over one additional hour of work per week. When age is controlled, the association drops by over 10 percent to one additional hour per week $[(1.02-1.14)/1.14 * 100 = -10.53]$. The results are consistent with our expectation that age is a confounder. We chose age because it is hard to argue that health influences age (except through altering the sample by attrition through death or very poor health).

## 10.2.3: Mediators

**Mediation** occurs when a third variable represents the mechanism through which an association between a predictor and outcome operates. The third variable is typically referred to as a **mediator**, an **intervening variable**, or a **proximal cause**.

Mediating relationships are often seen in elaborate conceptual models that do not simply suggest that two variables are related, but spell out how and why they are related. An empirical test that verifies not only the association between the two variables, but also

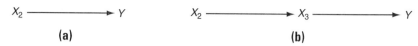

**(a)**  **(b)**

▓ **Figure 10.2**

the evidence for the intervening mechanisms, will provide more convincing support of the theory. Similarly, if two theories suggest two different mechanisms, we would ideally use an empirical model to test which explanatory framework is best reflected in the data.

Going back to our example of teenage mothers, we can consider mechanisms that explain why those who share a household with their own parents complete more years of schooling. One reason might be because these young mothers receive help with childcare from their parents. Table 10.4 uses hypothetical data to show how such a mediating relationship might be revealed with regression analysis. In Model 1, *Coresidence with Parents* is positively and significantly associated with *Years of Schooling*. In Model 2,

▓ **Table 10.4: Completely Mediating Association between Coresidence with Parents and Years of Schooling**

| | Dependent Variable: Years of Schooling | |
| --- | --- | --- |
| | Model 1 | Model 2 |
| Coresidence with Parents | 1.32* | −0.03 |
| | (0.08) | (0.08) |
| Hours of Parental Help with Child Care Per Week | — | 0.20* |
| | | (0.01) |
| Intercept | 10.48* | 7.96* |
| | (0.05) | (0.11) |

* $p < .05$ —indicates excluded from model.
*Source*: Hypothetical data created for illustration purposes.

when we add *Hours of Parental Help with Child Care Per Week* to the model, the association with *Coresidence with Parents* falls to close to zero and is insignificant.

Notice that the statistical pattern of results looks quite similar in Tables 10.1 and 10.4. A coefficient estimate for one variable that was significant in Model 1 becomes insignificant in Model 2. The statistical results do not tell us if the third variable is a confounder or a mediator. We determine its role in the model based on conceptual ideas about how the variable should operate and ideally based on careful measurement. For example, conceptually, we might consider the extent to which it is plausible for the parents' SES to be influenced by their daughter's teenage pregnancy (placing it in the middle intervening rather than prior confounding position). Methodologically, we might be careful to measure parents' SES when the daughter was born or in elementary school, so that it precedes the teenage pregnancy.

As with confounding, mediation may be partial rather than complete. We illustrate such partial mediation in Table 10.5.

▨ **Table 10.5:  Partially Mediating Association between Coresidence with Parents and Years of Schooling**

| | Dependent Variable: Years of Schooling | |
| --- | --- | --- |
| | Model 1 | Model 2 |
| Coresidence with Parents | 1.32* | 0.53* |
| | (0.08) | (0.09) |
| Hours of Parental Help with Child Care Per Week | — | 0.12* |
| | | (0.01) |
| Intercept | 10.48* | 8.55 |
| | (0.05) | (0.11) |

* p < .05 —indicates excluded from model.
*Source:* Hypothetical data created for illustration purposes.

In this case, the association between *Coresidence with Parents* and *Years of Schooling* drops substantially in size, but remains significant in Model 2. And, *Hours of Parental Help with Child Care Per Week* is less strongly associated with *Years of Schooling* than was the case in Table 10.4.

## 10.2.4:  A Note on Terminology: Mediation and Moderation

Students (and scholars) often confuse the terms mediation and **moderation**. The term moderation is another word used to describe the presence of an interaction. We say that $X_3$ moderates the relationship between $X_2$ and $Y$ when an interaction exists. Moderation is depicted visually as follows:

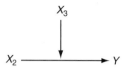

▦ **Figure 10.3**

Examining common definitions of these words may help you to remember their meaning. One definition of moderate in Webster's dictionary is "to become less violent, severe, intense, or rigorous." One definition of mediate is "to effect (a result) or convey (a message, gift, etc.) by or as if by an intermediary." Baron and Kenny (1986) provide a clear and accessible clarification of the distinctions between these terms.

### 10.2.5: Calculating Direct and Indirect Effects

The geneticist Sewall Wright developed a method called **path analysis** that often goes hand in hand with diagrams like those shown above to depict the causal pathways among variables. As Wright (1921, 557) wrote in a seminal paper on the topic:

> The degree of correlation between two variables can be calculated by well-known methods, but when it is found it gives merely the resultant of all connecting paths of influence. The present paper is an attempt to present a method of measuring the direct influence along each separate path in such a system and thus of finding the degree to which variation of a given effect is determined by each particular cause.

Figure 10.4(a) below depicts the "resultant of all connecting paths of influence" and Figure 10.4(b) one of the "separate paths" that Wright's technique was designed to measure.

(a)          (b)

▦ **Figure 10.4**

The association on the left is referred to as the **total effect** of $X_2$ on $Y$. The association in the top of Figure 10.4(b), depicted by the two arrows pointing from $X_2$ to $X_3$ and from $X_3$ to $Y$ is referred to as the **indirect effect** of $X_2$ on $Y$ through $X_3$. The bottom arrow in Figure 10.4(b), pointing from $X_2$ to $Y$, is referred to as the **direct effect** of $X_2$ on $Y$. The broader technique, referred to as path analysis, can incorporate much more complicated diagrams with multiple indirect pathways (e.g., indirect paths from $X_2$ to $Y$ through $X_4$, $X_5$, $X_6$, etc.) and additional stages of connections (e.g., indirect paths from $X_3$ to $Y$; see Box 10.1).

Figures 10.4(a) and (b) depict several regression models. We will define each of these formally so we can see how to calculate the direct, indirect, and total effects. We are going to use a new notational convention to make this clear. We will use $\hat{\beta}$ to denote a coefficient estimate for a multiple regression and we will use $b$ to denote a coefficient estimate from a bivariate regression. We will use subscripts to indicate which variable is the outcome and predictor, referring to $Y$ with the number 1, $X_2$ with the number 2, and $X_3$ with the number 3 and placing the outcome number first followed by the predictor in the subscript; for example, $b_{12}$ would denote a bivariate regression with $Y$ as the outcome and $X_2$ as the predictor.

We identify the regressions in Figures 10.4(a) and (b) using the arrows. Each arrow points from a predictor to an outcome. If an outcome has more than one arrow pointing to it, then the variables at the other end of the arrows are the predictors of a multiple regression. If an outcome has one arrow pointing to it, then the variable at the other end of that arrow is the single predictor in a bivariate regression. Notice that in Figure 10.4(b), $X_3$ is an outcome in one regression (an arrow points to it) and a predictor in another regression (an arrow points away from it).

■ **Box 10.1**

Many courses in path analysis and structural equation modeling are available, including through resources like summer courses at the University of Michigan (*http://www.icpsr.umich.edu/sumprog/*) and the University of Kansas (*http://www.quant.ku.edu/StatsCamps/ overview.html*) and likely at your local university. Numerous texts on the topics are available as well (e.g., Bollen 1989; MacKinnon 2008; Iacobucci 2008).

■ **Table 10.6: Regression Models Used in Estimating Indirect and Direct Effects**

| Figure 10.4(a) | $Y_i = b_1 + b_{12}X_{2i} + \hat{e}_{1i}$ | (10.1) |
|---|---|---|
| Figure 10.4(b) | $Y_i = \hat{\beta}_1 + \hat{\beta}_{12}X_{2i} + \hat{\beta}_{13}X_{3i} + \hat{\varepsilon}_{1i}$ | (10.2) |
| | $X_{3i} = b_3 + b_{32}X_{2i} + \hat{e}_{3i}$ | (10.3) |

We place these coefficients on the relevant paths below.

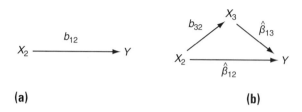

**(a)**                    **(b)**

■ **Figure 10.5**

The indirect effect of $X_2$ on $Y$ can be calculated by multiplying the coefficient estimates on the two paths associating $X_2$ to $Y$ through $X_3$; that is, the indirect effect is $b_{32} * \hat{\beta}_{13}$. The direct effect of $X_2$ on $Y$ is captured by the multiple regression coefficient estimate, $\hat{\beta}_{12}$. The total effect of $X_2$ on $Y$ can be calculated as the sum of the direct and indirect effects, $\hat{\beta}_{12} + b_{32} * \hat{\beta}_{13}$. This result will be equal to the total effect as calculated by the bivariate regression coefficient estimate, $b_{12}$.

■ **Table 10.7:  Calculating the Total, Direct, and Indirect Effects**

| | |
|---|---|
| Total | $b_{12} = \hat{\beta}_{12} + b_{32} * \hat{\beta}_{13}$ |
| Direct | $\hat{\beta}_{12}$ |
| Indirect | $b_{32} * \hat{\beta}_{13}$ |

### 10.2.6: Example 10.2

We will again use our NSFH example of men's hours of work, but now we will look at health limitations as a mediator of the effect of self-reported health. The limitations measure is the sum of six dichotomous items about whether or not the respondent perceives that physical or mental conditions limit his or her ability to: care for personal needs, such as dressing, eating or going to the bathroom; move about inside the house; work for pay; do day-to-day household tasks; climb a flight of stairs; and walk six blocks. The mediation of some of the effect of overall health on work hours through health limitations can be shown in figures, as follows. (For now, we will leave respondent's age out of the model, but we will show the full model below.)

**Figure 10.6**

The results of using SAS and Stata to estimate the three regressions shown in the diagrams are presented in Display B.10.2. We summarize them below:

**Table 10.8: Regression Models Used in Estimating Indirect and Direct Effects of Men's Health on their Hours of Work, with Mediation through Health Limitations**

| | |
|---|---|
| Figure 10.6(a) | $\widehat{hrwork}_i = 37.86 + 1.14 * health_i$ |
| Figure 10.6(b) | $\widehat{hrwork}_i = 38.52 + 1.008 * health_i - 5.656 * hlthlimit_i$ |
| | $\widehat{hlthlimit}_i = 0.12 - 0.0239 * health_i$ |

The values of the total, direct, and indirect effects can be calculated as follows:

**Table 10.9: Calculating the Total, Direct, and Indirect Effects of Men's Health on their Hours of Work, with Mediation through Health Limitations**

| | |
|---|---|
| Total | $b_{12} = 1.14$ |
| | $\hat{\beta}_{12} + b_{32} * \hat{\beta}_{113} = 1.008 + (-0.0239 * -5.656) = 1.14$ |
| Direct | $\hat{\beta}_{12} = 1.008$ |
| Indirect | $b_{32} * \hat{\beta}_{13} = -0.0239 * -5.656 = 0.135$ |

As expected, the sum of the direct and indirect effects equals the total effect.

## 10.2.7: Suppressor Variables

The example above is a case in which the indirect effect is of the same sign as the direct effect. As a consequence, the direct effect is smaller in magnitude than the total effect. It is also possible that the indirect effect may have the opposite sign as the direct effect. In these cases, the total effect may be close to zero, and adding the mediator variable to the model may reveal a significant association between the predictor of interest and the outcome. Mediators that have this type of effect are called **suppressor variables**.

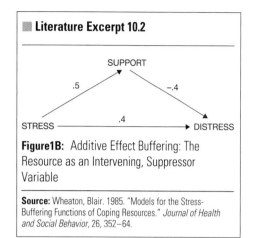

▨ **Literature Excerpt 10.2**

**Figure1B:** Additive Effect Buffering: The Resource as an Intervening, Suppressor Variable

**Source:** Wheaton, Blair. 1985. "Models for the Stress-Buffering Functions of Coping Resources." *Journal of Health and Social Behavior,* 26, 352–64.

Blair Wheaton (1985) provides an example of a suppressor variable in the field of medical sociology. Literature Excerpt 10.2 reproduces a figure that depicts a positive direct effect of stress on distress and a negative indirect effect of stress on distress through social support. As he notes:

Probably the central analytical fact about this model is that the indirect effect through social support operates in a direction opposite to the overall causal effect of stress on distress (suggesting the term *suppressor variable* is appropriate) . . . The point here is that the total causal effect of stress on distress (.2) is less than its direct effect (.4). It is the total effect, not just the direct effect, that reflects the impact of stress. Thus, it is clear in this case that support acts to buffer this overall impact (p.356, italics in the original).

The total effect of .2 is calculated based on the values on the arrows in Literature Excerpt 10.2, using the formulae we presented above (.4 + (5 * −.4) = .4 − .2 = .2).

Although researchers often think in terms of indirect effects that have the same sign as the direct effect, considering suppressor variables in our conceptual models and empirical work is important, especially because ignoring them can result in the overall association between a predictor of interest and outcome to be near zero (or of the "wrong sign").

### 10.2.8: A Note on the Meaning of Controlling

We can also use the regressions depicted above to help us to understand the meaning of controlling for other variables in multiple regression. In Equation 10.3, we regressed $X_3$ on $X_2$ as we calculated the indirect effect. The errors in this equation represent the variation in $X_3$ not explained by $X_2$. If we calculate these errors, and use them in a bivariate regression to predict $Y$, we will see that the resulting coefficient estimate equals $\hat{\beta}_{13}$ from Equation 10.2. Going through this process can make concrete the meaning of a multiple regression coefficient. When controlling for the other variable(s), we isolate the unique effect of one predictor on the outcome; that is, we answer the question: is the variation that the predictor does not share with other variables in the model related to the outcome?

We can use the `output` statement in SAS and the `predict` command in Stata to request the residuals from Equation 10.3 for our hours of work example. We then regress $Y$ on these residuals. The results are in the top two rows of Display B.10.3.

We can similarly regress $X_2$ on $X_3$ and obtain the errors, which represent the variation in $X_2$ not explained by $X_3$. If we regress $Y$ on these residuals, we will obtain the multiple regression coefficient estimate for $X_2$ from Equation 10.2. These results are shown in the bottom two rows of Display B.10.3.

Comparing the results in Display B.10.3 with those in the multiple regression shown in the middle panel of Display B.10.2 confirms that our residual regressions reproduce the multiple regression coefficients.

- In Display B.10.2, the coefficient for *Health* in the multiple regression (middle panel) is 1.008126. In Display B.10.3, the top panel shows the commands for creating the residuals from a regression of *Health* on *HlthLimit*. These residuals reflect the variation in *Health* not explained by *HlthLimit*. The results of the regression of *hrwork* on these residuals is shown in the top panel of Display B.10.3. The coefficient, 1.008126, matches the coefficient estimate for *Health* from the multiple regression in Display B.10.2.
- In Display B.10.2, the coefficient for *HlthLimit* in the multiple regression (middle panel) is –5.655881. In Display B.10.3, the bottom panel shows the commands for creating the residuals from a regression of *HlthLimit* on *Health*. These residuals reflect the variation in *HlthLimit* not explained by *Health*. The results of the regression of *hrwork* on these residuals is shown in the bottom panel of Display B.10.3. The coefficient, –5.655881, matches the coefficient estimate for *HlthLimit* from the multiple regression in Display B.10.2.

## 10.3: OMITTED VARIABLE BIAS

What if we believe part of the effect of health on hours of work operates through health limitations, but we do not have a measure of health limitations? How might we think through the potential effect of omitting the health limitations measure from our model? Or, what if we think that a portion of the association between health and hours of work is really due to age as a common cause, but we did not collect the study participants' ages? How might we think through the potential effect of omitting age from our model? **Omitted variable bias** is defined exactly the same as the indirect effect $(b_{32} * \hat{\beta}_{13})$, although because we typically do not have a measure of the omitted variable, we consider just the signs of the arrows on the direct and indirect pathways to anticipate the consequences of lacking this measure.

Table 10.10 can be used to assess the direction of the bias based on thinking through the signs on the pathways. The columns indicate the anticipated sign on the association between the included and excluded predictor variables. The rows indicate the anticipated sign of the association between the excluded predictor variable and the outcome. The product of the column and row signs are shown in the cells.

■ **Table 10.10: Expected Direction of Omitted Variable Bias**

| | | Direction of Association Between the *Included* and *Excluded Predictor* Variables | |
|---|---|---|---|
| | | Positive (+) | Negative (−) |
| Direction of Association Between *Excluded Predictor* Variable and *Outcome* Variable | Positive (+) | + + = positive | + − = negative |
| | Negative (−) | − + = negative | − − = positive |

So, if the sign of the association between the excluded variable is in the same direction with both the included predictor and outcome, then the omitted variable bias will be positive. If the sign of the association between the excluded variable and the included predictor and outcome are in different directions, then the omitted variable bias will be negative.

We can combine the expected direction of the omitted variable bias with the expected direction of the direct effect of the included variable to produce an assessment of whether we will likely understate or overstate the direct effect of the included variable due to omitted variable bias.

Typical indirect effects, like our example of health limitations mediating the association of health with hours of work, fall on the diagonal of Table 10.11. Omitting the indirect effect overstates the direct effect of the predictor of interest. This is sometimes referred to as biasing the effect away from zero. Suppression effects fall off the diagonal. The direct effect and indirect effect have opposite signs, and so omitting the suppressor variable leads to an understatement of the direct effect of the predictor of interest. This is sometimes referred to as bias toward zero. In cases in which the indirect effect of a suppressor variable is larger than the direct effect, the sign on the coefficient estimate for the predictor of interest will be in the wrong direction, relative to its expected direct effect.

■ **Table 10.11: Expected Misstatement of the Magnitude of the Direct Effect of Included Variable**

| | | Direction of *Omitted Variable Bias* | |
| --- | --- | --- | --- |
| | | Positive | Negative |
| Direction of *Direct Effect* of Included Variable | Positive | *positive + positive*<br>overstate<br>(bias away from zero) | *positive + negative*<br>understate<br>(bias toward zero)<br>or<br>incorrect direction<br>(wrong sign) |
| | Negative | *negative + positive*<br>understate<br>(bias toward zero)<br>or<br>incorrect direction<br>(wrong sign) | *negative + negative*<br>overstate<br>(bias away from zero) |

These tables can be helpful if a reviewer of your research raises a question about a variable that you left out of the model or if you realize too late that you should have measured an important concept. If the variable is conceptualized as having an indirect effect of the same sign as the direct effect, then omitting it overstates the direct effect of your predictor of interest. But, if the variable is conceptualized as having an indirect effect of opposite sign as the direct effect, then omitting it understates the direct effect of your predictor of interest.

Notice that the tables above can be used to think about the direction of omitted variable bias for confounders as well as mediators. For example, if a confounding excluded variable is expected to be associated in the same direction with the predictor of interest and the outcome, then the omitted variable bias will be positive. If the estimated effect of the predictor of interest with the outcome is also positive, then the omitted variable may be biasing the estimated effect upward.

## 10.4: SUMMARY

In this chapter, we examined how adding variables to a multiple regression model affects the coefficients and their standard errors. Multiple regression allows us to estimate the pathways that our theoretical frameworks suggest link variables and to adjust for confounders in our observational data. In many cases, adding a variable will lead to the coefficient estimate for our predictor of interest to decrease in magnitude, either because of adding a common cause it shares with the outcome (spuriousness) or because of adding an intervening variable that links it to the outcome (mediation). In some cases, adding a variable will lead the coefficient estimate of interest to increase in magnitude (suppression). It is useful to think carefully about potential suppressors, since omitting them can lead to an underestimation of the direct effect of our predictor of interest. Distinguishing between confounders and mediators cannot be determined by the empirical methods we study in this book. Conceptual rationales and measurement decisions determine their placement in the model (e.g., assuring that confounders precede and mediators follow predictors of interest logically and operationally). In cross-sectional data, the direction of arrows will often be debatable.

## KEY TERMS

Confounder (also Common Cause, Extraneous Variable)

Direct Effect

Indirect Effect

Mediation

Mediator (also Proximal Cause, Intervening Variable)

Moderation

Omitted Variable Bias

Path Analysis

Spurious Relationship

Suppressor Variables

Total Effect

# REVIEW QUESTIONS

**10.1** How can you calculate the direct and indirect effects in a path diagram in which the effect of $X_2$ on $Y$ is partially mediated by $X_3$?

**10.2** How can you calculate the total effect based on the direct and indirect effects?

**10.3** How can you draw examples of path diagrams in which omitted variable bias overstates the direct effect of $X_2$ on $Y$ away from zero in a *positive* direction, in which omitted variable bias overstates the direct effect of $X_2$ on $Y$ away from zero in a *negative* direction, and in which omitted variable bias understates the direct effect of $X_2$ on $Y$ (toward zero)?

# REVIEW EXERCISES

**10.1.** Refer to Brumbaugh and colleagues Table 3 presented in Literature Excerpt 10.1. Compare the coefficients for the *Female* and *Age* variables in Model 2 versus Model 3 (when marital, cohabitation, and parenthood histories are added to the model). Discuss whether you would interpret the changes in coefficients for these two variables as *mediation, suppression*, and/or *confounding*. What else might you need to know in order to make a determination?

**10.2.** Refer to Dooley and Prause's Table 3 presented in Literature Excerpt 6.1.

(a) Compare the coefficients for *Age* in Model 2 and Model 3. What do the results suggest is the possible direction of correlations among *Age, Weeks of Gestation*, and *Birth Weight*?

(b) Compare the coefficients for *Weight prior to pregnancy* in Model 3 and Model 4. What do the results suggest is the possible direction of correlations among *Weight prior to pregnancy, Weight gain during pregnancy*, and *Birth Weight*?

**10.3.** Suppose that a researcher anticipates that the association between stress and distress may be suppressed because persons who are exposed to stressors may elicit social support from their social networks and social support reduces distress.

(a) *Calculate the indirect effect* of stress on distress through social support using the results found in the table below.

(b) *Comment* on whether the researcher's expectation about the suppressor effect is confirmed.

(c) As you answer the question, be sure to *fill out the diagram,* by writing variable names in the boxes and coefficients in the circles.

| Predictor Variable | Outcome Variable | | |
|---|---|---|---|
| | Social Support | Distress | Distress |
| Stress | 0.52* | 1.80* | 0.61* |
| | (0.03) | (0.01) | (0.08) |
| Social Support | — | −2.29* | — |
| | | (0.01) | |
| Intercept | −26.08 | −79.76 | −19.98 |
| | (3.40) | (0.49) | (7.81) |

* $p < .05$

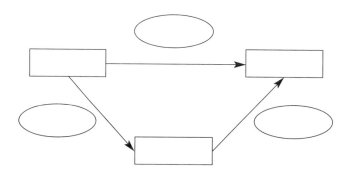

---

10.4. Consider the three regression model results presented below.

(a) Calculate the missing coefficient for the bivariate regression of SATMATH on FEMALE (cell labeled **A** below) based on the other coefficients listed in the table.

(b) As you answer the question, be sure to fill out the diagram by writing variable names in the boxes and coefficients in the circles.

| Predictor Variable | Outcome Variable | | |
|---|---|---|---|
| | SATMATH | SATMATH | NUMMATH |
| FEMALE | **A** | −98.16 | −2.51 |
| NUMMATH | — | 28.83 | — |
| Intercept | 473.51 | 358.30 | 4.00 |

* $p < .05$

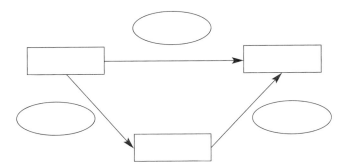

## CHAPTER EXERCISE

In this exercise, you will write a SAS and Stata batch program to estimate a mediated relationship.

To begin, prepare the shell of your batch program, including the commands to save your output and initial commands (e.g., SAS library, turning *more* off and closing *open logs* in Stata).

Start with the NOS 1996 dataset that you created in Chapter 4 and exclude cases with missing data using the following command on the data step in SAS:

```
if mngpctfem~=. & ftsize~=. & (n14=1 | n14=2) & (n15a=1 | n15a=2) & (n15b=1 | n15b=2)
                    & (n15c=1 | n15c=2) & (n15d=1 | n15d=2) & (n15e=1 | n15e=2);
```

and the use command in Stata:

```
if mngpctfem~=. & ftsize~=. & (n14==1 | n14==2) & (n15a==1 | n15a==2) & (n15b==1 | n15b==2) ///
                    & (n15c==1 | n15c==2) & (n15d==1 | n15d==2) & (n15e==1 | n15e==2)
```

Add the following code to your batch program to create a summary measure of the number of formal rules and regulations in the organization:

| **SAS** | **Stata** |
|---|---|
| if n14=1 \| n14=2 then Dn14= n14=1; | generate Dn14= n14==1 if n14==1 \| n14==2 |
| if n15a=1 \| n15a=2 then Dn15a=n15a=1; | generate Dn15a=n15a==1 if n15a==1 \| n15a==2 |
| if n15b=1 \| n15b=2 then Dn15b=n15b=1; | generate Dn15b=n15b==1 if n15b==1 \| n15b==2 |
| if n15c=1 \| n15c=2 then Dn15c=n15c=1; | generate Dn15c=n15c==1 if n15c==1 \| n15c==2 |
| if n15d=1 \| n15d=2 then Dn15d=n15d=1; | generate Dn15d=n15d==1 if n15d==1 \| n15d==2 |
| if n15e=1 \| n15e=2 then Dn15e=n15e=1; | generate Dn15e=n15e==1 if n15e==1 \| n15e==2 |
| formalsum=Dn14+Dn15a+Dn15b+ Dn15c+Dn15d+Dn15e; | generate formalsum=Dn14+Dn15a+Dn15b+/// Dn15c+Dn15d+Dn15e |

Notice that we are not restricting the analyses to organizations founded since 1950 for this assignment.

In all cases, conduct two-sided hypothesis tests. Use a 5 percent alpha unless otherwise indicated.

### 10.1 Total, Direct, and Indirect Effects

(a) SAS/Stata tasks:

(i) Regress *mngpctfem* on *ftsize* (we will refer to this as *Regression #1*).

(ii) Regress *mngpctfem* on *ftsize* and *formalsum* (we will refer to this as *Regression #2*).

(iii) Regress *formalsum* on *ftsize* (we will refer to this as *Regression #3*).

(b) Write-up tasks:

(i) Use Regression #2 and Regression #3 to calculate the total, direct, and indirect effects.

(ii) Comment on how the total effect calculated in "Question 10.1bi" relates to the coefficient in Regression #1.

## COURSE EXERCISE

**COURSE EXERCISE 10**

Identify at least one confounding and at least one mediating relationship that you might estimate based on the data set that you created in the course exercise to Chapter 4. Ideally, think about conceptual reasons why you might expect that one variable confounds or mediates the effect of another variable.

For the meditational relationship, estimate the regression models needed to calculate the total, direct, and indirect effects. Calculate each of these effects based on the results.

For the confounding relationship, show how the coefficient estimate for your predictor of interest changes when you add the confounding variable to the regression model. What is the direction of bias when this variable is omitted from the regression model? How is it related to the outcome and predictor of interest?

Using the commands shown in Display B.10.3, show how you can reproduce the multiple regression coefficients by regressing the outcome on the residuals.

# OUTLIERS, HETEROSKEDASTICITY, AND MULTICOLLINEARITY

## CHAPTER 11: OUTLIERS, HETEROSKEDASTICITY, AND MULTICOLLINEARITY

In this chapter we discuss several violations of model assumptions in OLS, and strategies for identifying and dealing with them. In particular, we consider how extreme data points can influence point estimates and how heteroskedasticity and multicollinearity can influence standard errors. Extreme data points can arise for various reasons, one of which is mis-specification of the form of the relationship between the predictor and outcome (e.g., as linear rather than nonlinear). Heteroskedasticity means that the assumption of constant conditional variance is violated. Multicollinearity is an extreme case of expected correlations among predictors. Indeed, some correlation among predictors is needed in order to benefit from multiple regression's ability to reduce bias: adding variables to a multiple regression model will not change the coefficient estimate of our predictor of interest unless the new variables are correlated at least to some degree with our predictor. At the same time, correlation among the predictors increases the standard errors of the coefficients. Very high correlation among predictor variables, or multicollinearity, will make our coefficient estimates quite imprecise.

## 11.1: OUTLIERS AND INFLUENTIAL OBSERVATIONS

A historical perspective by one of the developers of a widely used diagnostic for influential observations provides a nice introduction to this section:

> My own ideas on diagnostics grew out of theoretical work on robust estimation and applied work on econometric models. One model we were working on just did not agree with economic theory. After looking at residuals and some plots without noting

anything, I questioned the economic theory. The economists said our estimated model could not be right, so I checked all of the data again. One observation had been entered incorrectly several iterations back and it seemed to me that I should have had a better clue to that problem long before I questioned the economic theory. Leaving each observation out one-at-a-time was easy to talk about and after a few weeks of programming easy to do in practice (Welsch 1986, 404).

This quote highlights the intimate relationship between diagnosing influential observations and model testing, as well as how influential observations might arise (data entry error) and can be diagnosed (examining how excluding a single case affects the estimates).

### 11.1.1: Definition of Outliers and Influential Observations

An **outlier** is a value that falls far from the rest of the values on a variable. In a regression context, we typically are especially interested in outliers on the outcome variable. In some cases, outliers may go hand in hand with the outcome variable not being normally distributed. For instance, in our distance example, we know that the outcome variable of miles from the mother is highly skewed, with many zero values but some adults living thousands of miles from the mother. In a univariate context where the variable is expected to be normally distributed, outliers are often identified as values falling more than two standard deviations above or below the mean. Although we expect about 5 percent of observations to fall in this range by chance, a larger number of outliers or very extreme outliers may suggest non-normality or errors in the data.

An **influential observation** is an observation whose inclusion in or exclusion from the data file greatly changes the results of the regression analysis. These observations can be extreme because they fall relatively far from the mean across the predictor variables and because they fall relatively far from the conditional regression line (regression surface in multiple regression). It is extremely important to watch for such influential observations to avoid reporting results that depend on a single or small number of observations.

As illustrated in Literature Excerpt 11.1 an observation may be an outlier and not influential or influential but not an outlier. In Figure 1a, point $A'$ is an outlier on $Y$ but not influential. Point is $A''$ influential but not an outlier on $Y$. In Figure 1b, point $A'$ is both an outlier and influential. Figures 1c and 1d further visualize how the slope of the regression line might change when a single point is included or excluded. In Figure 1c, when point $A$ is excluded, the estimate of the slope is zero, but when $A$ is included, the slope is positive (the regression line is pulled upward by the influential observation). In Figure 1d, when point $A$ is excluded, the estimate of the slope is positive, but when $A$ is included, the slope is zero (the regression line is pulled downward by the influential observation).

▨ **Literature Excerpt 11.1**

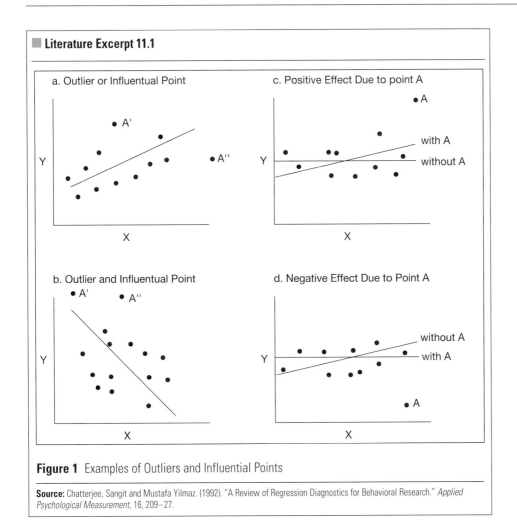

**Figure 1**  Examples of Outliers and Influential Points

**Source:** Chatterjee, Sangit and Mustafa Yilmaz. (1992). "A Review of Regression Diagnostics for Behavioral Research." *Applied Psychological Measurement*, 16, 209–27.

## 11.1.2: Possible Reasons for Outliers and Influential Observations

One reason for outliers and influential observations is data errors. Thus, it is always important to check for data entry and data management errors. We have emphasized in earlier chapters the importance of verifying that the distribution of values match codebooks in existing data, that missing data values are appropriately recoded, and that the distribution of values on created variables conform with desired results. Skipping these steps at the data management stage can lead to identification of outliers at the analysis stage (or if overlooked at the analysis stage, potentially reporting erroneous results). If you are using existing data, it is also helpful to stay in contact with the study staff, where possible (for example, signing up to listservs where data corrections and new data releases are announced). When you collect your own data, it is important to first carefully collect data (to reduce reporting errors from respondents), then to carefully enter the data (to reduce keystroke errors), and finally to examine the distribution of variables (to identify extreme values that should be verified against the raw data).

Another reason for outliers and influential observations is that the model is not specified correctly. When the relationship between a predictor and outcome is curvilinear, but a linear relationship between the predictor and outcome is estimated, diagnostics on the linear model may identify apparent influential observations that are no longer influential once a nonlinear model is estimated. The extreme cases may also conform to a different model than the other cases (perhaps suggesting an interaction or other subgroup analysis is needed).

The Lyons (2007) article that we examined in Chapter 1 provides an example. The table reproduced in Literature Excerpt 1.1b had columns labeled *Excluding Outliers*. In the text, Lyons notes that three influential observations were identified using some of the techniques we describe in the next section. These were three communities that had "unusually high levels of antiwhite hate crime given their affluence and racial (white) homogeneity" and were all located on Chicago's southwest side (p. 844). When these cases were omitted from the analysis, the expected association was revealed (antiwhite hate crime was higher in poorer areas with high residential mobility). As Lyon's notes "thorough explanation for these different patterns in the southwest warrants qualitative exploration beyond the scope of the present study" (p. 844) but he speculates about the potential reasons for these three cases conforming to a different model in the discussion. He notes that all three communities are areas with high antiblack hate crime that is explained well by his defended communities model. He speculates that the high antiwhite crime could be retaliatory for current antiblack crime, or due to historical racial tensions in these areas. He also speculates that whites may be more ready to define and report acts as racially motivated in these areas (p. 849).

Another alternative to excluding outlying cases or conducting subgroup analyses would be to transform the outcome, as we did using the log transformation in Chapter 9. In fact, two of our Literature Excerpts examining nonlinear relationships in Chapter 9 used a log transformation of the outcome (Glauber 2007; Pritchett and Summers 1996; Willson 2003 used another—square root—transformation because she found the log transformation too severe).

Yet another alternative to deal with outliers would be to move to a model that makes different assumptions about the distribution of the outcome; that is, values that seem unusual in a normal distribution may not be unusual for other distributions. In Chapter 12, we will provide a roadmap to models designed for dichotomous, ordinal, nominal, and count outcomes. In fact, our fourth Literature Excerpt in Chapter 9, Cornwell, Laumann, and Schumm 2008, used one of these models (Poisson regression) for their count outcome.

### 11.1.3: Identifying Outliers and Influential Observations

Many diagnostic measures have been developed with cutoff values to use as a general guide in identifying influential observations (see Chatterjee and Hadi 1986; Fox 1991; Belsey, Kuh, and Welsch 1980). We will cover five that are used frequently in the literature: (a) hat values, (b) studentized residuals, (c) Cook's distance, (d) DFFITS, and (e) DFBETAs.

The first two diagnostic measures do not measure influence, per se, but the potential for influence based on an outlying value on the outcome (studentized residuals) or the predictors (hat values). The latter three measure the change in the predicted values or coefficient estimates when a case is excluded from versus included in the estimation sample.

Each of these diagnostics is calculated for every case in the data set. Suggested cutoffs are available to identify extreme values for each diagnostic, although it is generally recommended that these be used along with graphs to identify extreme values in the context of a given data set (Chatterjee and Hadi 1986; Welsch 1986). We will illustrate how to use such cutoffs and graphs below.

The first diagnostic considers how unusual each observation is on the predictor variables and the second considers how far each observation falls from the fitted regression line (surface).

■ **Hat values** ($h_i$) measure the distance of the $X$-values for the $i$th case from the means of the $X$-values for all $n$ cases. Hat values fall between zero and 1, and a larger hat value indicates that the case falls far from the center (means) of the predictors in multidimensional space. The hat value is sometimes called a measure of **leverage** because observations that are distant from the center of the X variables have the potential to greatly alter the coefficient estimates from the model. For moderate to large sample sizes, $2(k-1)/n$ or $3(k-1)/n$ has been suggested as a cutoff for identifying high leverage (where $k$ is the number of predictors plus one for the intercept, and thus the cutoff is two or three times the number of predictors divided by the sample size; Chatterjee and Hadi 1986; Hoaglin and Kempthorne 1986).

▪ **Studentized residuals** are the estimated errors we are familiar with ($\hat{\varepsilon}_i = Y_i - \hat{Y}_i$) scaled by the root mean square error and hat value.

$$t_i^* = \frac{Y_i - \hat{Y}_i}{\hat{\sigma}_i \sqrt{1 - h_i}}$$

In the formula, $\hat{\sigma}_i$ is the mean square error for the regression model when the $i$th residual is excluded from the estimation sample and $h_i$ is the hat value. The studentized residuals follow a $t$-distribution with $n$-$k$ degrees of freedom (where $k$ is the number of predictors in the model plus one for the intercept). Values larger than 2 or 3 in absolute value are generally taken to be extreme, since about 95 percent of observations should fall within the range of $-2$ and 2.

▪ **Standardized residuals** may be calculated similarly.

$$t_i = \frac{Y_i - \hat{Y}_i}{\hat{\sigma} \sqrt{1 - h_i}}$$

Here, $\hat{\sigma}$ is the usual root mean square error for the regression model when all cases are included in the estimation sample.

Three other measures directly assess influence: how results from the regression differ when the observation is included or omitted.

▪ **Cook's distance** combines the standardized residuals and hat values in a single calculation:

$$D_i = \frac{t_i^2}{k-1} * \frac{h_i}{(1 - h_i)}$$

Because $t_i$ is squared in the numerator, Cook's distance is always positive in value. A cutoff of $4/n$ has been suggested to flag potentially influential observations.

▪ **DFFITS** is similar to Cook's distance, but calculated based on the studentized residual:

$$DFFIT_i = t_i^* \sqrt{h_i / (1 - h_i)}$$

Cook's distance and DFFITS can identify different values as influential, because the DFFITS measures the influence on both the coefficients and conditional variance (because $t_i^*$ uses $\hat{\sigma}_i$), while Cook's distance captures only the influence on the coefficients (because $t_i$ uses $\hat{\sigma}$).

The word DFFIT comes from "difference in fit," and can be remembered as an abbreviation of this phrase.[1]

For large data sets, the cutoff for high DFFITS is that the absolute value of the DFFITS is greater than $2\sqrt{(k-1)/n}$ where $k$ is equal to the number of parameters in the model plus 1 for the intercept. For smaller data sets, a cutoff of 1 or 2 has also been suggested (Chatterjee and Hadi 1986).[2]

■ **DFBETAS** provide a standardized measure of the difference between a parameter estimate when all $n$ cases are used to estimate the regression $(\hat{\beta}_k)$ and when the $i$th case is omitted from the sample prior to running the regression $(\hat{\beta}_{ki})$.

$$DFBETA_i = \frac{(\hat{\beta}_k - \hat{\beta}_{ki})}{\hat{\sigma}_{\hat{\beta}_k}}$$

Thus, DFBETA can be remembered as "difference in beta." A separate DFBETA is calculated for each predictor variable included in the model. For large data sets, when the absolute value of DFBETA is larger than $2\sqrt{n}$ there may be high influence (Chatterjee and Hadi 1986).

Table 11.1 summarizes the diagnostic measures and their cutoffs.

▨ **Table 11.1: Summary of Diagnostic Measures of Outlying and Influential Observations**

| Name | Definition | Cutoff Values | |
|---|---|---|---|
| Hat value | Measures how far observations fall from center of the predictor variables. Also known as leverage. | $2(k-1)/n$ | $3(k-1)/n$ |
| Studentized or standardized residuals | Scaled residuals from regression model. (Scaling factor differs slightly between two versions) | 2 | 3 |
| Cook's distance | Combined measure based on standardized residuals and hat values. | $4/n$ | |
| *DFFITS* | Capture changes in the "fit" of the regression, both the coefficients and conditional variance, when a case is excluded from the sample. | 1 or 2 | $2\sqrt{(k-1)/n}$ |
| *DFBETAS* (one for each predictor in model) | Capture how each coefficient changes when a case is excluded from the sample. | $2/\sqrt{n}$ | |

*Note:* $k$ is the number of predictors plus one for the intercept. $n$ is the sample size. Most measures can be negative or positive in sign, so take the absolute value (use abs () function in SAS and Stata) before evaluating their magnitude relative to the cutoffs.

### 11.1.4: Strategies for Addressing Outliers and Influential Observations

As noted above, a common approach to dealing with non-normality of the outcome variable within the context of OLS is to take the log of the outcome variable. Obviously, this approach won't work in all cases (e.g., variables with numerous zeros or negative values, or dichotomous outcomes). But in some cases it does a good job of normalizing the outcome variable and reducing outliers.

Another common approach to dealing with influential cases in OLS is to see how these cases differ from other cases and to refine the model accordingly. When the influential cases are few in number, not an obvious data entry error, or have no apparent substantive meaning, it is possible simply to examine how sensitive the findings are to the inclusion or exclusion of the outlying observations. In the case of the Lyons (2007) Literature Excerpt mentioned above, the three outlying cases hung together in a substantive way, but were too few in number to be separately analyzed, so they were excluded from the presented results but mentioned in the results and discussion sections.

### 11.1.5: Example 11.1

We will use our distance example to illustrate the process of identifying and addressing outliers in practice. As we noted above, although cutoffs exist for considering diagnostic values extreme, there is no hard-and-fast procedure for identifying and addressing outliers. There are many decision points where different researchers might make somewhat different choices. We try to convey this fluid process in our example. It is typically useful to use graphs of the variables and multiple diagnostic values to try to get a full picture of potential outliers and their actual influence. In practice, we are typically concerned with being sure that our results are not sensitive to the inclusion or exclusion of one or more influential observations (including those that are data entry errors) and that we have not mis-stated an association when it applies differently to some subset of the data. We may need to take several different approaches to looking for and addressing outliers before we are confident that these problems are not present. We would typically present the final result that we conclude best represents the data, but summarize in a footnote, paragraph, or appendix, what we did to come to this conclusion. In examining outliers (and heteroskedasticity in the next section) we will estimate a multiple regression similar to the one we estimated in Table 7.2, but allowing for the nonlinear associations of *g1yrschl* and *g2numsis* with *g1miles*, and including the *other* racial-ethnic group.

### Graphing the Variables

Graphing the variables in the data set, especially the outcome and central predictors, is an important general first step in any project. Graphs can also be specifically useful in identifying outliers. Univariate plots can be examined for extreme values that fall far from the remainder of the distribution. Histograms and box plots, and stem-and-leaf plots for small data sets, are useful for this purpose. Scatterplots can be used to plot the outcome against the predictor to see if some values fall far from the major cloud of data, suggesting potential influence.

## Univariate Plots

The `proc univariate` command in SAS provides a number of useful plots (including a histogram or stem-and-leaf plot and a box plot). We use the following command to request plots of the outcome (in natural units and logged form) and two central predictors (mother's schooling and number of sisters).

```
proc univariate plots;
   var g1miles logg1miles g1yrschl g2numsis;
run;
```

In Stata we can use the `stem` command for stem-and-leaf plots, the `histogram` command for histograms, and the `graph` command for box plots. We find box plots particularly helpful for examining potential outliers, although you should experiment with which graph works best for you.[3]

| Stem-and-Leaf | Histogram | Box Plot |
|---|---|---|
| stem g1miles | histogram g1miles | graph box g1miles |
| stem logg1miles | histogram logg1miles | graph box logg1miles |
| stem g1yrschl | histogram g1yrschl | graph box g1yrschl |
| stem g2numsis | histogram g2numsis | graph box g2numsis |

Recall that in a box plot, the top of the box is the 75th percentile and the bottom of the box is the 25th percentile. The horizontal line through the middle of the box is the median. Dots represent outliers, which are defined in the box plot as values more than 1.5 times the interquartile range above or below the box (recall the interquartile range is the difference between the value at the 75th and the 25th percentiles). Whiskers extend from the box to the largest and smallest values that are not outliers.

Figure 11.1 provides a box plot of a standard normal variable for reference. Notice that some outliers are present, even for a normal variable (because we expect 5 percent of the values to fall beyond about two standard deviations above and below the mean). In Figure 11.1, a few values are nearly four standard deviations above and below the mean.

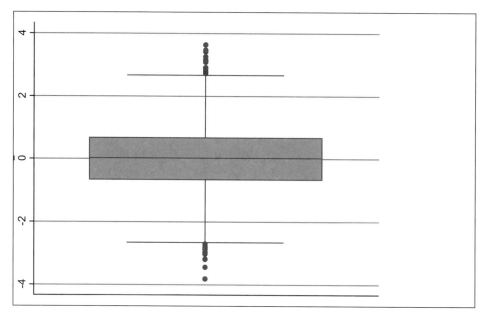

▨ **Figure 11.1:** Box Plot for a Standard Normal Variable.

Figure 11.2 presents the box plots of our four variables, using the Stata results. The boxes for *logg1miles* (Figure 11.2(b)) and *g1yrschl* (Figure 11.2(c)) look the most symmetric, and *logg1miles* has no outliers. *g1yrschl* has some outliers, especially the small values between zero and five years of schooling. In contrast *g1miles* and *g2numsis* both are skewed upward (skewed right) with some outliers falling far from the rest of the data (see Figures 11.2(a) and (d)). The top values on *g1miles* is one respondent who reports living 9,000 miles from his mother. The next closest value is 6,000 miles. On *g2numsis*, one respondent reports 20 and one respondent reports 22 sisters. The next closest value is 11. These results reinforce the benefit of the log transformation of the outcome for better approximating a normal distribution of the outcome in our case (compare Figures 11.2(a) and (b)) and suggest the potential for outliers and influential observations in the models that we estimated in earlier chapters with the nonlogged outcome and *g1yrschl* and *g2numsis* as continuous predictors.

*Scatterplots*
In SAS we can use the `proc gplot` statement to create scatterplots. For example, to graph *g1miles* on the Y axis and *g1yrschl* on the X axis, we would type:

```
proc gplot;
   plot g1miles*g1yrschl;
run;
```

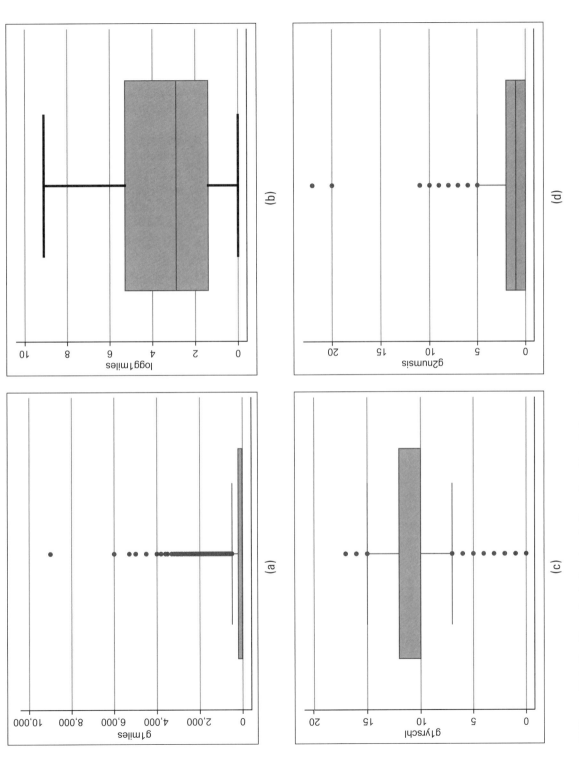

**Figure 11.2:** Box Plots of *g1miles, logg1miles, g1yrschl,* and *g2numsis.*

In Stata, we can create scatterplots using `graph twoway scatterplot` followed by the name of our outcome and a predictor variable. For example, `graph twoway scatterplot g1miles g1yrschl` would graph *g1miles* on the *Y* axis and *g1yrschl* on the *X* axis.[4]

We plotted the nonlogged and logged versions of *g1miles* against the mother's years of schooling and the adult's number of sisters, and show the results from Stata in Figure 11.3. The plots with *g1miles* in its natural units are on the left and those with *g1miles* in logged form are on the right. As shown by the red arrows, the case whose mother is reported to live 9,000 miles away (Figures 11.3(a) and (c)) and cases who report about 20 or more sisters (Figures 11.3(c) and (d)) stand out.

## Diagnostic Values

Graphs can also be used to examine the distribution of diagnostic values. Regardless of whether values fall above or below cutoffs, values that fall far from the rest of the data are worth examining. Sets of cases that fall above the cutoff can also be examined to see whether they share a common characteristic that might suggest how a model could be modified to reduce their influence and/or how sensitive the results are to their exclusion.

In SAS, we can request all of the diagnostics with options on the model statement (`/r influence`). By default, the table of values is simply listed in the ouput. But, we can use an `ods` statement to save the diagnostics in a data file to examine.[5] SAS names these statistics by default, with names that map clearly onto the names we used above, and we will use these names in Stata as well (e.g., the *DFBETAS* are named *DFB_* followed by the variable name).

```
proc reg;
  model logg1miles=g1yrschl sqg1yrschl g2numsis4 g2numsis56
                   g2numsis7p
                   amind mexamer white other female
                   g2earn10000 g1age g2age g2numbro
                   /r influence;
  ods output outputstatistics=influence;
run;
```

After creating the output data file, we can examine the values of the diagnostics using the `proc univariate` command.

```
proc univariate plots;
  var HatDiagonal RStudent CooksD DFFITS
    DFB_g1yrschl DFB_sqg1yrschl
    DFB_g2numsis4 DFB_g2numsis56 DFB_g2numsis7p;
run;
```

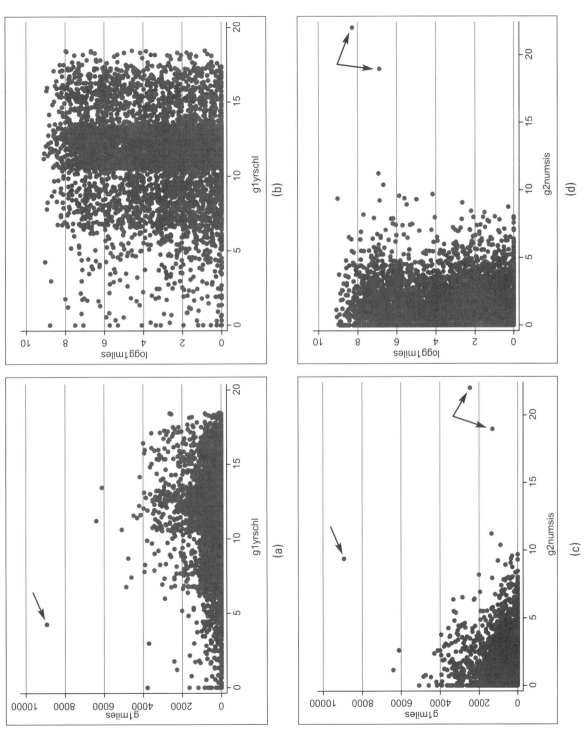

**Figure 11.3:** Scatterplots of *g1miles* and *logg1miles* with *g1yrschl* and *g2numsis*.

In Stata, we use the predict command to request each diagnostic be calculated, with options indicating which diagnostics. Stata adds a variable containing the diagnostic value to the current data set and allows us to choose names for these new variables. For simplicity, we use the same names as SAS chooses for its influence statistics.

```
regress logg1miles g1yrschl sqg1yrschl g2numsis4 ///
                    g2numsis56 g2numsis7p ///
                    amind mexamer white other female ///
                    g2earn10000 g1age g2age g2numbro
    predict HatDiagonal, hat
    predict RStudent, rstudent
    predict CooksD, cooksd
    predict DFFITS, dfits
    predict DFB_g1yrschl, dfbeta(g1yrschl)
    predict DFB_sqg1yrschl, dfbeta(sqg1yrschl)
    predict DFB_g2numsis4, dfbeta(g2numsis4)
    predict DFB_g2numsis56, dfbeta(g2numsis56)
    predict DFB_g2numsis7p, dfbeta(g2numsis7p)
```

We can use expressions in SAS and Stata to identify whether the values exceed the cutoffs. Display B.11.1 shows these calculations and summarizes the results. For most of the diagnostics, less than 5 percent of cases fall above the cutoff. The exceptions are leverage (*HatDiagonal*) with about 6 percent of cases above 3 and DFFITS with just over 5 percent above the cutoff.

Because these cutoffs are only rough guides, it is also helpful to graph the diagnostic values to see whether any fall far from the rest. We use proc univariate in SAS and the graph box command in Stata to create box plots of the results.

Figure 11.4 (see pp 376–7) presents box plots for the six diagnostic values. The hat value (Figure 11.4(a), with *Y* axis labeled *Leverage*), Cook's distance (Figure 11.4(c)), and DFFITS (Figure 11.4(d)) have values that clearly fall from the rest of the distribution (above about 0.10, above about 0.04, and below about –0.40, respectively). We use the list command to examine the cases that fall beyond these values. The syntax is list <variable list> if <expression> in Stata and proc print; var <variable list>; where <expression>; run; in SAS. We listed each of the variables in the regression model using an expression to identify the cases beyond the identified extreme value. For example, in Stata we typed:

```
list MCASEID g1miles g1yrschl sqg1yrschl g2numsis4 ///
    g2numsis56 g2numsis7p ///
```

```
       amind mexamer white other female g2earn10000 g1age ///
       g2age g2numbro ///
       if HatDiagonal>0.10
   list MCASEID g1miles g1yrschl sqg1yrschl g2numsis4 ///
       g2numsis56 g2numsis7p ///
       amind mexamer white other female g2earn10000 g1age ///
       g2age g2numbro ///
       if CooksD>0.04
   list MCASEID g1miles g1yrschl sqg1yrschl g2numsis4 ///
       g2numsis56 g2numsis7p ///
       amind mexamer white other female g2earn10000 g1age ///
       g2age g2numbro ///
       if DFFITS<-0.40
```

The results (not shown in a box) reveal that the same two cases are identified as extreme across these measures. Both of these cases have earnings of nearly $1 million/year. A box plot (not shown) revealed that earnings are highly skewed to the right. Although we are not focusing on earnings in this chapter, these results suggest that we might want to log or otherwise transform earnings to reduce outliers (and, as discussed above, it is common to log earnings, especially in substantive fields that intersect with economics).

In addition to examining cases whose diagnostic values are quite extreme relative to the rest of the sample, we also examined cases that fell above the cutoff on any of the diagnostics, creating a variable *anyhi*. For example, in Stata we would type:[6]

```
   generate anyhi=HatDiagonal3Hi==1 |RStudentHi==1| ///
       CooksDHi==1|DFFITSHi==1| ///
       DFB_g1yrschlHi==1 |DFB_sqg1yrschlHi==1| ///
       DFB_g2numsis4Hi==1|DFB_g2numsis56Hi==1| ///
       DFB_g2numsis7pHi==1
```

We created box plots separately for cases coded "0" and "1" on this *anyhi* dummy variable. The results are in Figure 11.5. In each panel, the graph on the left is a box plot for the cases coded 0 on *anyhi* and the graph on the right is a box plot for cases coded "1" on *anyhi*. The graphs on the right show that outlying and influential cases tend to have high values on *g1miles* (Figure 11.5(a)), both low and high values on *g1yrschl* (Figure 11.5(b)), and high values on *g2numsis* (Figure 11.5(c)).

We also summarized all of the predictor variables for the subset of the sample identified as an outlier or influential observation on any diagnostic measure versus

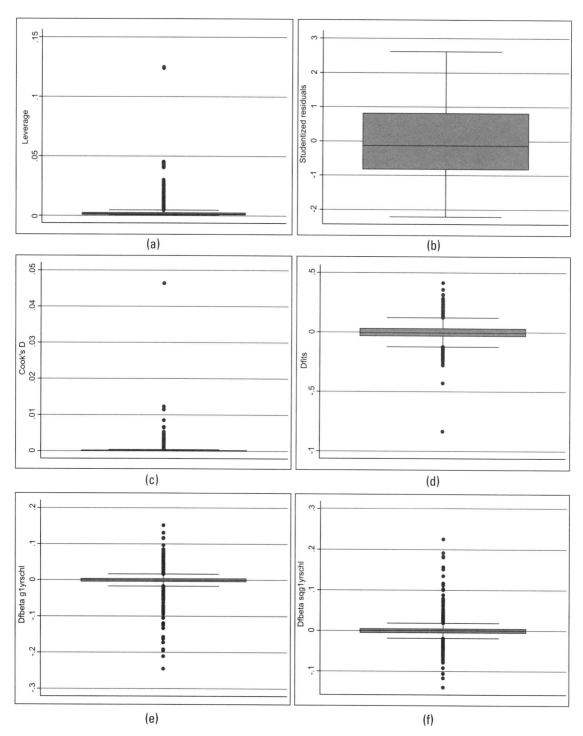

**Figure 11.4:** Box Plots of Diagnostic Values.

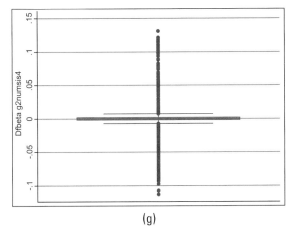

(g)

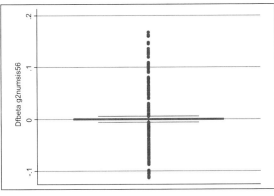

(h)

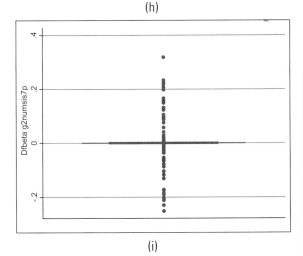

(i)

■ **Figure 11.4:** Box Plots of Diagnostic Values—Cont'd.

those not so identified (see Box 11.1). The results are listed in Display B.11.2, and summarized in Table 11.2 below.

The results show that the cases that are extreme on at least one diagnostic on average have mothers with fewer years of schooling (9.89 versus 11.65), more sisters (across dummies *g2numsis4*, *g2numsis56*, and *g2numsis7p*, 0.24 + 0.21 + 0.07 = 0.52 versus 0.01 + 0.01 + 0.00 = 0.02), and are more likely to be nonwhite (1-*white* is 1–0.51 = 0.49 versus 1–0.86 = 0.14).

In fact, all of the cases of American Indian and other ethnicities and all of the cases with seven or more sisters are identified as outliers or influential observations by at least one diagnostic measure (i.e., the mean is 0.00 for the subgroup *anyhi* = 0 for the variables *g2numsis7p*, *amind*, and *other*). Each of these groups is small, containing about 70 or less cases in the entire data set. When we examined the individual diagnostics for these cases (not shown), we saw that all have hat values that fall above the cutoff, reflecting their rareness in the data set. In addition, the DFBETA for *g2numsis7p* is above the cutoff for nearly every case with seven or more sisters, indicating that the coefficient

■ **Box 11.1**

In SAS, the diagnostics must be merged into the original data file to make these calculations, as shown at the top of Display B.11.2. Typically, when merging two files, you will use the `by` option and merge on an identifier (e.g., `by MCASEID`). Because we know that both data files are sorted in the same order in this case, we do not need a by variable on the merge.

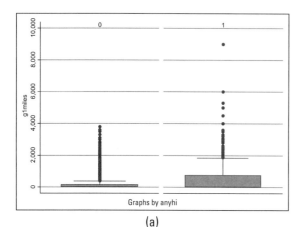

(a)

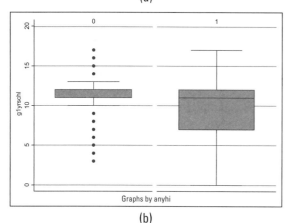

(b)

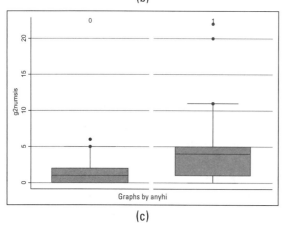

(c)

■ **Figure 11.5:** Box Pots of *g1 miles*, *g1 yrschl*, and *g2 numsis* for Cases that Do and Do Not Fall above the Cutoff on any Diagnostics.

■ **Table 11.2: Means for Predictor Variables Among Cases Without and With any Extreme Values Across Diagnostics**

|  | anyhi = 0 | anyhi = 1 |
|---|---|---|
|  | (*n* = 4,694) | (*n* = 778) |
| g1yrschl | 11.65 | 9.89 |
| sqg1yrschl | 141.97 | 118.37 |
| g2numsis4 | 0.01 | 0.24 |
| g2numsis56 | 0.01 | 0.21 |
| g2numsis7p | 0.00 | 0.07 |
| amind | 0.00 | 0.03 |
| mexamer | 0.01 | 0.15 |
| white | 0.86 | 0.51 |
| other | 0.00 | 0.09 |
| female | 0.60 | 0.63 |
| g2earn10000 | 3.04 | 3.24 |
| g1age | 59.94 | 60.95 |
| g2age | 34.26 | 34.47 |
| g2numbro | 1.30 | 2.21 |

estimate is highly sensitive to which of this small set of cases are in the sample. As we discuss below, one decision we might make in the face of these results would be to say that the NSFH data are not well suited for examining American Indian and the heterogeneous other race-ethnicities and families with numerous siblings, and to exclude these subgroups from analyses.

Before discussing decision-making in more detail, we examine an additional concrete way to examine the sensitivity of the results to a set of outlying and influential observations: re-estimating the regression model after excluding the cases that we have identified as having outliers and/or influential observations across the diagnostics.[7] Display B.11.3 shows the original multiple regression model estimated on the full sample. Display B.11.4 shows the multiple regression estimated on the partial sample, with outliers and influential observations excluded (*anyhi* = 0). We summarize the results below, in Table 11.3, for the full and partial samples (*anyhi* = 0).

Notice that the coefficients are not estimated in the partial sample for the three small groups mentioned above (American Indians, other race-ethnicities, and those with seven or more sisters) because all of the cases are coded a "1" on *anyhi* and thus are excluded from the partial sample (the notation—in the table indicates that they are excluded).

The coefficient estimates for the predictors of focus (mother's years of schooling and number of sisters) are fairly similar between the full and partial samples, although the

■ **Table 11.3: Coefficient Estimates and Standard Errors from Regressions Estimated on the Full Sample and the Partial Sample (With Outlying and/or Influential Observations Excluded)**

| | Coefficient | | Standard Error | |
|---|---|---|---|---|
| | Full | Partial | Full | Partial |
| g1yrschl | −0.05 | −0.01 | 0.05 | 0.08 |
| sqg1yrschl | 0.008** | 0.008* | 0.00 | 0.00 |
| g2numsis4 | −0.03 | −0.07 | 0.16 | 0.29 |
| g2numsis56 | −0.28 | −0.33 | 0.18 | 0.36 |
| g2numsis7p | 0.99** | — | 0.33 | — |
| amind | 1.15* | — | 0.48 | — |
| mexamer | −0.15 | −0.46 | 0.21 | 0.34 |
| white | 0.41** | 0.59** | 0.09 | 0.10 |
| other | 0.97** | — | 0.29 | — |
| female | −0.07 | 0.02 | 0.07 | 0.07 |
| g2earn10000 | 0.04** | 0.07** | 0.01 | 0.01 |
| g1age | 0.01+ | 0.01* | 0.01 | 0.01 |
| g2age | 0.01* | 0.02* | 0.01 | 0.01 |
| g2numbro | 0.04+ | 0.06* | 0.02 | 0.02 |
| _cons | 1.12 | 0.08 | 0.37 | 0.54 |

Sample size is 5,472 for full sample and 4,694 for partial sample, with cases above cutoffs for outliers or influential observations excluded.

— Excluded.

** $p < 0.01$, * $p < 0.05$, + $p < 0.10$, two-sided $p$-values.

standard errors are larger, especially for the dummy indicators of numbers of sisters. These rising standard errors reflect the smaller sample sizes in the partial sample (in the full and partial samples, respectively, there are 245 versus 62 cases with four sisters; and 200 versus 40 cases with five or six sisters).

We can also get a concrete picture of the effect of outlying and influential observations by plotting predicted values based on the coefficients estimated for the full and partial samples, using procedures similar to those introduced in Chapter 9. Figure 11.6 shows such graphs.[8]

The results are strikingly similar in the full and partial samples, even with the much reduced sample sizes, although obviously the cases with seven or more siblings are not in the partial data file. This gives us confidence in the general form of the association for these two predictors of focus. We would need to make a choice for a manuscript about which results to present. We could present the full sample results, and note that the result for seven or more sisters is quite sensitive to which cases are included. Or, we might decide to exclude cases with seven or more sisters, perhaps noting that a design that specifically oversampled adults from very large families would be needed, better to estimate distance for that group. A similar decision might be made to exclude adults of American Indian and other race-ethnicities, given that their relatively small representation in the sample makes it difficult to obtain accurate estimates for these groups.

## 11.2: HETEROSKEDASTICITY

As discussed in Chapter 5, the OLS model assumes that the variance of the errors are homoskedastic—constant across levels of $X$. If the error variances are in fact different at different levels of $X$, then they are referred to as **heteroskedastic**.

The OLS estimators are not biased under heteroskedasticity (i.e., the expected value of the estimate $\hat{\beta}_1$ equals the population parameter $\beta_1$). However, the OLS estimates are no longer the *best* (lowest variance) estimators. Rather, it is possible to use another estimator that is also unbiased but will have a lower variance than the OLS estimator. Depending on the form of heteroskedasticity, it is possible that the standard errors of some predictors are underestimated and that the standard errors for some predictor variables are overestimated in OLS (Hayes and Cai 2007).

### 11.2.1: Possible Reasons for Heteroskedasticity

If the reason for heteroskedasticity can be identified, then it is possible to correct for the nonconstant variance through various modeling techniques. For example, a common

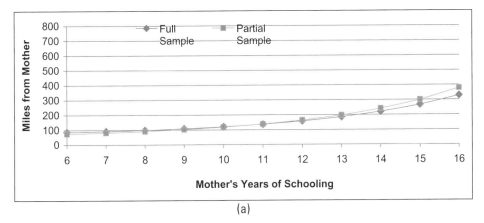

(a)

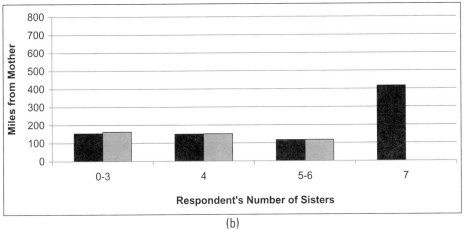

(b)

■ **Figure 11.6:** Graph of Predicted Values from Multiple Regression in Full Sample and Partial Sample (with Outliers and Influential Observations Excluded).

culprit in aggregate data is that the variances may depend on the size of the group (e.g., state, city, employer, school) on which calculations of means and proportions are made. The variance will be smaller for the groups with the largest size. Figure 11.7 provides an example of heteroskedasticity in state-level data in which violent crime rates were regressed on the percentage of the state's families with below-poverty incomes. The plot of the errors against the population size looks like a reverse fan, with the widest spread at the smallest population sizes. Another typical example is consumer expenditure data: if we regress expenditures on income, we often see increasing conditional variance with increasing income, explained by the larger discretionary income among persons with higher incomes (Downs and Rocke 1979).

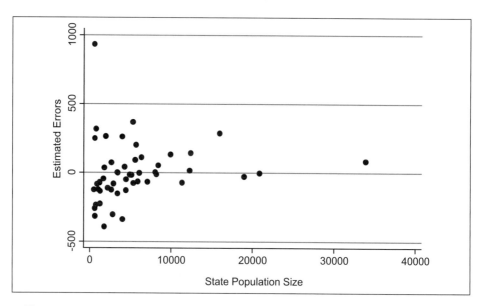

■ **Figure 11.7:** Heteroskedastic Errors from Regression of State Violent Crime Rate on Percentage of Families in Poverty in 2000.

Until the 1980s, nonconstant variance was commonly dealt with using strategies to test for heteroskedasticity, identify the source of the differences in conditional variances, and adjust for them. For example, a technique called *weighted least squares* allows researchers to weight each case in the calculation of the sum of squared residuals depending on the source of heteroskedasticity. In aggregate data, like our state-level example, cases with larger sample size would receive a lower weight. The problem with this approach is that at times there is not an obvious culprit producing nonconstant variance, and even when one culprit is identified, it is unknown whether all sources of heteroskedasticity have been addressed when an adjustment for one source of heteroskedasticity is made (Hayes and Cai 2007). Transformations of $Y$, such as the log transformation, may also reduce heteroskedasticity; however, again, such a transformation may not correct for all sources of heteroskedasticity.

### 11.2.2: Heteroskedasticity-Consistent Standard Errors

Today, general adjustments for heteroskedasticity of unknown form are possible and can be calculated in most statistical packages, including SAS and Stata. These **heteroskedasticity-consistent standard errors** are also sometimes referred to as **robust standard errors, heteroskedasticity robust standard errors, White standard errors,** or **Huber-White standard errors.** With recent developments, they are increasingly referred to with the letter HC followed by a number (e.g., HC0). These values are easier to calculate and broader in scope than earlier approaches to correcting for heteroskedasticity. As Hayes and Cai (2007, 711) put it:

> The appeal of this method lies in the fact that, unlike such methods as [Weighted Least Squares], it requires neither knowledge about nor a model of the functional form of the heteroskedasticity. It does not require the use of an arbitrary transformation of Y, and no intensive computer simulation is necessary.

In fact, all of the values needed to calculate the heteroskedasticity-consistent standard errors are available in a standard estimation of OLS.

In the bivariate context, we can write HC0 by substituting the value $\sigma^2_i$ for $\sigma^2$ in the formula for the standard error of the regression slope (Equation 5.8). We convert to a variance by squaring the result and multiplying the top and bottom by $\Sigma(X_i - \bar{X})^2$.

$$\sigma^2_{\hat{\beta}_1} = \frac{\sum(X_i - \bar{X})^2 \sigma^2_i}{\left[\sum(X_i - \bar{X})^2\right]^2}$$

Under homoskedasticity, because $\sigma^2_i = \sigma^2$ this formula reduces to the usual formula for the estimated variance of the OLS coefficient:

$$\sigma^2_{\hat{\beta}_1} = \frac{\sum(X_i - \bar{X})^2 \sigma^2_i}{\left[\sum(X_i - \bar{X})^2\right]^2} = \frac{\sigma^2 \sum(X_i - \bar{X})^2}{\left[\sum(X_i - \bar{X})^2\right]^2} = \frac{\sigma^2}{\sum(X_i - \bar{X})^2}$$

The heteroskedasticity-consistent variances can be estimated by using the estimated errors for each observation based on the sample regression equation (i.e., the $\hat{\varepsilon}^2_i$):

$$HC0 = \frac{\sum(X_i - \bar{X})^2 \hat{\varepsilon}^2_i}{\left[\sum(X_i - \bar{X})^2\right]^2}$$

The square root of this number is the heteroskedasticity-consistent standard error.

Recently, researchers increasingly use a slightly different calculation that adjusts for the hat values:

$$HC3 = \frac{\sum(X_i - \bar{X})^2 * (\hat{\varepsilon}^2_i / (1 - h_i)^2)}{\left[\sum(X_i - \bar{X})^2\right]^2}$$

Additional variants HC1, HC2, and HC4 are available, although recent simulations by Long and Ervin (2000) found HC3 to have superior performance. Importantly, their simulations showed that HC3 performed well in small as well as large samples and that HC3 was not inferior to the OLS standard errors when the homoskedasticity assumption held. Based on their results, the authors recommend that researchers routinely report HC3 standard errors, rather than the traditional OLS standard errors.

*T*-values for coefficients can be calculated in the usual fashion based on this heteroskedasticity-consistent standard error:

$$t = \frac{\hat{\beta}_1 - \beta_1^*}{\sqrt{HC3}}$$

where $\hat{\beta}_1$ is the standard OLS parameter estimate and $\beta_1^*$ is the hypothesized value, typically zero for our "straw man" hypothesis.

The formula for the heteroskedasticity-consistent standard error can be extended to multiple regression in a straightforward fashion and the procedure for calculating *t*-values is the same, although matrix algebra is required to write the formula (see Hayes and Cai 2007 and Long and Ervin 2000 for the matrix formulae).

### 11.2.3: Calculating Heteroskedasticity-Consistent Standard Errors in SAS and Stata

Stata's standard release calculates the various heteroskedasticity-consistent standard errors. Hayes and Cai (2007) wrote macros to calculate the various heteroskedasticity-consistent standard errors and to use them to calculate *t*-values and *p*-values in SAS. The web site for downloading the macros is provided in Appendix I (and via a link on the textbook web site *http://www.routledge.com/textbooks/9780415991544*).

### Example 11.2
Display B.11.5 presents the HC3 standard errors for the full sample multiple regression presented in Display B.11.3. We summarize the resulting coefficient estimates and standard errors from these two models (Display B.11.3 and Display B.11.5) in Table 11.4.

As expected, the coefficient estimates are the same in both models, but the summarized *p*-values sometimes differ because the standard errors differ. The coefficient for *amind* is now significant at $p < 0.01$ rather than $p < 0.05$. The coefficient for *g2numbro* is now significant at $p < 0.05$ rather approaching significance at $p < 0.10$. The standard

**Table 11.4: OLS and Heteroskedasticity-Consistent Standard Errors**

| | Coefficient | | Standard Error | | |
|---|---|---|---|---|---|
| | OLS | HC3 | OLS | HC3 | Difference |
| g1yrschl | −0.05 | −0.05 | 0.05 | 0.05 | 0.00 |
| sqg1yrschl | 0.008** | 0.008** | 0.00 | 0.00 | 0.00 |
| g2numsis4 | −0.03 | −0.03 | 0.16 | 0.16 | 0.00 |
| g2numsis56 | −0.28 | −0.28 | 0.18 | 0.18 | 0.00 |
| g2numsis7p | 0.99** | 0.99** | 0.33 | 0.37 | 0.04 |
| amind | 1.15* | 1.15** | 0.48 | 0.40 | −0.08 |
| mexamer | −0.15 | −0.15 | 0.21 | 0.20 | −0.01 |
| white | 0.41** | 0.41** | 0.09 | 0.09 | 0.00 |
| other | 0.97** | 0.97** | 0.29 | 0.32 | 0.03 |
| female | −0.07 | −0.07 | 0.07 | 0.07 | 0.00 |
| g2earn10000 | 0.04** | 0.04** | 0.01 | 0.01 | 0.00 |
| g1age | 0.01+ | 0.01+ | 0.01 | 0.01 | 0.00 |
| g2age | 0.01* | 0.01* | 0.01 | 0.01 | 0.00 |
| g2numbro | 0.04+ | 0.04* | 0.02 | 0.02 | 0.00 |
| _cons | 1.12 | 1.12 | 0.37 | 0.38 | 0.01 |

Sample size is 5,472.

OLS = Model with standard calculation of standard errors. HC3 = Model with heteroskedasticity-consistent standard errors using the $HC_3$ formula.

** $p < 0.01$, * $p < 0.05$, + $p < 0.10$, two-sided $p$-values.

errors change the most for variables that contain the smallest groups in the data set, although in some cases the standard errors for these variables are larger (other race-ethnicities and adults with seven or more sisters) and in some cases smaller (American Indian race-ethnicities) under the HC3 calculation.

## 11.3: MULTICOLLINEARITY

We end this chapter by considering multicollinearity. We have seen that the multiple regression model allows us to obtain estimates of the direct effect of a predictor variable on the outcome, parsing out the portion due to confounders and the flows through indirect pathways. It is also the case that when the predictor variables are themselves correlated, it can be difficult to separate out their unique effects on the outcome variable, depending on the degree of the correlation. At the extreme, if two predictor variables are perfectly correlated, then they are indistinguishable and we cannot disentangle their separate influences on the outcome variable. The situation of high correlation among predictor variables is referred to as **multicollinearity**.

## 11.3.1: Diagnosing Multicollinearity

Several approaches can be taken to diagnose multicollinearity. We will discuss three of them: (a) Variance Inflation Factors, (b) significant model $F$ but no significant individual coefficients, and (c) rising standard errors in models with controls.

### Variance Inflation Factors
We have already seen the formula which is used to define the **Variance Inflation Factor** (VIF). Recall that in a model with two predictors $X_2$ and $X_3$ the standard error and variance of the coefficients are:[9]

$$\text{Standard Error } \hat{\sigma}_{\hat{\beta}_{12}} = \sqrt{\frac{\hat{\sigma}^2}{\sum (X_{2i} - \overline{X}_2)^2 (1 - r_{23}^2)}} \quad \hat{\sigma}_{\hat{\beta}_{13}} = \sqrt{\frac{\hat{\sigma}^2}{\sum (X_{3i} - \overline{X}_3)^2 (1 - r_{23}^2)}}$$

$$\text{Variance } \hat{\sigma}_{\hat{\beta}_{12}}^2 = \frac{\hat{\sigma}^2}{\sum (X_{2i} - \overline{X}_2)^2 (1 - r_{23}^2)} \quad \hat{\sigma}_{\hat{\beta}_{13}}^2 = \frac{\hat{\sigma}^2}{\sum (X_{3i} - \overline{X}_3)^2 (1 - r_{23}^2)}$$

The VIF is defined as the right-hand term in the denominator:

$$\text{VIF} = \frac{1}{(1 - r_{23}^2)}$$

Substituting the VIF into the formula for the variance of the coefficient estimates makes the name clear, as the VIF is the factor by which the variance increases due to the correlation between the predictors.

$$\hat{\sigma}_{\hat{\beta}_{12}}^2 = \frac{\hat{\sigma}^2}{\sum (X_{2i} - \overline{X}_2)^2} * \text{VIF} \quad \hat{\sigma}_{\hat{\beta}_{13}}^2 = \frac{\hat{\sigma}^2}{\sum (X_{3i} - \overline{X}_3)^2} * \text{VIF}$$

Thus, when the two predictor variables are uncorrelated and VIF = 1, the VIF has no impact on the estimated variance of the coefficient. However, as the correlation between the two predictor variables increases, the VIF will get larger, and, all else equal, the estimated variance of the coefficient will increase as well.

Substituting the extreme values of the correlation into the VIF formula can also be instructive.

When the two predictor variables are uncorrelated, then $r^2_{23} = 0$ and thus:

$$VIF = \frac{1}{(1-0)} = \frac{1}{1} = 1$$

When the two predictor variables are perfectly correlated, then $r^2_{23} = 1$ and thus:

$$VIF = \frac{1}{(1-1)} = \frac{1}{0} = \infty$$

So, increasing correlation between the predictor variables will make the estimated variance of the coefficients, and thus their standard errors, larger. As a consequence, $t$-values will be smaller, and it will be harder to reject the null hypothesis that the coefficient is zero. As the correlation approaches the extreme of 1, the standard errors will "blow up," approaching infinity. In SAS and Stata, if two variables are perfectly correlated, the software will drop one variable from the model and report this problem in the results.

The formula for the VIF can be generalized to multiple regression models with more than two predictor variables:

$$VIF = \frac{1}{(1-R^2_j)}$$

where $R^2_j$ is the $R$-squared value from a regression of one of the predictor variables, call it $X_j$, on the remaining predictor variables. Note that when there are more than two predictor variables, the VIF is different for each predictor variable in the model.

Since we commonly report standard errors of coefficients, rather than variances of coefficients, it is often useful to take the square root of the VIF for interpretation. A VIF of 4 would be interpreted as follows: the estimated variance of the coefficient for this predictor variable is four times larger, and the standard error for the coefficient for this predictor variable is $\sqrt{4} = 2$ times larger, due to the variation that this predictor variable shares with the other predictor variables in the model, all else equal.

A cutoff value of 4 or 10 is sometimes given for regarding a VIF as high. But, it is important to evaluate the consequences of the VIF in the context of the other elements of the standard error, which may offset it (such as sample size; O'Brien 2007). We discuss below various remedies for multicollinearity.

### Significant Model *F* but No Significant Individual Coefficients

A telltale sign of high multicollinearity is having a significant model *F*-value but no significant *t*-values for individual coefficients in the model. Think about what this means: jointly, the coefficients in the model are significant (the rejection of the null for the model *F* tells us that at least one of the individual coefficients differs significantly from zero). But, no individual coefficient is significant. These contradictory findings are indicative of the fact that the predictors are so highly correlated in the model that it is impossible to isolate their individual impact on the outcome. However, the variance that they share is also shared by the outcome variable (together, they reduce the SSE). In fact, in this kind of situation, if we include only one of the collinear variables in the model, then that variable will be revealed as significant (and similarly for all of the collinear variables).

### Rising Standard Errors in Models with Controls

Another sign of multicollinearity relates to the VIF, but is diagnosed without actually calculating the VIFs: seeing the standard errors become substantially larger in models with additional controls. Watching for this sign of multicollinearity can be useful when examining published results that report standard errors.

### Example 11.3

We will first use a hypothetical example to demonstrate the three signs of multicollinearity. In this model, a teenage mother's years of schooling is predicted by her report during middle school of how far she thought she would go in school (*higrade7*) and how important school was to her (*impschl7*).

For instructional purposes, we will show how to calculate the VIFs by hand after regressing one predictor on the other predictor. And, we will also ask SAS and Stata to calculate the VIFs (with the `/vif` option in SAS and the `estat vif` command in Stata). Display B.11.6 shows the results.

First, notice that in the multiple regression predicting years of school (*yrschl*) based on *higrade7* and *impschl7* that the overall model *F* is significant ($F(2,997) = 9.66$, $p < 0.05$). But, neither variable has a large *t*-value (*t*-value of −1.10 for *higrade7*; *t*-value of 1.51 for *impschl7*). The VIFs are quite large at 105.89. (The VIF is shown in the last column of the `Parameter Estimates` table in SAS and in a new output table in Stata.) And, the regression of *higrade7* on *impschl7* in the bottom panel of the table shows that the R-squared is quite high, at 0.9906, reflecting the fact that we defined the variables to share nearly all of their variance. Plugging into the VIF formula gives:

$$VIF = \frac{1}{(1-0.9906)} = 106.38$$

Within rounding error, this result is equivalent to the value calculated by SAS and Stata.[10]

Display B.11.7 shows the regression of *yrschl* on *higrade7* and *impschl7* in bivariate regressions, revealing that each is significantly positively associated with the outcome when considered on its own. We put the values side by side in a table below to see the change in the coefficient estimates and standard errors between the bivariate and multiple regression.

■ **Table 11.5: Predicting Years of Schooling in Bivariate and Multiple Regression Models**

| | Dependent Variable: Years of Schooling | | |
|---|---|---|---|
| Expected Highest Grade (7th Grade Report) | 0.18* | — | −0.48 |
| | (0.043) | | (0.44) |
| Importance of School (7th Grade Report) | — | 1.80* | 6.56 |
| | | (0.42) | (4.35) |
| Intercept | 8.24 | 8.17 | 8.25 |
| | (0.65) | (0.64) | (0.65) |

* $p < .05$ —indicates excluded from model.

*Source:* Hypothetical data created for illustration purposes.

Notice that the standard errors are about 10 times larger in the multiple than the bivariate regression, in line with $\sqrt{VIF} = \sqrt{105.89} = 10.29$, so, even without seeing the VIFs or the overall *F*-test, we can identify a problem with multicollinearity in this table simply by noticing the very large increase in the standard errors between the bivariate and multiple regressions. By examining the standard errors in publications, you can watch for such problems (and this is one reason why it is so useful to report the standard errors in tables). Also, notice that the coefficient estimates have changed markedly between the bivariate and multiple regressions. The sign of the coefficient estimate for expected highest grade has even changed sign. This reflects the very imprecise estimate of the coefficient estimates in the multiple regression with such a high level of multicollinearity.

### Example 11.4

We will also use our multiple regression results from the NSFH hours of work example, with health, health limitations, and age as predictors, to demonstrate the three signs of multicollinearity. The results of estimating the full regression model and requesting VIFs in SAS and Stata are in Display B.11.8. We summarize them below, alongside the results from Displays B.10.1 and B.10.2 from Chapter 10.

▨ **Table 11.6: Predicting Mens' Hours of Work in the NSFH**

| | Dependent Variable: Hours of Work Per Week | | | |
| --- | --- | --- | --- | --- |
| | Model 1 (Display B.10.1) | Model 2 (Display B.10.1) | Model 3 (Display B.10.2) | Model 4 (Display B.11.8) |
| Self-Reported Health | 1.144* (0.31) | 1.017* (0.31) | 1.008* (0.31) | 0.896* (0.31) |
| Age (Years) | — | −0.105* (0.02) | — | −0.103* (0.02) |
| Health Limitations | — | — | −5.656* (2.45) | −5.125* (2.44) |
| Intercept | 37.86 (1.29) | 42.27 (1.51) | 38.52 (1.32) | 42.80 (1.53) |

\* $p < .05$ —indicates excluded from model.
*Source:* NSFH, Wave 1, Sample of Men, $n = 3,742$.

Notice that the standard errors for our predictors change very little across the models. This is consistent with the fact that the VIFs are all quite low (ranging from 1.01 to 1.04 for the three predictors; see again Display B.11.8).

Notice also that in this case the VIFs differ across the variables, since there are more than two predictors in the model. In Display B.11.9, we show one of the regressions of a predictor (*g2age*) on the other two predictors (*Health* and *HlthLimit*) that is needed to calculate the VIF for *g2age* (VIF = $1/(1–0.0073) = 1.01$).

## 11.3.2: Remedies for Multicollinearity

If multicollinearity appears to be a problem with your data, what can you do? A high VIF in and of itself may not require action, especially if you have a large sample size and substantial variation on the relevant predictor; that is, as we discussed in Chapters 5 and 6, the correlation among the predictors is only one of several components that contribute to the variance of a coefficient in a multiple regression. If we increase sample size, increase variation in $X$, or reduce conditional variance, then the estimated variance of the coefficient will decrease (all else constant). Thus, with a large sample size, ample variation in the predictors, and small conditional variance, it may be possible still to see significant $t$-ratios even with relatively high correlation among the predictors. It is important to consider these offsetting components when evaluating multicollinearity and choosing possible remedies for it. Still, if you see a significant overall model $F$

value with no significant *t*-values, or see that standard errors increase dramatically between the bivariate and multiple regression models, then you may need to take steps to address it. As you think about potential multicollinearity in your model, we recommend that you consider two key questions:

1. Are the predictor variables indicators of the same or different constructs?
2. How strongly are the predictor variables correlated in the real world as opposed to your sample (and why)?

### Same Construct/Strongly Correlated

The two predictors we looked at that had a severe multicollinearity problem in our hypothetical example—expected highest grade and importance of school—are an example of this kind of situation.

If the variables are really multiple indicators of the same construct, then you will likely want to combine the multiple indicators into a single measure. Sometimes researchers do this in an ad hoc fashion simply by picking the indicators that seem to measure the same thing and summing these variables. A better approach is to use confirmatory factor analysis or item response theory to verify that the variables measure the same thing and to produce the summary measure (Andrich 1988; Harrington 2008; Long 1983; Ostini and Nering 2006).[11]

Another common ad hoc approach to this problem is to use a single indicator at a time in the model. This has the benefit of allowing the researcher to see if the pattern of results is replicated across the multiple indicators of the same construct. However, the disadvantage is that to the extent that the variables really do indicate a single construct, then each individual indicator of the construct will have greater measurement error than would a combined measure developed through one of the approaches just discussed.

### Different Constructs/Strongly Correlated

If the predictors are strongly correlated but conceptually do not seem to indicate the same construct, then combining them into a single summary measure does not make sense. Similarly, if they measure distinct constructs, then including a single indicator at a time into the model will lead to omitted variable bias.

Often, high correlations among measures of different constructs arise because the two predictors go together so strongly in the population that, even though it is theoretically interesting to think of the two independently, it is hard in reality to think of varying one of the predictors independently of the other. In other words, ask yourself: in the population, is it possible to hold one of the predictors constant while varying the other predictor (or, will one predictor almost always change when you vary the other)? An example of this situation is a father's and mother's education. In the USA, spouses and partners tend to have similar educational attainment. However, we might develop a conceptual framework regarding why the mother's versus father's educational attainment might have different associations with an outcome (like child development).

So, what can you do in this situation? If an experimental study can be constructed, then the researcher can manipulate the levels of the key predictor variables so that they are uncorrelated in the sample. Of course, this may not be possible in many social science studies (e.g., we cannot ethically or practically take men and women of different educational levels and then partner them together). If an experimental study is not feasible, then a very large sample and substantial variation on each predictor (e.g., mother's and father's education) may be required to overcome the variance inflation created by the strong correlation between predictor variables.

### Different Constructs/Not Strongly Correlated

If the two variables measure distinct constructs and you don't think they are strongly correlated in the population, then consider whether your sampling strategy creates more correlation between the two variables than exists in the population. Sometimes a sample has been drawn for only a subgroup of the population and the correlation among key variables is higher in that subgroup than in the full population. For example, in a study that uses family income and neighborhood income as predictors of children's vocabulary scores, if the samples are drawn from economically homogeneous areas, then there will be higher correlation between the two measures of income than would be the case in a nationally representative sample. If a study aimed to separate the effects of each type of income, then a nationally representative data source might be needed.

### Same Construct/Not Strongly Correlated

If you think that the two variables indicate the same thing, then they should be correlated in the population. If you later realize that they would not be correlated in the population, then you would need to revisit your conceptual framework to consider whether they do indeed measure the same thing.

---

**SUMMARY**

**11**

### 11.4: SUMMARY

In this chapter, we discussed some common problems that can arise in estimating regression models (outliers and influential observations, heteroskedasticity, and multicollinearity). You should regularly check for these issues throughout a project and correct them where needed. A mis-specified model can go hand in hand with these problems; thus you must often go back and forth between specifying alternative conceptual models and checking for violations of model assumptions. For example, cases may be influential (greatly affect the regression estimates) when the form of the relationship is mis-specified as linear rather than nonlinear. Outlying cases may also influence the estimated form of the relationship.

We discussed using diagnostic measures to identify outlying and influential cases, and various strategies for dealing with them. We also presented contemporary formulae for estimating standard errors to adjust for heteroskedasticity, even when the form of such nonconstant variance is unknown. These heteroskedasticity-consistent standard errors can be easily estimated in SAS and Stata. We also introduced three strategies for identifying multicollinearity and suggested ways to think carefully about whether predictors indicate the same or distinct constructs and how they are correlated in the population (as opposed to your sample) when making decisions about how to address multicollinearity. Careful study planning can allow you to increase the components of the standard error that offset multicollinearity (e.g., larger sample size, more variation on $X$, lower conditional standard deviation).

## KEY TERMS

KEY TERMS
11

Cook's distance

DFBETA

DFFITS

Hat value
(also Leverage)

Heteroskedasticity

Heteroskedasticity-Consistent Standard Errors
(also Robust Standard Errors, Heteroskedasticity Robust Standard Errors, White Standard Errors, Huber–White Standard Errors)

Influential Observation

Multicollinearity

Outlier

Studentized Residual (also Standardized Residual)

Variance Inflation Factor

## REVIEW QUESTIONS

**11.1.** What is the difference between an outlier and an influential observation? How might you proceed if you identified some outliers and influential observations in your dataset?

**11.2.** What is the major difference in the formulae for heteroskedasticity-consistent standard errors and OLS standard errors? What are the differences in the assumptions for these standard errors?

**11.3.** What are three telltale signs of multicollinearity?

**11.4.** What is the formula for calculating the VIF and how does it relate to the formula for the standard error of the slope in multiple regression? In a regression model with two predictors, how does the VIF change as the correlation between the two predictors becomes smaller (approaches and then equals zero) and becomes larger (approaches and then equals one)?

## REVIEW EXERCISES

**11.1.** What other components of the standard error of the slope could compensate for a high VIF? In your response, be sure to *write the general formula* for the standard error of the slope in *multiple regression* and to be specific *about how each component* in this formula affects the standard error of the slope.

**11.2.** Write the *general equation* for the VIF.

(a) *Interpret the VIF* for the NUMKID variable in the results shown below.

(b) *Why* is the VIF for MARRY *identical* to the VIF for NUMKID in this model?

| Analysis of Variance | | | | | |
|---|---|---|---|---|---|
| Source | DF | Sum of Squares | Mean Square | F Value | Pr > F |
| Model | 2 | 167.26747 | 83.63373 | 0.46 | 0.6325 |
| Error | 7639 | 1394622 | 182.56604 | | |
| Corrected Total | 7641 | 1394789 | | | |

| | | | |
|---|---|---|---|
| Root MSE | 13.51170 | R-Square | 0.0001 |
| Dependent Mean | 40.80764 | Adj R-Sq | −0.0001 |
| Coeff Var | 33.11071 | | |

| Parameter Estimates | | | | | | |
|---|---|---|---|---|---|---|
| Variable | DF | Parameter Estimate | Standard Error | t-Value | Pr > \|t\| | Variance Inflation |
| Intercept | 1 | 40.83340 | 0.25175 | 162.20 | <.0001 | 0 |
| marry | 1 | 0.17007 | 0.31834 | 0.53 | 0.5932 | 1.04711 |
| NUMKID | 1 | −0.11647 | 0.13094 | −0.89 | 0.3738 | 1.04711 |

## CHAPTER EXERCISE

CHAPTER
EXERCISE

11

In this exercise, you will write SAS and Stata batch programs to examine outliers, heteroskedasticity, and multicollinearity.

To begin, prepare the shell of your batch program, including the commands to save your output and initial commands (e.g., SAS library, turning *more* off and closing open logs in Stata).

Start with the NOS 1996 dataset that you created in Chapter 4 and exclude cases with missing data using the following command on the data step in SAS:

```
if mngpctfem~=. & ftsize~=. & (n14=1 | n14=2)
              & (n15a=1 | n15a=2) & (n15b=1 | n15b=2)
              & (n15c=1 | n15c=2) & (n15d=1 | n15d=2)
              & (n15e=1 | n15e=2);
```

and the use command in Stata:

```
if mngpctfem~=. & ftsize~=. & (n14==1 | n14==2) ///
                & (n15a==1 | n15a==2) & (n15b==1 | n15b==2) ///
                & (n15c==1 | n15c==2) & (n15d==1 | n15d==2) ///
                & (n15e==1 | n15e==2)
```

Add the following code to your batch program to create a summary measure of the number of formal rules and regulations in the organization:

| SAS | Stata |
|---|---|
| `if n14=1 | n14=2 then Dn14=n14=1;` | `generate Dn14= n14==1 if n14==1 | n14==2` |
| `if n15a=1 | n15a=2 then Dn15a=n15a=1;` | `generate Dn15a=n15a==1 if n15a==1 | n15a==2` |
| `if n15b=1 | n15b=2 then Dn15b=n15b=1;` | `generate Dn15b=n15b==1 if n15b==1 | n15b==2` |
| `if n15c=1 | n15c=2 then Dn15c=n15c=1;` | `generate Dn15c=n15c==1 if n15c==1 | n15c==2` |
| `if n15d=1 | n15d=2 then Dn15d=n15d=1;` | `generate Dn15d=n15d==1 if n15d==1 | n15d==2` |
| `if n15e=1 | n15e=2 then Dn15e=n15e=1;` | `generate Dn15e=n15e==1 if n15e==1 | n15e==2` |
| `formalsum=Dn14+Dn15a+Dn15b+Dn15c+Dn15d+Dn15e;` | `generate formalsum=Dn14+Dn15a+Dn15b+Dn15c+Dn15d+Dn15e` |

*Notice that we are not restricting the analyses to organizations founded since 1950 for this assignment.*

*In all cases, conduct two-sided hypothesis tests. Use a 5 percent alpha unless otherwise indicated.*

## 11.1   Outliers and Influential Observations

(a) SAS/Stata tasks:

(i)   Regress the log of *mngpctfem* on the log of *ftsize* (refer back to Chapter Exercise from Chapter 9; be sure to account for zeros of negative values).

(ii)   Calculate the hat values, studentized residuals, Cook's Distance, DFFITS, and DFBETAs. (Note that you will have only one DFBETA because there is only one predictor in the model.)

(iii)   Create dummy variables to indicate which cases have extreme values on each of the five diagnostic measures (use the cutoffs shown in Display A.11).

(iv)   Create a dummy variable to indicate cases that are coded 1 on at least one of the five dummy variables you just created. Re-estimate the regression model for the subgroup with no values above the cutoffs on the five diagnostics and for the subgroup with at least one value above the cutoffs of the five diagnostics.

(v)   Summarize the *mngpctfem* and *ftsize* variables for the subgroup with no values above the cutoffs on the five diagnostics and for the subgroup with at least one value above the cutoffs of the five diagnostics.

(b) Write-up tasks:
    (i)   Describe the difference in the coefficient estimates in the regression results for the subgroups with and without outliers/influential observations. What do the results from the summarize command suggest about which cases have extreme values? How might you present and interpret these findings in a paper?

## 11.2   Heteroskedasticity-Consistent Standard Errors

(a) SAS/Stata tasks:
    (i)   Re-estimate the regression from Question 11.1 but request $HC_3$ heteroskedasticity-consistent standard errors.

(b) Write-up tasks:
    (i)   Discuss how the coefficient, standard error, and $t$-test for the results in Question 11.2 compared to the results for the corresponding model estimated on the full sample in Question 11.1.

## 11.3   Multicollinearity

(a) SAS/Stata tasks:
    (i)   Calculate the Pearson correlation between *ftsize* and *formalsum*.
    (ii)  Regress *mngpctfem* on *ftsize* and *formalsum*, and ask SAS and Stata to calculate the variance inflation factors.

(b) Write-up tasks:
    (i)   Show how to calculate the variance inflation factor based on the Pearson correlation.

## COURSE EXERCISE

**COURSE EXERCISE**
**11**

Choose one of the multiple regression models that you estimated in earlier chapters, including at least one continuous predictor variable.

Use diagnostic measures to identify potentially outlying or influential observations. Examine the cases with high values, and make a decision about how to proceed based on the results (e.g., do any with extreme values appear to be data entry errors? Do the results suggest that you should be running subgroup analyses? Do the results suggest that you should exclude a subset of cases from your models?).

Re-estimate your model with heteroskedasticity-consistent standard errors. How different are the results from those with the OLS standard errors assuming homoskedasticity?

Calculate the VIF for each of your predictor variables. Estimate bivariate regressions of your outcome on each of your predictors, and compare the standard errors to those in the multiple regression model (i.e., how much do the standard errors increase in the multiple regression versus the bivariate regressions?). Conduct the model $F$-test. If it is significant, is at least one of the individual $t$-tests significant? If any of these procedures indicate problematically high levels of multicollinearity in your model, think about which strategy for addressing multicollinearity best fits your situation. If you were to collect new data to examine your research questions, what would be important for avoiding problems with multicollinearity in your particular situation?

*Part 3*

# WRAPPING UP

*Chapter 12*

# PUTTING IT ALL TOGETHER AND THINKING ABOUT WHERE TO GO NEXT

## CHAPTER 12: PUTTING IT ALL TOGETHER AND THINKING ABOUT WHERE TO GO NEXT

In this final chapter, we will revisit the literature excerpts from Chapter 1 to help "put together" what we have learned about regression analysis. And, we will provide a roadmap to additional statistical topics that you may need or encounter as you read the literature and conduct your own projects as well as providing you with strategies for learning about these topics.

## 12.1: REVISITING LITERATURE EXCERPTS FROM CHAPTER 1

To help us link back to where we started, and put together what we have learned, we will begin by revisiting the three literature excerpts that we presented in Chapter 1. With more understanding under your belt now, you can interpret the findings we examined in Chapter 1 in some greater detail, and also better distinguish basic regression concepts that you now understand from advanced topics that you may want to study.

### 12.1.1: Revisiting Literature Excerpt 1.3

We will take the excerpts from Chapter 1 in reverse order, and begin with Literature Excerpt 1.3. Recall that this excerpt was publishing by Christine Li-Grining and examined correlates of preschoolers' "effortful control." Looking back at her Table 3, several features should be more familiar and interpretable now than they were when we first looked at this table in Chapter 1.

For example, in her column labels, she indicated that she listed the $R$-squared value, $B$ (unstandardized coefficient), SE B, and $\beta$ (standardized coefficient), all of which we now know how to interpret. She also reports a $\Delta R^2$ value, which is the change in $R$-squared and a complementary approach to capturing the additional variation explained by a subset of variables to the partial $F$-test that we covered in Chapter 6 (in fact, she reports the partial $F$-test for the set of five *Risk factors and child-mother interaction* in the text, which are $F(5,423) = 3.76$, $p < .01$ for *Delayed gratification* and $F(5,421) = 6.21$, $p < .001$ for *Executive control*, p. 215).[1]

Let's go ahead and interpret a few of her results in more detail.

■ We know that the $R$-squared values are the percentage of variation in the outcome explained by the set of *Child characteristics*. So, child age, gender, race, and negative emotionality explain 19 percent of the variation in *Delayed gratification* and 42 percent of the variation in *Executive control*.

- Comparing the row of coefficient estimates (*B*s) to the row of standard errors (SE *B*s) we can see that the coefficient estimates with asterisks are all at least twice their standard errors, and those without asterisks are all less than twice their standard errors, as expected.
- Standardized coefficients are provided that help us to interpret the substantive size of significant effects, but for dummy variables, partially standardized coefficients would be more appropriate, something we know now how to calculate. For example, partially standardizing the significant gender coefficient (circled in red in Literature Excerpt 1.3) would involve dividing by the standard deviation of the outcome, which is reported to be 0.72 (p. 213). The result is $-0.27/0.72 = -0.38$.[2] Thus, boys score over one quarter point lower than girls on delayed gratification, a difference of nearly four tenths of a standard deviation. This is a small to moderate effect, relative to Cohen's cutoffs, but seems more substantial relative to other predictors. For example, the difference between boys and girls is about three quarters the size of the standardized change associated with one more year of age ($0.35/0.72 = 0.49$).

Reading the full article, we would confront some terms that remain unfamiliar, and we will cover in the roadmap. For example, Li-Grining mentions that she applies probability weights, something common in large-scale surveys that use complex designs to oversample subgroups (p. 214).

## 12.1.2: Revisiting Literature Excerpt 1.2

Our second literature excerpt from Chapter 1 was by Stephen Vaisey, who examined how a sense of collective belonging developed within communes during the 1970s. Looking back at his Table 2 (Literature Excerpt 1.2b) we see that he presented unstandardized coefficients (*b*), standardized coefficients ($\beta$) and *t*-values (*t*) as well as *R*-squared and adjusted *R*-squared values.

Based on the *R*-squared values, we see that the set of variables in the model explain a substantial amount of the variation in the outcome—over three quarters. Three of the variables have coefficients that differ significantly from zero: authority, investment, and strength of moral order, with *t*-values larger than 2.00.[3] As we noted in Chapter 1, one of these significant variables has a sign which is the reverse of its simple correlation with the outcome (authority; see again Literature Excerpt 1.2a). We now know that it would be helpful if Vaisey had provided the coefficient estimates and standard errors from a bivariate regression, and the standard errors from the multiple regression, to help us to learn more about this sign reversal for authority. Indeed, an Appendix Table A2 in Vaisey's article suggests that multicollinearity may be a concern, with several correlations between predictor variables being 0.70 or larger. The variable authority is itself correlated above 0.70 with four other variables: *Spatiotemporal*, *Investment*, *Moral order*, and *Previous group*.

### 12.1.3: Revisiting Literature Excerpt 1.1

We will end with the Lyons article, which was the first excerpt that we considered in Chapter 1. We will not reinterpret his results in detail, because he in fact uses one of the advanced topics that we include in our roadmap next (negative binomial regression, appropriate for count outcomes like his number of hate crimes). But, we will use his results as a bridge to that roadmap, and to help us to see how much of the basic concepts of regression analysis that we learned in this book are transferable knowledge that you can build upon in learning such advanced techniques.

Indeed, even though he is using a different type of regression than OLS, much of what we see in his tables is consistent with what we learned for OLS. He presents a series of models in his Table 6 (Literature Excerpt 1.1a) and Table 8 (Literature Excerpt 1.1b) which allow us to see how the coefficient estimates and standard errors change as different variables are included in the model. He introduces several interactions which are product terms between two predictor variables, and he uses predicted values to interpret the significant interaction (his Table 7, in Literature Excerpt 1.1a).

Indeed, much of what you have learned about OLS regression translates to advanced topics in regression analysis. You have the foundation on which to build, depending on which advanced topics are particularly relevant in your field and for your research interests.

## 12.2: A ROADMAP TO STATISTICAL METHODS

In this section, we provide a roadmap to many commonly used techniques that complement or extend beyond the foundation we have provided. We have tried to cover many common techniques, although we have left some out to keep the roadmap manageable (e.g., we do not cover the *fuzzy set analysis* technique used by Vaisey in Literature Excerpt 1.2, or important topics related to regression modeling such as *multiple imputation, social network analysis*, and *meta-analysis*). In an attempt to make the roadmap useful but not overwhelming, we have focused on the comparative strengths of each technique. Table 12.1 summarizes the situations in which each is often used. We hope that this will help you to identify which techniques might be useful in your field or your work. We suggest strategies to locate courses and resources for learning more about any that you identify as relevant to you (indeed, entire courses are often devoted to each topic we consider in the roadmap!).

▨ **Table 12.1: Some Key Situations that Lend Themselves to Different Models**

**Categorical Dependent Variable**

| | |
|---|---|
| Logit/Probit | Dichotomous outcome |
| Ordered Logit/Probit | Ordinal outcome |
| Multinomial Logit | Nominal outcome, person characteristics as predictors |
| Conditional Logit | Nominal outcome, outcome category characteristics as predictors |
| Poisson Regression | Count outcome (e.g., number of hospital visits, number of symptoms, number of arrests); conditional variance constrained |
| Negative Binomial Regression | Count outcome (e.g., number of hospital visits, number of symptoms, number of arrests); conditional variance allows for "overdispersion" and "underdispersion" |
| Log-Linear Models | Finding a simple structure to explain the pattern of counts in a cross-classification table |

**Limited Dependent Variable**

| | |
|---|---|
| Censored Regression | All persons have $X$ values; $Y$ values missing for some persons |
| Truncated Regression | $X$ and $Y$ not observed for some persons |
| Sample Selection | Like censored and truncated problems, although the selection (why some persons have missing data) is modeled |
| Event History Analysis | Time until an event; duration or survival analysis |

**Systems of Equations**

| | |
|---|---|
| Structural Equation Modeling | Writing a theoretical model as a path of direct and indirect effects (mediation) and estimating multiple equations simultaneously, allowing for correlated errors |
| Multilevel Models | Nested structure (children clustered within classes clustered within schools; time points clustered within persons) |

**Measurement Theory**

| | |
|---|---|
| Factor Analysis and Item Response Theory | Identifying the dimensions (concepts) measured by items |

**Additional Topics**

| | |
|---|---|
| Sampling Theory | Samples other than simple random sample (weights to adjust for oversampling; adjustments for clustered samples) |
| First Difference and Fixed Effects Models | Using longitudinal data (or other clustered data) to adjust for stable characteristics of persons (or other units) |

### 12.2.1: Categorical Outcomes

We saw in Chapter 11 problems that can arise when a non-normal outcome variable produces outlying and influential cases in OLS regression. A number of models exist which are specifically designed for dichotomous, ordinal, and nominal outcomes (see Long 1997 and Long and Freese 2006 and Maddala 1983 for a detailed treatment of these types of models).

Before previewing these models, it is helpful to remind ourselves that whether a variable is nominal or ordinal or interval is not always clear cut, conceptually or in common practice. Often, the same variable can be (or is) treated in different ways. For example, we can think of (and measure) years of education as an interval variable, where one more always measures one more year spent in school. Or, we could think of (and measure) educational attainment as an ordinal variable, perhaps collapsing data into highest degree attained (e.g., junior high, high school, two-year college, four-year college, masters/JD, PhD/MD). If a researcher had focused on a sample that only covered a limited range of the educational distribution (e.g., say welfare recipients who primarily had a high school diploma or not) or had particular theoretical interest in a single contrast, education might be further collapsed to a dichotomous variable.

### Dichotomous Outcomes

The **logit** and **probit** models are appropriate for dichotomous outcomes. They assume that the outcome takes on two values, produced by a binomial distribution. They can be interpreted in terms of the probability of the category coded one on the outcome (e.g., how are earnings associated with the probability of being married?), although some extra effort are involved in translating to probability units for interpretation (see Long 1997 and Long and Freese 2006). The logit model can also be intereperted in terms of odds (e.g., how are earnings related to the odds of being married relative to not being married?), and such interpretations are common in some subfields (including health).

### Ordinal Outcomes

The **ordered logit** and **ordered probit** models are appropriate for outcome variables that have a natural order, but in which the distance between adjacent categories is unknown.

The standard linear regression model is appropriate for such outcome variables only if the true distances between outcome categories are about equal. If they are not, then results from the standard linear regression model applied to such outcome variables can be misleading (even though Likert scales are sometimes analyzed with OLS, like the *Health* outcome that we analyzed in Chapters 10 and 11). The ordered logit/probit models can be derived as a natural extension of logit and probit models.

## Nominal Categories

Common models for outcomes that do not have a natural order are the **multinomial logit** and the **conditional logit** models. The difference between these two models is that the multinomial logit uses characteristics of the individual to predict the probability of being in the outcome categories while the conditional logit uses characteristics of the outcome to predict the probability of being in the outcome category. For example, consider the study of an unmarried young adult's residential choice (e.g., living with their parents, living alone, or living with room-mates). With a multinomial logit, we might predict their choice based on their own characteristics, such as their gender and their school/work status. With a conditional logit, we might predict their choice based on characteristics of their options, such as the financial cost to them of living in each setting and their anticipated freedom to "do what they want" in each setting. The multinomial logit is more common in the social sciences than the conditional logit, although the conditional logit model fits with a number of interesting theoretical questions (but data are often less available on characteristics associated with the outcome categories). It is also possible to combine the models, if characteristics of both the individuals making choices and the alternatives are available (Agresti 2002).

### 12.2.2: Censored and Truncated Samples

There are also models for "limited dependent variables" which are continuous for much of their range, but censored or truncated in some way (see Long 1997 and Long and Freese 2006). These models fall into two major categories:

1. Censored dependent variables: all study participants have observations for all the predictor variables. But, for some participants, the value of the outcome variable is unknown. Often these unknown outcome variables are missing if the true value of the outcome falls below some censoring value, hence the term censored dependent variable. For example, wages might not be observed for persons earning below the minimum wage. Often, the outcome variable is assumed to be some value (e.g., zero on the censoring value) for the censored participants.
2. Truncated dependent variables: participants are excluded from the sample (have neither predictor nor outcome) based on their true value on the outcome.

These problems are similar to the problem of **sample selection** in which peoples' chances of participating in the study are related to their value on the outcome variable.

In all of these cases, if we estimate a standard linear regression model on the available data, then our parameter estimates will be biased.

There are three major approaches to these models:

1.  The **tobit** model adjusts for censoring using the probability of a person being censored. Censoring can happen "from above," "from below," or both.
2.  The **truncated regression model** adjusts the expected value from the regression model for the fact that we observe only a portion of the distribution.
3.  **Sample selection models** not only adjust for the censoring or truncation, but also model the mechanism of selection. The most well known of these approaches is **Heckman's selection model** (Heckman 1979; see also Stolzenberg and Relles 1997 and Bushway, Johnson, and Slocum 2007). In short, a probit model is estimated in which the probability of being in the sample is modeled. Then a transformation of the predicted value from this probit is included in the regression of the outcome of interest on the predictor variables in the truncated sample. Researchers also increasingly use propensity scores to adjust for measured differences between groups (Mueser, Troske, and Gorislavsky 2007).

### 12.2.3: Count Outcomes

Models for count outcomes share something in common with the models we have already considered—like the models for ordinal and nominal outcomes, they attempt to recognize more appropriately the measurement form of the outcome variable. Like the censored and truncated regression models, extensions of models for count outcomes adjust for an overabundance of zero counts in the data or for the failure to observe persons with zero counts.

The major models are: (a) the **Poisson** regression model, (b) the **Negative Binomial** regression model, (c) the zero modified count model, and (d) the truncated count model. Possion and negative binomial regression models have become especially common in social science when researchers analyze count outcomes.

Each of these adjusts for some cautions of the others: the Negative Binomial adjusts for a restriction that the conditional variance must equal the conditional mean in the Poisson regression model (allowing for something referred to as underdispersion and overdispersion; see again Literature Excerpt 1.1). The zero modified and truncated count models adjust for censoring and truncation, respectively.

### 12.2.4: Log-Linear Models

We have already considered a basic concept of log-linear models: We logged the outcome variable to allow for a nonlinear relationship with the predictor variables.

Log-linear models also rely on the log transformation but use the Poisson rather than normal distribution and move away from the dependent and independent variable structures. Instead, the objective is to analyze the counts in the cells of a cross-classification table (e.g., a cross-tabulation of race against education) to determine if a simple underlying structure can explain the pattern of counts (Agresti 2007).

### 12.2.5: Structural Equation Models

Structural equation models allow several regression models, with different outcomes, to be estimated simultaneously (Bollen 1989). This can allow us to capture better a complete theoretical model building on the basic concepts of path analysis that we learned in Chapter 10 (e.g., pathways of direct and indirect effects). Through this simultaneous estimation, we can also allow for errors across equations to be correlated. Structural equation modeling is sometimes associated with the computer package LISREL (Scientific Software International 2009), although many other packages for estimating SEM models are available (e.g., AMOS, EQS, MPlus).

### 12.2.6: Multilevel Models

In recent years, multilevel models, sometimes referred to as hierarchical linear models (HLMs), have become increasingly popular in the social sciences (Bickel 2007; Rabe-Hesketh and Skrondal 2005; Raudenbush and Bryk 2002). These models allow the appropriate analysis of data which has a natural nested structure, for example students clustered within classrooms clustered within schools or times of measurement clustered within persons. These models are important because they can adjust for the lack of independence within the clusters (i.e., OLS standard errors would typically be too small if the clustering were ignored). They also allow theoretically interesting cross-level effects to be estimated.

### 12.2.7: First Difference and Fixed Effects Models

We can also capitalize on clustered data, including longitudinal data with individuals clustered over time, to adjust for unmeasured characteristics of the units of study that might otherwise bias our results. One way to do this is with first difference or fixed effects models (Allison 2005; Wooldridge 2009). The intuition is the same as that which motivates the paired $t$-test that you may have learned about in introductory statistics: by taking a difference in measures for the same individual, we adjust for any pre-existing (time-constant) characteristics of the individual that affect their measure at both time points, and see if what remains can be explained by the predictor variable(s) of interest.

### 12.2.8: Survey Methods/Complex Data Analysis

All of the models we have dealt with in this book assume simple random samples. Often the samples used in the social sciences come instead from **complex samples.**

Most frequently, persons do not have a uniform chance of being part of the sample, but instead some persons are oversampled (e.g., the NSFH that we have analyzed in this book oversampled persons from certain racial/ethnic backgrounds and with certain kinds of household composition). This beneficially increases sample size for these subgroups, but simple descriptive statistics will not reflect the population unless this oversampling is accounted. In these cases, **sampling weights** are created to produce statistics that are representative of the original targeted population. Typically, these weights are the inverse of the probability of selection (e.g., if someone was sampled with probability of 0.5, then their weight is 2).

In addition, samples are often **clustered,** typically in multiple sample stages. Most typically, clustering occurs to save costs—most national samples first choose "primary sampling units" and then select smaller sampling units within these, and finally select households and/or individuals. For example, they might first sample cities and then census tracts within cities and then households. This lowers the costs of the survey by focusing the survey teams' efforts on smaller geographic areas. It also avoids the problem of not having a listing of all persons in the USA to sample from (e.g., once a census tract is chosen, the survey team goes out to "enumerate" (list) every living space in that census tract).

Most statistical packages allow for the adjustment of weights. However, debate exists about when they should be applied. This is because weighted estimates, although unbiased, are less efficient than unweighted estimates. Descriptive statistics should always be weighted to reflect the population (e.g., weighting would be used to get an estimate of the percentage of the population who are married, based on the sample). However, regression models are sometimes weighted and sometimes unweighted (DuMouchel and Duncan 1983; Korn and Graubard 1995; Reiter, Zanutto, and Hunter 2005).

Regression models estimated on complex samples should also account for the fact that persons who live in a cluster are often more alike than persons randomly sampled from the population. Because of this nonindependence, the standard errors typically should be larger than they are estimated to be based on standard statistical approaches (like OLS). Statistics appropriate for complex sampling have been developed and are available in specialized packages (like SUDAAN; Research Triangle Institute 2009; Some multipurposes packages, like Stata, also have built-in complex sampling commands). In addition, an extension of the heteroskedasticity-consistent robust standard errors that we discussed in Chapter 11 can be used to adjust standard errors for clustered samples.

## 12.2.9: Event History Analysis

Event history analysis (sometimes also called survival analysis or duration analysis) is appropriate when the outcome variable is the time until an event occurs (Blossfeld and Rohwer 2002; Yamaguchi 1991). For example: What predicts the length of time until a dating couple becomes engaged? What predicts the length of time until a company dissolves? What predicts the length of time until a person dies? These approaches are more appropriate than OLS for such questions because, for example, they adjust for the fact that some of the people who have not had the event at the time of the last interview may still have the event in the future.

## 12.2.10: Measurement Theory/Psychometrics

Errors in the measurement of variables can distort or cloud relationships. Ideally, we should be able to distinguish poor measurement of our key concepts from lack of relationships among the true concepts. Psychometric models allow us to devise better measures. As discussed in Chapter 11, commonly used techniques are exploratory and confirmatory factor analyses and item response theory models (Andrich 1988; Harrington 2008; Long 1983; Ostini and Nering 2006).

Factor analysis uses correlations among variables to identify whether particular sets of variables tap underlying dimensions. Confirmatory differs from exploratory factor analysis in that you would explicitly lay out in advance which variables map onto which dimensions. As an example, social support is often divided into affective (expressing caring, "being there," spending time together) and instrumental (loaning money, watching one another's children, providing a job tip) dimensions. Item response theory models also identify whether items capture particular dimensions, but are not focused on correlations. For example, the Rasch measurement model differs from the factor analysis approach in that it analyzes the joint order of persons and items assuming that items "lower" on the dimension are always answered positively by more persons than items "higher" on the dimension. For example, we might hypothesize that "providing a job tip" would be "lower" on the dimension of instrumental social support (observed more often) than "loaning money." Applying Rasch analyses and confirmatory factor models can greatly improve measures of key constructs, allowing models to test better for theoretical relationships.

## 12.3: A ROADMAP TO LOCATING COURSES AND RESOURCES

As noted above, we have painted broad brush strokes in this roadmap, and to actually understand and apply the techniques would require substantial additional reading and training. Here we offer some suggestions for locating courses and resources for doing so. We have also only "dipped our toe" into the capabilities of the SAS and Stata software, and provided a smattering of references to other advanced software. Thus, we also provide some suggestions about courses and resources for learning about these powerful statistical packages below.

### 12.3.1: Statistics

We have provided references to books and journal articles throughout the book that offer a starting point for reading about advanced topics. Most likely, though, you will want to take a course or otherwise interact with instructors and peers as you learn an advanced topic. One way to do so is to take courses at your own university or other universities in your area. In checking for courses, it is often helpful to look at course offerings in departments outside of your discipline (including the statistics department), since much of statistics training crosses disciplinary boundaries and the limited number of faculties in any one department generally cannot cover all advanced topics (although, if available, taking a course in or close to your discipline can help you to understand how to apply it to research questions in your field). You will likely want to check for courses in fields such as human development, education, public health, social work, and public policy (if available in your geographic area) as well as departments such as sociology, psychology, economics, and political science.

Short courses are also offered by organizations and individuals that allow you to immerse yourself in a topic, typically over a day or two or a week or two. For example, ICPSR has long offered summer programs in quantitative methods, ranging from basic introductory statistics to a number of the advanced topics in our roadmap (ICPSR 2009). The University of Kansas's Quantitative Psychology program offers regular five-day summer programs on a range of advanced topics (University of Kansas 2009). Some individual faculties regularly offer short courses as well. For example, University of Pennsylvania sociology professor Paul Allison offers regular courses at locations across the country on longitudinal data analysis, event history analysis, categorical data analysis, and missing data analysis through his *Statistical Horizons* company (Allison 2009).

### 12.3.2: Statistical Software

There are also numerous resources for learning statistical software, locally and nationally. You may want to start by checking with your local computing department (or those at other universities in your area) to see whether they offer basic courses in statistical programming. Most statistical software companies also offer their own training resources.

For example, Stata offers NetCourses, which are taught over the Internet on a variety of topics, as well as onsite training and affiliated short courses (Stata 2009b). SAS offers live web-based courses, e-learning, onsite training, and other training opportunities (SAS 2009). Specialized software companies also offer training (e.g., see again Scientific Software International 2009). Numerous excellent online resources and user communities also exist (one excellent example is the UCLA Statistical Computing Group, 2009 which offers detailed publicly accessible resources specific to SAS, Stata, and SPSS). Various listservs and online forums are also available.

## 12.4: SUMMARY

SUMMARY
**12**

In this chapter, we revisited the literature excerpts from Chapter 1 to reinforce what we have learned (and have yet to learn) about regression models. We provided a roadmap of advanced topics that relate to or build upon the foundation of regression techniques taught in this book. And, we offered suggestions about where and how to get additional training in statistics and statistical computing. We hope that these offer you a range of resources to turn to as you begin to apply regression modeling in your own research!

## KEY TERMS

KEY TERMS
**12**

Clustered

Complex Samples

Conditional Logit

Heckman's Selection Model

Logit

Multinomial Logit

Negative Binomial

Ordered Logit

Ordered Probit

Poisson

Probit

Sample Selection

Sampling Weights

*Appendix A*

# SUMMARY OF SAS AND STATA COMMANDS

The following displays summarize SAS and Stata commands used in the book.

The displays are organized by chapter, with each chapter's display summarizing the commands introduced in that chapter.

The displays are numbered with this appendix letter (A) followed by the chapter number and, where needed, a number representing the order where the commands were discussed in the chapter. For example Display A.4.1 summarizes the first set of SAS and Stata commands introduced in Chapter 4.

APPENDIX A

■ **Display A.3  Basic File Types in SAS and Stata**

| Type | Format | Content | Usual Extension | |
|------|--------|---------|-----------------|---|
| | | | **SAS** | **Stata** |
| **data** | plain text | **The information you want to analyze (but not yet in a format SAS or Stata can analyze)**<br>■ must be "read" by SAS or Stata before it can be used in analyses<br>■ organized in rows of observations and columns of variables<br>■ can be viewed by any text editor (e.g., Notepad, Textpad) or word processing software (e.g., Word Perfect, MS Word) | .dat<br>.raw<br>.txt | .dat<br>.raw<br>.txt |
| | SAS/Stata format | **The information you want to analyze (saved in a format SAS or Stata can analyze)**<br>■ can be viewed in SAS's spreadsheet utility called the "Table Editor" or Stata's spreadsheet called the "Data Browser"<br>■ cannot be viewed or used by other software (unless converted) | .sas7bdat | .dta |
| **batch program** | plain text | **The analyses you want SAS or Stata to conduct**<br>■ written in SAS or Stata syntax using SAS or Stata commands<br>■ can be written using any text editor (SAS and Stata also have built in editors)<br>■ can also be written or viewed using a word processor but do not save these files in the word processor's format or they will not be readable by SAS or Stata (they must "plain" or "ascii" text) files<br>■ be sure to **save** your batch programs frequently as you write them! | .sas | .do |
| **results** | plain text<br>rich text | **The results of the analyses SAS or Stata has conducted for you**<br>■ can be viewed in the SAS output window or Stata results window<br>   SAS has a separate "Log" window which reviews submitted statements, lists any messages and errors that occurred while SAS tried to run those statements, and summarizes how long it took to run the statements; this window is generally of temporary interest as you "debug" (look for errors in) a batch program<br>■ SAS' results (output window) can be saved in rich text format using the "ods" command (more on this in Chapter 4) so that the results can be viewed and edited in any word processor<br>   Stata combines the review of commands, error messages, and processing speed with the results in a single results window. Stata's results can be saved in a Stata-specific .smcl format, although we will use the plain text (with the .log extension) since they are easy to read in any software | .log<br>.rtf | .log<br>.smcl |

## ■ Display A.4.1.  Common Operators in SAS and Stata

| | Meaning | SAS Letters | SAS Symbols | Stata Symbols |
|---|---|---|---|---|
| Comparison Operators | | | | |
| | Equal to | EQ | ▭ | ▭ |
| | Not equal to | NE | ~= | ~= |
| | | | | != |
| | Greater than | GT | > | > |
| | Greater than or equal to | GE | >= | >= |
| | Less than | LT | < | < |
| | Less than or equal to | LE | <= | <= |
| Logical Operators | | | | |
| | And | AND | & | & |
| | Or | OR | \| | \| |
| | Not | NOT | ~ | ~ |
| | | | | ! |
| Arithmetic Operators | | | | |
| | Multiplication | n/a | * | * |
| | Division | n/a | / | / |
| | Addition | n/a | + | + |
| | Subtraction | n/a | − | − |

Notes:

■ The & symbol (read ampersand) is typically found above the number 7 on the keyboard.

■ On many keyboards, the | symbol (read vertical bar or pipe) is found on the key above the enter key (above the backward slash \). On the keyboard, it usually appears as two separate small lines, one above each other, although it appears as one solid vertical line on the screen.

■ The ~ symbol (read tilde) is usually found on the top left of the keyboard.

■ SAS can use either the letters or symbols. The letters may be easier to interpret at first, although using the symbol is more parallel to Stata.

■ There are some additional symbols allowed in each language (see the respective help pages for each package). For simplicity, we include only one in most cases.

■ To avoid unexpected results, put "or" expressions in parentheses and precede each variable name with an operator. For example, if (var1=6 | var1=7) not "if var1=6 | 7".

## ▪ Display A.4.2 Summary of SAS and Stata Commands for Reading and Saving Data and for Creating and Checking Variables

| | SAS | Stata |
|---|---|---|
| Locate formats | `libname library "<path>";` | `n/a` |
| Save results | `ods rtf body=<"path and filename.rtf">;` <br> `...` <br> `ods rtf close;` | `log using <filename>, replace text` <br> `...` <br> `log close` |
| Read and save data | `data <path and data filename>;` <br> `   set <path and data filename>` <br> `      (keep= <variable list>);` <br> `   if <expression>;` <br> `run;` | `use <variable list> using <data filename>///` <br> `   if <expression>` <br> `...` <br> `save <data filename>, replace` |
| Check raw data against codebook | `proc freq; tables <variable list>; run;` | `codebook <variable list>, tabulate(400)` |
| Crosstab created and original variables | `proc freq; tables <var1>*<var2> /missing; run;` | `tabulate <var1> <var2>, missing` |
| Create and modify variables | `if <expression> then <variable name> =` <br> `   expression;` | `generate <variable name> = <expression>///` <br> `   if <expression>` <br> `replace <variable name> = <expression> if <expression>` |
| Add comments | `/* comment */` <br> `* comment ;` | `/* comment */` <br> `* comment` <br> `// comment` |
| Notes | * You supply words in angle brackets < > <br> * Highlighted text is optional. <br> * In SAS, variable creation and modification must occur within the DATA Step (between the word DATA and RUN). <br> * SAS statements must end with a semi-colon. <br> * SAS is not case sensitive. You can type statements, data filenames, and variable names in lowercase and/or uppercase. | * Stata commands do not need a semi-colon at the end, but if you want to space a command over multiple lines you can use continuation comments (///) at the end of each continuing line (see examples above in "Read and Save Data" and "Create and Modify Variables" and in Display B.4.7). <br> * Stata is case sensitive. Commands must be typed in all lowercase. Variable names must be typed with lowercase and/or capitals to match how you named them or how they were named in the raw data. <br> * Add the following to the start of your batch program to facilitate debugging: <br><br> `version 10` <br> `set more off` <br> `capture drop _all` <br> `capture log close` <br><br> * Version 10 could be further specified with decimals (e.g., Version 10.1) to more precisely account for updates between major releases. |

■ Display A.5 **Summary of SAS and Stata Commands Introduced in Chapter Five (Basic Concepts of Bivariate Regression)**

| | SAS | Stata |
|---|---|---|
| Calculate sample means and standard deviations | `proc means;`<br>`  var <variable list>;`<br>`run;` | `summarize <variable list>` |
| Calculate Pearson correlation coefficients | `proc corr;`<br>`  var <variable list>;`<br>`run;` | `correlate <variable list>` |
| Estimate one bivariate regression model | `proc reg;`<br>`  model <depvar>=<indepvar>;`<br>`run;` | `regress <depvar> <indepvar>` |
| Estimate two bivariate regression models | `proc reg;`<br>`  model <depvar>=<indepvar1>;`<br>`  model <depvar>=<indepvar2>;`<br>`run;` | `regress <depvar> <indepvar1>`<br>`regress <depvar> <indepvar2>` |
| Request 95% confidence interval | `proc reg;`<br>`  model <depvar>=<indepvar> / clb;`<br>`run;` | `* in default output *` |
| Request standardized coefficients | `proc reg;`<br>`  model <depvar>=<indepvar> / stb;`<br>`run;` | `regress <depvar> <indepvar>, beta` |

■ Display A.6 **Summary of SAS and Stata Commands Introduced in Chapter Six (Basic Concepts of Multiple Regression)**

| | SAS | Stata |
|---|---|---|
| Estimate one multiple regression model | `proc reg;`<br>`  model <depvar>=<variable list>;`<br>`run;` | `regress <depvar> <variable list>` |
| Estimate "intercept-only" model | `proc reg;`<br>`  model hrchores=;`<br>`run;` | `regress hrchores` |
| Conduct general linear F test | `test <variable list>;`<br><br>`** in SAS, the variables in the variable list for the test command are separated by commas.`<br>`** conducts test based on most recently estimated regression model.` | `test <variable list>`<br><br>`** in Stata, the variables in the variable list are not separated by commas.`<br>`** conducts test based on most recently estimated regression model.` |

■ **Display A.7. Summary of SAS and Stata Commands Introduced in Chapter Seven (Dummy Variables)**

| | SAS | Stata |
|---|---|---|
| Succinct syntax to create dummy variable | `if <expression> then`<br>`<varname> = <true/false expression>;`<br><br>The true/false expression should evaluate to "true" for the cases that should be coded a '1' on the new dummy variable and should evaluate to "false" for the cases that should be coded a '0' on the new dummy variable. The shaded "if" qualifier can be used to be sure that cases with missing value codes are coded '.' missing on the new variable. | `generate <varname> = <true/false expression> if <expression>` |
| Request variance-covariance matrix of coefficients | `proc reg;`<br>`    model <depvar>=<variable list> /covb;`<br>`run;` | `regress <depvar> <variable list>`<br>`    estat vce` |
| Calculate linear combination of coefficients | N/A | `regress <depvar> <variable list>`<br>`    lincom <indepvar1> - <indepvar2>` |

# ■ Display A.8. Summary of SAS and Stata Commands Introduced in Chapter Eight (Interactions)

| | SAS | Stata |
|---|---|---|
| Create a product term to test for an interaction | `<interact> = <var1>*<var2>;` | `generate <interact> = <var1>*<var2>` |
| Test conditional effect of one variable within levels of the other variable | | |
| Re-estimating | `<var2C> = <var2> - <level2>;`<br><br>`<interactC> = <var1>*<var2C>;`<br><br>`proc reg;`<br>`model <depvar> = <var1> var2c interactC>;`<br>`run;` | `generate <var2C> = <var2> - <level2>`<br><br>`generate <interactC> = <var1>*<var2C>`<br><br>`regress <depvar> <var1> var2c interactC>` |
| test command | `test <var1> + <level2>*<interact> = 0;` | `test <var1> + <level2>*<interact> = 0` |
| lincom command | n/a | `lincom <var1> + <level2>*<interact>` |
| Conduct Chow test | `proc reg;`<br>`model <depvar> = <variable list> <groupvar>`<br>`<interaction list>;`<br>`test <group var> <interaction list>;`<br>`run;` | `regress <depvar> <variable list> //`<br>`<groupvar> <interaction list>`<br>`test <group*var> <interaction list>` |

Notes:

```
All variable creation in SAS must happen in the data step.
var1             = name of one variable in the interaction
var2             = name of the other variable in the interaction
level2           = level of other variable for calculating conditional effect of one variable (e.g., to
                   calculate the conditional effect of hours of work for families with two children, level2=2)
groupvar         = a variable defining two subgroups for a fully interacted model
variable list    = list of variables (other than groupvar) in a fully interacted model
interaction list = list of product terms created by multiplying together each variable with the groupvar
```

**APPENDIX A**

## ▪ Display A.9. Summary of SAS and Stata Commands Introduced in Chapter Nine (Nonlinear Relationships)

| | SAS | Stata |
|---|---|---|
| Create squared predictor | `<sqvar1> = <var1> * <var1>;` | `generate <sqvar1> = <var1> * <var1>` |
| Create logged predictor or outcome | `<logvar1> = log(var1);` | `generate <logvar1> = log(var1)` |
| Estimate quadratic model | `proc reg;`<br>`  model <depvar> = <var1> <sqvar1>;`<br>`run;` | `regress <depvar> <var1> <sqvar1>` |
| Estimate log-log model | `proc reg;`<br>`  model <logdepvar> = <logindepvar>;`<br>`run;` | `regress <logdepvar> <logindepvar>` |
| Estimate log-lin model | `proc reg;`<br>`  model <logdepvar> = <indepvar>;`<br>`run;` | `regress <logdepvar> <indepvar>` |
| Estimate lin-log model | `proc reg;`<br>`  model <depvar> = <logindepvar>;`<br>`run;` | `regress <depvar> <logindepvar>` |
| Calculate adjustment for predictions and *R*-squared in natural units | `/* Obtain predicted values, in log units */`<br>`proc reg;`<br>`  model <logdepvar> =<indepvar>;`<br>`  output out=predict(keep=<depvar> logYhat logResid)`<br>`         predicted=logYhat residual=logResid;`<br>`run;`<br><br>`/* Transform predicted values back to natural units */`<br>`data predict2;`<br>`  set predict;`<br>`  explogYhat=exp(logYhat);`<br>`  explogResid=exp(logResid);`<br>`run;`<br><br>`/* Coefficient from the following regression is`<br>`adjustment factor for predicted values */`<br>`proc means data=predict2;`<br>`  var explogResid;`<br>`run;`<br><br>`/* R-squared from the following regression is R-squared`<br>`in natural units of Y */`<br>`proc reg data=predict2;`<br>`  model <depvar>=explogYhat;`<br>`run;` | `/* Obtain predicted values, in log units */`<br><br>`regress <logdepvar> <indepvar>`<br>`predict logYhat`<br>`predict logResid, residuals`<br><br>`/* Transform predicted values back to natural units */`<br><br>`generate explogYhat=exp(logYhat)`<br>`generate explogResid=exp(logResid)`<br><br>`/* Coefficient from the following regression is`<br>`adjustment factor for predicted values */`<br>`summarize explogResid // adjustment`<br><br>`/* R-squared from the following regression is`<br>`R-squared in natural units of Y */`<br>`regress <depvar> explogYhat // R-squared` |

Note:

The log of zero or negative values is undefined. A positive value may be added to the variable prior to transformation to deal with such values (if these values are not numerous and the maximum value on the variable is not small). Care must be taken to add this value back following transformation of predicted values into their natural units.

**■ Display A.10.  Summary of SAS and Stata Commands Introduced in Chapter Ten (Indirect Effects and Omitted Variable Bias)**

|  | SAS | Stata |
|---|---|---|
| Estimating total effect | ```proc reg;<br>  model <depvar> = <indepvar>;<br>run;``` | ```regress <depvar> <indepvar>``` |
| Estimating direct effect | ```proc reg;<br>  model <depvar> = <indepvar> <medvar>;<br>run;``` | ```regress <depvar> <indepvar> <medvar>``` |
| Estimating extra regression for calculating indirect effect | ```proc reg;<br>  model <medvar> = <indepvar>;<br>run;``` | ```regress <medvar> <indepvar>``` |
| Notes:<br><br>```indepvar   =   predictor of interest```<br>```medvar     =   mediating variable``` | | |

■ Display A.11. Summary of SAS and Stata Commands Introduced in Chapter Eleven (Outliers, Heteroskedasticity, and Multicollinearity)

| | SAS | Stata |
|---|---|---|
| **Outlying and Influential Observations** | | |
| Graph individual variables | `proc univariate plots;`<br>`  var <variable list>;`<br>`run;` | `stem <variable name>`<br>`histogram <variable name>`<br>`graph box <variable name>` |
| Create scatterplots | `proc gplot;`<br>`  plot <var1>*<var2>;`<br>`run;` | `graph twoway scatterplot <var1> <var2>` |
| Calculate diagnostic measures | `proc reg;`<br>`  model <depvar> = <variable list> /r influence;`<br>`  ods output outputstatistics=<influence>;`<br>`run;`<br><br>Notes: In Stata, to calculate DFBETAS, the code "predict DFB_<indepvar1>, dfbeta(<indepvar1>)" should be repeated for each variable in the variable list. | `regress <depvar> <variable list>`<br>`  predict HatDiagonal, hat`<br>`  predict RStudent, rstudent`<br>`  predict CooksD, cooksd`<br>`  predict DFFITS, dfits`<br>`  predict DFB_<indepvar1>, dfbeta(<indepvar1>)`<br>`  predict DFB_<indepvar2>, dfbeta(<indepvar2>)`<br>`  predict DFB_<indepvar3>, dfbeta(<indepvar3>)` |
| Compare diagnostic measures to cutoffs | `data <influence2>;`<br>`  set <influence>;`<br>`  HatDiagonal3Hi =HatDiagonal         >3*(k-1)/n;`<br>`  RStudentHi     =abs(RStudent)      >2;`<br>`  CooksDHi       =CooksD             >4/n;`<br>`  DFFITSHi       =abs(DFFITS)        >2*sqrt((k-1)/n);`<br>`  DFB_indepvar1Hi =abs(DFB_indepvar1) >2/sqrt(n);`<br>`  DFB_indepvar2Hi =abs(DFB_indepvar2) >2/sqrt(n);`<br>`  DFB_indepvar3Hi =abs(DFB_indepvar3) >2/sqrt(n);`<br>`run;`<br><br>Notes: In the formula for the cutoffs, substitute the number of predictors in your model for k and substitute the sample size on which your regression is estimated for n. Repeat the command for DFB variables for each independent variable in your model. | `generate HatDiagonal3Hi =HatDiagonal         >3*(k-1)/n`<br>`generate RStudentHi     =abs(RStudent)      >2`<br>`generate CooksDHi       =CooksD             >4/n`<br>`generate DFFITSHi       =abs(DFFITS)        >2*sqrt((k-1)/n)`<br>`generate DFB_indepvar1Hi =abs(DFB_indepvar1) >2/sqrt(n)`<br>`generate DFB_indepvar2Hi =abs(DFB_indepvar2) >2/sqrt(n)`<br>`generate DFB_indepvar3Hi =abs(DFB_indepvar3) >2/sqrt(n)` |

(continued)

■ **Display A.11. Summary of SAS and Stata Commands Introduced in Chapter Eleven (Outliers, Heteroskedasticity, and Multicollinearity) (continued)**

| | SAS | Stata |
|---|---|---|
| **Outlying and Influential Observations (continued)** | | |
| Identify cases with any extreme values | ```anyhi = HatDiagonal3Hi=1 | RStudentHi=1                | CooksDHi=1 | DFFITSHi=1                | DFB_indepvar1Hi=1 | DFB_indepvar2Hi=1                | DFB_indepvar3Hi=1;``` | ```generate anyhi = HatDiagonal3Hi==1 | RStudentHi==1 ///                | CooksDHi==1 | DFFITSHi==1          ///                | DFB_indepvar1Hi==1 | DFB_indepvar2Hi==1 ///                | DFB_indepvar3Hi==1``` |
| Summarize and regress cases with and without extreme values | ```data <influence3>; merge <original file name> <influence2>; run; proc sort data=<influence3>; by <anyhi>; run; proc means data=<influence3>;    var <variable list>;    by <anyhi>; run; proc reg data=<influence3>;    model <depvar> = <variable list>;    by <anyhi>; run;``` | ```bysort <anyhi>: summarize <variable list> bysort <anyhi>: regress <depvar> <variable list>``` |
| List cases | ```proc print;    var <variable list>;    where <expression>; run;``` | ```list <variable list> if <expression>``` |
| **Heteroskedasticity** | | |
| Calculate heteroskedasticity-consistent standard errors | ```%include "c:\nsfh_distance\sas\hcreg.sas"; %HCREG(data = <filename>, dv = <depvar>, iv = <variable list>);``` | ```regress <depvar> <variable list>, vce(hc3)``` |
| **Multicollinearity** | | |
| Calculate variance inflation factors | ```proc reg;    model <depvar> = <variable list> /vif; run;``` | ```regress <depvar> <variable list> estat vif``` |

# EXAMPLES OF DATA CODING, AND OF THE SAS AND STATA INTERFACE, COMMANDS, AND RESULTS, BASED ON THE NATIONAL SURVEY OF FAMILIES AND HOUSEHOLDS

The following are summaries of data coding for the National Survey of Families and Households and of the SAS and Stata interface, commands, and results from the examples used throughout the text (using both the distance and hours of chores examples drawn from the NSFH).

The programs and results are also available on the course web site.

Each display is numbered with this appendix letter (B), the chapter number and a number to signify the order in which the display is referenced in the chapter. For example B.5.3 would be the third display referenced in Chapter 5.

**APPENDIX B**

■ **Display B.3.1 Summary of Variables in Analytic Data File for NSFH Distance Example**

| Created Variable Name | Original Variable Name | Description | Missing Data Codes | Notes |
|---|---|---|---|---|
| **Main Interview** **\* Household Composition** | | | | |
| — | MCASEID | Case Number | n/a | — |
| g2age | M2BP01 | Age Of Respondent | 97,98 | Years |
| — | M2DP01 | Sex Of Respondent | n/a | — |
| **Main Interview** **\* Social Background** | | | | |
| — | M484 | Which Group Describes R (Race) | 97,99 | — |
| — | M497A | Country/State Live When Born | 995,997 998,999 | 1–51, 990=U.S. 996=always lived here |
| g1yrschl | M502 | Highest Grade School Mother Completed | 98,99 | 25=GED |
| **Self-Administered** **Primary Respondent** **\* Attitudes** | | | | |
| — | E1301 | Mother Living Or Deceased | 7,8,9 | 1=still living |
| g1age | E1302 | Age of R's Mother | 96,97 98,99 | Years |
| g1miles | E1305 | How Far Away R Mother Lives | 9996,9997 9998,9999 | Miles 9994=Foreign Country 9995=Mother Live w/R |
| g2numbro | E1332A | Number Living Brothers R Has | 97,98,99 | 96=Inapplicable (zero) |
| g2numsis | E1332B | Number Living Sisters R Has | 97,98,99 | 96=Inapplicable (zero) |
| **Weights and** **Constructed Variables** | | | | |
| g2earn | IREARN | R Total Earnings | 9999997 9999998 9999999 | 1986 dollars |

■ Display B.4.1  Locating Data, Programs, and Results Files in Stata

(a) Writing a program in Stata's Do-File Editor.

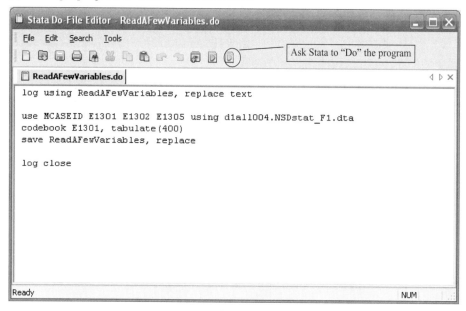

(b) Running programs and viewing results in Stata.

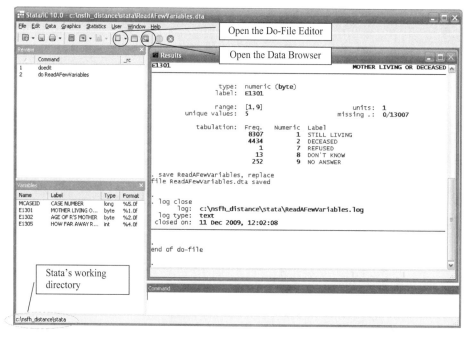

## ■ Display B.4.2 Locating Data, Programs, and Results Files in SAS

(a) Writing a program in SAS's enhanced editor and viewing messages

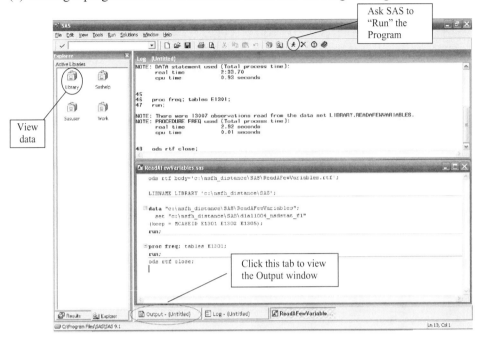

(b) Viewing data in SAS.

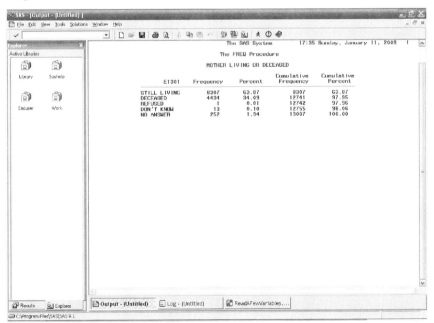

■ **Display B.4.3  Viewing Data in SAS and Stata**

(a) Viewing data in Stata.

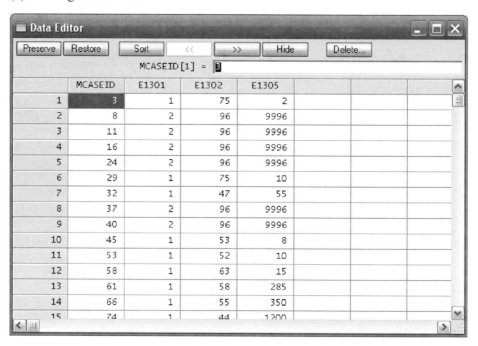

(b) Viewing data in SAS.

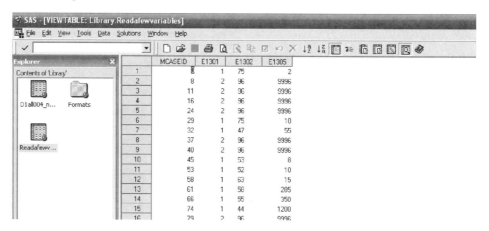

■ **Display B.4.4 Program to Read A Few Variables from the NSFH Wave 1 Data File**

| | SAS | Stata |
|---|---|---|
| **Program File** | ReadAFewVariables.sas | ReadAFewVariables.do |
| | ```ods rtf body="c:\nsfh_distance\SAS\ ReadAFewVariables.rtf"; LIBNAME LIBRARY "c:\nsfh_distance\SAS"; data "c:\nsfh_distance\SAS\ ReadAFewVariables"; set "c:\nsfh_distance\SAS\ diall004_nsdstat_f1" (keep = MCASEID E1301 E1302 E1305); run; proc freq; tables E1301; run; ods rtf close;``` | ```log using ReadAFewVariables, replace text use MCASEID E1301 E1302 E1305 using diall004.NSDstat_F1.dta codebook E1301, tabulate (400) save ReadAFewVariables, replace log close``` |
| **Results File** | ReadAFewVariables.rtf | ReadAFewVariables.log |

SAS Results File — **ReadAFewVariables.rtf**

**MOTHER LIVING OR DECEASED**

| E1301 | Frequency | Percent | Cumulative Frequency | Cumulative Percent |
|---|---|---|---|---|
| STILL LIVING | 8307 | 63.87 | 8307 | 63.87 |
| DECEASED | 4434 | 34.09 | 12741 | 97.95 |
| REFUSED | 1 | 0.01 | 12742 | 97.96 |
| DON'T KNOW | 13 | 0.10 | 12755 | 98.06 |
| NO ANSWER | 252 | 1.94 | 13007 | 100.00 |

Stata Results File — **ReadAFewVariables.log**

```
----------------------------------------------
E1301        MOTHER LIVING OR DECEASED
----------------------------------------------

           type:  numeric (byte)
          label:  E1301

          range:  [1,9]                 units:  1
  unique values:  5               missing .:  0/13007

     tabulation:  Freq.   Numeric  Label
                   8307         1  STILL LIVING
                   4434         2  DECEASED
                      1         7  REFUSED
                     13         8  DON'T KNOW
                    252         9  NO ANSWER
```

■ **Display B.4.5  Example of Checking New Variable Creation in Stata**

```
log using CheckingCreatedVariables, replace text
use MCASEID IREARN M502 using d1all004.NSDstat_F1.dta
generate g1yrschl=M502 if M502<98
replace g1yrschl=12 if M502==25

tabulate g1yrschl M502, missing
generate g2earn=IREARN*207.342/109.6 if IREARN<9999997
save CheckingCreatedVariables, replace
log close
```

|  | M502 (HIGHEST GRADE SCHOOL MOTHER COMPLETED) | | | | | | | | | | | |
| g1yrschl | 0 | 1 | 2 | 3 | 4 | | 15 | 16 | 17 | 25 | 98 | 99 | Total |
| 0 | 290 | 0 | 0 | 0 | 0 | . . . | 0 | 0 | 0 | 0 | 0 | 0 | 290 |
| 1 | 0 | 27 | 0 | 0 | 0 | . . . | 0 | 0 | 0 | 0 | 0 | 0 | 27 |
| 2 | 0 | 0 | 63 | 0 | 0 | . . . | 0 | 0 | 0 | 0 | 0 | 0 | 63 |
| 3 | 0 | 0 | 0 | 189 | 0 | . . . | 0 | 0 | 0 | 0 | 0 | 0 | 189 |
| 4 | 0 | 0 | 0 | 0 | 190 | . . . | 0 | 0 | 0 | 0 | 0 | 0 | 190 |
| 5 | 0 | 0 | 0 | 0 | 0 | . . . | 0 | 0 | 0 | 0 | 0 | 0 | 187 |
| 6 | 0 | 0 | 0 | 0 | 0 | . . . | 0 | 0 | 0 | 0 | 0 | 0 | 482 |
| 7 | 0 | 0 | 0 | 0 | 0 | . . . | 0 | 0 | 0 | 0 | 0 | 0 | 258 |
| 8 | 0 | 0 | 0 | 0 | 0 | . . . | 0 | 0 | 0 | 0 | 0 | 0 | 1,494 |
| 9 | 0 | 0 | 0 | 0 | 0 | . . . | 0 | 0 | 0 | 0 | 0 | 0 | 359 |
| 10 | 0 | 0 | 0 | 0 | 0 | . . . | 0 | 0 | 0 | 0 | 0 | 0 | 648 |
| 11 | 0 | 0 | 0 | 0 | 0 | . . . | 0 | 0 | 0 | 0 | 0 | 0 | 522 |
| 12 | 0 | 0 | 0 | 0 | 0 | . . . | 0 | 0 | 0 | 4 | 0 | 0 | 4,128 |
| 13 | 0 | 0 | 0 | 0 | 0 | . . . | 0 | 0 | 0 | 0 | 0 | 0 | 259 |
| 14 | 0 | 0 | 0 | 0 | 0 | . . . | 0 | 0 | 0 | 0 | 0 | 0 | 613 |
| 15 | 0 | 0 | 0 | 0 | 0 | . . . | 134 | 0 | 0 | 0 | 0 | 0 | 134 |
| 16 | 0 | 0 | 0 | 0 | 0 | . . . | 0 | 688 | 0 | 0 | 0 | 0 | 688 |
| 17 | 0 | 0 | 0 | 0 | 0 | . . . | 0 | 0 | 282 | 0 | 0 | 0 | 282 |
| . | 0 | 0 | 0 | 0 | 0 | . . . | 0 | 0 | 0 | 0 | 2,155 | 39 | 2,194 |
| Total | 290 | 27 | 63 | 189 | 190 | . . . | 134 | 688 | 282 | 4 | 2,155 | 39 | 13,007 |

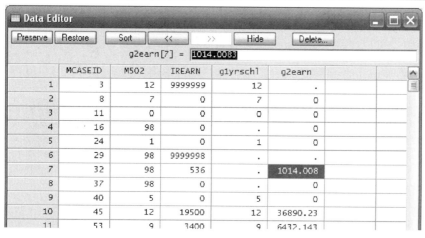

■ **Display B.4.6  Example of Checking New Variable Creation in SAS**

```
ods rtf body="c:\nsfh_distance\SAS\CheckingCreatedVariables.rtf";

LIBNAME LIBRARY "c:\nsfh_distance\SAS";

data "c:\nsfh_distance\SAS\CheckingCreatedVariables";
  set "c:\nsfh_distance\SAS\d1all004_nsdstat_f1"
(keep = MCASEID IREARN M502);

if M502<98 then g1yrschl=M502;
if M502=25 then g1yrschl=12;

if IREARN<9999997 then g2earn=IREARN*207.342/109.6;

run;

proc freq; tables g1yrschl*M502 /missing;
run;
ods rtf close;
```

| Frequency | M502 (HIGHEST GRADE SCHOOL MOTHER COMPLETED) | | | | | | | | | | | | |
| | **0** | **1** | **2** | **3** | **4** | **...** | **15** | **16** | **17** | **25** | **98** | **99** | **Total** |
|---|---|---|---|---|---|---|---|---|---|---|---|---|---|
| **.** | 0 | 0 | 0 | 0 | 0 | ... | 0 | 0 | 0 | 0 | 2155 | 39 | 2194 |
| **0** | 290 | 0 | 0 | 0 | 0 | ... | 0 | 0 | 0 | 0 | 0 | 0 | 290 |
| **1** | 0 | 27 | 0 | 0 | 0 | ... | 0 | 0 | 0 | 0 | 0 | 0 | 27 |
| **2** | 0 | 0 | 63 | 0 | 0 | ... | 0 | 0 | 0 | 0 | 0 | 0 | 63 |
| **3** | 0 | 0 | 0 | 189 | 0 | ... | 0 | 0 | 0 | 0 | 0 | 0 | 189 |
| **4** | 0 | 0 | 0 | 0 | 190 | ... | 0 | 0 | 0 | 0 | 0 | 0 | 190 |
| **5** | 0 | 0 | 0 | 0 | 0 | ... | 0 | 0 | 0 | 0 | 0 | 0 | 187 |
| **6** | 0 | 0 | 0 | 0 | 0 | ... | 0 | 0 | 0 | 0 | 0 | 0 | 482 |
| **7** | 0 | 0 | 0 | 0 | 0 | ... | 0 | 0 | 0 | 0 | 0 | 0 | 258 |
| **8** | 0 | 0 | 0 | 0 | 0 | ... | 0 | 0 | 0 | 0 | 0 | 0 | 1494 |
| **9** | 0 | 0 | 0 | 0 | 0 | ... | 0 | 0 | 0 | 0 | 0 | 0 | 359 |
| **10** | 0 | 0 | 0 | 0 | 0 | ... | 0 | 0 | 0 | 0 | 0 | 0 | 648 |
| **11** | 0 | 0 | 0 | 0 | 0 | ... | 0 | 0 | 0 | 0 | 0 | 0 | 522 |
| **12** | 0 | 0 | 0 | 0 | 0 | ... | 0 | 0 | 0 | 4 | 0 | 0 | 4128 |
| **13** | 0 | 0 | 0 | 0 | 0 | ... | 0 | 0 | 0 | 0 | 0 | 0 | 259 |
| **14** | 0 | 0 | 0 | 0 | 0 | ... | 0 | 0 | 0 | 0 | 0 | 0 | 613 |
| **15** | 0 | 0 | 0 | 0 | 0 | ... | 134 | 0 | 0 | 0 | 0 | 0 | 134 |
| **16** | 0 | 0 | 0 | 0 | 0 | ... | 0 | 688 | 0 | 0 | 0 | 0 | 688 |
| **17** | 0 | 0 | 0 | 0 | 0 | ... | 0 | 0 | 282 | 0 | 0 | 0 | 282 |
| **Total** | 290 | 27 | 63 | 189 | 190 | ... | 134 | 688 | 282 | 4 | 2155 | 39 | 13007 |

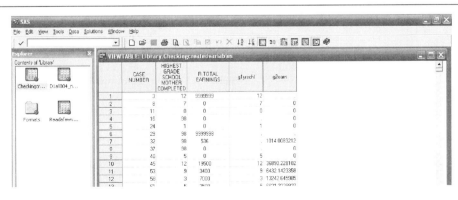

■ **Display B.4.7  Program to Create Full Analytic Data File**

| | SAS | Stata |
|---|---|---|
| Commands | (see code below) | (see code below) |

**SAS**

```
ods rtf body="c:\nsfh_distance\SAS\CreateData.rtf";

LIBNAME LIBRARY "c:\nsfh_distance\SAS";

data "c:\nsfh_distance\SAS\CreateData";
  set "c:\nsfh_distance\SAS\diall004_nsdstat_f1"
    (keep = MCASEID M2BP01 M2DP01 M484 M497A M502
    E1301 E1302 E1305 E1332A E1332B IREARN);

if    E1301=1 &
      (M497A<=51 | M497A==990 | M497A==996) &
      E1305>=9994 & E1305~=9995;

if M2BP01<97      then g2age=M2BP01;

if M502<98        then g1yrsch1=M502;
if M502=25        then g1yrsch1=12;

if E1302<96       then g1age=E1302;

if E1305<9996     then g1miles=E1305;

if E1332A<97      then g2numbro=E1332A;
if E1332A=96      then g2numbro=0;

if E1332B<97      then g2numsis=E1332B;
if E1332B=96      then g2numsis=0;
/* Adjust earnings from 1986 to 2007 dollars with the
   CPI ftp://ftp.bls.gov/pub/special.requests/
   pi/cpiai.txt */
if IREARN<999997 then g2earn=IREARN*207.342/109.6;

run;

proc freq;
run;

ods rtf close;
```

**Stata**

```
version 10
capture drop _all
capture log close
set more off

log using CreateData, replace text

use MCASEID M2BP01 M2DP01 M484 M497A M502        ///
    E1301 E1302 E1305 E1332A E1332B IREARN       ///
    using diall004.NSDstat_F1.dta                ///
    if E1301==1 &                                ///
    (M497A<=51 | M497A==990 | M497A==996) &       ///
    E1305~=9994 & E1305~=9995

generate g2age=M2BP01       if M2BP01<97

generate g1yrsch1=M502      if M502<98
replace g1yrsch1=12         if M502==25

generate g1age=E1302        if E1302<96

generate g1miles=E1305      if E1305<9996

generate g2numbro=E1332A    if E1332A<97
replace g2numbro=0          if E1332A==96

generate g2numsis=E1332B    if E1332B<97
replace g2numsis=0          if E1332B==96

/* Adjust earnings from 1986 to 2007 dollars with the
   CPI ftp://ftp.bls.gov/pub/special.requests/
   cpi/cpiai.txt*/
generate g2earn=IREARN*207.342/109.6 ///
   if IREARN<999997

codebook, tabulate(600)

save CreateData, replace

log close
```

## Display B.4.8 Example of Errors in SAS Editor Window

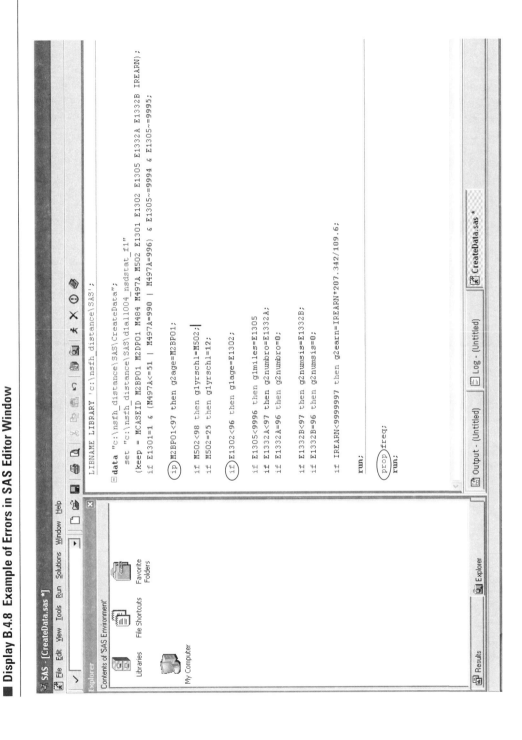

## ▪ Display B.4.9 Example of Errors in SAS Log Window

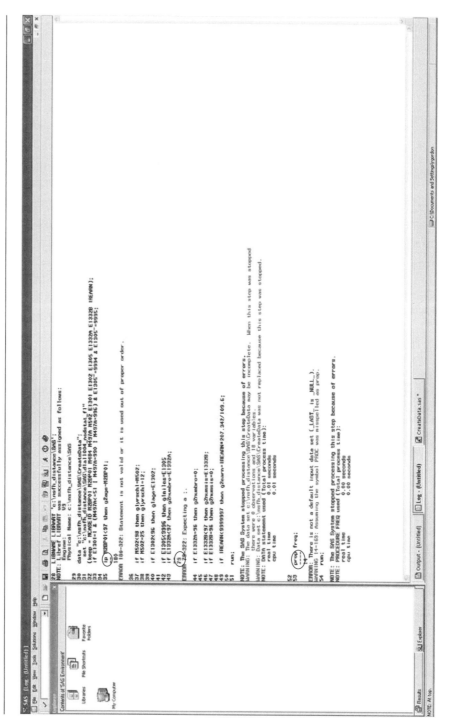

■ **Display B.4.10 Example of Error Message in Stata**

Stata/IC 10.0 - C:\nsfh_distance\stata\d1all004.NSDstat_F1.dta

File Edit Data Graphics Statistics User Window Help

Results

```
20-user Stata for Windows (network) perpetual license:
       Serial number: 81910526399
       Licensed to: Server Services
                    UIC - ACCC

Notes:
    1.  (/m# option or -set memory-) 1.00 MB allocated to data

. do c:\nsfh_distance\stata\CreateDataErrors.do

. use MCASEID M2BP01 M2DP01 M484 M497A M502 E1301 E1302 E1305 E1332A E1332B IRE
> N using d1all004.NSDstat_F1.dta if E1301==1 & (M497A<=51 | M497A==990 | M497A=
> 996) & E1305~=9994 & E1305~=9995

. genrate g2age=M2BP01 if M2BP01<97
unrecognized command:   genrate
r(199);

end of do-file

r(199);
```

College Station, Texas 77845 USA
800-STATA-PC          http://www.stata.com
979-696-4600          stata@stata.com
979-696-4601 (fax)

Command

Review
_rc
Command

Variables

| Name | Label | Type | Format |
|------|-------|------|--------|
| MCASEID | CASE NUMBER | long | %5.0f |
| M2BP01 | AGE OF RESPOND... | byte | %2.0f |
| M2DP01 | SEX OF RESPONDE... | byte | %1.0f |
| M484 | WHICH GROUP DE... | byte | %2.0f |
| M497A | COUNTRY/STATE L... | int | %3.0f |
| M502 | HIGHEST GRADE S... | byte | %2.0f |
| E1301 | MOTHER LIVING O... | byte | %1.0f |
| E1302 | AGE OF R'S MOTHER | byte | %2.0f |
| E1305 | HOW FAR AWAY R... | int | %4.0f |
| E1332A | NUMBER LIVING B... | byte | %2.0f |
| E1332B | NUMBER LIVING SI... | byte | %2.0f |
| IREARN | R TOTAL EARNINGS | long | %7.0f |

C:\nsfh_distance\stata

## ■ Display B.5.1 Regression of Distance from Mother on Respondent's Earnings

|  | SAS | Stata |
|---|---|---|
| Commands | ```
proc reg;
  Model g1miles=g2earn;
run;
``` | ```
regress g1miles g2earn
``` |

### Results

**SAS**

| Number of Observations Read | 6350 |
|---|---|
| Number of Observations Used | 6350 |

**Analysis of Variance**

| Source | DF | Sum of Squares | Mean Square | F Value | Pr > F |
|---|---|---|---|---|---|
| Model | 1 | 10850283 | 10850283 | 26.90 | <.0001 |
| Error | 6348 | 2560477708 | 403352 | | |
| Corrected Total | 6349 | 2571327990 | | | |

| Root MSE | 635.09989 | R-Square | 0.0042 |
|---|---|---|---|
| Dependent Mean | 283.47102 | Adj R-Sq | 0.0041 |
| Coeff Var | 224.04403 | | |

**Parameter Estimates**

| Variable | DF | Parameter Estimate | Standard Error | t Value | Pr > |t| |
|---|---|---|---|---|---|
| Intercept | 1 | 249.81157 | 10.27798 | 24.31 | <.0001 |
| g2earn | 1 | 0.00113 | 0.00021852 | 5.19 | <.0001 |

**Stata**

```
      Source |       SS       df       MS              Number of obs =    6350
-------------+------------------------------           F(  1,  6348) =   26.90
       Model |  10850282.2       1  10850282.2         Prob > F      =  0.0000
    Residual |  2.5605e+09    6348  403351.876         R-squared     =  0.0042
-------------+------------------------------           Adj R-squared =  0.0041
       Total |  2.5713e+09    6349  404997.321         Root MSE      =  635.1
```

```
     g1miles |      Coef.   Std. Err.      t    P>|t|     [95% Conf. Interval]
-------------+----------------------------------------------------------------
      g2earn |   .0011333   .0002185     5.19   0.000     .000705    .0015617
       _cons |   249.8116   10.27798    24.31   0.000    229.6632   269.9599
```

**Display B.5.2 Regression of Distance from Mother on Respondent's Earnings: Confidence Interval**

|  | SAS | Stata |
|---|---|---|
| Commands | `proc reg;`<br>`   Model g1miles=g2earn /clb;`<br>`run;` | `regress g1miles g2earn` |

### Results

**SAS**

| Number of Observations Read | 6350 |
|---|---|
| Number of Observations Used | 6350 |

**Analysis of Variance**

| Source | DF | Sum of Squares | Mean Square | F Value | Pr > F |
|---|---|---|---|---|---|
| Model | 1 | 10850283 | 10850283 | 26.90 | <.0001 |
| Error | 6348 | 2560477708 | 403352 | | |
| Corrected Total | 6349 | 2571327990 | | | |

| Root MSE | 635.09989 | R-Square | 0.0042 |
|---|---|---|---|
| Dependent Mean | 283.47102 | Adj R-Sq | 0.0041 |
| Coeff Var | 224.04403 | | |

**Parameter Estimates**

| Variable | DF | Parameter Estimate | Standard Error | t Value | Pr > |t| | 95% Confidence Limits | |
|---|---|---|---|---|---|---|---|
| Intercep | 1 | 249.81157 | 10.27798 | 24.31 | <.0001 | 229.66325 | 269.95989 |
| g2earn | 1 | 0.00113 | 0.00021852 | 5.19 | <.0001 | 0.00070498 | 0.00156 |

**Stata**

```
  Source |       SS       df       MS              Number of obs =    6350
---------+------------------------------           F( 1,  6348) =   26.90
   Model | 10850282.2      1 10850282.2           Prob > F      = 0.0000
Residual | 2.5605e+09   6348 403351.876           R-squared     = 0.0042
---------+------------------------------           Adj R-squared = 0.0041
   Total | 2.5713e+09   6349 404997.321           Root MSE      =   635.1

 g1miles |      Coef.   Std. Err.      t    P>|t|     [95% Conf. Interval]
---------+--------------------------------------------------------------
  g2earn |   .0011333    .0002185    5.19   0.000     .000705    .0015617
   _cons |   249.8116    10.27798   24.31   0.000     229.6632    269.9599
```

■ **Display B.5.3  Regression of Distance from Mother on Respondent's Earnings: Rescaled Earnings to $10,000 Units**

## Commands

### SAS

```
LIBNAME LIBRARY 'c:/nsfh_distance/SAS';
data Rescale10000;
  set "c:/nsfh_distance/SAS/
    CreateData";
  if g1miles~=.& g2earn~=.;
  g2earn10000=g2earn/10000;
run;

proc reg;
  Model g1miles=g2earn10000;
run;
```

### Stata

```
use CreateData if g1miles~=. & g2earn~=.
generate g2earn10000=g2earn/10000
regress g1miles g2earn10000
```

## Results

### SAS

| Number of Observations Read | 6350 |
|---|---|
| Number of Observations Used | 6350 |

Analysis of Variance

| Source | DF | Sum of Squares | Mean Square | F Value | Pr > F |
|---|---|---|---|---|---|
| Model | 1 | 10850283 | 10850283 | 26.90 | <.0001 |
| Error | 6348 | 2560477708 | 403352 | | |
| Corrected Total | 6349 | 2571327990 | | | |

| Root MSE | 635.09989 | R-Square | 0.0042 |
|---|---|---|---|
| Dependent Mean | 283.47102 | Adj R-Sq | 0.0041 |
| Coeff Var | 224.04403 | | |

Parameter Estimates

| Variable | DF | Parameter Estimate | Standard Error | t Value | Pr > |t| |
|---|---|---|---|---|---|
| Intercept | 1 | 249.81157 | 10.27798 | 24.31 | <.0001 |
| g2earn10000 | 1 | 11.33347 | 2.18517 | 5.19 | <.0001 |

### Stata

```
      Source |       SS       df       MS              Number of obs =    6350
-------------+------------------------------           F( 1,  6348) =   26.90
       Model |  10850282.2     1  10850282.2           Prob > F      =  0.0000
    Residual |  2.5605e+09  6348  403351.876           R-squared     =  0.0042
-------------+------------------------------           Adj R-squared =  0.0041
       Total |  2.5713e+09  6349  404997.321           Root MSE      =   635.1

------------------------------------------------------------------------------
     g1miles |      Coef.   Std. Err.      t    P>|t|     [95% Conf. Interval]
-------------+----------------------------------------------------------------
 g2earn10000 |   11.33347   2.185166     5.19   0.000     7.049807    15.61713
       _cons |   249.8116   10.27798    24.31   0.000     229.6632    269.9599
------------------------------------------------------------------------------
```

## ■ Display B.5.4 Standard Deviations of Distance from Mother and Respondent's Earnings

<table>
<tr><td></td><td style="text-align:center">SAS</td><td style="text-align:center">Stata</td></tr>
<tr>
<td>Commands</td>
<td>

```
proc means;
var g1miles g2earn;
run;
```

</td>
<td>

```
summarize g1miles g2earn
```

</td>
</tr>
<tr>
<td>Results</td>
<td>

| Variable | N | Mean | Std Dev | Minimum | Maximum |
|---|---|---|---|---|---|
| g1miles | 6350 | 283.4710236 | 636.3939981 | 1.0000000 | 9000.00 |
| g2earn | 6350 | 29699.16 | 36475.81 | 0 | 94590.28 |

</td>
<td>

```
Variable |   Obs      Mean     Std. Dev.   Min       Max
---------+-------------------------------------------------
 g1miles |  6350    283.471    636.394      1       9000
  g2earn |  6350   29699.16   36475.81      0    945903.3
```

</td>
</tr>
</table>

# ■ Display B.5.5 Regression of Distance from Mother on Respondent's Earnings: Standardized Coefficient using Rescaled Variables

| | SAS | Stata |
|---|---|---|
| **Commands** | | |

**SAS**

```
LIBNAME LIBRARY 'c:\nsfh_distance\SAS';
data RescalesD;
set "c:\nsfh_distance\SAS\CreateData";
if g1miles~=. & g2earn~=.;
g1milesSD=g1miles/636.394;
g2earnSD=g2earn/36475.81;
run;

proc reg;
    Model g1milesSD=g2earnSD;
run;
```

**Stata**

```
use CreateData if g1miles~=. & g2earn~=.
generate g1milesSD=g1miles/636.394
generate g2earnSD=g2earn/36475.81
regress g1milesSD g2earnSD
```

**Results**

**SAS**

| Number of Observations Read | 6350 |
|---|---|
| Number of Observations Used | 6350 |

Analysis of Variance

| Source | DF | Sum of Squares | Mean Square | F Value | Pr > F |
|---|---|---|---|---|---|
| Model | 1 | 26.79100 | 26.79100 | 26.90 | <.0001 |
| Error | 6348 | 6322.20896 | 0.99594 | | |
| Corrected Total | 6349 | 6348.99996 | | | |

| Root MSE | 0.99797 | R-Square | 0.0042 |
|---|---|---|---|
| Dependent Mean | 0.44543 | Adj R-Sq | 0.0041 |
| Coeff Var | 224.04403 | | |

Parameter Estimates

| Variable | DF | Parameter Estimate | Standard Error | t Value | Pr > |t| |
|---|---|---|---|---|---|
| Intercept | 1 | 0.39254 | 0.01615 | 24.31 | <.0001 |
| g2earnSD | 1 | 0.06496 | 0.01252 | 5.19 | <.0001 |

**Stata**

```
Source |       SS       df       MS              Number of obs =    6350
-------+------------------------------           F( 1, 6348)   =   26.90
 Model | 26.7909973     1  26.7909973            Prob > F      =  0.0000
Residual| 6322.20892  6348  .995937133           R-squared     =  0.0042
-------+------------------------------           Adj R-squared =  0.0041
 Total | 6348.99992  6349  .999999987            Root MSE      =  .99797

-------------------------------------------------------------------------
g1milesSD |   Coef.   Std. Err.    t    P>|t|   [95% Conf. Interval]
----------+--------------------------------------------------------------
 g2earnSD | .0649594  .0125246   5.19  0.000    .040407    .0895118
    _cons | .3925423  .0161503  24.31  0.000    .3608822   .4242024
-------------------------------------------------------------------------
```

■ **Display B.5.6 Regression of Distance from Mother on Respondent's Earnings: Standardized Coefficients Calculated by SAS and Stata**

|  | SAS | Stata |
|---|---|---|
| Commands | ```
proc reg;
  Model g1miles=g2earn /stb;
run;
``` | ```
regress g1miles g2earn, beta
``` |

**SAS Results**

| Number of Observations Read | 6350 |
|---|---|
| Number of Observations Used | 6350 |

Analysis of Variance

| Source | DF | Sum of Squares | Mean Square | F Value | Pr > F |
|---|---|---|---|---|---|
| Model | 1 | 10850283 | 10850283 | 26.90 | <.0001 |
| Error | 6348 | 256047708 | 403352 | | |
| Corrected Total | 6349 | 257132799 | | | |

| Root MSE | 635.09989 | R-Square | 0.0042 |
|---|---|---|---|
| Dependent Mean | 283.47102 | Adj R-Sq | 0.0041 |
| Coeff Var | 224.04403 | | |

Parameter Estimates

| Variable | DF | Parameter Estimate | Standard Error | t Value | Pr > |t| | Standardized Estimate |
|---|---|---|---|---|---|---|
| Intercept | 1 | 249.81157 | 10.27798 | 24.31 | <.0001 | 0 |
| g2earn | 1 | 0.00113 | 0.00021852 | 5.19 | <.0001 | 0.06496 |

**Stata Results**

```
Source |       SS       df       MS              Number of obs =    6350
-------+------------------------------           F(  1,  6348) =   26.90
 Model | 10850282.2     1   10850282.2           Prob > F      =  0.0000
Residual| 2.5605e+09   6348   403351.876         R-squared     =  0.0042
-------+------------------------------           Adj R-squared =  0.0041
 Total | 2.5713e+09   6349   404997.321          Root MSE      =   635.1

g1miles |   Coef.   Std. Err.     t    P>|t|                    Beta
--------+------------------------------------------------------------
 g2earn | .0011333  .0002185    5.19   0.000               .0649594
  _cons | 249.8116  10.27798    24.31   0.000
```

■ **Display B.5.7 Correlation of Distance from Mother and Respondent's Earnings**

| | SAS | Stata |
|---|---|---|
| Commands | `proc corr;`<br>`  var g1miles g2earn;`<br>`run;` | `correlate g1miles g2earn` |
| Results | Pearson Correlation Coefficients, N = 6350<br>Prob > \|r\| under H0: Rho=0<br><br>| | g1miles | g2earn |<br>\| g1miles \| 1.00000 \| 0.06496 <.0001 \|<br>\| g2earn \| 0.06496 <.0001 \| 1.00000 \| | `             |  g1miles   g2earn`<br>`-------------+------------------`<br>`     g1miles |   1.0000`<br>`      g2earn |   0.0650   1.0000` |

**Display B.5.8 Regression of Respondent's Earnings on Distance from Mother**

|  | SAS | Stata |
|---|---|---|
| Commands | `proc reg;`<br>`  Model g2earn=g1miles;`<br>`run;` | `regress g2earn g1miles` |

**Results**

**SAS**

| Number of Observations Read | 6350 |
|---|---|
| Number of Observations Used | 6350 |

Analysis of Variance

| Source | DF | Sum of Squares | Mean Square | F Value | Pr > F |
|---|---|---|---|---|---|
| Model | 1 | 35645009542 | 3564509542 | 26.90 | <.0001 |
| Error | 6348 | 8.411601E12 | 1325078994 |  |  |
| Corrected Total | 6349 | 8.447246E12 |  |  |  |

| Root MSE | 36402 | R-Square | 0.0042 |
|---|---|---|---|
| Dependent Mean | 29699 | Adj R-Sq | 0.0041 |
| Coeff Var | 122.56789 |  |  |

Parameter Estimates

| Variable | DF | Parameter Estimate | Standard Error | t Value | Pr > |t| |
|---|---|---|---|---|---|
| Intercept | 1 | 28844 | 500.08351 | 57.28 | <.0001 |
| g1miles | 1 | 3.72324 | 0.71786 | 5.19 | <.0001 |

**Stata**

```
      Source |       SS       df       MS              Number of obs =    6350
-------------+------------------------------           F(  1,  6348) =   26.90
       Model |  3.5645e+10     1   3.5645e+10           Prob > F      =  0.0000
    Residual |  8.4116e+12  6348  1.3251e+09           R-squared     =  0.0042
-------------+------------------------------           Adj R-squared =  0.0041
       Total |  8.4472e+12  6349  1.3305e+09           Root MSE      =  36402

------------------------------------------------------------------------------
      g2earn |      Coef.   Std. Err.      t    P>|t|     [95% Conf. Interval]
-------------+----------------------------------------------------------------
     g1miles |   3.723237   .717864     5.19   0.000     2.315981    5.130492
       _cons |   28643.73   500.0835    57.28   0.000     27663.4    29624.06
------------------------------------------------------------------------------
```

## ■ Display B.5.9 Regression of Rescaled Earnings on Rescaled Miles

|  | SAS | Stata |
|---|---|---|
| Commands | `proc reg;`<br>`  Model g2earnSD=g1milesSD;`<br>`run;` | `regress g2earnSD g1milesSD` |

### SAS — Results

| Number of Observations Read | 6350 |
|---|---|
| Number of Observations Used | 6350 |

**Analysis of Variance**

| Source | DF | Sum of Squares | Mean Square | F Value | Pr > F |
|---|---|---|---|---|---|
| Model | 1 | 26.79100 | 26.79100 | 26.90 | <.0001 |
| Error | 6348 | 6322.20826 | 0.99594 |  |  |
| Corrected Total | 6349 | 6348.99926 |  |  |  |

| Root MSE | 0.99797 | R-Square | 0.0042 |
|---|---|---|---|
| Dependent Mean | 0.81422 | Adj R-Sq | 0.0041 |
| Coeff Var | 122.56789 |  |  |

**Parameter Estimates**

| Variable | DF | Parameter Estimate | Standard Error | t Value | Pr > |t| |
|---|---|---|---|---|---|
| Intercept | 1 | 0.78528 | 0.01371 | 57.28 | <.0001 |
| g1milesSD | 1 | 0.06496 | 0.01252 | 5.19 | <.0001 |

### Stata — Results

```
      Source |       SS       df       MS              Number of obs =    6350
-------------+------------------------------           F(  1,  6348) =   26.90
       Model | 26.7909952      1   26.7909952          Prob > F      =  0.0000
    Residual | 6322.20842   6348  .995937055           R-squared     =  0.0042
-------------+------------------------------           Adj R-squared =  0.0041
       Total | 6348.99942   6349   .99999909           Root MSE      =  .99797

------------------------------------------------------------------------------
    g2earnSD |     Coef.   Std. Err.      t    P>|t|    [95% Conf. Interval]
-------------+----------------------------------------------------------------
   g1milesSD |  .0649594   .0125246     5.19   0.000    .040407    .0895118
       _cons |  .7852802    .01371     57.28   0.000    .7584039    .8121564
------------------------------------------------------------------------------
```

**APPENDIX B**

## Display B.6.1 Regression of Distance from Mother on Respondent's Earnings and Mother's Years of Schooling

| | SAS | Stata |
|---|---|---|
| Commands | `proc reg;`<br>`  Model g1miles=g2earn g1yrschl;`<br>`run;` | `regress g1miles g2earn g1yrschl` |

### Results

**SAS**

| Number of Observations Read | 5475 |
|---|---|
| Number of Observations Used | 5475 |

**Analysis of Variance**

| Source | DF | Sum of Squares | Mean Square | F Value | Pr > F |
|---|---|---|---|---|---|
| Model | 2 | 22291081 | 11145541 | 27.04 | <.0001 |
| Error | 5472 | 2255081902 | 412113 | | |
| Corrected Total | 5474 | 2277372983 | | | |

| Root MSE | 641.96022 | R-Square | 0.0098 |
|---|---|---|---|
| Dependent Mean | 291.22721 | Adj R-Sq | 0.0094 |
| Coeff Var | 220.43277 | | |

**Parameter Estimates**

| Variable | DF | Parameter Estimate | Standard Error | t Value | Pr > |t| |
|---|---|---|---|---|---|
| Intercept | 1 | 78.34818 | 34.89219 | 2.25 | 0.0248 |
| g2earn | 1 | 0.00102 | 0.0002297 | 4.39 | <.0001 |
| g1yrschl | 1 | 15.92156 | 2.97055 | 5.36 | <.0001 |

**Stata**

| Source | SS | df | MS | | |
|---|---|---|---|---|---|
| Model | 22291080.9 | 2 | 11145540.4 | Number of obs = 5475 | |
| Residual | 2.2551e+09 | 5472 | 412112.921 | $F(2, 5472)$ = 27.04 | |
| | | | | Prob > F = 0.0000 | |
| Total | 2.2774e+09 | 5474 | 416034.524 | R-squared = 0.0098 | |
| | | | | Adj R-squared = 0.0094 | |
| | | | | Root MSE = 641.96 | |

| g1miles | Coef. | Std. Err. | t | P>|t| | [95% Conf. Interval] | |
|---|---|---|---|---|---|---|
| g2earn | .001022 | .000233 | 4.39 | 0.000 | .0005653 | .0014788 |
| g1yrschl | 15.92156 | 2.970548 | 5.36 | 0.000 | 10.09811 | 21.74502 |
| _cons | 78.34818 | 34.89219 | 2.25 | 0.025 | 9.945613 | 146.7507 |

■ **Display B.6.2 Regression of Distance from Mother on Respondent's Earnings and Mother's Years of Schooling: Earnings Rescaled to $10,000 Units**

| | SAS | Stata |
|---|---|---|
| Commands | `proc reg;`<br>`  Model g1miles=g2earn10000 g1yrschl;`<br>`run;` | `regress g1miles g2earn10000 g1yrschl` |

**SAS Results**

| Number of Observations Read | 5475 |
|---|---|
| Number of Observations Used | 5475 |

**Analysis of Variance**

| Source | DF | Sum of Squares | Mean Square | F Value | Pr > F |
|---|---|---|---|---|---|
| Model | 2 | 22291081 | 11145541 | 27.04 | <.0001 |
| Error | 5472 | 2255081902 | 412113 | | |
| Corrected Total | 5474 | 2277372983 | | | |

| Root MSE | 641.96022 | R-Square | 0.0098 |
|---|---|---|---|
| Dependent Mean | 291.22721 | Adj R-Sq | 0.0094 |
| Coeff Var | 220.43277 | | |

**Parameter Estimates**

| Variable | DF | Parameter Estimate | Standard Error | t Value | Pr > |t| |
|---|---|---|---|---|---|
| Intercept | 1 | 78.34818 | 34.89219 | 2.25 | 0.0248 |
| g2earn10000 | 1 | 10.22048 | 2.32971 | 4.39 | <.0001 |
| g1yrschl | 1 | 15.92156 | 2.97055 | 5.36 | <.0001 |

**Stata Results**

| Source | SS | df | MS | | |
|---|---|---|---|---|---|
| Model | 22291080.9 | 2 | 11145540.5 | Number of obs = 5475 | |
| Residual | 2.2551e+09 | 5472 | 412112.921 | F( 2, 5472) = 27.04 | |
| | | | | Prob > F = 0.0000 | |
| Total | 2.2774e+09 | 5474 | 416034.524 | R-squared = 0.0098 | |
| | | | | Adj R-squared = 0.0094 | |
| | | | | Root MSE = 641.96 | |

| g1miles | Coef. | Std. Err. | t | P>|t| | [95% Conf. Interval] |
|---|---|---|---|---|---|---|
| g2earn10000 | 10.22048 | 2.329708 | 4.39 | 0.000 | 5.653329 | 14.78763 |
| g1yrschl | 15.92156 | 2.970548 | 5.36 | 0.000 | 10.09811 | 21.74502 |
| _cons | 78.34818 | 34.89219 | 2.25 | 0.025 | 9.945613 | 146.7507 |

■ **Display B.6.3 Regression of Distance from Mother on Respondent's Earnings and Mother's Years of Schooling: Schooling Rescaled to 4-Year Units**

| | SAS | Stata |
|---|---|---|
| Commands | ```
glyrschl4=glyrschl/4;
. . .
proc reg;
   Model glmiles=g2earn glyrschl4;
run;
``` | ```
generate glyrschl4=glyrschl/4
regress glmiles g2earn glyrschl4
``` |

**Results (SAS):**

| | |
|---|---|
| Number of Observations Read | 5475 |
| Number of Observations Used | 5475 |

**Analysis of Variance**

| Source | DF | Sum of Squares | Mean Square | F Value | Pr > F |
|---|---|---|---|---|---|
| Model | 2 | 22291081 | 11145541 | 27.04 | <.0001 |
| Error | 5472 | 2255081902 | 412113 | | |
| Corrected Total | 5474 | 2277372983 | | | |

| | | | |
|---|---|---|---|
| Root MSE | 641.96022 | R-Square | 0.0098 |
| Dependent Mean | 291.22721 | Adj R-Sq | 0.0094 |
| Coeff Var | 220.43277 | | |

**Parameter Estimates**

| Variable | DF | Parameter Estimate | Standard Error | t Value | Pr > |t| |
|---|---|---|---|---|---|
| Intercept | 1 | 78.34818 | 34.89219 | 2.25 | 0.0248 |
| g2earn | 1 | 0.00102 | 0.00023297 | 4.39 | <.0001 |
| glyrschl4 | 1 | 63.68624 | 11.88219 | 5.36 | <.0001 |

**Results (Stata):**

```
  Source |       SS       df       MS              Number of obs =    5475
---------+------------------------------           F(  2,  5472) =   27.04
   Model | 22291080.9      2  11145540.4           Prob > F      =  0.0000
Residual | 2.2551e+09   5472  412112.921           R-squared     =  0.0098
---------+------------------------------           Adj R-squared =  0.0094
   Total | 2.2774e+09   5474  416034.524           Root MSE      =  641.96

   glmiles |    Coef.    Std. Err.     t    P>|t|   [95% Conf. Interval]
-----------+----------------------------------------------------------------
    g2earn |  .001022    .000233     4.39  0.000    .0005653    .0014788
 glyrschl4 | 63.68624   11.88219     5.36  0.000    40.39242    86.98006
     _cons | 78.34818   34.89219     2.25  0.025    9.945613    146.7507
```

| Notes | The variable creation in SAS must take place in the DATA step. |
|---|---|

■ **Display B.6.4 Regression of Hours of Chores on Number of Children (Full NSFH Sample of 3,116 Employed Women)**

| | SAS | Stata |
|---|---|---|
| **Commands** | `proc reg;`<br>`Model hrchores=numkid;`<br>`run;` | `regress hrchores numkid` |

**Results**

**SAS**

| Number of Observations Read | 3116 |
|---|---|
| Number of Observations Used | 3116 |

**Analysis of Variance**

| Source | DF | Sum of Squares | Mean Square | F Value | Pr > F |
|---|---|---|---|---|---|
| Model | 1 | 152453 | 152453 | 215.16 | <.0001 |
| Error | 3114 | 2206415 | 708.54674 | | |
| Corrected Total | 3115 | 2358868 | | | |

| Root MSE | 26.61854 | R-Square | 0.0646 |
|---|---|---|---|
| Dependent Mean | 34.45250 | Adj R-Sq | 0.0643 |
| Coeff Var | 77.26156 | | |

**Parameter Estimates**

| Variable | DF | Parameter Estimate | Standard Error | t Value | Pr > |t| |
|---|---|---|---|---|---|
| Intercept | 1 | 28.69373 | 0.61767 | 46.45 | <.0001 |
| numkid | 1 | 6.38361 | 0.43519 | 14.67 | <.0001 |

**Stata**

```
      Source |       SS       df       MS              Number of obs =    3116
-------------+------------------------------           F( 1,  3114) =  215.16
       Model | 152453.437      1  152453.437           Prob > F      =  0.0000
    Residual | 2206414.53   3114  708.546735           R-squared     =  0.0646
-------------+------------------------------           Adj R-squared =  0.0643
       Total | 2358867.97   3115  757.260986           Root MSE      = 26.619

    hrchores |      Coef.   Std. Err.      t    P>|t|     [95% Conf. Interval]
-------------+----------------------------------------------------------------
      numkid |   6.383609    .435193    14.67   0.000     5.530315    7.236903
       _cons |   28.69373    .6176739    46.45   0.000     27.48264    29.90482
```

■ **Display B.6.5 Intercept-only Model for Hours of Chores (Full NSFH Sample of 3,116 Employed Women)**

|  | SAS | Stata |
|---|---|---|
| Commands | ```
proc reg;
  Model hrchores=;
run;
``` | ```
regress hrchores
``` |

**SAS Results**

| Number of Observations Read | 3116 |
|---|---|
| Number of Observations Used | 3116 |

**Analysis of Variance**

| Source | DF | Sum of Squares | Mean Square | F Value | Pr > F |
|---|---|---|---|---|---|
| Model | 0 | 0 | . | . | . |
| Error | 3115 | 2358868 | 757.26099 | | |
| Corrected Total | 3115 | 2358868 | | | |

| Root MSE | 27.51838 | R-Square | 0.0000 |
|---|---|---|---|
| Dependent Mean | 34.45250 | Adj R-Sq | 0.0000 |
| Coeff Var | 79.87337 | | |

**Parameter Estimates**

| Variable | DF | Parameter Estimate | Standard Error | t Value | Pr > |t| |
|---|---|---|---|---|---|
| Intercept | 1 | 34.45250 | 0.49297 | 69.89 | <.0001 |

**Stata Results**

```
Source |       SS       df       MS              Number of obs =    3116
-------+------------------------------           F(  0,  3115) =    0.00
 Model |        0        0        .              Prob > F      =       .
Residual| 2358867.97   3115  757.260986          R-squared     =  0.0000
-------+------------------------------           Adj R-squared =  0.0000
 Total | 2358867.97   3115  757.260986           Root MSE      =  27.518

---------------------------------------------------------------------------
hrchores|    Coef.   Std. Err.      t    P>|t|   [95% Conf. Interval]
-------+-------------------------------------------------------------------
  _cons | 34.4525   .4929741    69.89   0.000    33.48592    35.41909
---------------------------------------------------------------------------
```

# Display B.6.6 Regression of Hours of Chores on Number of Children and Hours of Paid Work (Full NSFH Sample of 3,116 Employed Women)

| | SAS | Stata |
|---|---|---|
| Commands | `proc reg;`<br>`  Model hrchores=numkid hrwork;`<br>`run;` | `regress hrchores numkid hrwork` |

**Results**

**SAS**

| Number of Observations Read | 3116 |
|---|---|
| Number of Observations Used | 3116 |

**Analysis of Variance**

| Source | DF | Sum of Squares | Mean Square | F Value | Pr > F |
|---|---|---|---|---|---|
| Model | 2 | 179987 | 89993 | 128.58 | <.0001 |
| Error | 3113 | 2178881 | 699.92965 | | |
| Corrected Total | 3115 | 2358868 | | | |

| Root MSE | 26.45618 | R-Square | 0.0763 |
|---|---|---|---|
| Dependent Mean | 34.45250 | Adj R-Sq | 0.0757 |
| Coeff Var | 76.79031 | | |

**Parameter Estimates**

| Variable | DF | Parameter Estimate | Standard Error | t Value | Pr > |t| |
|---|---|---|---|---|---|
| Intercept | 1 | 36.24366 | 1.35126 | 26.82 | <.0001 |
| numkid | 1 | 6.15556 | 0.43406 | 14.18 | <.0001 |
| hrwork | 1 | −0.21040 | 0.03355 | −6.27 | <.00011 |

**Stata**

```
Source |       SS       df       MS              Number of obs =    3116
-------+------------------------------           F( 2,  3113) =  128.58
 Model | 179986.973      2   89993.4863          Prob > F      = 0.0000
Residual| 2178881      3113  699.929649          R-squared     = 0.0763
-------+------------------------------           Adj R-squared = 0.0757
 Total | 2358867.97   3115   757.260986          Root MSE      = 26.456

hrchores |   Coef.    Std. Err.     t     P>|t|    [95% Conf. Interval]
numkid   |  6.155562  .4340641    14.18   0.000    5.304481    7.006642
hrwork   |  -.2104033 .035466     -6.27   0.000    -.276179    -.1446275
 _cons   |  36.24366  1.351263    26.82   0.000    33.5942     38.89311
```

## Display B.6.7 Regression of Hours of Chores on Hours of Paid Work (Full NSFH Sample of 3,116 Employed Women)

|  | SAS | Stata |
|---|---|---|
| Commands | `proc reg;`<br>`  Model hrchores=hrwork;`<br>`run;` | `regress hrchores hrwork` |

### SAS Results

| Number of Observations Read | 3116 |
|---|---|
| Number of Observations Used | 3116 |

**Analysis of Variance**

| Source | DF | Sum of Squares | Mean Square | F Value | Pr > F |
|---|---|---|---|---|---|
| Model | 1 | 39226 | 39226 | 52.66 | <.0001 |
| Error | 3114 | 2319642 | 744.90747 | | |
| Corrected Total | 3115 | 2358868 | | | |

| Root MSE | 27.29299 | R-Square | 0.0166 |
|---|---|---|---|
| Dependent Mean | 34.45250 | Adj R-Sq | 0.0163 |
| Coeff Var | 79.21919 | | |

**Parameter Estimates**

| Variable | DF | Parameter Estimate | Standard Error | t Value | Pr > |t| |
|---|---|---|---|---|---|
| Intercept | 1 | 43.18768 | 1.29926 | 33.24 | <.0001 |
| hrwork | 1 | −0.25025 | 0.03449 | −7.26 | <.0001 |

### Stata Results

```
    Source |       SS       df       MS              Number of obs =    3116
-----------+------------------------------           F(  1,  3114) =   52.66
     Model |  39226.104      1    39226.104           Prob > F      =  0.0000
  Residual | 2319641.87   3114   744.907472           R-squared     =  0.0166
-----------+------------------------------           Adj R-squared =  0.0163
     Total | 2358867.97   3115   757.260986           Root MSE      =  27.293

-----------+------------------------------------------------------------------
  hrchores |     Coef.   Std. Err.      t    P>|t|    [95% Conf. Interval]
-----------+------------------------------------------------------------------
    hrwork | -.2502534   .0344861    -7.26   0.000    -.3178712   -.1826357
     _cons |  43.18768   1.299256    33.24   0.000     40.6402     45.73517
-----------+------------------------------------------------------------------
```

# ■ Display B.6.8 Regression of Distance from Mother on Respondent's Earnings, Mother's Years of Schooling, Respondent and Mother's Age, and Respondent's Number of Brothers and Sisters

## SAS

### Commands

```
LIBNAME LIBRARY 'c:\sfh_distance\SAS';
data CreateData;
  set "c:\sfh_distance\SAS\CreateData";
  if g1miles~=. & g2earn~=. & g1yrschl~=.
  & g1age~=. & g2age~=. & g2numbro~=. &
  g2numsis~=.;
run;
proc reg;
  Model g1miles=g2earn g1yrschl g1age g2age
    g2numbro g2numsis;
  test g1age, g2age, g2numbro, g2numsis;
run;
```

### Results

Number of Observations Used: 5475

Analysis of Variance

| Source | DF | Sum of Squares | Mean Square | F Value | Pr > F |
|---|---|---|---|---|---|
| Model | 6 | 38919757 | 6486626 | 15.85 | <.0001 |
| Error | 5468 | 2238453227 | 409373 | | |
| Corrected Total | 5474 | 2277372983 | | | |

| Root MSE | 639.82287 | R-Square | 0.0171 |
|---|---|---|---|
| Dependent Mean | 291.22721 | Adj R-Sq | 0.0160 |
| Coeff Var | 219.69886 | | |

Parameter Estimates

| Variable | DF | Parameter Estimate | Standard Error | t Value | Pr > |t| |
|---|---|---|---|---|---|
| Intercept | 1 | -275.28450 | 67.89939 | -4.05 | <.0001 |
| g2earn | 1 | 0.00082868 | 0.00023716 | 3.49 | 0.0005 |
| g1yrschl | 1 | 21.20263 | 3.13213 | 6.77 | <.0001 |
| g1age | 1 | 4.55911 | 1.44023 | 3.17 | 0.0016 |
| g2age | 1 | -0.13125 | 1.65018 | -0.08 | 0.9366 |
| g2numbro | 1 | 4.48160 | 5.96686 | 0.75 | 0.4526 |
| g2numsis | 1 | 16.94382 | 6.00851 | 2.82 | 0.0048 |

Test 1 Results for Dependent Variable g1miles

| Source | DF | Mean Square | F Value | Pr > F |
|---|---|---|---|---|
| Numerator | 4 | 4157169 | 10.15 | <.0001 |
| Denominator | 5468 | 409373 | | |

## Stata

### Commands

```
use CreateData if g1miles~=. & g2earn~=.       ///
  & g1yrschl~=. & g1age~=.                      ///
  & g2age~=. & g2numbro~=. & g2numsis~=.

regress g1miles g2earn g1yrschl g1age g2age    ///
  g2numbro g2numsis

test g1age g2age g2numbro g2numsis
```

### Results

```
                                        Number of obs =    5475
                                        F( 6,  5468) =     15.85
                                        Prob > F      =    0.0000
                                        R-squared     =    0.0171
                                        Adj R-squared =    0.0160
                                        Root MSE      =    639.82
```

| Source | SS | df | MS |
|---|---|---|---|
| Model | 38919756.6 | 6 | 6486626.11 |
| Residual | 2.2385e+09 | 5468 | 409373.304 |
| Total | 2.2774e+09 | 5474 | 416034.524 |

| g1miles | Coef. | Std. Err. | t | P>|t| | [95% Conf. Interval] | |
|---|---|---|---|---|---|---|
| g2earn | .0008287 | .0002372 | 3.49 | 0.000 | .003638 | .0012936 |
| g1yrschl | 21.20263 | 3.132135 | 6.77 | 0.000 | 15.0624 | 27.34286 |
| g1age | 4.559107 | 1.440233 | 3.17 | 0.002 | 1.735677 | 7.382537 |
| g2age | -.131249 | 1.650181 | -0.08 | 0.937 | -3.366261 | 3.103763 |
| g2numbro | 4.481604 | 5.96686 | 0.75 | 0.453 | -7.215816 | 16.17902 |
| g2numsis | 16.94382 | 6.008505 | 2.82 | 0.005 | 5.164763 | 28.72289 |
| _cons | -275.2845 | 67.89939 | -4.05 | 0.000 | -408.3943 | -142.1747 |

```
. test g1age g2age g2num

 (1)  g1age = 0
 (2)  g2age = 0
 (3)  g2numbro = 0
 (4)  g2numsis = 0

       F(  4,  5468) =   10.15
            Prob > F =    0.0000
```

■ **Display B.7.1 Codebook for M484 in Analytic Data File**

| | Commands | Results |
|---|---|---|
| **SAS** | `proc freq; run;` | |

**WHICH GROUP DESCRIBES R (RACE)**

| M484 | Frequency | Percent | Cumulative Frequency | Cumulative Percent |
|---|---|---|---|---|
| BLACK | 1165 | 16.88 | 1165 | 16.88 |
| WHITE/NOT HISPANIC | 5384 | 78.02 | 6549 | 94.90 |
| MEX/CHICANO/MEX.AM | 218 | 3.16 | 6767 | 98.06 |
| PUERTO RICAN | 40 | 0.58 | 6807 | 98.64 |
| CUBAN | 5 | 0.07 | 6812 | 98.71 |
| OTHER HISPANIC | 34 | 0.49 | 6846 | 99.20 |
| AMERICAN INDIAN | 31 | 0.45 | 6877 | 99.65 |
| ASIAN | 20 | 0.29 | 6897 | 99.94 |
| OTHER | 1 | 0.01 | 6898 | 99.96 |
| NO ANSWER | 3 | 0.04 | 6901 | 100.00 |

**Stata**

| Commands | `codebook` |
|---|---|

```
M484                                  WHICH GROUP DESCRIBES R (RACE)
-------------------------------------------------------------------

                 type:  numeric (byte)
                label:  M484

                range:  [1,99]                    units:  1
        unique values:  10                    missing .:  0/6901

          tabulation:  Freq.   Numeric  Label
                        1165      1      BLACK
                        5384      2      WHITE/NOT
                                         HISPANIC
                         218      3      MEX/CHICANO/MEX.AM
                          40      4      PUERTO RICAN
                           5      5      CUBAN
                          34      6      OTHER HISPANIC
                          31      7      AMERICAN INDIAN
                          20      8      ASIAN
                           1      9      OTHER
                           3     99      NO ANSWER
```

## ■ Display B.7.2 Creating and Checking *aframer* Dummy Variable

|  | SAS | Stata |
|---|---|---|
| Commands | ```
LIBNAME LIBRARY "c:\nsfh_distance\SAS";

data CheckAfrAmer;

  set "c:\nsfh_distance\SAS\CreateData";

  if M484=1 then aframer=1;
  if M484~=1 then aframer=0;
  if M484>=97 then aframer=.;

run;

proc freq;
  tables M484*aframer /NOCOL NOROW
    NOPERCENT;
run;
``` | ```
use CreateData

generate aframer=1 if M484==1
replace aframer=0 if M484~=1
replace aframer=. if M484>=97

tabulate M484 aframer, missing
``` |

**Results — SAS**

Table of M484 by aframer

| M484(WHICH GROUP DESCRIBES R (RACE)) | aframer | | |
|---|---|---|---|
| Frequency | 0 | 1 | Total |
| BLACK | 0 | 1165 | 1165 |
| WHITE/NOT HISPANIC | 5384 | 0 | 5384 |
| MEX/CHICANO/MEX.AM | 218 | 0 | 218 |
| PUERTO RICAN | 40 | 0 | 40 |
| CUBAN | 5 | 0 | 5 |
| OTHER HISPANIC | 34 | 0 | 34 |
| AMERICAN INDIAN | 31 | 0 | 31 |
| ASIAN | 20 | 0 | 20 |
| OTHER | 1 | 0 | 1 |
| NO ANSWER | 0 | 0 | 0 |
| Total | 5733 | 1165 | 6898 |

Frequency Missing = 3

**Results — Stata**

| WHICH GROUP DESCRIBES R (RACE) | aframer | | |
|---|---|---|---|
|  | 0 | 1 | . | Total |
| BLACK | 0 | 1,165 | 0 | 1,165 |
| WHITE/NOT HISPANIC | 5,384 | 0 | 0 | 5,384 |
| MEX/CHICANO/MEX.AM | 218 | 0 | 0 | 218 |
| PUERTO RICAN | 40 | 0 | 0 | 40 |
| CUBAN | 5 | 0 | 0 | 5 |
| OTHER HISPANIC | 34 | 0 | 0 | 34 |
| AMERICAN INDIAN | 31 | 0 | 0 | 31 |
| ASIAN | 20 | 0 | 0 | 20 |
| OTHER | 1 | 0 | 0 | 1 |
| NO ANSWER | 0 | 0 | 3 | 3 |
| Total | 5,733 | 1,165 | 3 | 6,901 |

■ **Display B.7.3 Creating and Checking** *other* **Dummy Variable**

|  | SAS | Stata |
|---|---|---|
| Commands | ```
LIBNAME LIBRARY "c:\nsfh_distance\SAS";

data CheckOther;

set "c:\nsfh_distance\SAS\CreateData";

if (M484>=4 & M484<=6) | M484=8 | M484=9
   then other=1;
if (M484>=1 & M484<=3) | M484=7 then
   other=0;
if M484>=97 then other=.;

run;

proc freq;
tables M484*other /NOCOL NOROW
   NOPERCENT;
run;
``` | ```
use CreateData

generate other=1 if (M484>=4 & M484<=6)  ///
   | M484==8 | M484==9

replace other=0 if (M484>=1 & M484<=3) | M484==7
replace other=. if M484>=97

tabulate M484 other, missing
``` |

**SAS Results**

Table of M484 by other

| M484(WHICH GROUP DESCRIBES R (RACE)) Frequency | other 0 | other 1 | Total |
|---|---|---|---|
| BLACK | 1165 | 0 | 1165 |
| WHITE/NOT HISPANIC | 5384 | 0 | 5384 |
| MEX/CHICANO/MEX.AM | 218 | 0 | 218 |
| PUERTO RICAN | 0 | 40 | 40 |
| CUBAN | 0 | 5 | 5 |
| OTHER HISPANIC | 0 | 34 | 34 |
| AMERICAN INDIAN | 31 | 0 | 31 |
| ASIAN | 0 | 20 | 20 |
| OTHER | 0 | 1 | 1 |
| NO ANSWER | 0 | 0 | 0 |
| Total | 6798 | 100 | 6898 |

Frequency Missing = 3

**Stata Results**

| WHICH GROUP DESCRIBES R (RACE) | other 0 | other 1 | other . | Total |
|---|---|---|---|---|
| BLACK | 1,165 | 0 | 0 | 1,165 |
| WHITE/NOT HISPANIC | 5,384 | 0 | 0 | 5,384 |
| MEX/CHICANO/MEX.AM | 218 | 0 | 0 | 218 |
| PUERTO RICAN | 0 | 40 | 0 | 40 |
| CUBAN | 0 | 5 | 0 | 5 |
| OTHER HISPANIC | 0 | 34 | 0 | 34 |
| AMERICAN INDIAN | 31 | 0 | 0 | 31 |
| ASIAN | 0 | 20 | 0 | 20 |
| OTHER | 0 | 1 | 0 | 1 |
| NO ANSWER | 0 | 0 | 3 | 3 |
| Total | 6,798 | 100 | 3 | 6,901 |

APPENDIX **B**

# ■ Display B.7.4 Regression of *glmiles* on Gender Dummies in the NSFH

**Regression of *glmiles* on *female* in the NSFH.**

| | SAS | Stata |
|---|---|---|
| Variable Creation | `female=(M2DP01=2);` | `generate female=(M2DP01==2)` |
| Regression Model | `proc reg;`<br>`  model glmiles=female;`<br>`run;` | `regress glmiles female` |

Select Results — SAS:

| Parameter Estimates | | | | | |
|---|---|---|---|---|---|
| Variable | DF | Parameter Estimate | Standard Error | t Value | Pr > \|t\| |
| Intercept | 1 | 323.75559 | 13.74255 | 23.56 | <.0001 |
| female | 1 | -59.38793 | 17.70771 | -3.35 | 0.0008 |

Select Results — Stata:

```
glmiles |      Coef.   Std. Err.       t    P>|t|
 female | -59.38793   17.70771    -3.35   0.001
   cons | 323.7556    13.74255    23.56   0.000
```

**Regression of *glmiles* on *female* in the NSFH.**

| | SAS | Stata |
|---|---|---|
| Variable Creation | `male=(M2DP01=1);` | `generate male=(M2DP01==1)` |
| Regression Model | `proc reg;`<br>`  model glmiles=male;`<br>`run;` | `regress glmiles male` |

Select Results — SAS:

| Parameter Estimates | | | | | |
|---|---|---|---|---|---|
| Variable | DF | Parameter Estimate | Standard Error | t Value | Pr > \|t\| |
| Intercept | 1 | 264.36766 | 11.16715 | 23.67 | <.0001 |
| male | 1 | 59.38793 | 17.70771 | 3.35 | 0.0008 |

Select Results — Stata:

```
glmiles |      Coef.   Std. Err.       t    P>|t|
 female | 59.38793    17.70771     3.35   0.001
   cons | 264.3677    11.16715    23.67   0.000
```

Comments: *Don't forget: In SAS, variable creation must occur in the data step!*

**APPENDIX B**

■ **Display B.7.5** *t*-test of `g1miles` by *M2DP01* in the **NSFH**

| | SAS | Stata |
|---|---|---|
| t-test | `proc ttest;`<br>`  class M2DP01;`<br>`  var g1miles;`<br>`run;` | `ttest g1miles, by(M2DP01)` |

**SAS — Statistics**

| Variable | M2DP01 | N | Lower CL Mean | Mean | Upper CL Mean | Std Err |
|---|---|---|---|---|---|---|
| g1miles | MALE | 2148 | 294.35 | 323.76 | 353.16 | 14.996 |
| g1miles | FEMALE | 3253 | 243.9 | 264.37 | 284.84 | 10.44 |
| g1miles | Diff (1–2) | | 24.674 | 59.388 | 94.102 | 17.708 |

**SAS — T-Tests**

| Variable | Method | Variances | DF | t Value | Pr > |t| |
|---|---|---|---|---|---|
| g1miles | Pooled | Equal | 5399 | 3.35 | 0.0008 |
| g1miles | Satterthwaite | Unequal | 4097 | 3.25 | 0.0012 |

**Stata**

```
Two-sample t-test with equal variances
Group    |    Obs      Mean     Std. Err.        [95% Conf. Interval]
---------+--------------------------------------------------------------
MALE     |   2148   323.7556   14.99641          294.3466   353.1646
FEMALE   |   3253   264.3677   10.44012          243.8978   284.8375
---------+--------------------------------------------------------------
combined |   5401   287.9865    8.674792         270.9804   304.9926
---------+--------------------------------------------------------------
diff     |          59.38793   17.70771          24.67367   94.10218

diff = mean(MALE) - mean(FEMALE)                        t =   3.3538
Ho: diff = 0                          degrees of freedom =   5399

Ha: diff < 0              Ha: diff != 0              Ha: diff > 0
Pr(T < t) = 0.9996    Pr(|T| > |t|) = 0.0008    Pr(T > t) = 0.0004
```

Select Results

■ Display B.7.6 Regression of *glmiles* on *amind*, *mexamer*, **and** *white* in the NSFH

|  | SAS | Stata |
|---|---|---|
| **Variable Creation** | `LIBNAME LIBRARY 'c:\nsfh_distance\SAS';`<br>`data CreateData;`<br>`  set "c:\nsfh_distance\SAS\CreateData";`<br>`  if glmiles~=. & g2earn~=. & glyrschl~=. &`<br>`  & glage~=. & g2age~=. & g2numbro~=. &`<br>`  g2numsis~=.`<br>`  & ((M484>=1 & M484<=3) | M484==7) ;`<br><br>`  if M484<97 then aframer=(M484=1) ;`<br>`  if M484<97 then white=  (M484=2) ;`<br>`  if M484<97 then mexamer= (M484=3) ;`<br>`  if M484<97 then amind=  (M484=7) ;`<br><br>`run;` | `use CreateData if glmiles~=. & g2earn~=. & glyrschl~=.   ///`<br>`            & glage~=. & g2age~=. & g2numbro~=.   ///`<br>`            & g2numsis~=. & ((M484>=1 & M484<=3) | M484==7)`<br><br>`generate aframer= (M484==1) if M484<97`<br>`generate white=   (M484==2) if M484<97`<br>`generate mexamer= (M484==3) if M484<97`<br>`generate amind=   (M484==7) if M484<97` |
| **Regression Model** | `proc reg;`<br>`  model glmiles=amind mexamer white;`<br>`run;` | `regress glmiles amind mexamer white` |
| **Select Results** | (see tables below) | (see output below) |

**Stata output:**

```
      Source |       SS       df       MS              Number of obs =    5401
-------------+------------------------------          F(  3,  5397) =    2.42
       Model |  2946571.39      3  982190.464          Prob > F      =  0.0643
    Residual |  2.1918e+09   5397  406116.082          R-squared     =  0.0013
-------------+------------------------------          Adj R-squared =  0.0008
       Total |  2.1948e+09   5400  406436.123          Root MSE      =  637.27

     glmiles |      Coef.   Std. Err.      t    P>|t|     [95% Conf. Interval]
-------------+----------------------------------------------------------------
       amind |   157.0283   129.4861     1.21   0.225    -96.81669    410.8732
     mexamer |  -8.254855   54.48378    -0.15   0.880    -115.0651    98.55535
       white |   56.71736   24.77161     2.29   0.022     8.154998    105.2797
       _cons |   240.9717   22.84732    10.55   0.000     196.1818    285.7617
```

**SAS output:**

| Number of Observations Read | 5401 |
|---|---|
| Number of Observations Used | 5401 |

Analysis of Variance

| Source | DF | Sum of Squares | Mean Square | F Value | Pr > F |
|---|---|---|---|---|---|
| Model | 3 | 2946571 | 982190 | 2.42 | 0.0643 |
| Error | 5397 | 2191808493 | 406116 |  |  |
| Corrected Total | 5400 | 2194755064 |  |  |  |

| Root MSE | 637.27238 | R-Square | 0.0013 |
|---|---|---|---|
| Dependent Mean | 287.98648 | Adj R-Sq | 0.0008 |
| Coeff Var | 221.28552 |  |  |

Parameter Estimates

| Variable | DF | Parameter Estimate | Standard Error | t Value | Pr > |t| |
|---|---|---|---|---|---|
| Intercept | 1 | 240.97172 | 22.84732 | 10.55 | <.0001 |
| amind | 1 | 157.02828 | 129.48607 | 1.21 | 0.2253 |
| mexamer | 1 | -8.25485 | 54.48378 | -0.15 | 0.8796 |
| white | 1 | 56.71736 | 24.77161 | 2.29 | 0.0221 |

**■ Display B.7.7 Regression of** glmiles **on Race-Ethnicity Dummies in the NSFH, with Different Reference Categories**

## American Indian is Reference Category

|  | SAS | Stata |
|---|---|---|
| Commands | `model glmiles=aframer mexamer white;` | `regress glmiles aframer mexamer white` |

SAS — Select Results — Parameter Estimates:

| Variable | DF | Parameter Estimate | Standard Error | t Value | Pr > \|t\| |
|---|---|---|---|---|---|
| Intercept | 1 | 398.00000 | 127.45448 | 3.12 | 0.0018 |
| aframer | 1 | -157.02828 | 129.48607 | -1.21 | 0.2253 |
| mexamer | 1 | -165.28313 | 136.71549 | -1.21 | 0.2267 |
| white | 1 | -100.31092 | 127.81344 | -0.78 | 0.4326 |

Stata — Select Results:

| glmiles | Coef. | Std. Err. | t | P>\|t\| |
|---|---|---|---|---|
| aframer | -157.0283 | 129.4861 | -1.21 | 0.225 |
| mexamer | -165.2831 | 136.7155 | -1.21 | 0.227 |
| white | -100.3109 | 127.8134 | -0.78 | 0.433 |
| _cons | 398 | 127.4545 | 3.12 | 0.002 |

## Mexican American is Reference Category

|  | SAS | Stata |
|---|---|---|
| Commands | `model glmiles=amind aframer white;` | `regress glmiles amind aframer white` |

SAS — Select Results — Parameter Estimates:

| Variable | DF | Parameter Estimate | Standard Error | t Value | Pr > \|t\| |
|---|---|---|---|---|---|
| Intercept | 1 | 232.71687 | 49.46193 | 4.70 | <.0001 |
| amind | 1 | 165.28313 | 136.71549 | 1.21 | 0.2267 |
| aframer | 1 | 8.25485 | 54.48378 | 0.15 | 0.8796 |
| white | 1 | 64.97221 | 50.37971 | 1.29 | 0.1972 |

Stata — Select Results:

| glmiles | Coef. | Std. Err. | t | P>\|t\| |
|---|---|---|---|---|
| amind | 165.2831 | 136.7155 | 1.21 | 0.227 |
| aframer | 8.254855 | 54.48378 | 0.15 | 0.880 |
| white | 64.97221 | 50.37971 | 1.29 | 0.197 |
| _cons | 232.7169 | 49.46193 | 4.70 | 0.000 |

## White is Reference Category

|  | SAS | Stata |
|---|---|---|
| Commands | `model glmiles=amind mexamer aframer;` | `regress glmiles amind mexamer aframer` |

SAS — Select Results — Parameter Estimates:

| Variable | DF | Parameter Estimate | Standard Error | t Value | Pr > \|t\| |
|---|---|---|---|---|---|
| Intercept | 1 | 297.68908 | 9.57250 | 31.10 | <.0001 |
| amind | 1 | 100.31092 | 127.81344 | 0.78 | 0.4326 |
| mexamer | 1 | -64.97221 | 50.37971 | -1.29 | 0.1972 |
| aframer | 1 | -56.71736 | 24.77161 | -2.29 | 0.0221 |

Stata — Select Results:

| glmiles | Coef. | Std. Err. | t | P>\|t\| |
|---|---|---|---|---|
| amind | 100.3109 | 127.8134 | 0.78 | 0.433 |
| mexamer | -64.97221 | 50.37971 | -1.29 | 0.197 |
| aframer | -56.71736 | 24.77161 | -2.29 | 0.022 |
| _cons | 297.6891 | 9.572497 | 31.10 | 0.000 |

■ **Display B.7.8 Regression of** *glmiles* **on** *amindmex* **and** *white* **in the NSFH**

| | SAS | Stata |
|---|---|---|
| Variable Creation | `if M484<97 then amindmex= (M484=7 | M484=3);` | `generate amindmex= (M484==7 | M484==3) if M484<97` |
| Regression Model | `proc reg;`<br>`  model glmiles=amindmex white;`<br>`run;` | `regress glmiles amindmex white` |

**SAS — Select Results**

| Number of Observations Read | 5401 |
|---|---|
| Number of Observations Used | 5401 |

Analysis of Variance

| Source | DF | Sum of Squares | Mean Square | F Value | Pr > F |
|---|---|---|---|---|---|
| Model | 2 | 2353002 | 1176501 | 2.90 | 0.0553 |
| Error | 5398 | 2192402062 | 406151 | | |
| Corrected Total | 5400 | 2194755064 | | | |

| Root MSE | 637.29962 | R-Square | 0.0011 |
|---|---|---|---|
| Dependent Mean | 287.98648 | Adj R-Sq | 0.0007 |
| Coeff Var | 221.29498 | | |

Parameter Estimates

| Variable | DF | Parameter Estimate | Standard Error | t Value | Pr > |t| |
|---|---|---|---|---|---|
| Intercept | 1 | 240.97172 | 22.84830 | 10.55 | <0001 |
| amindmex | 1 | 13.37906 | 51.46347 | 0.26 | 0.7949 |
| white | 1 | 56.71736 | 24.77267 | 2.29 | 0.0221 |

**Stata — Select Results**

| | | | | Number of obs | = | 5401 |
|---|---|---|---|---|---|---|
| | | | | F( 2, 5398) | = | 2.90 |
| | | | | Prob > F | = | 0.0553 |
| | | | | R-squared | = | 0.0011 |
| | | | | Adj R-squared | = | 0.0007 |
| | | | | Root MSE | = | 637.3 |

| Source | SS | df | MS |
|---|---|---|---|
| Model | 2353001.59 | 2 | 1176500.79 |
| Residual | 2.1924e+09 | 5398 | 406150.808 |
| Total | 2.1948e+09 | 5400 | 406436.123 |

| glmiles | Coef. | Std. Err. | t | P>|t| | [95% Conf. Interval] |
|---|---|---|---|---|---|
| amindmex | 13.37906 | 51.46347 | 0.26 | 0.795 | -87.51011 114.2682 |
| white | 56.71736 | 24.77267 | 2.29 | 0.022 | 8.152923 105.2818 |
| _cons | 240.9717 | 22.8483 | 10.55 | 0.000 | 196.1798 285.7636 |

## ■ Display B.7.9 Using the Test Command to Test the Difference in Means among the Included Categories

| | SAS | Stata |
|---|---|---|
| Regression Model and Test Commands | `proc reg;`<br>`  model g1miles=amind mexamer white;`<br>`  test amind=mexamer;`<br>`  test amind=white;`<br>`  test mexamer=white;`<br>`run;` | `regress g1miles amind mexamer white`<br>`test amind=mexamer`<br>`test amind=white`<br>`test mexamer=white` |

Select Results:

Test 1 Results for Dependent Variable g1miles

| Source | DF | Mean Square | F Value | Pr > F |
|---|---|---|---|---|
| Numerator | 1 | 593570 | 1.46 | 0.2267 |
| Denominator | 5397 | 406116 | | |

Test 2 Results for Dependent Variable g1miles

| Source | DF | Mean Square | F Value | Pr > F |
|---|---|---|---|---|
| Numerator | 1 | 250146 | 0.62 | 0.4326 |
| Denominator | 5397 | 406116 | | |

Test 3 Results for Dependent Variable g1miles

| Source | DF | Mean Square | F Value | Pr > F |
|---|---|---|---|---|
| Numerator | 1 | 675452 | 1.66 | 0.1972 |
| Denominator | 5397 | 406116 | | |

Stata results:

```
. test amind=mexamer

 (1)  amind - mexamer = 0

       F(  1,  5397) =    1.46
            Prob > F =  0.2267

. test amind=white

 (1)  amind - white = 0

       F(  1,  5397) =    0.62
            Prob > F =  0.4326

. test mexamer=white

 (1)  mexamer - white = 0

       F(  1,  5397) =    1.66
            Prob > F =  0.1972
```

■ **Display B.7.10 The Variance–Covariance Matrix of the Estimated Regression Coefficients**

| | SAS | Stata |
|---|---|---|
| New Commands | ```
proc reg;
  model g1miles=amind mexamer white /covb;
run;
``` | ```
regress g1miles amind mexamer white
estat vce
``` |
| Select Results | **Covariance of Estimates**<br><br>| Variable | Intercept | amind | mexamer | white |<br>|---|---|---|---|---|<br>| Intercept | 522.0001049 | -522.0001049 | -522.0001049 | -522.0001049 |<br>| amind | -522.0001049 | 16766.643371 | 522.00010494 | 522.00010494 |<br>| mexamer | -522.0001049 | 522.00010494 | 2968.4825245 | 522.00010494 |<br>| white | -522.0001049 | 522.00010494 | 522.00010494 | 613.63279484 | | ```
Covariance matrix of coefficients of regress model

    e(V) |      amind      mexamer        white         cons
---------+------------------------------------------------------
   amind | 16766.643
 mexamer |   522.0001     2968.4825
   white |   522.0001      522.0001    613.63279
   _cons |  -522.0001     -522.0001    -522.0001     522.0001
``` |
| Comments | In SAS, we request the variance–covariance matrix of the estimated regression coefficients with an **option** following the model statement. The option is **covb** which can be remembered as a shorthand for "**covariance**" of the "**betas**." As with all options, the word **covb** needs to be separated with a **forward slash** "/" from the last predictor variable listed in the model statement. | In Stata, we request the variance–covariance matrix of the estimated regression coefficients with a **command** following the regression command. The command is **estat** (a general command for "postestimation statistics"). The subcommand **vce** asks Stata to list the variance covariance **estimates**. |
```

■ **Display B.7.11 The Stata** `lincom` **Command**

| | Stata |
|---|---|
| New Commands | `regress glmiles amind mexamer white`<br>`lincom mexamer-amind`<br>`lincom white-amind`<br>`lincom white-mexamer` |
| Select Results | `. lincom mexamer-amind`<br><br>`(1) - amind + mexamer = 0`<br><br>`  glmiles |      Coef.   Std. Err.      t    P>|t|     [95% Conf. Interval]`<br>`---------+-------------------------------------------------------------`<br>`     (1) | -165.2831   136.7155   -1.21   0.227    -433.3007   102.7344`<br><br>`. lincom white-amind`<br><br>`(1) - amind + white = 0`<br><br>`  glmiles |      Coef.   Std. Err.      t    P>|t|     [95% Conf. Interval]`<br>`---------+-------------------------------------------------------------`<br>`     (1) | -100.3109   127.8134   -0.78   0.433    -350.8769   150.255`<br><br>`. lincom white-mexamer`<br><br>`(1) - mexamer + white = 0`<br><br>`  glmiles |      Coef.   Std. Err.      t    P>|t|     [95% Conf. Interval]`<br>`---------+-------------------------------------------------------------`<br>`     (1) |  64.97221   50.37971    1.29   0.197    -33.79235   163.7368` |
| Comments | In Stata, we use the **lincom** command to calculate and test the significance of linear combinations of coefficients. As with the **test** command, we use the variable name to indicate to Stata which coefficient is of interest. We will generally use simple linear combinations that take the sum or difference of two coefficients, although more complicated linear combinations are possible (see **help lincom** in Stata). |

■ **Display B.7.12 Regression of** *g1miles* **on** *amind, mexamer, white,* **and** *female* **in the NSFH**

| | SAS | Stata |
|---|---|---|
| Regression Model and Test Commands | `proc reg;`<br>`  model g1miles=amind mexamer white female`<br>`                g2earn g2age g2numbro`<br>`                g2numsis g1yrschl g1age;`<br>`  test amind=mexamer;`<br>`  test amind=white;`<br>`  test mexamer=white;`<br>`run;` | `regress g1miles amind mexamer white female ///`<br>`        g2earn g2age g2numbro g2numsis ///`<br>`        g1yrschl g1age`<br><br>`test amind=mexamer`<br><br>`test amind=white`<br><br>`test mexamer=white` |

**SAS — Parameter Estimates**

| Variable | DF | Parameter Estimate | Standard Error | t Value | Pr>|t| |
|---|---|---|---|---|---|
| Intercept | 1 | -293.50824 | 71.86251 | -4.08 | <.0001 |
| amind | 1 | 185.14082 | 128.53760 | 1.44 | 0.1498 |
| mexamer | 1 | 61.84987 | 55.31577 | 1.12 | 0.2636 |
| white | 1 | 35.06974 | 25.22134 | 1.39 | 0.1644 |
| female | 1 | -33.79229 | 18.79857 | -1.80 | 0.0723 |
| g2earn | 1 | 0.00066760 | 0.00025048 | 2.67 | 0.0077 |
| g2age | 1 | -0.03834 | 1.65004 | -0.02 | 0.9815 |
| g2numbro | 1 | 5.20427 | 5.96187 | 0.87 | 0.3827 |
| g2numsis | 1 | 18.56337 | 6.03220 | 3.08 | 0.0021 |
| g1yrschl | 1 | 21.11644 | 3.27596 | 6.45 | <.0001 |
| g1age | 1 | 4.60571 | 1.44147 | 3.20 | 0.0014 |

**Stata output**

Number of obs = 5401
F( 10, 5390) = 10.38
Prob > F = 0.0000
R-squared = 0.0189
Adj R-squared = 0.0171
Root MSE = 632.06

| Source | SS | df | MS |
|---|---|---|---|
| Model | 41483650.7 | 10 | 4148365.07 |
| Residual | 2.1533e+09 | 5390 | 399493.769 |
| Total | 2.1948e+09 | 5400 | 406436.123 |

| g1miles | Coef. | Std. Err. | t | P>|t| | [95% Conf. Interval] | |
|---|---|---|---|---|---|---|
| amind | 185.1408 | 128.5376 | 1.44 | 0.150 | -66.84484 | 437.1265 |
| mexamer | 61.84987 | 55.31577 | 1.12 | 0.264 | -46.59139 | 170.2911 |
| white | 35.06974 | 25.22134 | 1.39 | 0.164 | -14.37427 | 84.51376 |
| female | -33.7923 | 18.79857 | -1.80 | 0.072 | -70.64509 | 3.060504 |
| g2earn | .0006676 | .0002505 | 2.67 | 0.008 | .0001766 | .0011586 |
| g2age | -.038338 | 1.650044 | -0.02 | 0.981 | -3.273091 | 3.196415 |
| g2numbro | 5.204269 | 5.961866 | 0.87 | 0.383 | -6.483399 | 16.89194 |
| g2numsis | 18.56337 | 6.032196 | 3.08 | 0.002 | 6.737823 | 30.38891 |
| g1yrschl | 21.11644 | 3.275965 | 6.45 | 0.000 | 14.69423 | 27.53866 |
| g1age | 4.605706 | 1.441466 | 3.20 | 0.001 | 1.77985 | 7.431563 |
| _cons | -293.5082 | 71.86251 | -4.08 | 0.000 | -434.3878 | -152.6287 |

(1) amind - mexamer = 0

F( 1, 5390) = 0.82
Prob > F = 0.3657

(1) amind - white = 0

F( 1, 5390) = 1.40
Prob > F = 0.2373

(1) mexamer - white = 0

F( 1, 5390) = 0.26
Prob > F = 0.6092

**SAS — Select Results**

Test 1 Results for Dependent Variable g1miles

| Source | DF | Mean Square | F Value | Pr > F |
|---|---|---|---|---|
| Numerator | 1 | 326914 | 0.82 | 0.3657 |
| Denominator | 5390 | 399494 | | |

Test 2 Results for Dependent Variable g1miles

| Source | DF | Mean Square | F Value | Pr > F |
|---|---|---|---|---|
| Numerator | 1 | 558148 | 1.40 | 0.2373 |
| Denominator | 5390 | 399494 | | |

Test 3 Results for Dependent Variable g1miles

| Source | DF | Mean Square | F Value | Pr > F |
|---|---|---|---|---|
| Numerator | 1 | 104437 | 0.26 | 0.6092 |
| Denominator | 5390 | 399494 | | |

**APPENDIX B**

## ■ Display B.7.13 Means for Interval Predictors

| | SAS | Stata |
|---|---|---|
| Commands | `proc means;`<br>`  var g2earn g2age g2numbro g2numsis`<br>`      g1yrschl g1age;`<br>`run;` | `summarize g2earn g2age g2numbro g2numsis ///`<br>`          g1yrschl g1age` |

Select Results

SAS:

| Variable | N | Mean | Std Dev | Minimum | Maximum |
|---|---|---|---|---|---|
| g2earn | 5401 | 30752.28 | 37626.16 | 0 | 945903.28 |
| g2age | 5401 | 34.3527125 | 10.0265673 | 16.0000000 | 76.0000000 |
| g2numbro | 5401 | 1.4328828 | 1.5560708 | 0 | 19.0000000 |
| g2numsis | 5401 | 1.3817812 | 1.5511505 | 0 | 22.0000000 |
| g1yrschl | 5401 | 11.4141826 | 2.9210752 | 0 | 17.0000000 |
| g1age | 5401 | 60.1349750 | 11.3189828 | 29.0000000 | 95.0000000 |

Stata:

| Variable | Obs | Mean | Std. Dev. | Min | Max |
|---|---|---|---|---|---|
| g2earn | 5401 | 30752.28 | 37626.16 | 0 | 945903.3 |
| g2age | 5401 | 34.35271 | 10.02657 | 16 | 76 |
| g2numbro | 5401 | 1.432883 | 1.566071 | 0 | 19 |
| g2numsis | 5401 | 1.381781 | 1.551151 | 0 | 22 |
| g1yrschl | 5401 | 11.41418 | 2.921075 | 0 | 17 |
| g1age | 5401 | 60.13498 | 11.31898 | 29 | 95 |

■ **Display B.8.1 Regression of** *hrchores* **on** *married*, *female*, **and** *fem_marr* **in the NSFH**

| | SAS | Stata |
|---|---|---|
| Variable Creation | `fem_marr=female*married;` | `generate fem_marr=female*married` |
| Regression Model | `proc reg;`<br>`  model hrchores=married female fem_marr;`<br>`run;` | `regress hrchores married female fem_marr` |

**SAS — Select Results**

| Number of Observations Read | 6054 |
|---|---|
| Number of Observations Used | 6054 |

**Analysis of Variance**

| Source | DF | Sum of Squares | Mean Square | F Value | Pr > F |
|---|---|---|---|---|---|
| Model | 3 | 297807 | 99269 | 169.97 | <.0001 |
| Error | 6050 | 3553423 | 584.03690 | | |
| Corrected Total | 6053 | 3831230 | | | |

| Root MSE | 24.16686 | R-Square | 0.0777 |
|---|---|---|---|
| Dependent Mean | 27.72382 | Adj R-Sq | 0.0773 |
| Coeff Var | 87.17001 | | |

**Parameter Estimates**

| Variable | DF | Parameter Estimate | Standard Error | t Value | Pr > |t| |
|---|---|---|---|---|---|
| Intercept | 1 | 22.31421 | 0.69677 | 32.03 | <.0001 |
| married | 1 | -2.60489 | 0.90427 | -2.88 | 0.0040 |
| female | 1 | 10.19964 | 0.91505 | 11.15 | <.0001 |
| fem_marr | 1 | 6.63953 | 1.25582 | 5.29 | <.0001 |

**Stata — Select Results**

| Source | SS | df | MS |
|---|---|---|---|
| Model | 297806.971 | 3 | 99268.9903 |
| Residual | 3553423.25 | 6050 | 584.036902 |
| Total | 3831230.23 | 6053 | 632.947336 |

```
Number of obs =     6054
F( 3,  6050)  =   169.97
Prob > F      =   0.0000
R-squared     =   0.0777
Adj R-squared =   0.0773
Root MSE      =   24.167
```

| hrchores | Coef. | Std. Err. | t | P>|t| | [95% Conf. Interval] |
|---|---|---|---|---|---|
| married | -2.604886 | .9042679 | -2.88 | 0.004 | -4.377573 | -.8321986 |
| female | 10.19964 | .9150482 | 11.15 | 0.000 | 8.405821 | 11.99346 |
| fem_marr | 6.63953 | 1.255823 | 5.29 | 0.000 | 4.177669 | 9.101391 |
| _cons | 22.31421 | .6967666 | 32.03 | 0.000 | 20.9483 | 23.68013 |

■ **Display B.8.2 Regression of** *hrchores* **on** *married*, *female*, **and** *fem_marr* **in the NSFH: Reversing Reference Category on Conditioning Variable**

| | SAS | Stata |
|---|---|---|
| | | **Conditional Effect of Being Married for Women** |
| Variable | `if female~=. then male=female=0;`<br>`male_marr=male*married;` | `generate male=female==0 if female~=.`<br>`generate male_marr=male*married` |
| Regression | `proc reg;`<br>`  model hrchores=married male male_marr;`<br>`run;` | `regress hrchores married male male_marr` |

Stata Select Results:

```
      Source |       SS           df       MS              Number of obs =    6054
-------------+----------------------------------            F(3, 6050)    =  169.97
       Model | 297806.971          3   399268.9903         Prob > F      =  0.0000
    Residual | 353423.25        6050   584.036902          R-squared     =  0.0777
-------------+----------------------------------            Adj R-squared =  0.0773
       Total | 383230.23        6053   632.947336          Root MSE      =  24.167

    hrchores |      Coef.   Std. Err.      T    P>|t|     [95% Conf. Interval]
-------------+----------------------------------------------------------------
     married |   4.034644   .8714309     4.63   0.000     2.326329    5.742959
        male |  -10.19964    .915482   -11.15   0.000    -11.99346   -8.405821
   male_marr |   -6.63953   1.255823    -5.29   0.000    -9.101391   -4.177669
       _cons |   32.51386   .5931521    54.82   0.000     31.35107    33.67664
```

SAS Select Results — Parameter Estimates:

| Variable | DF | Parameter Estimate | Standard Error | t Value | Pr > |t| |
|---|---|---|---|---|---|
| Intercept | 1 | 32.51386 | 0.59315 | 54.82 | <.0001 |
| married | 1 | 4.03464 | 0.87143 | 4.63 | <.0001 |
| male | 1 | -10.19964 | 0.91505 | -11.15 | <.0001 |
| male_marr | 1 | -6.63953 | 1.25582 | -5.29 | <.0001 |

(Continued on next page)

■ **Display B.8.2 Regression of** hrchores **on** married, female, **and** fem_marr **in the NSFH: Reversing Reference Category on Conditioning Variable (Continued)**

| | SAS | Stata |
|---|---|---|
| | **Conditional Effect of Being Female for Married Adults** | |
| Variable | ```
if married~=. then unmarried=married=0;
fem_unmarr=female*unmarried;
``` | ```
generate unmarried=married=0 if married~=.
generate fem_unmarr=female*unmarried
``` |
| Regression | ```
proc reg;
  model hrchores=unmarried female fem_unmarr;
run;
``` | ```
regress hrchores unmarried female fem_unmarr
``` |
| Select Results | **Parameter Estimates** | ```
Source |     SS       df     MS          Number of obs =    6054
                                         F( 3,  6050)  =  169.97
  Model | 297806.971    3  9926.9903    Prob > F      =  0.0000
Residual| 3533423.25 6050 584.036902    R-squared     =  0.0777
                                        Adj R-squared =  0.0773
  Total | 3831230.23 6053 632.947336    Root MSE      =  24.167

 hrchores |   Coef.  Std. Err.    T   P>|t| [95% Conf. Interval]
unmarried | 2.604886 .9042679  2.88  0.004  .8321986  4.377573
   female | 16.83917 .8601041 19.58  0.000  15.15306 18.52528
fem_unmarr| -6.63953 1.255823 -5.29  0.000 -9.101391 -4.177669
    _cons | 19.70933 .5763824 34.19  0.000  18.57941 20.83924
``` |

| Variable | DF | Parameter Estimate | Standard Error | t Value | Pr > \|t\| |
|---|---|---|---|---|---|
| Intercept | 1 | 19.70933 | 0.57638 | 34.19 | <.0001 |
| unmarried | 1 | 2.60489 | 0.90427 | 2.88 | 0.0040 |
| female | 1 | 16.83917 | 0.86010 | 19.58 | <.0001 |
| fem_unmarr | 1 | -6.63953 | 1.25582 | -5.29 | <.0001 |
```

**■ Display B.8.3 Regression of *hrchores* on *married, female,* and *fem_marr* in the NSFH: General Linear F-test**

|  | SAS | Stata |
|---|---|---|
| Regression Model | ```proc reg;\n  model hrchores=married female fem_marr;\n  test married+fem_marr=0;\n  test female+fem_marr=0;\nrun;``` | ```regress hrchores married female fem_marr\ntest married+fem_marr=0\ntest female+fem_marr=0``` |

SAS Select Results:

**Parameter Estimates**

| Variable | DF | Parameter Estimate | Standard Error | t Value | Pr > |t| |
|---|---|---|---|---|---|
| Intercept | 1 | 22.31421 | 0.69677 | 32.03 | <.0001 |
| married | 1 | -2.60489 | 0.90427 | -2.88 | 0.0040 |
| female | 1 | 10.19964 | 0.91505 | 11.15 | <.0001 |
| fem_marr | 1 | 6.63953 | 1.25582 | 5.29 | <.0001 |

**Test 1 Results for Dependent Variable hrchores**

| Source | DF | Mean Square | F Value | Pr > F |
|---|---|---|---|---|
| Numerator | 1 | 12519 | 21.44 | <.0001 |
| Denominator | 6050 | 584.03690 | | |

**Test 2 Results for Dependent Variable hrchores**

| Source | DF | Mean Square | F Value | Pr > F |
|---|---|---|---|---|
| Numerator | 1 | 223862 | 383.30 | <.0001 |
| Denominator | 6050 | 584.03690 | | |

Stata Select Results:

```
     Source |       SS       df       MS              Number of obs =    6054
                                                      F( 3,  6050)  =  169.97
      Model | 297806.971      3  99268.9903           Prob > F      =  0.0000
   Residual | 3533423.25   6050  584.036902           R-squared     =  0.0777
                                                      Adj R-squared =  0.0773
      Total | 3831230.23   6053  632.947336           Root MSE      =  24.167

   hrchores |    Coef.   Std. Err.     T    P>|t|    [95% Conf. Interval]
    married | -2.604886  .9042679   -2.88  0.004   -4.377573  -.8321986
     female |  10.19964  .9150482   11.15  0.000    8.405821   11.99346
   fem_marr |   6.63953  1.255823    5.29  0.000    4.177669   9.101391
      _cons |  22.31421  .6967666   32.03  0.000    20.9483    23.68013

 (1) married + fem_marr = 0
      F(1,  6050) =   21.44
         Prob > F =    0.0000

 (1) female + fem_marr = 0
      F(1,  6050) =  383.30
         Prob > F =    0.0000
```

# ■ Display B.8.4 Regression of *hrchores* on *married*, *female*, **and** *fem_marr* **in the NSFH: Linear Combination of Coefficients**

## SAS

**Regression Model**

```
proc reg;
  model hrchores=married female fem_marr / covb;
run;
```

**Select Results**

| Number of Observations Read | 6054 |
|---|---|
| Number of Observations Used | 6054 |

### Analysis of Variance

| Source | DF | Sum of Squares | Mean Square | F Value | Pr > F |
|---|---|---|---|---|---|
| Model | 3 | 297807 | 99269 | 169.97 | <.0001 |
| Error | 6050 | 3553423 | 584.03690 | | |
| Corrected Total | 6053 | 3831230 | | | |

| Root MSE | 24.16686 | R-Square | 0.0777 |
|---|---|---|---|
| Dependent Mean | 27.72382 | Adj R-Sq | 0.0773 |
| Coeff Var | 87.17001 | | |

### Parameter Estimates

| Variable | DF | Parameter Estimate | Standard Error | t Value | Pr > |t| |
|---|---|---|---|---|---|
| Intercept | 1 | 22.31421 | 0.69677 | 32.03 | <.0001 |
| married | 1 | -2.60489 | 0.90427 | -2.88 | 0.0040 |
| female | 1 | 10.19964 | 0.91505 | 11.15 | <.0001 |
| fem_marr | 1 | 6.63953 | 1.25582 | 5.29 | <.0001 |

### Covariance of Estimates

| Variable | Intercept | married | female | fem_marr |
|---|---|---|---|---|
| Intercept | 0.4854837087 | -0.485483709 | -0.485483709 | 0.4854837087 |
| married | -0.485483709 | 0.817700376 | 0.4854837087 | -0.817700376 |
| female | -0.485483709 | 0.4854837087 | 0.8373131675 | -0.837313167 |
| fem_marr | 0.4854837087 | -0.817700376 | -0.837313167 | 1.5770922228 |

## Stata

**Regression Model**

```
regress hrchores married female fem_marr
estat vce
```

**Select Results**

```
      Source |       SS       df       MS              Number of obs =    6054
-------------+------------------------------           F(  3,  6050) =  169.97
       Model | 297806.971      3   99268.9903          Prob > F      =  0.0000
    Residual | 3533423.25   6050  584.036902           R-squared     =  0.0777
-------------+------------------------------           Adj R-squared =  0.0773
       Total | 3831230.23   6053  632.947336           Root MSE      =  24.167

    hrchores |    Coef.    Std. Err.      t    P>|t|    [95% Conf. Interval]
-------------+----------------------------------------------------------------
     married | -2.604886   .9042679    -2.88   0.004   -4.377573   -.8321986
      female |  10.19964   .9150482    11.15   0.000    8.405821    11.99346
    fem_marr |   6.63953   1.255823     5.29   0.000    4.177669    9.101391
       _cons |  22.31421   .6967666    32.03   0.000    20.9483     23.68013

       estat vce

Covariance matrix of coefficients of regress model

        e(V) |    married     female    fem_marr       _cons
-------------+------------------------------------------------
     married |  .81770038
      female |  .48548371   .83731317
    fem_marr | -.81770038  -.83731317   1.5770922
       _cons | -.48548371  -.48548371   .48548371   .48548371
```

■ **Display B.8.5 Regression of** *hrchores* **on** *married, female,* **and** *fem_marr* **in the NSFH: Stata** *lincom* **Command**

| | SAS | Stata |
|---|---|---|
| Regression Model | n/a | regress hrchores married female fem_marr<br>  lincom married+fem_marr<br>  lincom female+fem_marr<br><br>      Source \|      SS       df      MS                Number of obs =    6054<br>    ----------+------------------------------        F( 3,  6050) =   169.97<br>       Model \| 297806.971       3  99268.9903        Prob > F     =   0.0000<br>    Residual \| 3533423.25    6050  584.036902        R-squared    =   0.0777<br>    ----------+------------------------------        Adj R-squared =  0.0773<br>       Total \| 3831230.23    6053  632.947336        Root MSE     =   24.167<br><br>    hrchores \|   Coef.   Std. Err.      T    P>\|t\|  [95% Conf. Interval]<br>    ---------+--------------------------------------------------------------<br>     married \| -2.604886  .9042679   -2.88   0.004   -4.377573   -.8321986<br>      female \|  10.19964   .9150482   11.15   0.000    8.405821    11.99346<br>    fem_marr \|   6.63953  1.255823    5.29   0.000    4.177669    9.101391<br>       _cons \|  22.31421  .6967666   32.03   0.000     20.9483    23.68013<br><br>    (1)  married + fem_marr = 0<br><br>    hrchores \|   Coef.   Std. Err.      T    P>\|t\|  [95% Conf. Interval]<br>    ---------+--------------------------------------------------------------<br>         (1) \|  4.034644  .8714309    4.63   0.000    2.326329    5.742959<br><br>           lincom female+fem_marr<br><br>    (1)  female + fem_marr = 0<br><br>    hrchores \|   Coef.   Std. Err.      T    P>\|t\|  [95% Conf. Interval]<br>    ---------+--------------------------------------------------------------<br>         (1) \|  16.83917  .8601041   19.58   0.000    15.15306    18.52528 |
| Select Results | n/a | |

**APPENDIX B**

## ■ Display B.8.6 Regression of *hrchores* on *hrwork*, *female* and *fem_hrwork* in the NSFH

| | SAS | Stata |
|---|---|---|
| Variable Creation | `fem_hrwork=female*hrwork;` | `generate fem_hrwork= female*hrwork` |
| Regression Model | `proc reg;`<br>`  model hrchores=hrwork female fem_hrwork;`<br>`run;` | `regress hrchores hrwork female fem_hrwork` |

**SAS — Select Results**

| Number of Observations Read | 6054 |
|---|---|
| Number of Observations Used | 6054 |

Analysis of Variance

| Source | DF | Sum of Squares | Mean Square | F Value | Pr > F |
|---|---|---|---|---|---|
| Model | 3 | 328274 | 109425 | 188.99 | <.0001 |
| Error | 6050 | 3502956 | 579.00099 | | |
| Corrected Total | 6053 | 3831230 | | | |

| Root MSE | 24.06244 | R-Square | 0.0857 |
|---|---|---|---|
| Dependent Mean | 27.72382 | Adj R-Sq | 0.0852 |
| Coeff Var | 86.79338 | | |

Parameter Estimates

| Variable | DF | Parameter Estimate | Standard Error | t Value | Pr > |t| |
|---|---|---|---|---|---|
| Intercept | 1 | 26.12448 | 1.42428 | 18.34 | <.0001 |
| hrwork | 1 | -0.12432 | 0.03142 | -3.96 | <.0001 |
| female | 1 | 16.99004 | 1.83148 | 9.28 | <.0001 |
| fem_hrwork | 1 | -0.12560 | 0.04382 | -2.87 | 0.0042 |

**Stata — Select Results**

```
Source |       SS       df       MS              Number of obs =    6054
-------+------------------------------           F( 3,  6050) =  188.99
 Model | 328274.257      3  109424.752           Prob > F      = 0.0000
Residual| 3502955.97   6050  579.000987          R-squared     = 0.0857
-------+------------------------------           Adj R-squared = 0.0852
 Total | 3831230.23   6053  632.947336           Root MSE      = 24.062

  hrchores |      Coef.   Std. Err.      T    P>|t|    [95% Conf. Interval]
-----------+----------------------------------------------------------------
    hrwork | -.1243169   .0314199   -3.96   0.000   -.1859111   -.0627226
    female | 16.99004    1.831485    9.28   0.000    13.39968    20.5804
fem_hrwork | -.1256042   .0438182   -2.87   0.004   -.2115036   -.0397048
     _cons | 26.12448    1.424276   18.34   0.000    23.33239    28.91657
```

■ Display B.8.7 Regression of *hrchores* on *hrwork*, *female* and *fem_hrwork* in the NSFH: **Conditional Regression Equations for Effect of Hours of Paid Work among Women**

| SAS | Stata |
|---|---|
| **Approach #1: Re-estimating Regression** | |
| `male_hrwork=male*hrwork;` <br> `...` <br> `proc reg;` <br> `  model hrchores=hrwork male male_hrwork;` <br> `run;` | `generate male_hrwork=male*hrwork` <br> `regress hrchores hrwork male male_hrwork` |

Parameter Estimates (SAS)

| Variable | DF | Parameter Estimate | Standard Error | t Value | Pr > \|t\| |
|---|---|---|---|---|---|
| Intercept | 1 | 43.11452 | 1.15142 | 37.44 | <.0001 |
| hrwork | 1 | −0.24992 | 0.03054 | −8.18 | <.0001 |
| male | 1 | −16.99004 | 1.83148 | −9.28 | <.0001 |
| male_hrwork | 1 | 0.12560 | 0.04382 | 2.87 | 0.0042 |

Stata output:

```
hrchores |      Coef. Std. Err.      T    P>|t|  [95% Conf. Interval]
hrwork|-.2499211 .0305422  -8.18   0.000  -.3097947  -.1900475
     male|-16.99004 1.831485  -9.28   0.000  -20.5804  -13.39968
male_hrwork| .1256042 .0438182   2.87   0.004   .0397048   .2115036
    _cons| 43.11452 1.151423  37.44   0.000   40.85732   45.37172
```

| SAS | Stata |
|---|---|
| **Approach #2: F-Test** | |
| `proc reg;` <br> `  model hrchores=hrwork male female fem_hrwork;` <br> `  test hrwork+fem_hrwork=0;` <br> `run;` | `regress hrchores hrwork female fem_hrwork` <br> `  test hrwork+fem_hrwork=0` |

Test 1 Results for Dependent Variable hrchores (SAS)

| Source | DF | Mean Square | F Value | Pr > F |
|---|---|---|---|---|
| Numerator | 1 | 38769 | 66.96 | <.0001 |
| Denominator | 6050 | 579.00099 | | |

Stata output:

```
(1) hrwork + fem_hrwork = 0

  F(1, 6050) = 66.96
    Prob > F = 0.0000
```

| SAS | Stata |
|---|---|
| **Approach #3: Linear Combination** | |
| n/a (or use /covb and hand calculate) | `regress hrchores hrwork female fem_hrwork` <br> `  lincom hrwork+fem_hrwork` |
| n/a | |

Stata output:

```
hrchores |      Coef.    Std. Err.      T  P>|t|  [95% Conf. Interval]
(1)|-.2499211  .0305422  -8.18  0.000  -.3097947  -.1900475
```

APPENDIX B

■ **Display B.8.8. Regression of _hrchores_ on _hrwork_, _female_ and _fem_hrwork_ in the NSFH: Conditional Effect of Gender within Hours of Paid Work (Approach #1: Re-Estimation).**

| | SAS | Stata |
|---|---|---|
| Variable Creation and Regression Model | ```
hrworkC10=hrwork-10;
hrworkC20=hrwork-20;
hrworkC30=hrwork-30;
hrworkC40=hrwork-40;
hrworkC50=hrwork-50;

fem_hrwC10=female*hrworkC10;
fem_hrwC20=female*hrworkC20;
fem_hrwC30=female*hrworkC30;
fem_hrwC40=female*hrworkC40;
fem_hrwC50=female*hrworkC50;
...

proc reg;
  model hrchores=hrworkC10 female fem_hrwC10;
  model hrchores=hrworkC20 female fem_hrwC20;
  model hrchores=hrworkC30 female fem_hrwC30;
  model hrchores=hrworkC40 female fem_hrwC40;
  model hrchores=hrworkC50 female fem_hrwC50;
run;
``` | ```
generate hrworkC10=hrwork-10
generate hrworkC20=hrwork-20
generate hrworkC30=hrwork-30
generate hrworkC40=hrwork-40
generate hrworkC50=hrwork-50

generate fem_hrwC10=female*hrworkC10
generate fem_hrwC20=female*hrworkC20
generate fem_hrwC30=female*hrworkC30
generate fem_hrwC40=female*hrworkC40
generate fem_hrwC50=female*hrworkC50

regress hrchores hrworkC10 female fem_hrwC10
regress hrchores hrworkC20 female fem_hrwC20
regress hrchores hrworkC30 female fem_hrwC30
regress hrchores hrworkC40 female fem_hrwC40
regress hrchores hrworkC50 female fem_hrwC50
``` |

**SAS**

Parameter Estimates

| Variable | DF | Parameter Estimate | Standard Error | t Value | Pr > \|t\| |
|---|---|---|---|---|---|
| Intercept | 1 | 24.88131 | 1.12982 | 22.02 | <.0001 |
| hrworkC10 | 1 | -0.12432 | 0.03142 | -3.96 | <.0001 |
| female | 1 | 15.73400 | 1.42960 | 11.01 | <.0001 |
| fem_hrwC10 | 1 | -0.12560 | 0.04382 | -2.87 | 0.0042 |

Parameter Estimates

| Variable | DF | Parameter Estimate | Standard Error | t Value> | Pr > \|t\| |
|---|---|---|---|---|---|
| Intercept | 1 | 23.63814 | 0.84964 | 27.82 | <.0001 |
| hrworkC20 | 1 | -0.12432 | 0.03142 | -3.96 | <.0001 |
| female | 1 | 14.47796 | 1.05697 | 13.70 | <.0001 |
| fem_hrwC20 | 1 | -0.12560 | 0.04382 | -2.87 | 0.0042 |

**Stata**

```
hrchores |      Coef. Std. Err.      T    P>|t|    [95% Conf. Interval]
hrworkC10 |-.1243169  .0314199    -3.96   0.000   -.1859111   -.0627226
   female | 15.734    1.429602    11.01   0.000    12.93147    18.53653
fem_hrwC10|-.1256042  .0438182    -2.87   0.004   -.2115036   -.0397048
    _cons | 24.88131  1.129823    22.02   0.000    22.6646     27.09617

hrchores |      Coef. Std. Err.      T    P>|t|    [95% Conf. Interval]
hrworkC20 |-.1243169  .0314199    -3.96   0.000   -.1859111   -.0627226
   female | 14.47796  1.056974    13.70   0.000    12.40591    16.55
fem_hrwC20|-.1256042  .0438182    -2.87   0.004   -.2115036   -.0397048
    _cons | 23.63814  .8496356    27.82   0.000    21.97256    25.30373
```

(Continued on next page)

**Display B.8.8. Regression of** *hrchores* **on** *hrwork,* *female* **and** *fem_hrwork* **in the NSFH: Conditional Effect of Gender within Hours of Paid Work (Approach #1: Re-Estimation). (Continued)**

Select Results

```
  hrchores |      Coef.   Std. Err.      T    P>|t|   [95% Conf. Interval]
-----------+----------------------------------------------------------------
 hrworkC30 | -.1243169  .0314199   -3.96   0.000   -.1859111   -.0627226
    female |  13.22192  .7580464   17.44   0.000    11.73588    14.70796
fem_hrwC30 | -.1256042  .0438182   -2.87   0.004   -.2115036   -.0397048
     _cons |  22.39498  .6039071   37.08   0.000     21.2111    23.57885
```

```
  hrchores |      Coef.   Std. Err.      T    P>|t|   [95% Conf. Interval]
-----------+----------------------------------------------------------------
 hrworkC40 | -.1243169  .0314199   -3.96   0.000   -.1859111   -.0627226
    female |  11.96587  .6450443   18.55   0.000    10.70136    13.23039
fem_hrwC40 | -.1256042  .0438182   -2.87   0.004   -.2115036   -.0397048
     _cons |  21.15181  .4527355   46.72   0.000    20.26428    22.03933
```

```
  hrchores |      Coef.   Std. Err.      T    P>|t|   [95% Conf. Interval]
-----------+----------------------------------------------------------------
 hrworkC50 | -.1243169  .0314199   -3.96   0.000   -.1859111   -.0627226
    female |  10.70983  .8009605   13.37   0.000    9.139664      12.28
fem_hrwC50 | -.1256042  .0438182   -2.87   0.004   -.2115036   -.0397048
     _cons |  19.90864  .4922231   40.41   0.000    18.94292    20.87435
```

Parameter Estimates

| Variable | DF | Parameter Estimate | Standard Error | t Value | Pr > |t| |
|---|---|---|---|---|---|
| Intercept | 1 | 22.39498 | 0.60391 | 37.08 | <.0001 |
| hrworkC30 | 1 | -0.12432 | 0.03142 | -3.96 | <.0001 |
| female | 1 | 13.22192 | 0.75805 | 17.44 | <.0001 |
| fem_hrwC30 | 1 | -0.12560 | 0.04382 | -2.87 | 0.0042 |

Parameter Estimates

| Variable | DF | Parameter Estimate | Standard Error | t Value | Pr > |t| |
|---|---|---|---|---|---|
| Intercept | 1 | 21.15181 | 0.45274 | 46.72 | <.0001 |
| hrworkC40 | 1 | -0.12432 | 0.03142 | -3.96 | <.0001 |
| female | 1 | 11.96587 | 0.64504 | 18.55 | <.0001 |
| fem_hrwC40 | 1 | -0.12560 | 0.04382 | -2.87 | 0.0042 |

Parameter Estimates

| Variable | DF | Parameter Estimate | Standard Error | t Value | Pr > |t| |
|---|---|---|---|---|---|
| Intercept | 1 | 19.90864 | 0.49262 | 40.41 | <.0001 |
| hrworkC50 | 1 | -0.12432 | 0.03142 | -3.96 | <.0001 |
| female | 1 | 10.70983 | 0.80096 | 13.37 | <.0001 |
| fem_hrwC50 | 1 | -0.12560 | 0.04382 | -2.87 | 0.0042 |

APPENDIX
**B**

■ **Display B.8.9. Regression of** *hrchores* **on** *hrwork*, *female* **and** *fem_hrwork* **in the NSFH: Conditional Effect of Gender within Hours of Paid Work (Approach #2:** *F*-test).

|  | SAS | Stata |
|---|---|---|
| Regression Model | ```
proc reg;
  model hrchores=hrwork female fem_hrwork;
  test female+10*fem_hrwork=0;
  test female+20*fem_hrwork=0;
  test female+30*fem_hrwork=0;
  test female+40*fem_hrwork=0;
  test female+50*fem_hrwork=0;
run;
``` | ```
regress hrchores hrwork female fem_hrwork
  test female+10*fem_hrwork=0
  test female+20*fem_hrwork=0
  test female+30*fem_hrwork=0
  test female+40*fem_hrwork=0
  test female+50*fem_hrwork=0
``` |

Select Results (SAS):

**Test 1 Results for Dependent Variable hrchores**

| Source | DF | Mean Square | F Value | Pr > F |
|---|---|---|---|---|
| Numerator | 1 | 70134 | 121.13 | <.0001 |
| Denominator | 6050 | 579.00099 | | |

**Test 2 Results for Dependent Variable hrchores**

| Source | DF | Mean Square | F Value | Pr > F |
|---|---|---|---|---|
| Numerator | 1 | 108634 | 187.62 | <.0001 |
| Denominator | 6050 | 579.00099 | | |

**Test 3 Results for Dependent Variable hrchores**

| Source | DF | Mean Square | F Value | Pr > F |
|---|---|---|---|---|
| Numerator | 1 | 176148 | 304.23 | <.0001 |
| Denominator | 6050 | 579.00099 | | |

**Test 4 Results for Dependent Variable hrchores**

| Source | DF | Mean Square | F Value | Pr > F |
|---|---|---|---|---|
| Numerator | 1 | 199246 | 344.12 | <.0001 |
| Denominator | 6050 | 579.00099 | | |

**Test 5 Results for Dependent Variable hrchores**

| Source | DF | Mean Square | F Value | Pr > F |
|---|---|---|---|---|
| Numerator | 1 | 103520 | 178.79 | <.0001 |
| Denominator | 6050 | 579.00099 | | |

Select Results (Stata):

```
(1) female + 10 fem_hrwork = 0
       F( 1, 6050) = 121.13
            Prob > F = 0.0000

(1) female + 20 fem_hrwork = 0
       F( 1, 6050) = 187.62
            Prob > F = 0.0000

(1) female + 30 fem_hrwork = 0
       F( 1, 6050) = 304.23
            Prob > F = 0.0000

(1) female + 40 fem_hrwork = 0
       F( 1, 6050) = 344.12
            Prob > F = 0.0000

(1) female + 50 fem_hrwork = 0
       F( 1, 6050) = 178.79
            Prob > F = 0.0000
```

■ **Display B.8.10 Regression of** *hrchores* **on** *hrwork,* *female* **and** *fem_hrwork* **in the NSFH: Conditional Effect of Gender within Hours of Paid Work (Approach #3: Linear Combinations)**

| | SAS | Stata |
|---|---|---|
| Regression Model | n/a | `regress hrchores hrwork female fem_hrwork`<br>`  lincom female+10*fem_hrwork`<br>`  lincom female+20*fem_hrwork`<br>`  lincom female+30*fem_hrwork`<br>`  lincom female+40*fem_hrwork`<br>`  lincom female+50*fem_hrwork` |
| Select Results | n/a | (see below) |

```
(1) female + 10 fem_hrwork = 0

hrchores |   Coef.  Std. Err.    T   P>|t|  [95% Conf. Interval]
---------+
     (1) | 15.734   1.429602  11.01  0.000   12.93147  18.53653

(1) female + 20 fem_hrwork = 0

hrchores |   Coef.  Std. Err.    T   P>|t|  [95% Conf. Interval]
---------+
     (1) |14.47796  1.056974  13.70  0.000   12.40591     16.55

(1) female + 30 fem_hrwork = 0

hrchores |   Coef.  Std. Err.    T   P>|t|  [95% Conf. Interval]
---------+
     (1) |13.22192  .7580464  17.44  0.000   11.73588  14.70796

(1) female + 40 fem_hrwork = 0

hrchores |   Coef.  Std. Err.    T   P> t   [95% Conf. Interval]
---------+
     (1) |11.96587  .6450443  18.55  0.000   10.70136  13.23039

(1) female + 50 fem_hrwork = 0

hrchores |   Coef.  Std. Err.    T   P>|t|  [95% Conf. Interval]
---------+
     (1) |10.70983  .8009605  13.37  0.000   9.139664     12.28
```

■ **Display B.8.11 Regression of** *hrchores* **on** *married, hrwork, numkid, aframer,* **and** *mexamer* **in the NSFH: Total Sample**

| | SAS |
|---|---|
| Regression Model | ```proc reg;
    model hrchores=married hrwork numkid
        aframer mexamer;
run;``` |

| Number of Observations Read | 5820 |
|---|---|
| Number of Observations Used | 5820 |

**Analysis of Variance**

| Source | DF | Sum of Squares | Mean Square | F Value | Pr > F |
|---|---|---|---|---|---|
| Model | 5 | 283291 | 56658 | 95.30 | <.0001 |
| Error | 5814 | 3456552 | 594.52212 | | |
| Corrected Total | 5819 | 3739843 | | | |

| Root MSE | 24.38282 | R-Square | 0.0757 |
|---|---|---|---|
| Dependent Mean | 27.79553 | Adj R-Sq | 0.0750 |
| Coeff Var | 87.72210 | | |

**Parameter Estimates**

| Variable | DF | Parameter Estimate | Standard Error | t Value | Pr > |t| |
|---|---|---|---|---|---|
| Intercept | 1 | 36.06816 | 0.99963 | 36.08 | <.0001 |
| married | 1 | -3.06821 | 0.68093 | -4.51 | <.0001 |
| hrwork | 1 | -0.29439 | 0.02169 | -13.58 | <.0001 |
| numkid | 1 | 4.36420 | 0.32221 | 13.54 | <.0001 |
| aframer | 1 | 5.22127 | 0.86797 | 6.02 | <.0001 |
| mexamer | 1 | 8.91881 | 1.64745 | 5.41 | <.0001 |

| | Stata |
|---|---|
| | `regress hrchores married hrwork numkid aframer mexamer` |

| Source | SS | df | MS | | |
|---|---|---|---|---|---|
| Model | 283291.065 | 5 | 56658.2129 | Number of obs = 5820 |
| Residual | 3456551.62 | 5814 | 594.522122 | F( 5, 5814) = 95.30 |
| | | | | Prob > F = 0.0000 |
| | | | | R-squared = 0.0757 |
| | | | | Adj R-squared = 0.0750 |
| Total | 3739842.68 | 5819 | 642.695082 | Root MSE = 24.383 |

| hrchores | Coef. | Std. Err. | t | P>|t| | [95% Conf. Interval] |
|---|---|---|---|---|---|
| married | -3.068214 | .6809263 | -4.51 | 0.000 | -4.403083 -1.733345 |
| hrwork | -.2943945 | .0216851 | -13.58 | 0.000 | -.3369054 -.2518836 |
| numkid | 4.364196 | .322211 | 13.54 | 0.000 | 3.732542 4.995849 |
| aframer | 5.221268 | .8679679 | 6.02 | 0.000 | 3.519727 6.922808 |
| mexamer | 8.918807 | 1.647446 | 5.41 | 0.000 | 5.689201 12.14841 |
| _cons | 36.06816 | .9996316 | 36.08 | 0.000 | 34.10851 38.02781 |

Select Results

| | SAS | Stata |
|---|---|---|
| Regression Model | `proc sort; by female; run;`<br>`proc reg;`<br>`  model hrchores=married hrwork numkid`<br>`                 aframer mexamer;`<br>`  by female;`<br>`run;` | `bysort female: regress hrchores married hrwork ///`<br>`                       numkid aframer mexamer` |

**Males**

```
  Source |       SS       df       MS              Number of obs =    2849
---------+------------------------------           F(  5,  2843) =   17.75
   Model | 35661.778        5 7132.3556            Prob > F      = 0.0000
Residual |1142398.38     2843 401.828484           R-squared     = 0.0303
---------+------------------------------           Adj R-squared = 0.0286
   Total |1178060.16     2848 413.644718           Root MSE      = 20.046

hrchores |   Coef.    Std. Err.     t    P>|t|   [95% Conf. Interval]
 married | -3.184859  .8664546   -3.68  0.000  -4.883803 -1.485916
  hrwork | -.109475   .0269866   -4.06  0.000  -.1623902 -.0565598
  numkid |  1.445786  .4002116    3.61  0.000   .610514   2.23052
 aframer |  6.041829  1.084095    5.57  0.000   3.916138  8.16752
 mexamer |  5.768344  1.849813    3.12  0.002   2.141233  9.395456
   _cons | 25.18235   1.303018   19.33  0.000  22.62739  27.73731
```

Select Results

| | | Parameter Estimates | | | |
|---|---|---|---|---|---|
| Variable | DF | Parameter Estimate | Standard Error | t Value | Pr>\|t\| |
| Intercept | 1 | 25.18235 | 1.30302 | 19.33 | <.0001 |
| married | 1 | -3.18486 | 0.86645 | -3.68 | 0.0002 |
| hrwork | 1 | -0.10947 | 0.02699 | -4.06 | <.0001 |
| numkid | 1 | 1.44579 | 0.40021 | 3.61 | 0.0003 |
| aframer | 1 | 6.04183 | 1.08409 | 5.57 | <.0001 |
| mexamer | 1 | 5.76834 | 1.84981 | 3.12 | 0.0018 |

**Females**

```
  Source |       SS       df       MS              Number of obs =    2971
---------+------------------------------           F(  5,  2965) =   59.11
   Model | 207458.514       5 41491.7028           Prob > F      = 0.0000
Residual |2081280.21    2965 701.949481           R-squared     = 0.0906
---------+------------------------------           Adj R-squared = 0.0891
   Total |2288738.73    2970 770.6191              Root MSE      = 26.494

hrchores |   Coef.    Std. Err.     t    P>|t|   [95% Conf. Interval]
 married |  1.297806  1.010148    1.28  0.199  -.6828562  3.278467
  hrwork | -.2151042  .0344651   -6.24  0.000  -.2826822 -.1475262
  numkid |  6.240049  .4831701   12.91  0.000   5.292666  7.187432
 aframer |  4.063293  1.256962    3.23  0.001   1.598687  6.527899
 mexamer | 14.26334   2.638858    5.41  0.000   9.089161  19.43752
   _cons | 34.69164   1.506367   23.03  0.000  31.738    37.64527
```

Select Results

| | | Parameter Estimates | | | |
|---|---|---|---|---|---|
| Variable | DF | Parameter Estimate | Standard Error | t Value | Pr>\|t\| |
| Intercept | 1 | 34.69164 | 1.50637 | 23.03 | <.0001 |
| married | 1 | 1.29781 | 1.01015 | 1.28 | 0.1990 |
| hrwork | 1 | -0.21510 | 0.03447 | -6.24 | <.0001 |
| numkid | 1 | 6.24005 | 0.48317 | 12.91 | <.0001 |
| aframer | 1 | 4.06329 | 1.25696 | 3.23 | 0.0012 |
| mexamer | 1 | 14.26334 | 2.63886 | 5.41 | <.0001 |

■ **Display B.8.13 Regression of** *hrchores* **on** *married, hrwork, numkid, aframer,* **and** *mexamer* **in the NSFH: Fully Interacted Model**

| | SAS | Stata |
|---|---|---|
| Regression | ```
proc reg;
  model hrchores=married hrwork numkid
                 aframer mexamer female
                 fem_marr fem_hrwork
                 fem_numkid fem_aframer
                 fem_mexamer;

  test female,fem_marr,fem_hrwork,fem_numkid,
       fem_aframer,fem_mexamer;
run;
``` | ```
regress hrchores married hrwork numkid aframer mexamer ///
                 female fem_marr fem_hrwork fem_numkid ///
                 fem_aframer fem_mexamer

test female fem_marr fem_hrwork fem_numkid ///
     fem_aframer fem_mexamer
``` |

**Stata output:**

```
      Source |       SS       df       MS              Number of obs =    5820
-------------+------------------------------           F( 11,  5808) =   84.54
       Model |  516164.093    11  46924.0084           Prob > F      =  0.0000
    Residual |  3223678.59  5808  555.04108           R-squared     =  0.1380
-------------+------------------------------           Adj R-squared =  0.1364
       Total |  3739842.68  5819  642.695082           Root MSE      =  23.559

------------------------------------------------------------------------------
    hrchores |      Coef.   Std. Err.      t    P>|t|     [95% Conf. Interval]
-------------+----------------------------------------------------------------
     married |  -3.184859   1.018329    -3.13   0.002    -5.181163   -1.188556
      hrwork |   -.109475   .0317168    -3.45   0.001    -.1716517   -.0472982
      numkid |   1.445786   .4703616     3.07   0.002     .5237018     2.36787
     aframer |   6.041829   1.274117     4.74   0.000     3.544085    8.539573
     mexamer |   5.768344   2.174053     2.65   0.008     1.506391     10.0303
      female |   9.509286   2.034569     4.67   0.000     5.520772     13.4978
    fem_marr |   4.482665   1.357879     3.30   0.001     1.820716    7.144615
  fem_hrwork |  -.1056292    .041044    -2.39   0.017    -.1920904   -.0191681
  fem_numkid |   4.794263   .6370517     7.53   0.000     3.545404    6.043122
 fem_aframer |  -1.978536   1.694894    -1.17   0.243    -5.301159    1.344087
 fem_mexamer |   8.494996   3.198859     2.66   0.008     2.224041    14.76595
       _cons |   25.18235   1.531414    16.44   0.000     22.18021    28.18449
------------------------------------------------------------------------------

.test female fem_marr fem_hrwork fem_numkid fem_aframer fem_mexamer
 (1)  female = 0
 (2)  fem_marr = 0
 (3)  fem_hrwork = 0
 (4)  fem_numkid = 0
 (5)  fem_aframer = 0
 (6)  fem_mexamer = 0
       F(  6,  5808) =   69.93
```

**Select Results — SAS**

Parameter Estimates

| Variable | DF | Parameter Estimate | Standard Error | t Value | Pr > |t| |
|---|---|---|---|---|---|
| Intercept | 1 | 25.18235 | 1.53141 | 16.44 | <.0001 |
| married | 1 | -3.18486 | 1.01833 | -3.13 | 0.0018 |
| hrwork | 1 | -0.10947 | 0.03172 | -3.45 | 0.0006 |
| numkid | 1 | 1.44579 | 0.47036 | 3.07 | 0.0021 |
| aframer | 1 | 6.04183 | 1.27412 | 4.74 | <.0001 |
| mexamer | 1 | 5.76834 | 2.17405 | 2.65 | 0.0080 |
| female | 1 | 9.50929 | 2.03457 | 4.67 | <.0001 |
| fem_marr | 1 | 4.48267 | 1.35788 | 3.30 | 0.0010 |
| fem_hrwork | 1 | -0.10563 | 0.04410 | -2.39 | 0.0167 |
| fem_numkid | 1 | 4.79426 | 0.63705 | 7.53 | <.0001 |
| fem_aframer | 1 | -1.97854 | 1.69489 | -1.17 | 0.2431 |
| fem_mexamer | 1 | 8.49500 | 3.19886 | 2.66 | 0.0079 |

Test 1 Results for Dependent Variable hrchores

| Source | DF | Mean Square | F Value | Pr > F |
|---|---|---|---|---|
| Numerator | 6 | 38812 | 69.93 | <.0001 |
| Denominator | 5808 | 555.04108 | | |

■ **Display B.8.14 Regression of** *hrchores* **on** *married,* *hrwork,* *numkid,* *aframer,* **and** *mexamer* **in the NSFH: Conditional Regression for Women from Fully Interacted Model**

| | Stata |
|---|---|
| Regression Model | ```regress hrchores married hrwork numkid aframer mexamer ///                female fem_marr fem_hrwork fem_numkid fem_aframer fem_mexamer  lincom married+fem_marr lincom hrwork+fem_hrwork lincom numkid+fem_numkid lincom aframer+fem_aframer lincom mexamer+fem_mexamer lincom _cons+female``` |
| Select Results | ```( 1)  married + fem_marr = 0   hrchores |    Coef.   Std. Err.     t    P>|t|  [95% Conf. Interval]        (1) | 1.297806  .8982444   1.44  0.149  -.4630881   3.058699   ( 1)  hrwork + fem_hrwork = 0   hrchores |    Coef.    Std. Err.    t    P>|t|  [95% Conf. Interval]        (1) | -.2151042  .0306471  -7.02  0.000  -.2751839  -.1550244   ( 1)  numkid + fem_numkid = 0   hrchores | Coef.  Std. Err.    t    P>|t|  [95% Conf. Interval]        (1) | 6.240049  .4296449  14.52  0.000  5.397785   7.082313   ( 1)  aframer + fem_aframer = 0   hrchores |    Coef.   Std. Err.     t    P>|t|  [95% Conf. Interval]        (1) | 4.063293  1.117717   3.64  0.000   1.872152   6.254434   ( 1)  mexamer + fem_mexamer = 0   hrchores |     Coef.   Std. Err.     t    P>|t|  [95% Conf. Interval]        (1) | 14.26334  2.344528   6.08  0.000   9.663272  18.86341   ( 1)  female + _cons = 0   hrchores |    Coef.  Std. Err.     t    P>|t|  [95% Conf. Interval]        (1) | 34.69164  1.339493  25.90  0.000  32.06573  37.31754``` |

■ **Display B.8.15 Regression of** *hrchores* **on** *hrwork* **and** *numkid* **in the NSFH: Interaction Model**

| | SAS | Stata |
|---|---|---|
| Variable Creation | `numk_hrw=numkid*hrwork;` | `generate numk_hrw=numkid*hrwork` |
| Regression Model | `proc reg;`<br>`  model hrchores=hrwork numkid numk_hrw;`<br>`run;` | `regress hrchores hrwork numkid numk_hrw` |

**SAS — Select Results**

| Number of Observations Read | 6054 |
|---|---|
| Number of Observations Used | 6054 |

**Analysis of Variance**

| Source | DF | Sum of Squares | Mean Square | F Value | Pr > F |
|---|---|---|---|---|---|
| Model | 3 | 258894 | 86298 | 146.15 | <.0001 |
| Error | 6050 | 3572336 | 590.46874 | | |
| Corrected Total | 6053 | 3831230 | | | |

| Root MSE | 24.29956 | R-Square | 0.0676 |
|---|---|---|---|
| Dependent Mean | 27.72382 | Adj R-Sq | 0.0671 |
| Coeff Var | 87.64868 | | |

**Parameter Estimates**

| Variable | DF | Parameter Estimate | Standard Error | t Value | Pr > |t| |
|---|---|---|---|---|---|
| Intercept | 1 | 31.89119 | 1.11564 | 28.59 | <.0001 |
| hrwork | 1 | -0.19367 | 0.02678 | -7.23 | <.0001 |
| numkid | 1 | 9.06985 | 0.80394 | 11.28 | <.0001 |
| numk_hrw | 1 | -0.12751 | 0.01924 | -6.63 | <.0001 |

**Stata — Select Results**

```
Source |      SS           df       MS            Number of obs   =      6054
-------+----------------------------------        F( 3,  6050)    =    146.15
 Model |258894.349          3   86298.1165        Prob > F        =    0.0000
Residual|3572335.88      6050   590.46874         R-squared       =    0.0676
-------+----------------------------------        Adj R-squared   =    0.0671
 Total |3831230.23      6053   632.947336         Root MSE        =      24.3

hrchores |   Coef.    Std. Err.      t     P>|t|     [95% Conf. Interval]
---------+--------------------------------------------------------------
 hrwork  | -.193674    .026785    -7.23    0.000    -.2461821    -.1411658
 numkid  | 9.069852    .8039375   11.28    0.000     7.493849     10.64586
numk_hrw | -.127512    .0192373   -6.63    0.000     -.165224    -.0897999
   _cons | 31.89119    1.115636   28.59    0.000     29.70415     34.07824
```

■ **Display B.8.16 Regression of** *hrchores* **on** *hrwork* **and** *numkid* **in the NSFH: Conditional Effects**

| | Conditional Effects of Hours of Work | Conditional Effects of Number of Children |
|---|---|---|
| Regression Model | ```
lincom hrwork+numk_hrw*0
lincom hrwork+numk_hrw*1
lincom hrwork+numk_hrw*2
lincom hrwork+numk_hrw*3
lincom hrwork+numk_hrw*4
``` | ```
lincom numkid+numk_hrw*10
lincom numkid+numk_hrw*20
lincom numkid+numk_hrw*30
lincom numkid+numk_hrw*40
lincom numkid+numk_hrw*50
``` |
| Select Results | ```
(1) hrwork = 0

hrchores |    Coef.   Std. Err.       t    P>|t|   [95% Conf. Interval]
     (1) | -.193674   .026785     -7.23   0.000   -.2461821   -.1411658

(1) hrwork + numk_hrw = 0

hrchores |    Coef.   Std. Err.       t    P>|t|   [95% Conf. Interval]
     (1) | -.321186   .02145     -14.97   0.000   -.3632355   -.2791364

(1) hrwork + 2 numk_hrw = 0

hrchores |    Coef.   Std. Err.       t    P>|t|   [95% Conf. Interval]
     (1) | -.4486979  .030707    -14.61   0.000   -.5088945   -.3885014

(1) hrwork + 3 numk_hrw = 0

hrchores |    Coef.   Std. Err.       t    P>|t|   [95% Conf. Interval]
     (1) | -.5762099  .0465391   -12.38   0.000   -.6674431   -.4849768

(1) hrwork + 4 numk_hrw = 0

hrchores |    Coef.   Std. Err.       t    P>|t|   [95% Conf. Interval]
     (1) | -.7037219  .0642573   -10.95   0.000   -.826891    -.5777547
``` | ```
(1) numkid + 10 numk_hrw = 0

hrchores |    Coef.   Std. Err.       t    P>|t|   [95% Conf. Interval]
     (1) | 7.794732   .6292909    12.39   0.000    6.561098    9.028367

(1) numkid + 20 numk_hrw = 0

hrchores |    Coef.   Std. Err.       t    P>|t|   [95% Conf. Interval]
     (1) | 6.519613   .4687363    13.91   0.000    5.600723    7.438503

(1) numkid + 30 numk_hrw = 0

hrchores |    Coef.   Std. Err.       t    P>|t|   [95% Conf. Interval]
     (1) | 5.244493   .3426887    15.30   0.000    4.572701    5.916285

(1) numkid + 40 numk_hrw = 0

hrchores |    Coef.   Std. Err.       t    P>|t|   [95% Conf. Interval]
     (1) | 3.969373   .2986177    13.29   0.000    3.383976    4.55477

(1) numkid + 50 numk_hrw = 0

hrchores |    Coef.   Std. Err.       t    P>|t|   [95% Conf. Interval]
     (1) | 2.694253   .3673208     7.33   0.000    1.974174    3.414333
``` |

APPENDIX **B**

■ **Display B.9.1 Models Relating Distance to Mother's Years of Schooling: Outcome in Natural Units**

| | SAS | Stata |
|---|---|---|
| Variable Creation | `logg1yrschl=log(g1yrschl+1);`<br>`sqg1yrschl=g1yrschl*g1yrschl;` | `generate logg1yrschl=log(g1yrschl+1)`<br>`generate sqg1yrschl=g1yrschl*g1yrschl` |
| Regression Model | `proc reg;`<br>`  model g1miles=g1yrschl;`<br>`  model g1miles=logg1yrschl;`<br>`  model g1miles=g1yrschl sqg1yrschl;`<br>`run;` | `regress g1miles g1yrschl`<br>`regress g1miles logg1yrschl`<br>`regress g1miles g1yrschl sqg1yrschl` |

Select Results — SAS:

Analysis of Variance

| Source | DF | Sum of Squares | Mean Square | F Value | Pr > F |
|---|---|---|---|---|---|
| Model | 1 | 14307140 | 14307140 | 34.59 | <.0001 |
| Error | 5470 | 226259142 | 413638 | | |
| Corrected Total | 5471 | 277906282 | | | |

Parameter Estimates

| Variable | DF | Parameter Estimate | Standard Error | t Value | Pr > |t| |
|---|---|---|---|---|---|
| Intercept | 1 | 92.86328 | 34.81817 | 2.67 | 0.0077 |
| g1yrschl | 1 | 17.39260 | 2.95732 | 5.88 | <.0001 |

Select Results — Stata:

| Source | SS | df | MS |
|---|---|---|---|
| Model | 14307139.9 | 1 | 14307139.9 |
| Residual | 2.2626e+09 | 5470 | 413637.869 |
| Total | 2.2769e+09 | 5471 | 416177.35 |

```
                                          Number of obs =    5472
                                          F( 1,  5470)  =   34.59
                                          Prob > F      =  0.0000
                                          R-squared     =  0.0063
                                          Adj R-squared =  0.0061
                                          Root MSE      =  643.15
```

| g1miles | Coef. | Std. Err. | t | P>|t| | [95% Conf. Interval] | |
|---|---|---|---|---|---|---|
| g1yrschl | 17.3926 | 2.95732 | 5.88 | 0.000 | 11.59508 | 23.19013 |
| _cons | 92.86328 | 34.81817 | 2.67 | 0.008 | 24.60582 | 161.1207 |

(Continued on next page)

**■ Display B.9.1 Models Relating Distance to Mother's Years of Schooling: Outcome in Natural Units (Continued)**

```
  Source |       SS       df       MS                     Number of obs =    5472
---------+----------------------------------             F(  1,  5470) =   17.04
   Model | 7070413.39        1  7070413.39               Prob > F      =  0.0000
Residual | 2.2698e+09     5470  414960.854               R-squared     =  0.0031
---------+----------------------------------             Adj R-squared =  0.0029
   Total | 2.2769e+09     5471  416177.35               Root MSE      =  644.17

------------------------------------------------------------------------------
   glmiles |   Coef.   Std. Err.    t    P>|t|    [95% Conf. Interval]
-----------+------------------------------------------------------------------
logg1yrschl | 107.6181  26.07152   4.13  0.000    56.50752   158.7286
      _cons |  24.65057  65.1465    0.38  0.705   -103.0625   152.3636
------------------------------------------------------------------------------
```

```
  Source |       SS       df       MS                     Number of obs =    5472
---------+----------------------------------             F(  2,  5469) =   23.65
   Model | 19520918.9        2  9760459.43               Prob > F      =  0.0000
Residual | 2.2574e+09     5469  412760.169               R-squared     =  0.0086
---------+----------------------------------             Adj R-squared =  0.0082
   Total | 2.2769e+09     5471  416177.35               Root MSE      =  642.46

------------------------------------------------------------------------------
   glmiles |   Coef.    Std. Err.    t    P>|t|    [95% Conf. Interval]
-----------+------------------------------------------------------------------
 g1yrschl  | -25.69624  12.47849  -2.06  0.040   -50.15905  -1.233428
sqg1yrschl |  2.094442   .589306   3.55  0.000    .9391676   3.249716
     _cons | 293.7769   66.37331   4.43  0.000    163.6588   423.895
------------------------------------------------------------------------------
```

Select Results (continued)

**Analysis of Variance**

| Source | DF | Sum of Squares | Mean Square | F Value | Pr > F |
|---|---|---|---|---|---|
| Model | 1 | 7070413 | 7070413 | 17.04 | <.0001 |
| Error | 5470 | 2269835869 | 414961 | | |
| Corrected Total | 5471 | 2276906282 | | | |

**Parameter Estimates**

| Variable | DF | Parameter Estimate | Standard Error | t Value | Pr > |t| |
|---|---|---|---|---|---|
| Intercept | 1 | 24.65058 | 65.14650 | 0.38 | 0.7052 |
| logg1yrschl | 1 | 107.61806 | 26.07152 | 4.13 | <.0001 |

**Analysis of Variance**

| Source | DF | Sum of Squares | Mean Square | F Value | Pr > F |
|---|---|---|---|---|---|
| Model | 2 | 19520919 | 9760459 | 23.65 | <.0001 |
| Error | 5469 | 2257385363 | 412760 | | |
| Corrected Total | 5471 | 2276906282 | | | |

**Parameter Estimates**

| Variable | DF | Parameter Estimate | Standard Error | t Value | Pr > |t| |
|---|---|---|---|---|---|
| Intercept | 1 | 293.77688 | 66.37331 | 4.43 | <.0001 |
| g1yrschl | 1 | -25.69624 | 12.47849 | -2.06 | 0.0395 |
| sqg1yrschl | 1 | 2.09444 | 0.58931 | 3.55 | 0.0004 |

**APPENDIX B**

■ **Display B.9.2 Models Relating Distance to Mother's Years of Schooling: Logged Outcome**

## SAS

**Variable Creation**

```
logg1miles=log(g1miles);
```

**Regression Model**

```
proc reg;
    model logg1miles=g1yrschl;
    model logg1miles=logg1yrschl;
    model logg1miles=g1yrschl sqg1yrschl;
run;
```

**Select Results**

Analysis of Variance

| Source | DF | Sum of Squares | Mean Square | F Value | Pr > F |
|---|---|---|---|---|---|
| Model | 1 | 631.71941 | 631.71941 | 110.61 | <.0001 |
| Error | 5470 | 31239 | 5.71104 | | |
| Corrected Total | 5471 | 31871 | | | |

Parameter Estimates

| Variable | DF | Parameter Estimate | Standard Error | t Value | Pr > \|t\| |
|---|---|---|---|---|---|
| Intercept | 1 | 1.97424 | 0.12938 | 15.26 | <.0001 |
| g1yrschl | 1 | 0.11557 | 0.01099 | 10.52 | <.0001 |

Analysis of Variance

| Source | DF | Sum of Squares | Mean Square | F Value | Pr > F |
|---|---|---|---|---|---|
| Model | 1 | 404.56553 | 404.56553 | 70.33 | <.0001 |
| Error | 5470 | 31467 | 5.75257 | | |
| Corrected Total | 5471 | 31871 | | | |

Parameter Estimates

| Variable | DF | Parameter Estimate | Standard Error | t Value | Pr > \|t\| |
|---|---|---|---|---|---|
| Intercept | 1 | 1.27593 | 0.24256 | 5.26 | <.0001 |
| logg1yrschl | 1 | 0.81406 | 0.09707 | 8.39 | <.0001 |

Analysis of Variance

| Source | DF | Sum of Squares | Mean Square | F Value | Pr > F |
|---|---|---|---|---|---|
| Model | 2 | 721.84162 | 360.92081 | 63.37 | <.0001 |
| Error | 5469 | 31149 | 5.69561 | | |
| Corrected Total | 5471 | 31871 | | | |

Parameter Estimates

| Variable | DF | Parameter Estimate | Standard Error | t Value | Pr > \|t\| |
|---|---|---|---|---|---|
| Intercept | 1 | 2.80955 | 0.24656 | 11.40 | <.0001 |
| g1yrschl | 1 | -0.06357 | 0.04635 | -1.37 | 0.1703 |
| sqg1yrschl | 1 | 0.00871 | 0.00219 | 3.98 | <.0001 |

## Stata

**Variable Creation**

```
generate logg1miles=log(g1miles)
```

**Regression Model**

```
regress logg1miles g1yrschl
regress logg1miles logg1yrschl
regress logg1miles g1yrschl sqg1yrschl
```

**Select Results**

```
      Source |       SS       df       MS              Number of obs =    5472
-------------+------------------------------           F(  1,  5470) =  110.61
       Model | 631.719424     1  631.719424           Prob > F      =  0.0000
    Residual | 31239.4079  5470  5.71104349           R-squared     =  0.0198
-------------+------------------------------           Adj R-squared =  0.0196
       Total | 31871.1273  5471  5.82546652           Root MSE      =  2.3898

  logg1miles |      Coef.   Std. Err.      t    P>|t|     [95% Conf. Interval]
-------------+----------------------------------------------------------------
    g1yrschl |   .1155714   .0109887    10.52   0.000     .0940292    .1371136
       _cons |   1.974236   .1293759    15.26   0.000     1.720608    2.227864
```

```
      Source |       SS       df       MS              Number of obs =    5472
-------------+------------------------------           F(  1,  5470) =   70.33
       Model | 404.565542     1  404.565542           Prob > F      =  0.0000
    Residual | 31466.5618  5470  5.75257071           R-squared     =  0.0127
-------------+------------------------------           Adj R-squared =  0.0125
       Total | 31871.1273  5471  5.82546652           Root MSE      =  2.3985

  logg1miles |      Coef.   Std. Err.      t    P>|t|     [95% Conf. Interval]
-------------+----------------------------------------------------------------
 logg1yrschl |   .8140615   .0970719     8.39   0.000     .6237619    1.004361
       _cons |   1.275927   .2425596     5.26   0.000     .8004137    1.75144
```

```
      Source |       SS       df       MS              Number of obs =    5472
-------------+------------------------------           F(  2,  5469) =   63.37
       Model | 721.841633     2  360.920816           Prob > F      =  0.0000
    Residual | 31149.2857  5469  5.69560901           R-squared     =  0.0226
-------------+------------------------------           Adj R-squared =  0.0223
       Total | 31871.1273  5471  5.82546652           Root MSE      =  2.3865

  logg1miles |      Coef.   Std. Err.      t    P>|t|     [95% Conf.Interval]
-------------+----------------------------------------------------------------
    g1yrschl |  -.0635735   .0463536    -1.37   0.170    -.154445    .027298
  sqg1yrschl |   .0087078   .0021891     3.98   0.000     .0044163    .0129993
       _cons |   2.809548   .2465555    11.40   0.000     2.326202    3.292895
```

■ **Display B.9.3 Calculating** $\widehat{\exp(\varepsilon)}$ **and Approximate *R*-squareds: Example for** *log–sq* **model**

| | SAS | Stata |
|---|---|---|
| Calculate Predictions | ```
proc reg;
  model logglmiles=glyrschl sqglyrschl;
  output out=predict(keep=glmiles logYhat
  logResid)predicted=logYhat residual=logResid;
run;
data predict2;
  set predict;
  explogYhat=exp(logYhat);
  explogResid=exp(logResid);
run;
``` | ```
regress logglmiles glyrschl sqglyrschl
predict logYhat
predict logResid, residuals
generate explogYhat=exp(logYhat)
generate explogResid=exp(logResid)
``` |
| Calculate $\widehat{\exp(\varepsilon)}$ | ```
proc means data=predict2;
  var explogResid;
run;
``` | ```
summarize explogResid
``` |
| Mean Results | **Analysis Variable: explogResid** <br><br> | N | Mean | Std Dev | Minimum | Maximum | <br> 5472 | 10.7577435 | 26.1452245 | 0.0143307 | 606.5467606 | | Variable \| Obs Mean Std. Dev. Min Max <br> explogResid\| 5472 10.75774 26.14523 .0143307 606.5466 |
| Estimate R-squared | ```
proc reg data=predict2;
  model glmiles=explogYhat;
run;
``` | ```
regress glmiles explogYhat
``` |
| Regression Results | Root MSE 642.76634  R Square 0.0075 <br> Dependent Mean 291.14912  Adj R-Sq 0.0073 <br> Coeff Var 220.76877 <br><br> **Parameter Estimates** <br> Variable \| DF \| Parameter Estimate \| Standard Error \| t Value \| Pr>\|t\| <br> Intercept \| 1 \| 160.35487 \| 22.17333 \| 7.23 \| <.0001 <br> explogYhat \| 1 \| 4.52556 \| 0.70585 \| 6.41 \| <.0001 | Source \| SS df MS  Number of obs = 5472 <br> F( 1, 5470) = 41.11 <br> Model \| 16983584.2 1 16983584.2  Prob > F = 0.0000 <br> Residual \| 2.2599e+9 5470 413148.574  R-squared = 0.0075 <br>  Adj R-squared = 0.0073 <br> Total \| 2.2769e+09 5471 416177.35  Root MSE = 642.77 <br><br> glmiles \| Coef. Std. Err. t P>\|t\| [95% Conf. Interval] <br> explogYhat \| 4.52556 .7058474 6.41 0.000 3.141819 5.909302 <br> _cons \| 160.3549 22.17332 7.23 0.000 116.8864 203.8234 |

**APPENDIX B**

## ■ Display B.9.4 Using a Dummy Variable Specification for Number of Sisters

### Create Variables

**SAS**
```
g2numsis1=g2numsis==1;
g2numsis2=g2numsis==2;
g2numsis3=g2numsis==3;
g2numsis4=g2numsis==4;
g2numsis5=g2numsis==5;
g2numsis6=g2numsis==6;
g2numsis7p=g2numsis>=7;
```

**Stata**
```
generate g2numsis1=g2numsis==1;
generate g2numsis2=g2numsis==2;
generate g2numsis3=g2numsis==3;
generate g2numsis4=g2numsis==4;
generate g2numsis5=g2numsis==5;
generate g2numsis6=g2numsis==6;
generate g2numsis7p=g2numsis>=7;
```

### Estimate Regression

**SAS**
```
proc reg data=nonlinear;
  model logg1miles=g2numsis1 g2numsis2
  g2numsis3 g2numsis4 g2numsis5
  g2numsis6 g2numsis7p;
run;
```

**Stata**
```
regress logg1miles g2numsis1 g2numsis2 g2numsis3 ///
  g2numsis4 g2numsis5 g2numsis6 g2numsis7p
```

### Regression Results

**SAS**

| Number of Observations Read | 5472 |
|---|---|
| Number of Observations Used | 5472 |

Analysis of Variance

| Source | DF | Sum of Squares | Mean Square | F Value | Pr > F |
|---|---|---|---|---|---|
| Model | 7 | 109.96480 | 15.70926 | 2.70 | 0.0085 |
| Error | 5464 | 31761 | 5.81280 | | |
| Corrected Total | 5471 | 31871 | | | |

| Root MSE | 2.41098 | R-Square | 0.0035 |
|---|---|---|---|
| Dependent Mean | 3.29182 | Adj R-Sq | 0.0022 |
| Coeff Var | 73.24148 | | |

Parameter Estimates

| Variable | DF | Parameter Estimate | Standard Error | t Value | Pr > \|t\| |
|---|---|---|---|---|---|
| Intercept | 1 | 3.28466 | 0.05584 | 58.82 | <.0001 |
| g2numsis1 | 1 | 0.10992 | 0.08191 | 1.34 | 0.1797 |
| g2numsis2 | 1 | -0.04876 | 0.09454 | -0.52 | 0.6060 |
| g2numsis3 | 1 | 0.03772 | 0.12230 | 0.31 | 0.7578 |
| g2numsis4 | 1 | -0.18858 | 0.16384 | -1.15 | 0.2498 |
| g2numsis5 | 1 | -0.45919 | 0.21950 | -2.09 | 0.0365 |
| g2numsis6 | 1 | -0.58571 | 0.29153 | -2.01 | 0.0446 |
| g2numsis7p | 1 | 0.71156 | 0.33281 | 2.14 | 0.0326 |

**Stata**

| Source | SS | df | MS |
|---|---|---|---|
| Model | 109.964799 | 7 | 15.709257 |
| Residual | 31761.1625 | 5464 | 5.81280427 |
| Total | 31871.1273 | 5471 | 5.82546652 |

```
Number of obs =   5472
F( 7,  5464)  =   2.70
Prob > F      = 0.0085
R-squared     = 0.0035
Adj R-squared = 0.0022
Root MSE      =  2.411
```

| logg1miles | Coef. | Std. Err. | t | P>\|t\| | [95% Conf. Interval] |
|---|---|---|---|---|---|
| g2numsis1 | .109918 | .0819074 | 1.34 | 0.180 | -.0506532 .2704891 |
| g2numsis2 | -.0487583 | .0945361 | -0.52 | 0.606 | -.2340868 .1365702 |
| g2numsis3 | .0377221 | .1222995 | 0.31 | 0.758 | -.2020336 .2774778 |
| g2numsis4 | -.188578 | .163842 | -1.15 | 0.250 | -.5097735 .1326175 |
| g2numsis5 | -.4591877 | .2194971 | -2.09 | 0.036 | -.8894893 -.028886 |
| g2numsis6 | -.5857075 | .2915286 | -2.01 | 0.045 | -1.15722 -.0141953 |
| g2numsis7p | .7115629 | .3328107 | 2.14 | 0.033 | .0591213 1.364004 |
| _cons | 3.284659 | .0558431 | 58.82 | 0.000 | 3.175184 3.394134 |

**Display B.3.3 Using a Dummy Variable Specification for Number of Sisters, Compared...**

## Create Variables

SAS:
```
g2numsis56 =g2numsis=5 | g2numsis=6;
```

Stata:
```
gen g2numsis56=g2numsis==5 | g2numsis==6
```

## Estimate Regression

SAS:
```
proc reg data=nonlinear;
   model logg1miles=g2numsis4 g2numsis56
         g2numsis7p;
run;

test g2numsis4=g2numsis56;
test g2numsis4=g2numsis7p;
test g2numsis56=g2numsis7p;
run;
```

Stata:
```
regress logg1miles g2numsis4 g2numsis56 g2numsis7p
test g2numsis4=g2numsis56
test g2numsis4=g2numsis7p
test g2numsis56=g2numsis7p
```

## Regression Results

### SAS

Parameter Estimates

| Variable | DF | Parameter Estimate | Standard Error | t Value | Pr>|t| |
|---|---|---|---|---|---|
| Intercept | 1 | 3.31437 | 0.03419 | 96.95 | <.0001 |
| g2numsis4 | 1 | −0.21829 | 0.15777 | −1.38 | 0.1665 |
| g2numsis56 | 1 | −0.53382 | 0.17386 | −3.07 | 0.0021 |
| g2numsis7p | 1 | 0.68185 | 0.32985 | 2.07 | 0.0388 |

Test 1 Results for Dependent Variable logg1miles

| Source | DF | Mean Square | F Value | Pr > F |
|---|---|---|---|---|
| Numerator | 1 | 10.96229 | 1.89 | 0.1697 |
| Denominator | 5468 | 5.81202 | | |

Test 2 Results for Dependent Variable logg1miles

| Source | DF | Mean Square | F Value | Pr > F |
|---|---|---|---|---|
| Numerator | 1 | 35.85169 | 6.17 | 0.0130 |
| Denominator | 5468 | 5.81202 | | |

Test 3 Results for Dependent Variable logg1miles

| Source | DF | Mean Square | F Value | Pr > F |
|---|---|---|---|---|
| Numerator | 162.83736 | 10.81 | 0.0010 | |
| Denominator | 5468 | 5.81202 | | |

### Stata

```
    Source |       SS       df       MS              Number of obs =    5472
-----------+------------------------------           F(  3,  5468) =    5.22
     Model | 90.9882392      3  30.3294131           Prob > F      =  0.0013
  Residual | 31780.1391   5468  5.81202251           R-squared     =  0.0029
-----------+------------------------------           Adj R-squared =  0.0023
     Total | 31871.1273   5471  5.82546652           Root MSE      =  2.4108

------------------------------------------------------------------------------
logg1miles |      Coef.   Std. Err.      t    P>|t|     [95% Conf. Interval]
-----------+------------------------------------------------------------------
 g2numsis4 |  -.2182923   .1577696    -1.38   0.167    -.5275836    .0909989
g2numsis56 |  -.5338165   .1738644    -3.07   0.002    -.8746599   -.1929731
g2numsis7p |   .6618486   .3298466     2.07   0.039     .0352181    1.32479
     _cons |   3.314373   .0341865    96.95   0.000     3.247354    3.381392
------------------------------------------------------------------------------

. test g2numsis4=g2numsis56

 (1)  g2numsis4 - g2numsis56 = 0

       F(  1,  5468) =    1.89
            Prob > F =    0.1697

. test g2numsis4=g2numsis7p

 (1)  g2numsis4 - g2numsis7p = 0

       F(  1,  5468) =    6.17
            Prob > F =    0.0130

. test g2numsis56=g2numsis7p

 (1)  g2numsis56 - g2numsis7p = 0

       F(  1,  5468) =   10.81
            Prob > F =    0.0010
```

**APPENDIX B**

■ Display B.10.1 Regression of Mens' Hours of Work on their Self-Reported Health, with and without Controls for Age

| | SAS | Stata |
|---|---|---|
| Commands | `proc reg;`<br>`  model hrwork=Health;`<br>`  model hrwork=Health g2age;`<br>`run;` | `regress hrwork Health`<br>`regress hrwork Health g2age` |

**SAS Results**

Parameter Estimates

| Variable | DF | Parameter Estimate | Standard Error | t Value | Pr > \|t\| |
|---|---|---|---|---|---|
| Intercept | 1 | 37.86048 | 1.28924 | 29.37 | <.0001 |
| Health | 1 | 1.14351 | 0.30536 | 3.74 | 0.0002 |

Parameter Estimates

| Variable | DF | Parameter Estimate | Standard Error | t Value | Pr > \|t\| |
|---|---|---|---|---|---|
| Intercept | 1 | 42.26619 | 1.51318 | 27.93 | <.0001 |
| Health | 1 | 1.01681 | 0.30504 | 3.33 | 0.0009 |
| g2age | 1 | −0.10497 | 0.01907 | −5.50 | <.0001 |

**Stata Results**

```
      Source |       SS       df       MS              Number of obs =    3742
-------------+------------------------------           F(  1,  3740) =   14.02
       Model | 2833.01506      1 2833.01506            Prob > F      =  0.0002
    Residual | 755569.781   3740 202.024006            R-squared     =  0.0037
-------------+------------------------------           Adj R-squared =  0.0035
       Total | 758402.796   3741 202.727291            Root MSE      =  14.214

------------------------------------------------------------------------------
      hrwork |      Coef.   Std. Err.      t    P>|t|     [95% Conf. Interval]
-------------+----------------------------------------------------------------
      Health |   1.143507   .3053625     3.74   0.000     .5448133    1.7422
       _cons |   37.86048   1.289238    29.37   0.000     35.3328    40.38816
------------------------------------------------------------------------------
```

```
      Source |       SS       df       MS              Number of obs =    3742
-------------+------------------------------           F(  2,  3739) =   22.22
       Model | 8907.66109      2 4453.83055            Prob > F      =  0.0000
    Residual | 749495.135   3739 200.453366            R-squared     =  0.0117
-------------+------------------------------           Adj R-squared =  0.0112
       Total | 758402.796   3741 202.727291            Root MSE      =  14.158

------------------------------------------------------------------------------
      hrwork |      Coef.   Std. Err.      t    P>|t|     [95% Conf. Interval]
-------------+----------------------------------------------------------------
      Health |   1.016814   .3050426     3.33   0.001     .4187475    1.61488
       g2age |   -.104973   .0190688    -5.50   0.000    -.1423593   -.0675867
       _cons |   42.26619   1.513182    27.93   0.000     39.29944   45.23293
------------------------------------------------------------------------------
```

**■ Display B.10.2 Regression Models Used in Estimating Indirect and Direct Effects of Men's Health on their Hours of Work, with Mediation through Health Limitations**

## SAS

**Commands**

```
proc reg;
    model hrwork=Health;
    model hrwork=Health HlthLimit;
    model HlthLimit=Health;
run;
```

**Results**

Parameter Estimates

| Variable | DF | Parameter Estimate | Standard Error | t Value | Pr > |t| |
|---|---|---|---|---|---|
| Intercept | 1 | 37.86048 | 1.28924 | 29.37 | <.0001 |
| Health | 1 | 1.14351 | 0.30536 | 3.74 | 0.0002 |

Parameter Estimates

| Variable | DF | Parameter Estimate | Standard Error | t Value | Pr > |t| |
|---|---|---|---|---|---|
| Intercept | 1 | 38.52018 | 1.31985 | 29.19 | <.0001 |
| Health | 1 | 1.00813 | 0.31078 | 3.24 | 0.0012 |
| HlthLimit | 1 | −5.65588 | 2.45198 | −2.31 | 0.0211 |

Parameter Estimates

| Variable | DF | Parameter Estimate | Standard Error | t Value | Pr > |t| |
|---|---|---|---|---|---|
| Intercept | 1 | 0.11664 | 0.00859 | 13.57 | <.0001 |
| Health | 1 | −0.02394 | 0.00204 | −11.76 | <.0001 |

## Stata

**Commands**

```
regress hrwork Health
regress hrwork Health HlthLimit
regress HlthLimit Health
```

**Results**

```
      Source |       SS       df       MS              Number of obs =    3742
                                                       F( 1,  3740)  =   14.02
       Model |2833.01506      1  2833.01506            Prob > F      =  0.0002
    Residual |755569.781   3740  202.024006            R-squared     =  0.0037
                                                       Adj R-squared =  0.0035
       Total |758402.796   3741  202.727291            Root MSE      =  14.214

      hrwork |     Coef.   Std. Err.      t    P>|t|    [95% Conf. Interval]
      Health |  1.143507   .3053625    3.74   0.000    .5448133   1.7422
       _cons |  37.86048   1.289238   29.37   0.000    35.3328    40.38816
```

```
      Source |       SS       df       MS              Number of obs =    3742
                                                       F( 2,  3739)  =    9.68
       Model |3906.67985      2  1953.33993            Prob > F      =  0.0001
    Residual |754496.117   3739  201.790884            R-squared     =  0.0052
                                                       Adj R-squared =  0.0046
       Total |758402.796   3741  202.727291            Root MSE      =  14.205

      hrwork |     Coef.   Std. Err.      t    P>|t|    [95% Conf. Interval]
      Health |  1.008126   .3107785    3.24   0.001    .3988139   1.617438
   HlthLimit | -5.655881    .451979   -2.31   0.021   -10.46323   -.848535
       _cons |  38.52018   1.319854   29.19   0.000    35.93248   41.10789
```

```
      Source |       SS       df       MS              Number of obs =    3742
                                                       F( 1,  3740)  =  138.32
       Model |1.24132056      1  1.24132056            Prob > F      =  0.0000
    Residual |33.5635748   3740  .008974218            R-squared     =  0.0357
                                                       Adj R-squared =  0.0354
       Total |34.8048953   3741  .009303634            Root MSE      =  .09473

   HlthLimit |     Coef.   Std. Err.      t    P>|t|    [95% Conf. Interval]
      Health | -.0239363   .0020352  -11.76   0.000   -.0279265   -.019946
       _cons |  .1166407   .0085927   13.57   0.000    .0997938   .1334875
```

**APPENDIX B**

## ■ Display B.10.3 Understanding Controlling for with Regression on Residuals for $X_3$

### SAS

**Commands**

```
proc reg data=HoursOfChoresMediate;
   model Health=HlthLimit;
   output out=prede2i (keep=hrwork e2i)
   r=e2i;
run;

proc reg data=prede2i;
   model hrwork=e2i;
run;
```

**Results**

Parameter Estimates

| Variable | DF | Parameter Estimate | Standard Error | t Value | Pr>|t| |
|---|---|---|---|---|---|
| Intercept | 1 | 42.60930 | 0.23246 | 183.30 | <.0001 |
| e2i | 1 | 1.00813 | 0.31110 | 3.24 | 0.0012 |

**Commands**

```
proc reg;
   model HlthLimit=Health;
   output out=prede3i(keep=hrwork e3i)
   r=e3i;
run;

proc reg data=prede3i;
   model hrwork=e3i;
run;
```

**Results**

Parameter Estimates

| Variable | DF | Parameter Estimate | Standard Error | t Value | Pr>|t| |
|---|---|---|---|---|---|
| Intercept | 1 | 42.60930 | 0.23262 | 183.17 | <.0001 |
| e3i | 1 | -5.65588 | 2.45625 | -2.30 | 0.0214 |

### Stata

**Commands**

```
regress Health HlthLimit
predict e2i, residuals
regress hrwork e2i
```

**Results**

```
Source |      SS          df      MS           Number of obs =    3742
-------+------------------------------         F( 1, 3740)   =   10.50
 Model | 2123.38643       1  2123.38643        Prob > F      =  0.0012
Residual| 756279.41     3740  202.213746       R-squared     =  0.0028
-------+------------------------------         Adj R-squared =  0.0025
 Total | 758402.796    3741  202.727291        Root MSE      =   14.22

hrwork |    Coef.   Std. Err.      t    P>|t|   [95% Conf. Interval]
-------+-------------------------------------------------------------
   e2i | 1.008126   .311104     3.24   0.001    .3981759   1.618076
 _cons | 42.6093    .2324628  183.30   0.000    42.15353   43.06507
```

**Commands**

```
regress HlthLimit Health
predict e3i, residuals
regress hrwork e3i
```

**Results**

```
Source |      SS          df      MS           Number of obs =    3742
-------+------------------------------         F( 1, 3740)   =    5.30
 Model | 1073.66477       1  1073.66477        Prob > F      =  0.0214
Residual| 757329.132    3740  202.49442        R-squared     =  0.0014
-------+------------------------------         Adj R-squared =  0.0011
 Total | 758402.796    3741  202.727291        Root MSE      =   14.23

hrwork |    Coef.   Std. Err.      t    P>|t|   [95% Conf. Interval]
-------+-------------------------------------------------------------
   e3i | -5.655881  2.456249    -2.30  0.021    -10.4716   -.8401623
 _cons | 42.6093    .2326241  183.17   0.000    42.15322   43.06538
```

| | SAS | Stata |
|---|---|---|
| **Request Diagnostics** | ```proc reg data=miles; model loggimiles=glyrschl sqglyrschl g2numsis4 g2numsis56 g2numsis7p amind mexamer white other female g2earn10000 g1age g2age g2numbro /r influence; ods output outputstatistics=influence; run;``` | ```regress loggimiles glyrschl sqglyrschl g2numsis4 /// g2numsis56 g2numsis7p amind /// mexamer white other female /// g2earn10000 g1age g2age /// g2numbro  predict HatDiagonal, hat predict RStudent, rstudent predict CooksD, cooksd predict DFFITS, dfits predict DFB_glyrschl, dfbeta(glyrschl) predict DFB_sqglyrschl, dfbeta(sqglyrschl) predict DFB_g2numsis4, dfbeta(g2numsis4) predict DFB_g2numsis56, dfbeta(g2numsis56) predict DFB_g2numsis7p, dfbeta(g2numsis7p)``` |
| **Use Cutoffs to Identify High Values** | ```data influence2; set influence; HatDiagonal3Hi  =HatDiagonal>3*14/5472; RStudentHi=     abs(RStudent)>2; CooksDHi=       CooksD>4/5472; DFFITSHi=     abs(DFFITS)>2*sqrt(14/5472); DFB_glyrschlHi=     abs(DFB_glyrschl)>2/sqrt(5472); DFB_sqglyrschlHi=     abs(DFB_sqglyrschl)>2/sqrt(5472); DFB_g2numsis4Hi=     abs(DFB_g2numsis4)>2/sqrt(5472); DFB_g2numsis56Hi=     abs(DFB_g2numsis56)>2/sqrt(5472); DFB_g2numsis7pHi=     abs(DFB_g2numsis7p)>2/sqrt(5472); run;``` | ```generate HatDiagonal3Hi   =HatDiagonal>3*14/5472 generate RStudentHi       =abs(RStudent)>2 generate CooksDHi         =CooksD>4/5472 generate DFFITSHi         =abs(DFFITS)>2*sqrt(14/5472) generate DFB_glyrschlHi   =abs(DFB_glyrschl)>2/sqrt(5472) generate DFB_sqglyrschlHi =abs(DFB_sqglyrschl)>2/sqrt(5472) generate DFB_g2numsis4Hi  =abs(DFB_g2numsis4)>2/sqrt(5472) generate DFB_g2numsis56Hi =abs(DFB_g2numsis56)>2/sqrt(5472) generate DFB_g2numsis7pHi =abs(DFB_g2numsis7p)>2/sqrt(5472)``` |
| **Sum High Values** | ```proc means data=influence2; var HatDiagonal3Hi RStudentHi CooksDHi DFFITSHi DFB_glyrschlHi DFB_sqglyrschlHi DFB_g2numsis4Hi DFB_g2numsis56Hi DFB_g2numsis7pHi; run;``` | ```summarize HatDiagonal3Hi RStudentHi CooksDHi DFFITSHi /// DFB_glyrschlHi DFB_sqglyrschlHi DFB_g2numsis4Hi /// DFB_g2numsis56Hi DFB_g2numsis7pHi``` |

**SAS output table**

| Variable | N | Mean | Std Dev | Minimum | Maximum |
|---|---|---|---|---|---|
| HatDiagonal3Hi | 5472 | 0.0575658 | 0.2329418 | 0 | 1.0000000 |
| RStudentHi | 5472 | 0.0135234 | 0.1155117 | 0 | 1.0000000 |
| CooksDHi | 5472 | 0.0467836 | 0.2111944 | 0 | 1.0000000 |
| DFFITSHi | 5472 | 0.0506213 | 0.2192433 | 0 | 1.0000000 |
| DFB_glyrschlHi | 5472 | 0.0383772 | 0.1921227 | 0 | 1.0000000 |
| DFB_sqglyrschlHi | 5472 | 0.0498904 | 0.2177383 | 0 | 1.0000000 |
| DFB_g2numsis4Hi | 5472 | 0.0314327 | 0.1745001 | 0 | 1.0000000 |
| DFB_g2numsis56Hi | 5472 | 0.0279605 | 0.1648748 | 0 | 1.0000000 |
| DFB_g2numsis7pHi | 5472 | 0.0098684 | 0.0988576 | 0 | 1.0000000 |

**Stata output table**

| Variable | Obs | Mean | Std. Dev. | Min | Max |
|---|---|---|---|---|---|
| HatDiagona~i | 5472 | .0575658 | .2329418 | 0 | 1 |
| RStudentHi | 5472 | .0135234 | .1155117 | 0 | 1 |
| CooksDHi | 5472 | .0467836 | .2111944 | 0 | 1 |
| DFFITSHi | 5472 | .0506213 | .2192433 | 0 | 1 |
| DFB_glyrsc~i | 5472 | .0383772 | .1921227 | 0 | 1 |
| DFB_sqglyr~i | 5472 | .0498904 | .2177383 | 0 | 1 |
| DFB_g2nu~4Hi | 5472 | .0314327 | .1745001 | 0 | 1 |
| DFB_g2nu~6Hi | 5472 | .0279605 | .1648748 | 0 | 1 |
| DFB_g2nu~pHi | 5472 | .0098684 | .0988576 | 0 | 1 |

# ■ Display B.11.2 Summary Statistics by *anyhi*

## SAS

**Commands**

```
data influence3;
  merge miles influence2;
run;
proc sort data=influence3; by anyhi; run;
proc means data=influence3;
  var glyrschl sqglyrschl g2numsis4 g2numsis56
      g2numsis7p amind mexamer white other
      female g2earn10000 g1age g2age g2numbro;
  by anyhi;
run;
```

**Results**

anyhi=0

| Variable | N | Mean | Std Dev | Minimum | Maximum |
|---|---|---|---|---|---|
| glyrschl | 4694 | 11.6516830 | 2.4922425 | 3.0000000 | 17.0000000 |
| sqglyrschl | 4694 | 141.9716660 | 56.9249794 | 9.0000000 | 289.0000000 |
| g2numsis4 | 4694 | 0.0132084 | 0.1141782 | 0 | 1.0000000 |
| g2numsis56 | 4694 | 0.0085215 | 0.0919277 | 0 | 1.0000000 |
| g2numsis7p | 4694 | 0 | 0 | 0 | 0 |
| amind | 4694 | 0 | 0 | 0 | 0 |
| mexamer | 4694 | 0.0102258 | 0.1006152 | 0 | 1.0000000 |
| white | 4694 | 0.8596080 | 0.3474303 | 0 | 1.0000000 |
| other | 4694 | 0 | 0 | 0 | 0 |
| female | 4694 | 0.5988496 | 0.4901836 | 0 | 1.0000000 |
| g2earn10000 | 4694 | 3.0421728 | 2.9186966 | 0 | 25.1610274 |
| g1age | 4694 | 59.9443971 | 11.1694472 | 29.0000000 | 95.0000000 |
| g2age | 4694 | 34.2645931 | 9.9245113 | 16.0000000 | 76.0000000 |
| g2numbro | 4694 | 1.3046442 | 1.3951270 | 0 | 9.0000000 |

anyhi=1

| Variable | N | Mean | Std Dev | Minimum | Maximum |
|---|---|---|---|---|---|
| glyrschl | 778 | 9.8856041 | 4.5470257 | 0 | 17.0000000 |
| sqglyrschl | 778 | 118.3740360 | 85.1018177 | 0 | 289.0000000 |
| g2numsis4 | 778 | 0.2352185 | 0.4244082 | 0 | 1.0000000 |
| g2numsis56 | 778 | 0.2056555 | 0.404398 | 0 | 1.0000000 |
| g2numsis7p | 778 | 0.0694087 | 0.2543114 | 0 | 1.0000000 |
| amind | 778 | 0.0321337 | 0.1764685 | 0 | 1.0000000 |
| mexamer | 778 | 0.1516710 | 0.3589324 | 0 | 1.0000000 |
| white | 778 | 0.5102828 | 0.5002158 | 0 | 1.0000000 |
| other | 778 | 0.0912596 | 0.2881632 | 0 | 1.0000000 |
| female | 778 | 0.6272494 | 0.4838476 | 0 | 1.0000000 |
| g2earn10000 | 778 | 3.2419257 | 6.8925063 | 0 | 94.5903285 |
| g1age | 778 | 60.9515568 | 12.0444342 | 30.0000000 | 95.0000000 |
| g2age | 778 | 34.4704370 | 10.5143048 | 17.0000000 | 73.0000000 |

## Stata

**Commands**

```
bysort anyhi: summarize glyrschl sqglyrschl g2numsis4 g2numsis56
                        g2numsis7p amind
                        mexamer white other
                        female g2earn10000 g1age
                        g2age g2numbro
```

**Results**

-> anyhi = 0

| Variable | Obs | Mean | Std. Dev. | Min | Max |
|---|---|---|---|---|---|
| glyrschl | 4694 | 11.65168 | 2.492242 | 3 | 17 |
| sqglyrschl | 4694 | 141.9717 | 56.92498 | 9 | 289 |
| g2numsis4 | 4694 | .0132084 | .1141782 | 0 | 1 |
| g2numsis56 | 4694 | .0085215 | .0919277 | 0 | 1 |
| g2numsis7p | 4694 | 0 | 0 | 0 | 0 |
| amind | 4694 | 0 | 0 | 0 | 0 |
| mexamer | 4694 | .0102258 | .1006152 | 0 | 1 |
| white | 4694 | .859608 | .3474303 | 0 | 1 |
| other | 4694 | 0 | 0 | 0 | 0 |
| female | 4694 | .5988496 | .4901836 | 0 | 1 |
| g2earn10000 | 4694 | 3.042173 | 2.918697 | 0 | 25.16103 |
| g1age | 4694 | 59.9444 | 11.16945 | 29 | 95 |
| g2age | 4694 | 34.26459 | 9.924511 | 16 | 76 |
| g2numbro | 4694 | 1.304644 | 1.395127 | 0 | 9 |

-> anyhi = 1

| Variable | Obs | Mean | Std. Dev. | Min | Max |
|---|---|---|---|---|---|
| glyrschl | 778 | 9.885604 | 4.547026 | 0 | 17 |
| sqglyrschl | 778 | 118.374 | 85.10182 | 0 | 289 |
| g2numsis4 | 778 | .2352185 | .4244082 | 0 | 1 |
| g2numsis56 | 778 | .2056555 | .404398 | 0 | 1 |
| g2numsis7p | 778 | .0694087 | .2543114 | 0 | 1 |
| amind | 778 | .0321337 | .1764685 | 0 | 1 |
| mexamer | 778 | .151671 | .3589324 | 0 | 1 |
| white | 778 | .5102828 | .5002158 | 0 | 1 |
| other | 778 | .0912596 | .2881632 | 0 | 1 |
| female | 778 | .6272494 | .4838476 | 0 | 1 |
| g2earn10000 | 778 | 3.241926 | 6.892506 | 0 | 94.59033 |
| g1age | 778 | 60.95116 | 12.0443 | 30 | 95 |
| g2age | 778 | 34.47044 | 10.5143 | 17 | 73 |
| g2numbro | 778 | 2.209512 | 2.194192 | 0 | 19 |

■ **Display B.11.3 Multiple Regression on Full Sample**

|  | SAS | Stata |
|---|---|---|
| Commands | ```proc reg data=miles;   model logg1miles=g1yrschl sqg1yrschl        g2numsis4 g2numsis56        g2numsis7p amind        mexamer white other        female g2earn10000        g1age g2age g2numbro; run;``` | ```regress logg1miles g1yrschl sqg1yrschl g2numsis4   ///                    g2numsis56 g2numsis7p   ///                    amind mexamer white other   ///                    female g2earn10000   ///                    g1age g2age g2numbro``` |

**SAS Results**

Parameter Estimates

| Variable | DF | Parameter Estimate | Standard Error | t Value | Pr > \|t\| |
|---|---|---|---|---|---|
| Intercept | 1 | 1.11740 | 0.36508 | 3.06 | 0.0022 |
| g1yrschl | 1 | -0.04726 | 0.04877 | -0.97 | 0.3326 |
| sqg1yrschl | 1 | 0.00822 | 0.00225 | 3.66 | 0.0003 |
| g2numsis4 | 1 | -0.02737 | 0.15664 | -0.17 | 0.8613 |
| g2numsis56 | 1 | -0.27900 | 0.17650 | -1.58 | 0.1140 |
| g2numsis7p | 1 | 0.98506 | 0.32826 | 3.00 | 0.0027 |
| amind | 1 | 1.14527 | 0.47983 | 2.39 | 0.0170 |
| mexamer | 1 | -0.14627 | 0.21035 | -0.70 | 0.4868 |
| white | 1 | 0.41083 | 0.09437 | 4.35 | <.0001 |
| other | 1 | 0.96906 | 0.29356 | 3.30 | 0.0010 |
| female | 1 | -0.07250 | 0.06968 | -1.04 | 0.2982 |
| g2earn10000 | 1 | 0.03740 | 0.00932 | 4.01 | <.0001 |
| g1age | 1 | 0.01005 | 0.00539 | 1.87 | 0.0621 |
| g2age | 1 | 0.01442 | 0.00612 | 2.35 | 0.0186 |
| g2numbro | 1 | 0.04218 | 0.02152 | 1.96 | 0.0500 |

**Stata Results**

| Source | SS | df | MS |
|---|---|---|---|
| Model | 1530.79748 | 14 | 109.342677 |
| Residual | 30340.3298 | 5457 | 5.55989185 |
| Total | 31871.1273 | 5471 | 5.82546652 |

```
Number of obs =    5472
F( 14,  5457) =   19.67
Prob > F      =  0.0000
R-squared     =  0.0480
Adj R-squared =  0.0456
Root MSE      =  2.3579
```

| logg1miles | Coef. | Std. Err. | t | P>\|t\| | [95% Conf. | Interval] |
|---|---|---|---|---|---|---|
| g1yrschl | -.0472574 | .048767 | -0.97 | 0.333 | -.1428602 | .0483454 |
| sqg1yrschl | .0082153 | .0022461 | 3.66 | 0.000 | .003812 | .0126187 |
| g2numsis4 | -.0273686 | .156636 | -0.17 | 0.861 | -.3344377 | .2797005 |
| g2numsis56 | -.279005 | .1765034 | -1.58 | 0.114 | -.625022 | .0670121 |
| g2numsis7p | .985055 | .3282596 | 3.00 | 0.003 | .3415353 | 1.628575 |
| amind | 1.145266 | .47983 | 2.39 | 0.017 | .2046076 | 2.085924 |
| mexamer | -.1462737 | .2103489 | -0.70 | 0.487 | -.5586414 | .266094 |
| white | .4108335 | .0943698 | 4.35 | 0.000 | .225831 | .5958359 |
| other | .9690582 | .293564 | 3.30 | 0.001 | .3935557 | 1.544561 |
| female | -.0724955 | .0696832 | -1.04 | 0.298 | -.2091024 | .0641113 |
| g2earn10000 | .0374032 | .0093177 | 4.01 | 0.000 | .0191369 | .0556696 |
| g1age | .0100507 | .0053863 | 1.87 | 0.062 | -.0005086 | .0206099 |
| g2age | .0144181 | .0061237 | 2.35 | 0.019 | .0024131 | .026423 |
| g2numbro | .0421847 | .0215228 | 1.96 | 0.050 | -8.53e-06 | .0843779 |
| _cons | 1.117402 | .365079 | 3.06 | 0.002 | .4017015 | 1.833102 |

■ **Display B.11.4 Multiple Regression on Partial Sample (with Outliers and Influential Observations Excluded)**

**Commands**

```
proc reg data=influence3;
model logg1miles=g1yrschl sqg1yrschl g2numsis4
    g2numsis56 mexamer white female
    g2earn10000 g1age g2age g2numbro;
by anyhi;
run;
```

```
bysort anyhi: regress logg1miles g1yrschl sqg1yrschl   ///
    g2numsis4 g2numsis56                               ///
    mexamer white female g2earn10000                   ///
    g1age g2age g2numbro
```

**Results**

anyhi=0

**Parameter Estimates**

| Variable | DF | Parameter Estimate | Standard Error | t Value | Pr > |t| |
|---|---|---|---|---|---|
| Intercept | 1 | 0.08020 | 0.53581 | 0.15 | 0.8810 |
| g1yrschl | 1 | -0.01033 | 0.08020 | -0.13 | 0.8975 |
| sqg1yrschl | 1 | 0.00783 | 0.00348 | 2.25 | 0.0245 |
| g2numsis4 | 1 | -0.07464 | 0.28542 | -0.26 | 0.7937 |
| g2numsis56 | 1 | -0.32919 | 0.35814 | -0.92 | 0.3580 |
| mexamer | 1 | -0.46111 | 0.33773 | -1.37 | 0.1722 |
| white | 1 | 0.59151 | 0.09960 | 6.00 | <.0001 |
| female | 1 | 0.02331 | 0.07320 | 0.32 | 0.7502 |
| g2earn10000 | 1 | 0.07116 | 0.01265 | 5.63 | <.0001 |
| g1age | 1 | 0.01328 | 0.00572 | 2.32 | 0.0202 |
| g2age | 1 | 0.01640 | 0.00644 | 2.55 | 0.0110 |
| g2numbro | 1 | 0.05895 | 0.02381 | 2.48 | 0.0133 |

```
-> anyhi = 0

    Source |       SS       df       MS              Number of obs =    4694
-----------+------------------------------          F( 11, 4682)  =   33.25
     Model | 1807.83975    11  164.349068           Prob > F      =  0.0000
  Residual | 23144.4657  4682  4.94328615           R-squared     =  0.0725
-----------+------------------------------          Adj R-squared =  0.0703
     Total | 24952.3055  4693  5.31691999           Root MSE      =  2.2234

 logg1miles |      Coef.   Std. Err.      t    P>|t|     [95% Conf. Interval]
   g1yrschl | -.0103337    .080199   -0.13   0.897    -.1675614    .1468941
 sqg1yrschl |  .0078314   .0034796    2.25   0.024    -.0010096    .0146531
  g2numsis4 | -.0746374   .2854243   -0.26   0.794    -.6342034    .4849286
 g2numsis56 | -.3291949   .3581387   -0.92   0.358    -1.031315    .3729254
    mexamer |  -.461109   .3377253   -1.37   0.172     -1.12321    .2009917
      white |  .5915109   .0995972    6.00   0.000      .398214    .7848078
     female |  .0233067   .0731999    0.32   0.750    -.1201995     .166813
g2earn10000 |  .0711568    .012648    5.63   0.000     .0463607    .0959528
      g1age |  .0132813   .0057172    2.32   0.020     .0020728    .0244897
      g2age |  .0164005   .0064434    2.55   0.011     .0037683    .0290326
   g2numbro |  .0589484   .0238129    2.48   0.013      .012264    .1056328
      _cons |  .0802039   .5358109    0.15   0.881    -.9702378    1.130646
```

■ **Display B.11.5 Multiple Regression for Full Sample with Heteroskedasticity-Consistent Standard Errors (HC3)**

| | SAS | Stata |
|---|---|---|
| Commands | ```%include "c:/nsfh_distance/sas/hcreg.sas"; %HCREG(data = miles, dv = logglmiles, iv = glyrschl sqglyrschl g2numsis4 g2numsis56 g2numsis7p amind mexamer white other female g2earnl0000 glage g2age g2numbro);``` | ```regress logglmiles glyrschl sqglyrschl g2numsis4 /// g2numsis56 g2numsis7p amind mexamer white other /// female g2earnl0000 glage g2age g2numbro /// , vce(hc3)``` |

Stata results:

```
Linear regression                    Number of obs =     5472
                                     F( 14,  5457) =    19.55
                                     Prob > F      =   0.0000
                                     R-squared     =   0.0480
                                     Root MSE      =   2.3579
```

| logglmiles | Coef. | Robust Std. Err. | t | P>|t| | [95% Conf. | Interval] |
|---|---|---|---|---|---|---|
| glyrschl | -.0472574 | .052199 | -0.91 | 0.365 | -.1495883 | .0550735 |
| sqglyrschl | .0082153 | .002876 | 3.44 | 0.001 | .003348 | .0128959 |
| g2numsis4 | -.0273686 | .155564 | -0.18 | 0.860 | -.332236 | .2775989 |
| g2numsis56 | -.279005 | .1843165 | -1.51 | 0.130 | -.6403388 | .0823289 |
| g2numsis7p | .985055 | .3685039 | 2.67 | 0.008 | .2626404 | 1.70747 |
| amind | 1.145266 | .4044861 | 2.83 | 0.005 | .3523117 | 1.93822 |
| mexamer | -.1462737 | .2047908 | -0.71 | 0.475 | -.5477454 | .2551979 |
| white | .4108335 | .0948416 | 4.33 | 0.000 | .2249061 | .5967608 |
| other | .9690582 | .3205739 | 3.02 | 0.003 | .3406054 | 1.597511 |
| female | -.0722955 | .0721381 | -1.00 | 0.315 | -.213915 | .0689239 |
| g2earn10000 | .0374032 | .0118305 | 3.16 | 0.002 | .0142108 | .0605957 |
| glage | .0100507 | .0053679 | 1.87 | 0.061 | -.0004725 | .0205738 |
| g2age | .0144181 | .0060699 | 2.38 | 0.018 | .0025186 | .0263175 |
| g2numbro | .0421847 | .021343 | 1.98 | 0.048 | .0003438 | .0840256 |
| _cons | 1.117402 | .382864 | 2.92 | 0.004 | .368358 | 1.867968 |

SAS results:

Heteroscedasticity-Consistent Regression Results

| OPUT | Coeff | SE(HC) | t | P > |t| |
|---|---|---|---|---|
| CONSTANT | 1.1174 | 0.3829 | 2.9185 | 0.0035 |
| G1YRSCHL | -0.0473 | 0.0522 | -0.9053 | 0.3653 |
| SQG1YRSCHL | 0.0082 | 0.0024 | 3.4409 | 0.0006 |
| G2NUMSIS4 | -0.0274 | 0.1556 | -0.1759 | 0.8604 |
| G2NUMSIS56 | -0.2790 | 0.1843 | -1.5137 | 0.1302 |
| G2NUMSIS7P | 0.9851 | 0.3685 | 2.6731 | 0.0075 |
| AMIND | 1.1453 | 0.4045 | 2.8314 | 0.0047 |
| MEXAMER | -0.1463 | 0.2048 | -0.7143 | 0.4751 |
| WHITE | 0.4108 | 0.0948 | 4.3318 | 0.0000 |
| OTHER | 0.9691 | 0.3206 | 3.0229 | 0.0025 |
| FEMALE | -0.0725 | 0.0721 | -1.0050 | 0.3150 |
| G2EARN10000 | 0.0374 | 0.0118 | 3.1616 | 0.0016 |
| G1AGE | 0.0101 | 0.0054 | 1.8724 | 0.0612 |
| G2AGE | 0.0144 | 0.0061 | 2.3753 | 0.0176 |
| G2NUMBRO | 0.0422 | 0.0213 | 1.9765 | 0.0481 |

Select Results

## ■ Display B.11.6 Calculation of Variance Inflation Factors in *Hypothetical Data Set*

### SAS

**Commands**

```
proc reg; model yrschl=higrade7 impschl7 /vif;
run;
proc reg; model higrade7=impschl7; run;
```

**Results**

Analysis of Variance

| Variable | DF | Sum of Squares | Mean Square | F Value | Pr > F |
|---|---|---|---|---|---|
| Model | 2 | 34.68552 | 17.34276 | 9.66 | <.0001 |
| Error | 997 | 1790.51348 | 1.79590 | | |
| Corrected Total | 999 | 1825.19900 | | | |

Parameter Estimates

| Variable | DF | Parameter Estimate | Standard Error | t Value | Pr > |t| | Variance Inflation |
|---|---|---|---|---|---|---|
| Intercept | 1 | 8.25317 | 0.64637 | 12.77 | <.0001 | 0 |
| higrade7 | 1 | -0.48197 | 0.43731 | -1.10 | 0.2707 | 105.88741 |
| impschl7 | 1 | 6.56431 | 4.34626 | 1.51 | 0.1313 | 105.88741 |

| Root MSE | 0.09700 | R-Square | 0.9906 |
|---|---|---|---|
| Dependent Mean | 15.17502 | Adj R-Sq | 0.9905 |
| Coeff Var | 0.63922 | | |

Parameter Estimates

| Variable | DF | Parameter Estimate | Standard Error | t Value | Pr > |t| |
|---|---|---|---|---|---|
| Intercept | 1 | 0.16405 | 0.04650 | 3.53 | 0.0004 |
| impschl7 | 1 | 9.89148 | 0.03057 | 323.54 | <.0001 |

### Stata

**Commands**

```
regress yrschl higrade7 impschl7
estat vif
regress higrade7 impschl7
```

**Results**

| Source | SS | df | MS |
|---|---|---|---|
| Model | 34.6855227 | 2 | 17.3427614 |
| Residual | 1790.51348 | 997 | 1.79590118 |
| Total | 1825.199 | 999 | 1.82702603 |

```
Number of obs =    1000
F( 2,  997)   =    9.66
Prob > F      =  0.0001
R-squared     =  0.0190
Adj R-squared =  0.0170
Root MSE      =  1.3401
```

| yrschl | Coef. | Std. Err. | t | P>|t| | [95% Conf. Interval] | |
|---|---|---|---|---|---|---|
| higrade7 | -.4819722 | .4373145 | -1.10 | 0.271 | -1.340135 | .3761902 |
| impschl7 | 6.564307 | 4.346258 | 1.51 | 0.131 | -1.964557 | 15.09317 |
| _cons | 8.253167 | .6463679 | 12.77 | 0.000 | 6.984769 | 9.521565 |

| Variable | VIF | 1/VIF |
|---|---|---|
| higrade7 | 105.89 | 0.009444 |
| impschl7 | 105.89 | 0.009444 |
| Mean VIF | 105.89 | |

| Source | SS | df | MS |
|---|---|---|---|
| Model | 984.958958 | 1 | 984.958958 |
| Residual | 9.39063078 | 999 | .00940945 |
| Total | 994.3495589 | 999 | .995344934 |

```
Number of obs =    1000
F( 1,  998)   =  0.0000
Prob > F      =  0.0000
R-squared     =  0.9906
Adj R-squared =  0.9905
Root MSE      =   .097
```

| higrade7 | Coef. | Std. Err. | t | P>|t| | [95% Conf. Interval] | |
|---|---|---|---|---|---|---|
| impschl7 | 9.891479 | .0305727 | 323.54 | 0.000 | 9.831484 | 9.951473 |
| _cons | .164046 | .0464975 | 3.53 | 0.000 | .0728021 | .25529 |

**■ Display B.11.7 Bivariate Regression of Years of Schooling and Educational Expectations in Hypothetical Data set**

| | SAS | Stata |
|---|---|---|
| Commands | `proc reg; model yrschl=higrade7; run;`<br>`proc reg; model yrschl=impschl7; run;` | `regress yrschl higrade7`<br>`regress yrschl impschl7` |

**Results (SAS — first regression)**

Parameter Estimates

| Variable | DF | Parameter Estimate | Standard Error | t Value | Pr > \|t\| |
|---|---|---|---|---|---|
| Intercept | 1 | 8.23941 | 0.64672 | 12.74 | <.0001 |
| higrade7 | 1 | 0.17539 | 0.04253 | 4.12 | <.0001 |

**Results (SAS — second regression)**

Parameter Estimates

| Variable | DF | Parameter Estimate | Standard Error | t Value | Pr > \|t\| |
|---|---|---|---|---|---|
| Intercept | 1 | 8.17410 | 0.64244 | 12.72 | <.0001 |
| impschl7 | 1 | 1.79689 | 0.42242 | 4.25 | <.0001 |

**Results (Stata — first regression)**

```
      Source |       SS       df       MS              Number of obs =    1000
-------------+------------------------------           F(  1,   998) =   17.01
       Model |  30.5888705     1  30.5888705           Prob > F      =  0.0000
    Residual |  1794.61013   998  1.79820654           R-squared     =  0.0168
-------------+------------------------------           Adj R-squared =  0.0158
       Total |    1825.199   999  1.82702603           Root MSE      =   1.341

------------------------------------------------------------------------------
      yrschl |      Coef.   Std. Err.      t    P>|t|     [95% Conf. Interval]
-------------+----------------------------------------------------------------
    higrade7 |    .175393   .0425256     4.12   0.000     .091932     .2588428
       _cons |   8.239408   .6467184    12.74   0.000    6.970324     9.508492
------------------------------------------------------------------------------
```

**Results (Stata — second regression)**

```
      Source |       SS       df       MS              Number of obs =    1000
-------------+------------------------------           F(  1,   998) =   18.10
       Model |  32.5041055     1  32.5041055           Prob > F      =  0.0000
    Residual |  1792.69489   998  1.79628747           R-squared     =  0.0178
-------------+------------------------------           Adj R-squared =  0.0168
       Total |    1825.199   999  1.82702603           Root MSE      =  1.3403

------------------------------------------------------------------------------
      yrschl |      Coef.   Std. Err.      t    P>|t|     [95% Conf. Interval]
-------------+----------------------------------------------------------------
    impschl7 |   1.796889   .4224157     4.25   0.000    .9679642     2.625814
       _cons |   8.174101   .6424435    12.72   0.000    6.913406     9.434796
------------------------------------------------------------------------------
```

APPENDIX **B**

**APPENDIX B**

## ■ Display B.11.8 Variance Inflation Factors in NSFH Example

|  | SAS | Stata |
|---|---|---|
| Commands | ```proc reg;    model hrwork=Health HlthLimit g2age /vif;  run;``` | ```regress hrwork Health HlthLimit g2age  estat vif``` |
| Results | *(see tables below)* | *(see tables below)* |

**SAS — Parameter Estimates**

| Variable | DF | Parameter Estimate | Standard Error | t Value | Pr > \|t\| | Variance Inflation |
|---|---|---|---|---|---|---|
| Intercept | 1 | 42.79665 | 1.53352 | 27.91 | <.0001 | 0 |
| Health | 1 | 0.89608 | 0.31030 | 2.89 | 0.0039 | 1.04161 |
| HlthLimit | 1 | -5.12471 | 2.44470 | -2.10 | 0.0361 | 1.03865 |
| g2age | 1 | -0.10337 | 0.01908 | -5.42 | <.0001 | 1.00734 |

**Stata**

| Source | SS | df | MS |
|---|---|---|---|
| Model | 9787.71371 | 3 | 3262.57124 |
| Residual | 748615.083 | 3738 | 200.271558 |
| Total | 758402.796 | 3741 | 202.727291 |

```
Number of obs =    3742
F( 3, 3738)   =   16.29
Prob > F      =  0.0000
R-squared     =  0.0129
Adj R-squared =  0.0121
Root MSE      =  14.152
```

| hrwork | Coef. | Std. Err. | t | P>\|t\| | [95% Conf. Interval] |
|---|---|---|---|---|---|
| Health | .8960821 | .310296 | 2.89 | 0.004 | .2877161 1.504448 |
| HlthLimit | -5.124713 | 2.444696 | -2.10 | 0.036 | -9.917782 -.316449 |
| g2age | -.1033697 | .0190755 | -5.42 | 0.000 | -.1407692 -.0659703 |
| _cons | 42.79665 | 1.533518 | 27.91 | 0.000 | 39.79003 45.80326 |

estat vif

| Variable | VIF | 1/VIF |
|---|---|---|
| Health | 1.04 | 0.960053 |
| HlthLimit | 1.04 | 0.962785 |
| g2age | 1.01 | 0.992709 |
| Mean VIF | 1.03 | |

■ **Display B.11.9  Results for Handcalculating Variance Inflation Factor for** *g1age* **in NSFH Example**

| | SAS | Stata |
|---|---|---|
| Commands | `proc reg;`<br>`   model g2age=Health Hlthlimit;`<br>`run;` | `regress g2age Health HlthLimit` |

**SAS Results**

| Root MSE | 12.13265 | R-Square | 0.0073 |
|---|---|---|---|
| Dependent Mean | 36.95778 | Adj R-Sq | 0.0068 |
| Coeff Var | 32.82841 | | |

Parameter Estimates

| Variable | DF | Parameter Estimate | Standard Error | t Value | Pr > |t| |
|---|---|---|---|---|---|
| Intercept | 1 | 41.37054 | 1.12728 | 36.70 | <.0001 |
| Health | 1 | -1.08391 | 0.26543 | -4.08 | <.0001 |
| HlthLimit | 1 | 5.13852 | 2.09422 | 2.45 | 0.0142 |

**Stata Results**

```
      Source |       SS       df       MS
-------------+------------------------------
       Model | 4042.11187     2  2021.05593
    Residual | 550385.217  3739  147.201181
-------------+------------------------------
       Total | 554427.329  3741  148.202975
```

```
Number of obs =    3742
F( 2,  3739)  =   13.73
Prob > F      =  0.0000
R-squared     =  0.0073
Adj R-squared =  0.0068
Root MSE      =  12.133
```

```
       g2age |      Coef.   Std. Err.      t    P>|t|     [95% Conf. Interval]
-------------+----------------------------------------------------------------
      Health | -1.083913   .2654336    -4.08   0.000    -1.604322   -.5635043
   HlthLimit |  5.138519   2.094216     2.45   0.014     1.032602    9.244437
       _cons |  41.37054   1.127277    36.70   0.000     39.16041    43.58068
```

# SCREENSHOTS OF DATA SET DOCUMENTATION

The following are screenshots from data set archives and documentation files.

Each is numbered to refer to this appendix letter (C), the chapter number where it is referenced and its order of reference in the chapter. For example, Display C.2.1 would be the first screenshot referred to in Chapter Two.

Note that because many of these screenshots were taken from active web sites they may not exactly match what you see if you visit the current web site. (In fact, the ICPSR web site was revised while this book was in press.)

## ▧ Display C.2.1  ICPSR Browse Screen, after Selecting "Family and Gender" and Sorting by Relevance

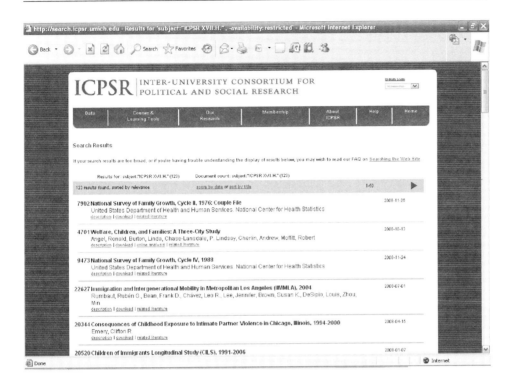

■ **Display C.2.2  General Search**

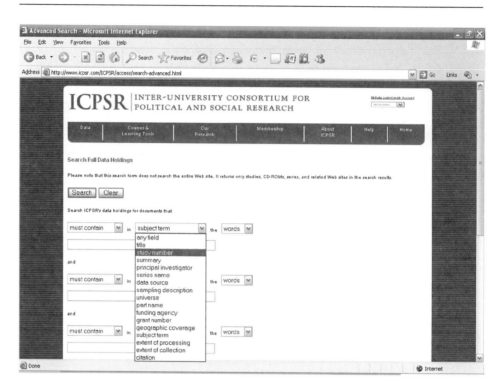

## ■ Display C.2.3 Searching Variable Lists Through Survey Documentation Analysis

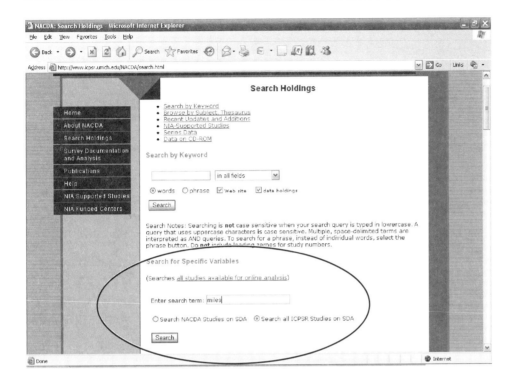

■ **Display C.2.4  Example of Variable Matches from Survey Documentation Analysis**

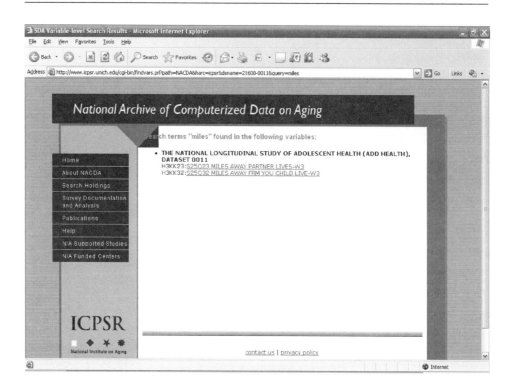

■ **Display C.2.5** **Details about Variable HSKK32 from Survey Documentation Analysis**

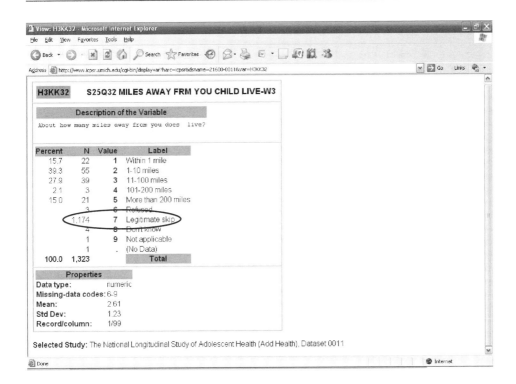

Selected Study: The National Longitudinal Study of Adolescent Health (Add Health), Dataset 0011

■ **Display C.2.6  Advanced Search Page for Bibliography**

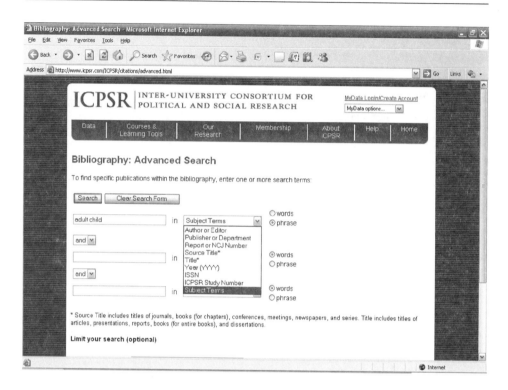

▨ **Display C.2.7  Results from Searching for "Miles" in Panel Study of Income Dynamics Variables**

### ◼ Display C.2.8  Details about Panel Study of Income Dynamics "Miles" Question

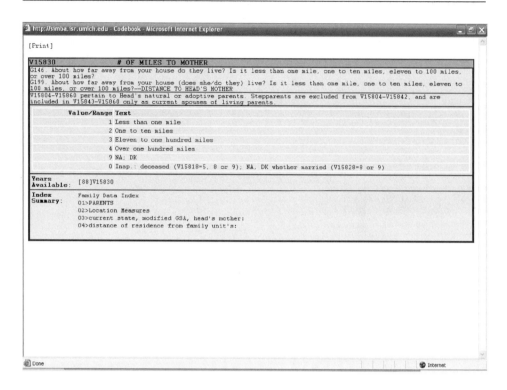

■ **Display C.3.1  The BADGIR Utility**

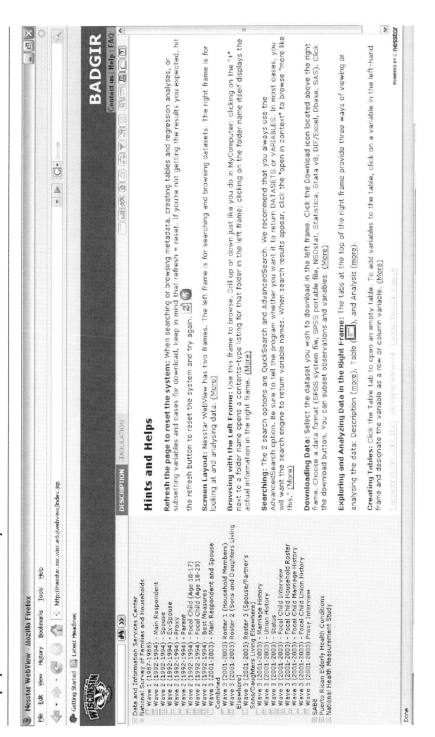

## ■ Display C.3.2 Identification of Relevant Variables in NSFH Content Outline

VIII. OUTLINE OF THE CONTENT OF THE
NATIONAL SURVEY OF FAMILIES AND HOUSEHOLDS

INTERVIEW WITH PRIMARY RESPONDENT

Household Composition
    A. Household composition
        1. Household composition – age sex, marital status

Social and Economic Characteristics

    A. Social background
        1. Race
        2. Religious preference and activity
        3. Recent residential movement
        4. Parent's occupation and education
        5. Family's receipt of public assistance during R's youth

SELF-ADMINISTERED QUESTIONNAIRE: PRIMARY RESPONDENT

SE-13 Parents, Relatives, and General Attitudes
    (all respondents)

    A. Information about mother
        1. Current age or age at death
        2. Health
        3. Quality of relationship with mother (Global)
        4. Current residence
        5. Contact with mother

    D. Brothers and sisters
        1. Number
        2. Quality of relationship

Source: Sweet, James, Larry Bumpass and Vaughn Call. 1988. The Design and Content of the National Survey of Families and Households. *NSFH Working Paper No. 1*. Available at *http://www.ssc.wisc.edu/cde/nsfhwp/nsfh1.pdf*

## ▨ Display C.3.3  Questionnaires and Skip Maps

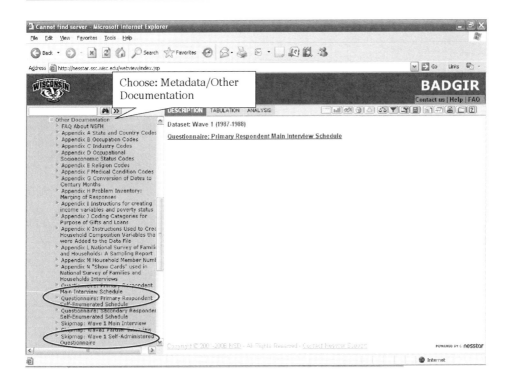

▪ **Display C.3.4  Example Skip Map: Mother's Age and Distance**

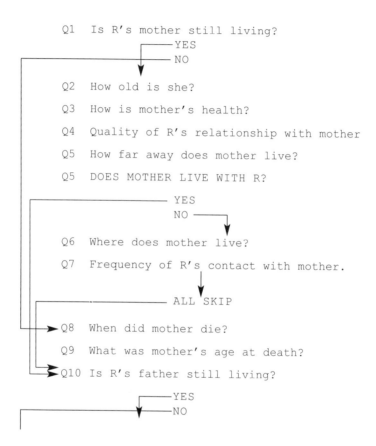

```
SE - 13 PARENTS, SIBS, AND ATTITUDES
* Includes: All Respondents
* SE Questionnaire pages 57-64
* Main Interview instructions on page 157


Q1  Is R's mother still living?
                              YES
                              NO

Q2  How old is she?

Q3  How is mother's health?

Q4  Quality of R's relationship with mother

Q5  How far away does mother live?

Q5  DOES MOTHER LIVE WITH R?

                              YES
                         NO

Q6  Where does mother live?

Q7  Frequency of R's contact with mother.


                              ALL SKIP

Q8  When did mother die?

Q9  What was mother's age at death?

Q10 Is R's father still living?

                              YES
                              NO
```

Source: Self-Enumerated Questionnaire Skip Map. Available at
*ftp://elaine.ssc.wisc.edu/pub/nsfh/T1se.pdf*

▓ **Display C.3.5  Example Questionnaire: Mother's Age and Distance**

1. Is your mother: (circle one)

    1 Still living          OR          2 Deceased
        :                                   :
                                            .

2. How old is she?                 8. In what year did she die?

    ____ Years old                      _____
                                          (year)

3. How would you describe your     9. What was her age when
   mother's health?                   she died?

    1 Very poor                        ____ Years old

    2 Poor

    3 Fair                             (GO TO THE NEXT PAGE)

    4 Good

    5 Excellent

4. How would you describe your relationship with your mother?

            1    2    3    4    5    6    7
    VERY POOR :____ ____ ____ ____ ____ ____ ____: EXCELLENT

5. About how far away does she live?    (IF YOUR MOTHER LIVES HERE WITH
                                         YOU, GO TO THE NEXT PAGE.)

    ____ Miles  IF YOU DO NOT        CITY _____
            KNOW, WRITE IN HER:      STATE _____

6. Does she live in:

    1 Her own home or apartment

    2 With a son or daughter

    3 In a nursing home

    4 Someplace else: (please specify) _____

| 7. During the past 12 months, about how often did you: | NOT AT ALL | ABOUT ONCE A YEAR | SEVERAL TIMES A YEAR | 1-3 TIMES A MONTH | ABOUT ONCE A WEEK | SEVERAL TIMES A WEEK |
|---|---|---|---|---|---|---|
| a. See your mother | 1 | 2 | 3 | 4 | 5 | 6 |

Source: Primary Respondent Self-Enumerated Schedule. Available at
*ftp://elaine.ssc.wisc.edu/pub/nsfh/i1se.001*

**APPENDIX C**

## Display C.3.6 Example Codebook: Mother Living or Deceased

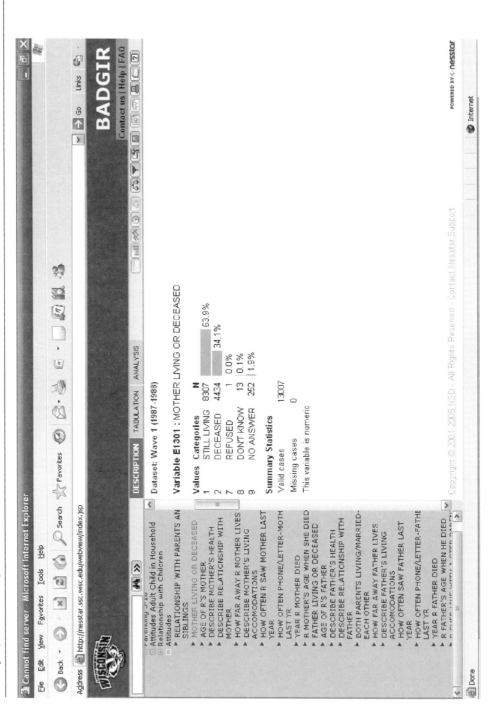

■ **Display C.3.7  Example Codebook: Mother's Age**

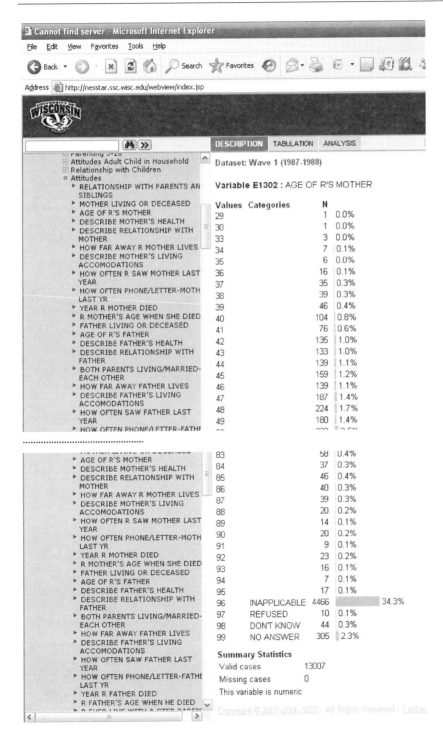

## ■ Display C.3.8 Example Codebook: Mother's Distance

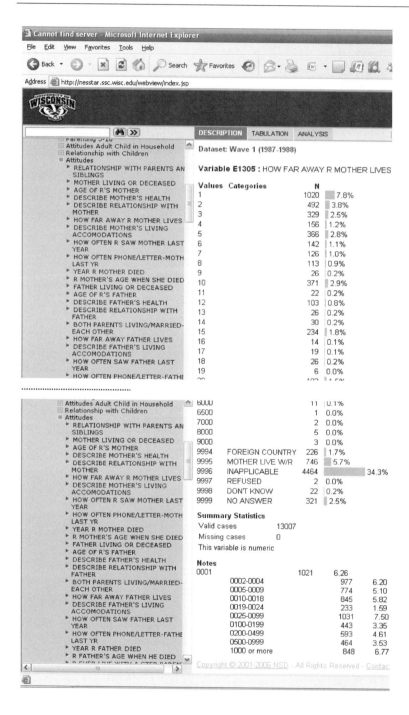

▨ **Display C.7  BADGIR Codebook Screenshots for M2DP01 and M484**

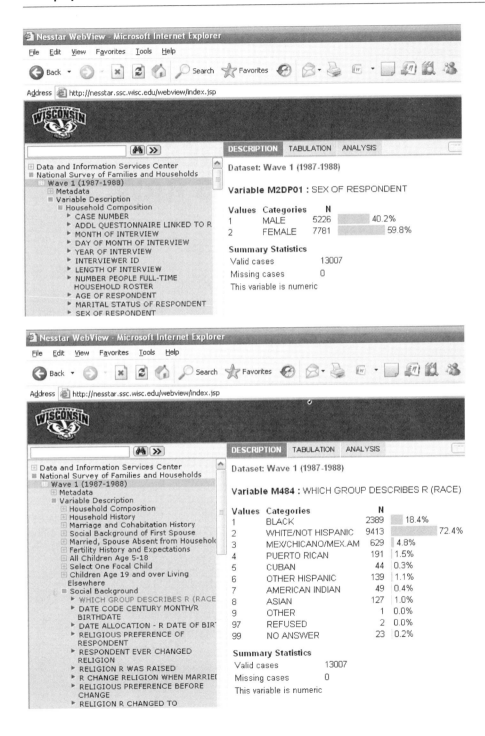

# ACCESSING THE NATIONAL SURVEY OF FAMILIES AND HOUSEHOLDS RAW DATA FILE

The archives we covered in Chapter 2 make data available over the Internet, so they are readily accessible. But, not all data are available in the same formats. Thus, the steps you will need to follow to download the data and read them into SAS and Stata will vary from archive to archive. Most archives make data available for download already formatted to be read directly into at least one statistical package. But, some will require you to translate the raw data from plain text format using a batch program. In most cases, archives provide a template batch program to accomplish this task which you will need to modify slightly. In this Appendix, we walk through the process of downloading the NSFH data. Appendix E provides notes about accessing the NOS data set used in the Chapter Exercises. If you locate your own data, for the course exercises or for a project or thesis, you may need to follow a somewhat different process.

## Downloading the Data in Stata Format

If you did not already, you want to begin by preparing a new project folder called *c:\nsfh_distance* on your hard drive to store the raw data files. Within this folder, create two subfolders, called *SAS* and *Stata*, where you will store each version of the downloaded raw data and the relevant batch programs and output.

Next, go to the BADGIR web site to access the data. Display D.1 shows the BADGIR screen for downloading the data. Clicking on the listed file called *d1all004* provides general information about the data file. Circled in red are the filename, *d1all004.NSDstat*, and number of observations (13,007) and variables (4,355). The *Download* icon is circled in black. When you click on this icon, you will be prompted to login or register. The registration screen requires you to affirm appropriate use of the data, including that you will not try to identify respondents, and that you will appropriately cite the data and notify the Center for Demography of Health and Aging at the University of Wisconsin and contact your local IRB office if you plan to use data for research purposes (The latter is not needed if you only download the data to complete course exercises. But, if you decide to write a research paper based on the data, then you should contact your local IRB regarding their policy on existing public use data sets and share your work with the Center for Demography of Health and Aging at the University of Wisconsin.)

BADGIR allows the data to be downloaded in numerous formats, as shown in Display D.2. This includes three versions of Stata (Versions 6, 7, and 8) and one SAS format. Similar to word processing software, Stata's formats often change with new releases to accommodate new features. They are "backward compatible" but not "forward compatible," meaning that data saved in older formats can be read with a newer release of Stata, but newer formats cannot be read by older releases of Stata. Stata 10 can read any of the earlier formats available from BADGIR, so we will select the most recent (version 8) for downloading. For SAS, the note circled in dashed black on Display D.1 tells us that BADGIR will produce SAS statements to read the plain text data (rather than a *.sas7bdat* file directly in SAS format). We will discuss below how to make three small modifications to the downloaded batch SAS batch program in order to convert the raw data file from plain text to SAS format.

The Stata Version 8 file is downloaded from BADGIR as a compressed file, with a *.zip* extension. When you double-click on this *.zip* file, your computer should associate the extension with a utility that will reveal its contents (a *.dta* Stata data file and a file that ends with *.missRecode*).

You should extract or copy these files to your Stata project folder (*c:\nsfh_distance\Stata*), and see something like the following:

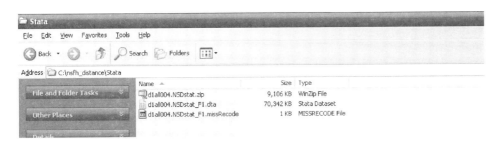

Notice that the Stata Data Set is much larger than the *.zip* file (size of over 70,000 versus 9,000 kbyte), which is why the compressed version is provided for downloading. The *.missRecode* file can be ignored. (Notice that the *.missRecode* extension is not meaningful to your computer—it does not know what software to use to open the file.

In fact, it is a plain text file containing a single sentence: "Recoding of missing values exported to Stata.")

By default, Stata accesses only a certain amount of computer resources. On our computer, Stata allocates 10 MB for the data when we double click a *.dta* file, enough to open a moderately sized data set. Above, we saw that our data file was over 70,000 kbyte (70 Mb). Our data set is thus too large for the defaults. The number of variables also exceeds the limit of 2,047 for Stata/IC. Because of this, we read just a few variables and subset of cases from the data file in our example batch programs (in Displays B.4.4 and B.4.7).

If your computer has Stata/SE or Stata/MP you can open the full *d1all004.NSDstat_F1.dta* data set. But, first you need to tell Stata to allocate more of the computer's memory. We know we need more than 70 Mb for our data file, perhaps 100 Mb. The command to tell Stata to do this is `set memory 100m`. If we put this single line into a plain text file called "*profile.do*", then Stata will allocate this much memory every time we double-click on our NSFH Stata data set *d1all004.NSDstat_F1.dta*. Note that this *profile.do* file should be stored in the same directory as the *.dta file*. An alternative approach is to permanently allocate more memory to Stata on your computer with the command "`set memory 100m, permanently`." This is a good alternative if you have Stata installed on your personal computer.

## Creating the SAS Raw Data File using StatTransfer

Especially since BADGIR does not provide the data file directly in SAS format, one easy way to obtain the SAS format is to use file conversion software to translate the data from Stata format to SAS format. StatTransfer and DBMS/Copy are widely used for this purpose (Hilbe 1996), although DBMS/Copy has been discontinued and does not directly read the latest versions of Stata. If you have access to such file conversion software, they offer a straightforward interface to select the type and location of the file to convert (in this case, the Stata Version 8 file *c:\nsfh_distance\Stata\ d1all004.NSDstat_F1.dta*) and the file to create (in this case, a SAS Version 9 file for Windows to be stored in *c:\nsfh_distance\SAS\d1all004_NSDstat_F1.sas7bdat*). Notice that we renamed the file slightly to make it easier to read in SAS (changing the first `.` to `_`) and match the name in our examples in Display B.4.4 and B.4.7.

## Creating the SAS Raw Data File with the BADGIR Utility

Let's now go back to BADGIR and download the plain text raw data and batch program to read it into SAS format. You will need to use this option if you do not have access to file conversion software like StatTransfer.

After choosing SAS format in BADGIR (see again Display D.2), we download a new compressed *.zip* file, save it to the SAS folder in our project folder (*C:\nsfh_distance\SAS*), and uncompress its contents to the same location. As expected, the compressed file contains the plain text ASCII raw data (with *.txt* extension) and the SAS batch program (with *.sas* extension).

We now need to open the *.sas* file in the enhanced editor window in SAS (launch SAS and choose File/Open after clicking in the enhanced editor window). The SAS batch program contains over 36,000 lines of code, and almost certainly looks overwhelmingly complicated to you. But, someone else invested the time to write and proof these 36,000 lines of syntax! All we need to do is to make a few simple changes so that the software knows where to find the *.txt* file and save the files on our computer.

Display D.3 shows the batch program, with the enhanced editor window in SAS maximized to show more lines of text. The first two lines, circled in red, are two of the three places where we need to make changes to the file. The remainder of the batch program is all prewritten, and should be left untouched (for example, the remaining visible lines provide value labels for two of the file's variables).

### First and Second Change
We need to first replace the empty quotes in the first two lines with our project folder's location, as follows:

```
LIBNAME LIBRARY ' ';
LIBNAME OUT ' ';
```

to

```
LIBNAME LIBRARY 'c:\nsfh_distance\SAS';
LIBNAME OUT 'c:\nsfh_distance\SAS';
```

This batch program illustrates the advantage of the SAS *libname*, discussed in Chapter 4. Although *libnames* can seem abstract to students, especially when first learning SAS, they provide an easy way to tailor a batch program to run on various computers. All we need to do is substitute between the quotes the location of the project folder on our computer.

### Third Change

The third change we need to make is embedded in the middle of the long batch program. We can use SAS's search utility to find it (see Display D.4 for the code we are searching). Select *Edit/Find* from the menus or type *Ctrl-F*. Type the word "INFILE" in the search box. You should make the following change, adding the path to our project folder so that SAS knows where to find the downloaded plain text (*.txt*) raw data file.

```
INFILE 'd1all004_NSDstat_F1.txt' LRECL = 7946;
```

to

```
INFILE 'c:\nsfh_distance\SAS\d1all004_NSDstat_F1.txt' LRECL
= 7946;
```

APPENDIX D

Now, you need to save these changes to the batch program. We strongly encourage you to save the file with a new name. We chose the name *d1all004_NSDstat_F1R.sas* for the new name, adding *R* to the end of the original filename to indicate *Revised*. By only slightly changing the filename, it is easy to connect it to the original (and we left the original file intact, in case we receive any errors messages when we run the batch program).

You are now ready to try running the batch program (e.g., click the icon of the little running person circled in black in Display D.4). The results, shown in Display D.5, report that the raw data file was appropriately read from our project folder. And, the correct number of cases is listed (13,007). If you don't see this message, scroll up in the Log Window to look for error messages, which will be in red or can be located by searching for the word ERROR in the Log Window. Check that you don't have typos in the path and filenames that you added to the batch program, and that you did not accidentally delete one of the quotes or other portions of the text (compare the spots where you made changes to the original file).

After the batch program runs successfully, you should see two new files in your project folder: (a) the data file *d1all004_nsdstat_f1.sas7bdat*, and (b) a file called *formats.sas7bcat* which stores the labels for the variables.

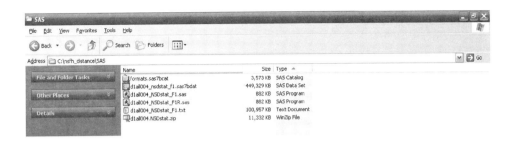

We included the following line at the beginning of our batch programs (see again Display B.4.4 and B.4.7) so that SAS will be able to locate the *formats.sas7bcat* file and associate the correct value and variable labels to the variables: LIBNAME LIBRARY 'c:\nsfh_distance\SAS'; Although we will not focus on the formats in this book, this statement is needed for your batch program to run properly if you created the SAS raw data file by running the BADGIR SAS batch program. Otherwise, you will receive an error message.

■ **Display D.1  Notes on NSFH Wave 1 Data Set in BADGIR**

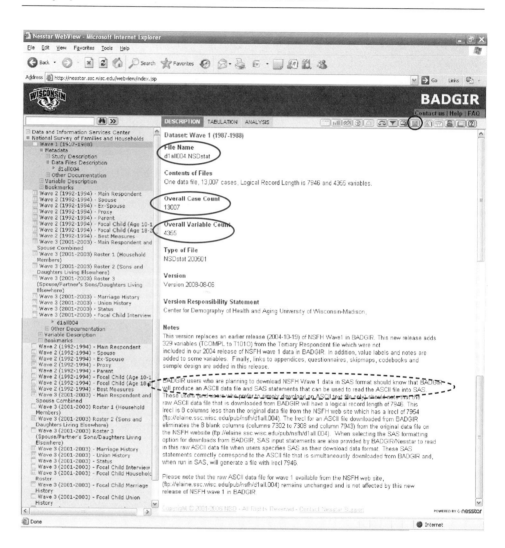

■ Display D.2 **BADGIR Download Screen for NSFH Wave 1 Data Set**

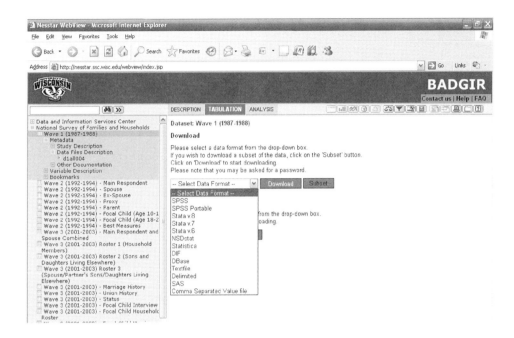

■ **Display D.3 SAS Editor Window: Updating LIBNAMES**

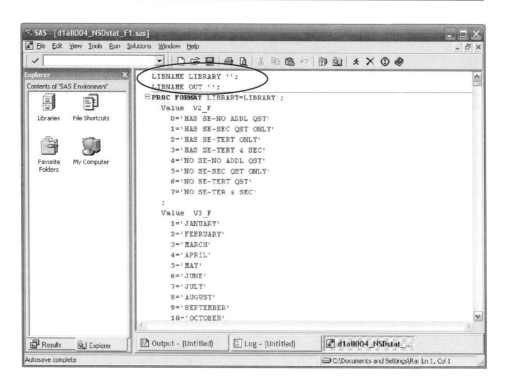

■ Display D.4  **SAS Editor Window: Updating Raw Data File Location**

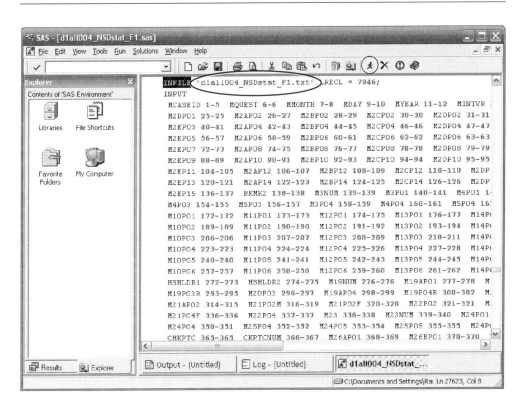

■ **Display D.5** Log window after submitting SAS batch program to read NSFH Wave 1 Data

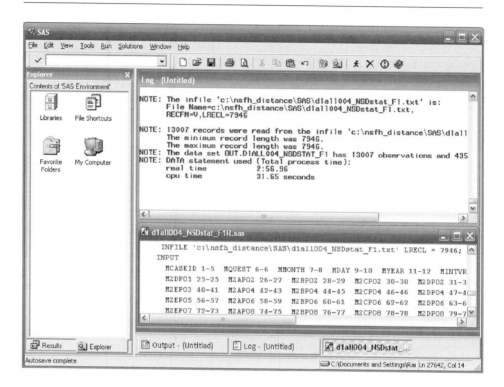

*Appendix E*

# ACCESSING THE NATIONAL ORGANIZATIONS SURVEY DATA

The NOS 1996 data can be downloaded from ICPSR in SPSS portable format or as ASCII text file with SAS or SPSS code for reading the ASCII files.

We will offer two ways to translate these data into SAS and Stata format: (1) use file conversion software and (2) read the SAS file and use a SAS procedure to write the Stata version of the data.

***Option 1:*** If you have access to file conversation software, such as StatTransfer, you can directly translate the SPSS portable file into a SAS and/or Stata data file. Be sure to choose SPSS portable file as the type for your input file, and Stata 10 and/or SAS version 9 for Windows for the output data files.

***Option 2:*** In the second approach, we will read the data by modifying the SAS code to execute on our computer and save a SAS format file. We will also include a PROC EXPORT command in the SAS batch program to save the data in Stata format. The steps are:

1. Remove the comments around the PROC FORMAT command (in line 51 and line 3217; find by searching for * SAS PROC FORMAT;—open comment is just after this—and * SAS DATA, INFILE, INPUT STATEMENTS;—close comment is just before this).
2. Remove the comments around the FORMAT statement (in line 4671 and 5341; search for * SAS FORMAT STATEMENT;—open comment is just after this—and the last line of the file—close comment is just before this).
3. In the first line of the batch program, add the statement LIBNAME LIBRARY; "c:\nos1996\sas";

4.  Search again for `* SAS PROC FORMAT;` Modify

    ```
    PROC FORMAT;
    ```

    to

    ```
    PROC FORMAT LIBRARY=LIBRARY;
    ```

5.  Search again for `* SAS DATA, INFILE, INPUT STATEMENTS;` Modify

    ```
    DATA;
    INFILE "physical-filename" LRECL=2951;
    ```

    to

    ```
    DATA "c:\nos1996\sas\nos_03190_0001_data.sas7bdat";
    INFILE "c:\nos1996\sas\03190-0001-data.txt" LRECL=2951;
    ```

6.  At the end of the file, add the following statement:

    ```
    PROC EXPORT outfile= "c:\nos1996\stata\nos_03190_0001_data.dta";
    RUN;
    ```

SAS does not translate formats when creating the Stata file with the "`outfile`" command; the error messages (that the SAS formats could not be found or loaded) can be ignored.

# USING SAS AND STATA'S ONLINE DOCUMENTATION

In Chapter 12, we provide a number of ideas and references for learning more about SAS and Stata. But, here we provide basic information about using the online documentation for each.

In Stata, typing `help <command name>` in the command line, and hitting return, will open a window with information about the purpose and syntax rules for the command. We provide a portion of Stata's help file for the `use` command in Display F.1 as an example. (Note that clicking on the word `varlist` (which will be blue on your screen) in the help window for `help use` would open another window that gives help about `varlist` defining it as "a list of variable names with blanks between.")

In Display F.1, we circled the aspects of the command that we used in Chapter 4, but you can see that there is more that you might want or need to know related to this command if you continue programming with Stata.

You can also get help about a particular command, or browse the help files, by choosing Help from the menus. The first portion of Help/Contents, for example, provides information about the basic structure of Stata's commands. Browsing this section may be helpful to you, now or (if it still feels too overwhelming at this stage) when we get to the end of the book.

The electronic help files in Stata are succinct summaries of the command. But, they are not as comprehensive as the Stata harcopy manuals (which provide excellent background information related to each command and typically offer several detailed examples).[1] In contrast, SAS's complete set of documentation is available electronically from the Help menu. This can make it somewhat harder to find a simple description of a command in SAS online files. But, it means that once you find what you need, you will have a wealth of information at your fingertips. We show an example in Display F.2 of entering the word "freq" into the SAS help Index. We can use the other tabs to Search for keywords. And, we can browse the Contents for help as well (most of the commands we use in this book are found in Base SAS and SAS/STAT).

■ **Display F.1 Example Stata Help Page for** use

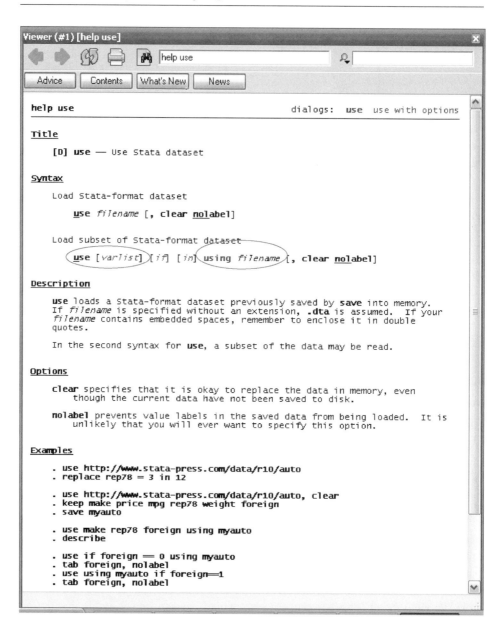

## ■ Display F.2  Example SAS Help Page for PROC FREQ

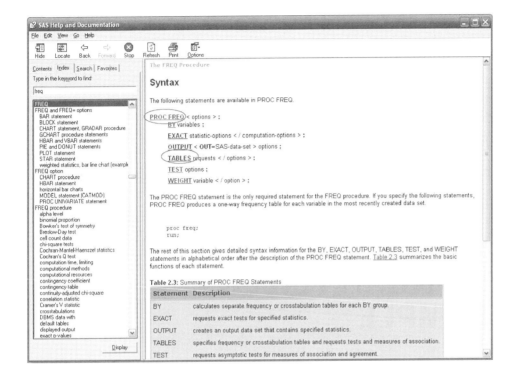

*Appendix G*

# EXAMPLE OF HAND-CALCULATING THE INTERCEPT, SLOPE, AND CONDITIONAL STANDARD DEVIATION USING STYLIZED SAMPLE

| | hrchores | numkid | $Y_i - \bar{Y}$ | $X_i - \bar{X}$ | $\Sigma(X_i - \bar{X})(Y_i - \bar{Y})$ | $\Sigma(X_i - \bar{X})^2$ | $\hat{Y}_i$ | $Y_i - \hat{Y}_i$ | $\Sigma(Y_i - \hat{Y}_i)^2$ |
|---|---|---|---|---|---|---|---|---|---|
| 1 | 21 | 0 | −15.04 | −2 | 30.08 | 4 | 23.78 | −2.78 | 7.7284 |
| 2 | 28 | 0 | −8.04 | −2 | 16.08 | 4 | 23.78 | 4.22 | 17.8084 |
| 3 | 25 | 0 | −11.04 | −2 | 22.08 | 4 | 23.78 | 1.22 | 1.4884 |
| 4 | 21 | 0 | −15.04 | −2 | 30.08 | 4 | 23.78 | −2.78 | 7.7284 |
| 5 | 32 | 0 | −4.04 | −2 | 8.08 | 4 | 23.78 | 8.22 | 67.5684 |
| 6 | 19 | 0 | −17.04 | −2 | 34.08 | 4 | 23.78 | −4.78 | 22.8484 |
| 7 | 22 | 0 | −14.04 | −2 | 28.08 | 4 | 23.78 | −1.78 | 3.1684 |
| 8 | 18 | 0 | −18.04 | −2 | 36.08 | 4 | 23.78 | −5.78 | 33.4084 |
| 9 | 21 | 0 | −15.04 | −2 | 30.08 | 4 | 23.78 | −2.78 | 7.7284 |
| 10 | 22 | 0 | −14.04 | −2 | 28.08 | 4 | 23.78 | −1.78 | 3.1684 |
| 11 | 24 | 1 | −12.04 | −1 | 12.04 | 1 | 29.91 | −5.91 | 34.9281 |
| 12 | 35 | 1 | −1.04 | −1 | 1.04 | 1 | 29.91 | 5.09 | 25.9081 |
| 13 | 25 | 1 | −11.04 | −1 | 11.04 | 1 | 29.91 | −4.91 | 24.1081 |
| 14 | 41 | 1 | 4.96 | −1 | −4.96 | 1 | 29.91 | 11.09 | 122.9881 |
| 15 | 40 | 1 | 3.96 | −1 | −3.96 | 1 | 29.91 | 10.09 | 101.8081 |
| 16 | 34 | 1 | −2.04 | −1 | 2.04 | 1 | 29.91 | 4.09 | 16.7281 |
| 17 | 38 | 1 | 1.96 | −1 | −1.96 | 1 | 29.91 | 8.09 | 65.4481 |
| 18 | 25 | 1 | −11.04 | −1 | 11.04 | 1 | 29.91 | −4.91 | 24.1081 |

|  | hrchores | numkid | $Y_i - \bar{Y}$ | $X_i - \bar{X}$ | $\Sigma(X_i - \bar{X})(Y_i - \bar{Y})$ | $\Sigma(X_i - \bar{X})^2$ | $\hat{Y}_i$ | $Y_i - \hat{Y}_i$ | $\Sigma(Y_i - \hat{Y}_i)^2$ |
|---|---|---|---|---|---|---|---|---|---|
| 19 | 26 | 1 | −10.04 | −1 | 10.04 | 1 | 29.91 | −3.91 | 15.2881 |
| 20 | 27 | 1 | −9.04 | −1 | 9.04 | 1 | 29.91 | −2.91 | 8.4681 |
| 21 | 32 | 2 | −4.04 | 0 | 0 | 0 | 36.04 | −4.04 | 16.3216 |
| 22 | 37 | 2 | 0.96 | 0 | 0 | 0 | 36.04 | 0.96 | 0.9216 |
| 23 | 47 | 2 | 10.96 | 0 | 0 | 0 | 36.04 | 10.96 | 120.1216 |
| 24 | 29 | 2 | −7.04 | 0 | 0 | 0 | 36.04 | −7.04 | 49.5616 |
| 25 | 34 | 2 | −2.04 | 0 | 0 | 0 | 36.04 | −2.04 | 4.1616 |
| 26 | 33 | 2 | −3.04 | 0 | 0 | 0 | 36.04 | −3.04 | 9.2416 |
| 27 | 33 | 2 | −3.04 | 0 | 0 | 0 | 36.04 | −3.04 | 9.2416 |
| 28 | 33 | 2 | −3.04 | 0 | 0 | 0 | 36.04 | −3.04 | 9.2416 |
| 29 | 37 | 2 | 0.96 | 0 | 0 | 0 | 36.04 | 0.96 | 0.9216 |
| 30 | 37 | 2 | 0.96 | 0 | 0 | 0 | 36.04 | 0.96 | 0.9216 |
| 31 | 44 | 3 | 7.96 | 1 | 7.96 | 1 | 42.17 | 1.83 | 3.3489 |
| 32 | 45 | 3 | 8.96 | 1 | 8.96 | 1 | 42.17 | 2.83 | 8.0089 |
| 33 | 38 | 3 | 1.96 | 1 | 1.96 | 1 | 42.17 | −4.17 | 17.3889 |
| 34 | 50 | 3 | 13.96 | 1 | 13.96 | 1 | 42.17 | 7.83 | 61.3089 |
| 35 | 53 | 3 | 16.96 | 1 | 16.96 | 1 | 42.17 | 10.83 | 117.2889 |
| 36 | 36 | 3 | −0.04 | 1 | −0.04 | 1 | 42.17 | −6.17 | 38.0689 |
| 37 | 50 | 3 | 13.96 | 1 | 13.96 | 1 | 42.17 | 7.83 | 61.3089 |
| 38 | 40 | 3 | 3.96 | 1 | 3.96 | 1 | 42.17 | −2.17 | 4.7089 |
| 39 | 34 | 3 | −2.04 | 1 | −2.04 | 1 | 42.17 | −8.17 | 66.7489 |
| 40 | 36 | 3 | −0.04 | 1 | −0.04 | 1 | 42.17 | −6.17 | 38.0689 |
| 41 | 41 | 4 | 4.96 | 2 | 9.92 | 4 | 48.3 | −7.3 | 53.29 |
| 42 | 48 | 4 | 11.96 | 2 | 23.92 | 4 | 48.3 | −0.3 | 0.09 |
| 43 | 44 | 4 | 7.96 | 2 | 15.92 | 4 | 48.3 | −4.3 | 18.49 |
| 44 | 53 | 4 | 16.96 | 2 | 33.92 | 4 | 48.3 | 4.7 | 22.09 |
| 45 | 53 | 4 | 16.96 | 2 | 33.92 | 4 | 48.3 | 4.7 | 22.09 |
| 46 | 40 | 4 | 3.96 | 2 | 7.92 | 4 | 48.3 | −8.3 | 68.89 |
| 47 | 45 | 4 | 8.96 | 2 | 17.92 | 4 | 48.3 | −3.3 | 10.89 |
| 48 | 52 | 4 | 15.96 | 2 | 31.92 | 4 | 48.3 | 3.7 | 13.69 |
| 49 | 44 | 4 | 7.96 | 2 | 15.92 | 4 | 48.3 | −4.3 | 18.49 |
| 50 | 60 | 4 | 23.96 | 2 | 47.92 | 4 | 48.3 | 11.7 | 136.89 |

Average: 36.04    2

Sum: 613    100    0    1614.23

Slope    6.13

Intercept    23.78

RMSE    5.799119904

# USING EXCEL TO CALCULATE AND GRAPH PREDICTED VALUES

The following Appendices show how to calculate and plot predicted values in Excel. The examples are primarily drawn from Chapter 8 (interaction models), but the concepts can be extended to plotting regression results from additive models. We also demonstrate their use with the nonlinear models from Chapter 9.

The Appendices are organized with this appendix letter (H) followed by the chapter number and the order in which they are discussed in the chapter.

Illustrations within each Appendix are referred to as displays, and are similarly given this appendix letter (H) followed by the chapter number and the order in which they are discussed in the Appendix.

### ■ Appendix H.8.1  Basic Steps of Creating Graphs in Excel

We will use the following basic steps across all of the interaction models examined in Chapter 8:

Step 1.   Copy the variable names and coefficient estimates from the results to Excel.

Step 2.   Make a cross-tabulation of the variables from the interaction, listing the values we would like to use in predictions.

Step 3.   If needed, choose values for the other variables in the model.

Step 4.   Write a formula for the prediction equation, using the copied coefficient estimates and selected values of the predictor variables.

Step 5.   Insert a chart (bar chart for interactions of two dummy variables, or line graph for interactions involving one or more interval variables).

Step 6.   Edit the chart as desired.

## Steps 1–4

Begin by opening Microsoft Excel. Excel will open with three worksheets (Sheet 1, Sheet 2, and Sheet 3) and the cursor in the top left cell (Column A, Row 1). We recommend that you save the worksheet in your project folder using a name that signals the purpose and contents of the file (e.g., *c:\hrchores\fem_marr.xlsx*). And, rename the first worksheet (e.g., to *PredictedValues*).[1]

For Step 1, it is easiest to cut-and-paste the results using an editor that will allow for selection of a rectangle. This is possible, for example, in the SAS output and editor windows, in Microsoft Word, or in some third-party text editors, such as TextPad or UltraEdit. In SAS and Microsoft Word, press the *Alt-* key before beginning the selection.[2] Start at the top left of the rectangle, and end at the bottom right corner. Display H.8.1 shows the difference between a regular selection and rectangle selection. You can then copy and paste this rectangle into Excel. Click on a cell in Excel and paste, and the values will be placed in a column of four separate cells. We clicked on cell B3 before pasting so we can also cut-and-paste the variable names into Column A (similarly select the rectangle containing the column of names in SAS) and add some labels. See Display H.8.2.

For Step 2, we next type in the levels of the predictors for which we would like to make predictions. In this case, the values are just 0 and 1 on each of our dummy variables. We can place these values anywhere within the worksheet. We put them slightly below our coefficient estimates, with labels so that we can refer to one set of predictions as the levels of the *married* variable and the other as the levels of the *female* variable (see top panel of Display H.8.3). Step 3 is not needed for this model, because we have no other variables in the model. In Appendix H.8.3, we will see how to choose values for other variables.

At Step 4, we type the prediction equation in Excel, referencing the cells that contain coefficient estimates and levels of the predictors. To do this, we click in the cell representing unmarried, males (cell C12 in our worksheet) and then we click in the formula bar (circled in red in Display H.8.3). Any equation can be typed in the formula bar after an equals sign, and Excel will calculate the result for you (e.g., Type "=2+2" in the formula bar, being sure to begin with the equals sign. When you hit enter, you should see the result 4).

Excel makes it easy to make calculations based on values contained in various cells of a worksheet. If we type the equals sign into the formula bar, and then click on any cell, Excel will enter a reference for that cell. For example, if we type the equals sign in the formula bar and then click on cell B3, the formula bar will contain =B3. We can use operators (+ for addition and * for multiplication) and click on the relevant cells to create our prediction equation. The result is shown in the bottom of Display H.8.3. Notice that Excel uses colors for the cell references to help us prove that our equation is correct. (The full colors are not shown in this book, but will be visible on your computers screen.)

We reproduce the equation here to make clear its relationship to Equation 8.1.

$$=B3+B4*C11+B5*B12+B6*C11*B12$$

The equation takes the four coefficient estimates, from cells B3 to B6, and sums them after multiplying by the relevant value of the predictor variables. Specifically, the equation begins with Cell B3 which contains the intercept (22.31). Next, it adds the product of Cell B4 (which contains the coefficient estimate for the first dummy, *married*, of –2.60) and Cell C11 (which contains a value of the first dummy, *married*, in this case 0). Next, it adds the product of Cell B5 (which contains the coefficient estimate for the second dummy, *female*, of 10.20) and Cell B12 (which contains a value of the second dummy, *female*, in this case 0). Finally, it adds the product of Cell B6 (which contains the coefficient estimate for the product term, *fem_marr*, of 6.64) and Cells C11 and B12 (which contain the values of the two dummy variables).

After you hit enter, Excel will calculate the results of the equation. Because we referenced the value of 0 for *married* and 0 for *female* for our first prediction, the result should be the predicted value for unmarried males. The result calculated by Excel of 22.31 matches our hand calculation from Table 8.8.

We could similarly type and click to create the equation for each of the three remaining cells. But, a great functionality of Excel is the ability to copy an equation so that it can be recalculated with different values. Normally, Excel uses *relative* cell references to do this. Display H.8.4 shows what happens when we click on cell C12, type Ctrl-C for copy and then click on cell D12 and type Ctrl-P for paste. Because the new cell is one position to the right of the old cell, all of the references in the equation have shifted one position to the right. This is what we desire for the value of *marry* but not for the rest of the values.

In Excel, we can use an *absolute* rather than a relative reference by placing a dollar sign ($) in front of any column letter or row number. The bottom panel of Display H.8.4 reproduces Excel help on absolute references and the top right panel shows how we used this approach to refer to the estimates for the coefficients by placing a $ before their column letters and row numbers. We also made the references to the predictor values partially absolute by placing a $ before the row number for the *married* reference and before the column letter for the *female* reference. Now, when we copy the formula to Cell D12 and click on it to see the cells boxed in color, the references look correct (see top/right screenshot in Display H.8.4). Finally, we copied and pasted the equation across all four cells and verify that Excel calculated all of the same predicted values as we calculated by hand in Table 8.8 (see top of Display H.8.5).

## Steps 5–6

We are now ready for Step 5, inserting a chart. Excel provides numerous chart styles, but we will use a simple bar chart. To do so, we first highlight the box of Cells from B11 at the top left to D13 at the bottom right (see again top panel of Display H.8.5). Then, we choose the `Insert` tab and click on `Column` and choose a 2-D Clustered Column from the top left (see red circles at top of Display H.8.5). This creates a basic bar chart (see bottom panel of Display H.8.5).

Finally, at Step 6, we can use the Chart Tools, shown in the toolbars in the bottom panel of Display H.8.5 to improve the look of the chart. For example, we clicked on Move Chart Location (circled with a black dotted line in Display H.8.5) to switch the chart to its own worksheet tab. We clicked on Layout tab (circled in red in Display H.8.5) to add titles to the chart and axes. We clicked on the gray chart style (circled in black in Display H.8.5) to change the fill of the bars to grayscale from color. We clicked on the Home tab (circled in black in Display H.8.5) to increase the font size. Finally, we clicked on Select Data (circled with a red dotted line in Display H.8.5) to label the series. Display H.8.6 shows how clicking on Select Data opens several boxes which allow us to select the words rather than values to label the legend series and horizontal axis.

### ▪ Display H.8.1  Using Alt-Key to Select a Rectangle in SAS

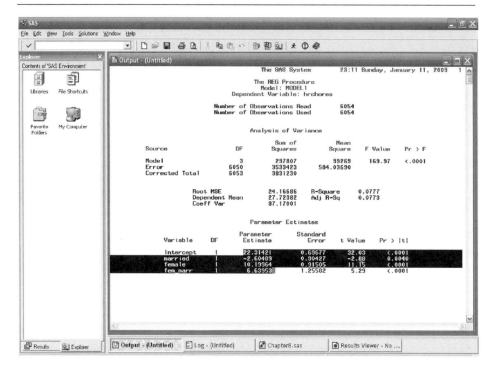

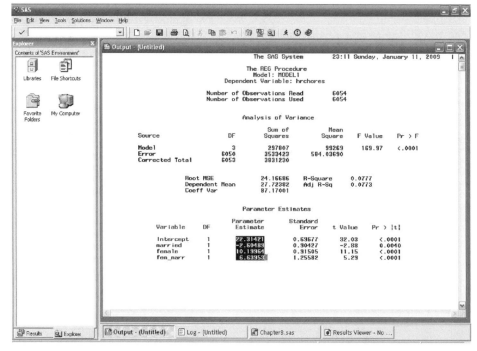

■ **Display H.8.2  Pasting Coefficients into Excel**

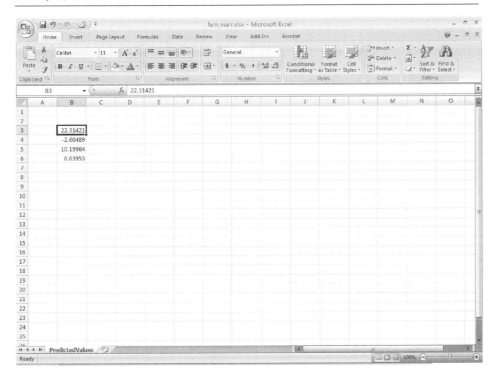

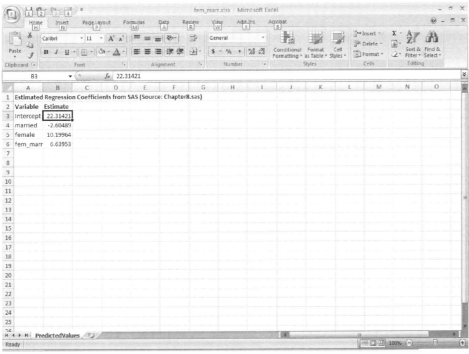

## ▪ Display H.8.3 Adding Levels of Predictors and Prediction Equation

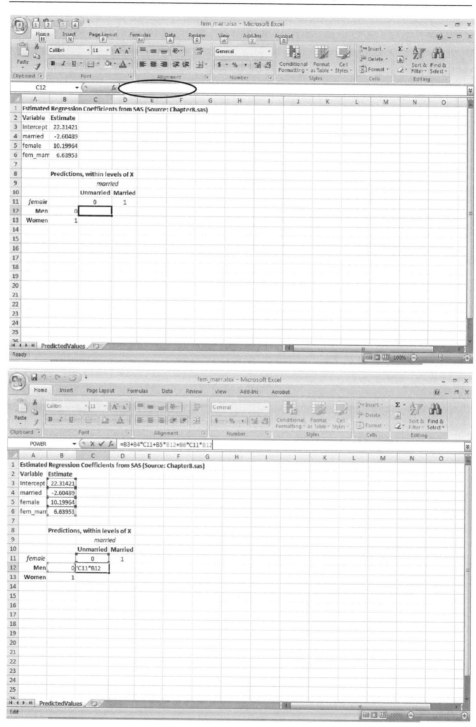

## ■ Display H.8.4  Using Fixed References

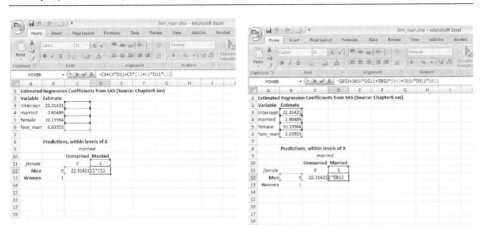

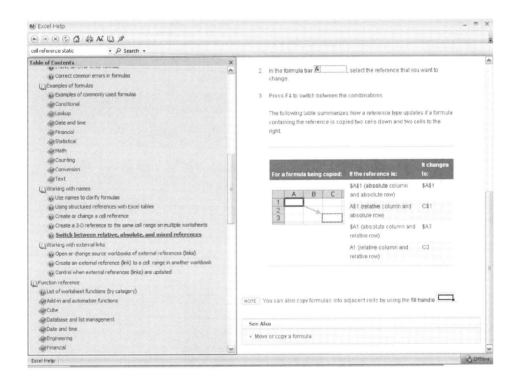

■ **Display H.8.5  Inserting a Chart**

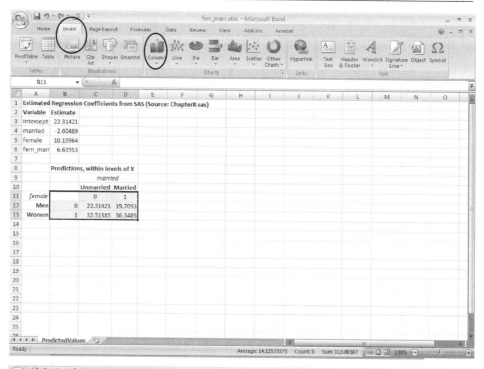

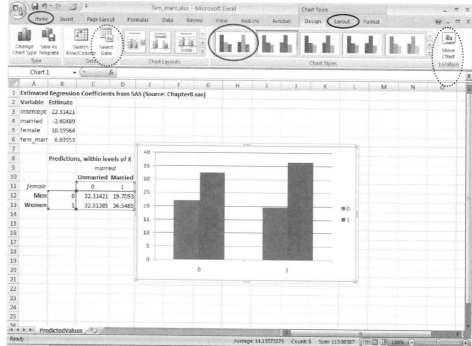

## ■ Display H.8.6 Labeling Series

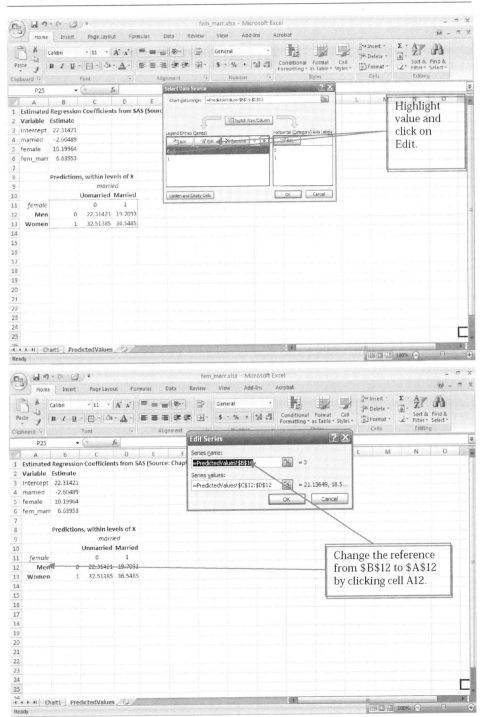

### ■ Appendix H.8.2  Graphing a Dummy by Interval Interaction and Calculating Conditional Effects (in Unstandardized, Semi-Standardized, and Standardized Form)

We can follow the basic steps introduced in Appendix H.8.1 to create a graph of a dummy by interval variable. The two main differences are that (a) we will provide more than two values for the interval variable for calculating predictions and (b) we will choose a Line chart from the Insert menu.

Display H.8.7 shows the values from the regression coefficients from Display B.8.6 cut-and-paste into Excel. Note that the *Predictions, within levels of X* are calculated similarly as in Appendix H.8.1, except that we now use five levels of the interval *hrwork* variable. The chart is also created similarly to the chart shown in Appendix H.8.1, except that we began by selecting a Line chart from the menus.

Display H.8.8 shows how we can use the formula bar to calculate differences between predicted values, both in their raw units and relative to the standard deviation of the outcome.

Display H.8.9 shows how we can also calculate the conditional effects of *hrwork* based on the coefficients, and use the standard deviation of the predictor (*hrwork*) and outcome (*hrchores*) to completely standardize these conditional effects.

■ **Display H.8.7  Presenting Results of a Dummy by Interval Variable Interaction in Excel**

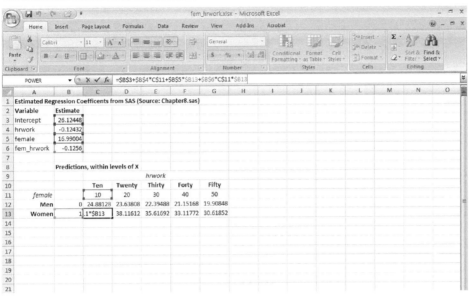

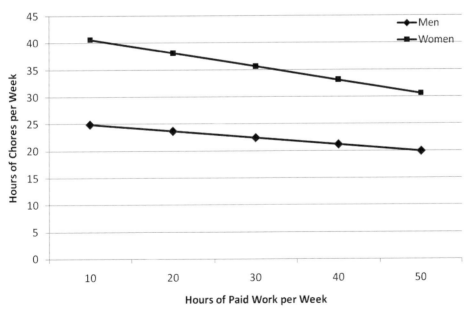

**Predicted Values for Hours of Chores
with Interaction between Hours of Paid Work and Gender**

■ **Display H.8.8 Calculating Differences in Predictions and Effect Size in Excel**

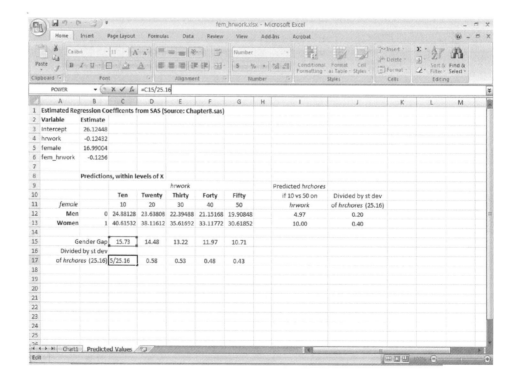

## ■ Display H.8.9  Calculating Standardized Effect of Interval Variable in Excel

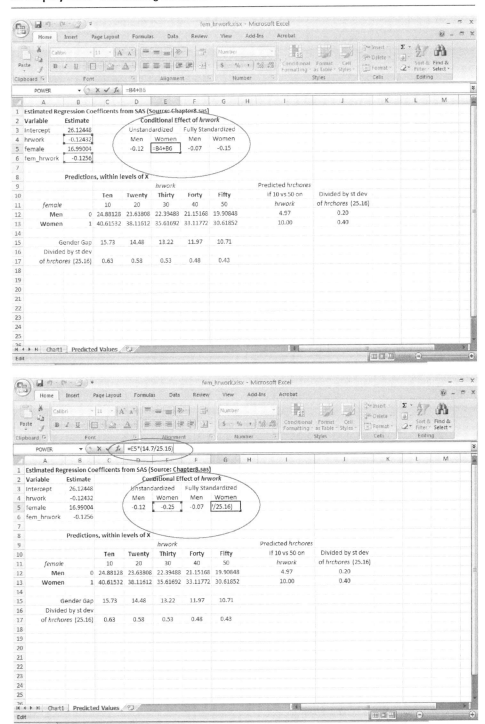

## ■ Appendix H.8.3  Graphing Results from a Fully Interacted Model

We will recalculate the results from Figure 8.2, which plotted the interaction between gender and hours of paid work, using the fully interacted model to illustrate how to make predictions with additional variables in the model, beyond those in the interaction.

Display H.8.10 shows the means for the predictor variables in the subsample included in the fully interacted model. Display H.8.11 shows where we added these mean values to our Excel spreadsheet and updated our equation.

The complete equation for cell C21 is now:

=$B$3+$B$5*C$19+$B$9*$B21+$B$11*C$19*$B21

+$B$4*$I$4+$B$6*$I$6+$B$7*$I$7+$B$8*$I$8

+$B$10*$B21*$I$4+$B$12*$B21*$I$6+$B$13*$B21*$I$7+$B$14*$B21*$I$8

For example $B$4*$I$4 multiplies the coefficient estimate for marital status (−3.18486) by the mean value of marital status (0.5297251); and, $B$10*$B21*$I$4 multiplies the coefficient estimate for the interaction of gender and marital status (4.48267) by gender (0 or 1, depending on the row) and the mean value of marital status (0.5297251).

Clearly, these equations become lengthy with numerous variables in the model, and in Chapter 9 we illustrate how to ask SAS and Stata to calculate predicted values (see also Long and Freese 2006 for their excellent *spost* utilities for calculating and graphing predicted values in Stata).

**Display H.8.10 Means of** *married, hrwork, numkid, aframer,* **and** *mexamer*

| | SAS | Stata |
|---|---|---|
| Regression Model | `proc means;`<br>`  var married hrwork numkid aframer mexamer;`<br>`run;` | `summarize married hrwork numkid aframer mexamer` |

Select Results — SAS

| Variable | N | Mean | Std Dev | Minimum | Maximum |
|---|---|---|---|---|---|
| married | 5820 | 0.5297251 | 0.4991585 | 0 | 1.0000000 |
| hrwork | 5820 | 38.9218213 | 14.7683344 | 0 | 93.0000000 |
| numkid | 5820 | 0.8159794 | 1.0449523 | 0 | 4.0000000 |
| aframer | 5820 | 0.1713058 | 0.3768084 | 0 | 1.0000000 |
| mexamer | 5820 | 0.0398625 | 0.1956530 | 0 | 1.0000000 |

Select Results — Stata

| Variable | Obs | Mean | Std. Dev. | Min | Max |
|---|---|---|---|---|---|
| married | 5820 | .5297251 | .4991585 | 0 | 1 |
| hrwork | 5820 | 38.92182 | 14.76833 | 0 | 93 |
| numkid | 5820 | .8159794 | 1.044952 | 0 | 4 |
| aframer | 5820 | .1713058 | .3768084 | 0 | 1 |
| mexamer | 5820 | .0398625 | .195653 | 0 | 1 |

■ **Display H.8.11 Predicted Values from Fully Interacted Model, Illustrating Interaction between Hours of Paid Work and Gender**

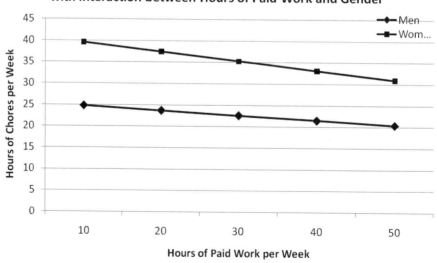

**Predicted Values for Hours of Chores with Interaction between Hours of Paid Work and Gender**

## ■ Appendix H.8.4  Graphing Results from an Interaction between Two Interval Variables

Display H.8.12 shows how we extend the procedures used for dummy by dummy and dummy by interval interactions for an interaction between two interval variables.

We first cut-and-pasted the coefficient estimates from the model shown in Display B.8.15 into Excel. We then added the multiple levels of number of children and copied the formula for calculating predictions across the cells.

We also added calculations of the unstandardized, semi-standardized (divided by the standard deviation of the outcome), and completely standardized (multiplied by the standard deviation of the predictor and divided by the standard deviation of the outcome) coefficients.

We plotted the results as a Line chart, modifying the default layout to increase readability (including deleting the lines for families for one and with three children, to reduce the amount of information and make it easier to compare the steepness of lines for the middle number of children, two, and the extremes, none and four).

■ **Display H.8.12  Presenting Interaction between Two Interval Variables in Excel**

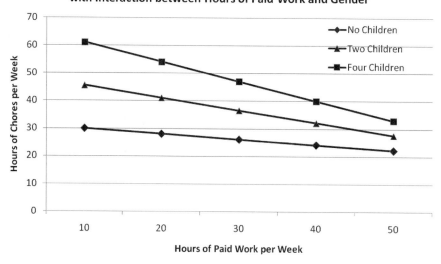

POWER    =$B$3+$B$4*$B12+$B$5*C$11+$B$6*$B12*C$11

| | A | B | C | D | E | F | G |
|---|---|---|---|---|---|---|---|
| 1 | Estimated Regression Coefficents from SAS (Source: Chapter8.sas) | | | | | | |
| 2 | Variable | Estimate | | | | | |
| 3 | Intercept | 31.89119 | | | | | |
| 4 | numkid | 9.06985 | | | | | |
| 5 | hrwork | -0.19367 | | | | | |
| 6 | numk_hrw | -0.12751 | | | | | |
| 7 | | | | | | | |
| 8 | Predictions, within levels of X | | | | | | |
| 9 | | | | *hrwork* | | | |
| 10 | | *numkid* | Ten | Twenty | Thirty | Forty | Fifty |
| 11 | | | 10 | 20 | 30 | 40 | 50 |
| 12 | | 0 | 2*C$11 | 28.01779 | 26.08109 | 24.14439 | 22.20769 |
| 13 | | 1 | 37.74924 | 34.53744 | 31.32564 | 28.11384 | 24.90204 |
| 14 | | 2 | 45.54399 | 41.05709 | 36.57019 | 32.08329 | 27.59639 |
| 15 | | 3 | 53.33874 | 47.57674 | 41.81474 | 36.05274 | 30.29074 |
| 16 | | 4 | 61.13349 | 54.09639 | 47.05929 | 40.02219 | 32.98509 |
| 17 | | | | | | | |
| 18 | | Unstandardized | 7.79 | 6.52 | 5.24 | 3.97 | 2.69 |
| 19 | | Semi-Standardized | 0.31 | 0.26 | 0.21 | 0.16 | 0.11 |
| 20 | Completely Standardized | | 0.25 | 0.21 | 0.17 | 0.13 | 0.09 |

10 Hour Increase in *hrwork*

| Unstandardized | Partially Standardized | Completely Standardized |
|---|---|---|
| -1.94 | -0.08 | -0.11 |
| -3.21 | -0.13 | -0.19 |
| -4.49 | -0.18 | -0.26 |
| -5.76 | -0.23 | -0.34 |
| -7.04 | -0.28 | -0.41 |

**Predicted Values for Hours of Chores with Interaction between Hours of Paid Work and Gender**

Legend: ◆ No Children  ▲ Two Children  ■ Four Children

Y-axis: Hours of Chores per Week (0 to 70)
X-axis: Hours of Paid Work per Week (10 to 50)

## ■ Appendix H.9  Graphing Results for Nonlinear Relationships

This appendix illustrates how Excel can be used to calculate and graph predicted values from various models that capture nonlinear relationships between the predictor and outcome variables.

Display H.9.1 shows predictions from the six alternative models for relating mother's years of schooling to the distance she lives from her adult child discussed in Chapter 9: the lin-lin, lin-log, lin-sq, log-lin, log-log, and log-sq models.

Display H.9.2 shows predictions from a flexible dummy variable model which relates numbers of sisters to distance using seven dummy variables.

Display H.9.3 shows results similar to Display H.9.2, but collapses to three dummy variables.

■ **Display H.9.1  Nonlinear Models for Distance Predicted by Mother's Years of Schooling**

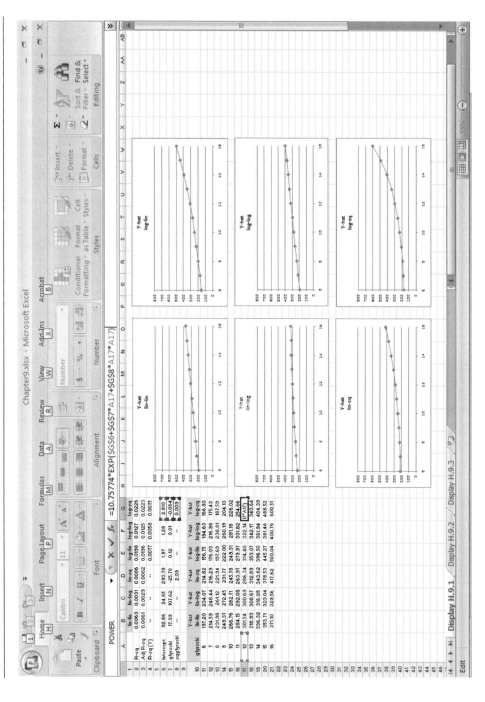

■ Display H.9.2  Predicted Values and Graph of Dummy Variable Specification for Number of Sisters

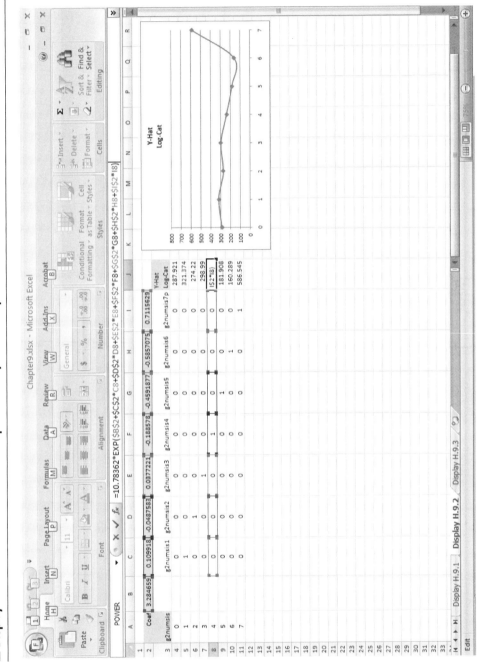

■ **Display H.9.3 Predicted Values and Graph of Dummy Variable Specification for Number of Sisters, Collapsed Model**

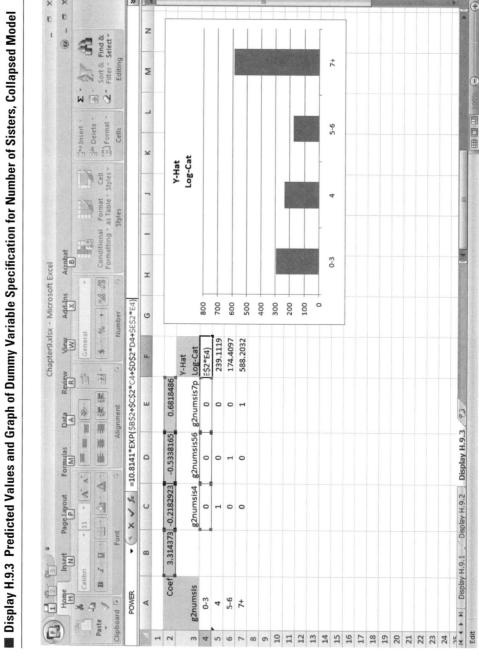

*Appendix I*

# USING HAYES-CAI SAS MACRO FOR HETEROSKEDASTICITY-CONSISTENT STANDARD ERRORS

This appendix provides instructions for using the SAS macro written by Hayes and Cai (2007) to estimate heteroskedasticity-consistent standard errors.

The SAS macro, called *hcreg.sas* should be saved to your computer from the link on their web site: *http://www.comm.ohio-state.edu/ahayes/SPSS%20programs/HCSEp.htm.* (Do not try to type it in or cut-and-paste it!)

To use the macro, you must first execute its syntax. This can be done readily with the `%include` command in SAS. When we place a filename after this command, SAS executes the code in that file.

We downloaded the macros and saved them in our *c:\nsfh_distance\SAS* folder so that our command reads:

```
%include "c:\nsfh_distance\sas\hcreg.sas";
```

The syntax for using the new command created by the macro is:

```
%HCREG(data = <filename>, dv = <depvar>, iv = <variable list>);
```

where *filename* is the name of a SAS data file, *dv* is the dependent variable, and *iv* is a list of variables separated by spaces. By default, the HC3 heteroskedasticity-consistent standard errors are calculated.

For our model regressing distance on mother's years of schooling we would use:

```
%HCREG(data = miles, dv = g1miles, iv = g1yrschl);
```

The filename should be a one-word name, so we use the temporary data set called *miles*.

# NOTES

## Chapter 2

2.1 These summary files and public use microdata files for the census are also available through the ICPSR archive.

2.2 Students can check if their university is a member and allows direct data access at: *http://www.icpsr.org/ICPSR/membership/ors.html*. Students at universities that do not allow direct data access should contact their organizational representative to get access.

2.3 These resources should be available through your local library. See *http://www.annualreviews.org/* and *http://thomsonreuters.com/products_services/scientific/Web_of_Science*.

2.4 At the time of this writing, ICPSR was developing a similar tool to allow for searching variable names of all datasets, not only those that are available for online analysis.

2.5 Students can search for the data set in ICPSR to read the study summary. See *http://www.icpsr.umich.edu/cocoon/ICPSR-STUDY/21600.xml*.

## Chapter 3

3.1 It also is possible to use Stata's `cmdlog` command to capture the commands you type interactively for later editing.

3.2 The self-administered questionnaire, in Display C.3.4, does not allow respondents to circle *don't know* or *refused*; thus it is sensible that most are recorded as *no answer*. The few recorded as *don't know* or *refused* may have written something on the questionnaire that allowed this coding or may have been among the small subset of respondents who were administered the survey via telephone.

3.3 The number coded as *inapplicable* is larger than the number who reported the mother to be deceased (4,434 in Display C.3.6) but smaller than the number who also did not provide a valid response to the first question ($266 = 1 + 13 + 252$ in Display C.3.6). With paper and pencil administration, especially in these self-administered questionnaires, skip errors are possible. These could be further explored after the raw data are downloaded.

3.4  As we will discuss when recoding this variable, most distances among US locations should be at most 5,000 miles, suggesting some respondents with larger values may have reported an incorrect value or may have provided a value for mothers living outside the USA. Such errors are especially likely in paper-and-pencil administrations.

3.5  Multiple imputation is increasingly used by researchers to deal with missing data but beyond the scope of this textbook (see Allison 2001 for an accessible treatment). We focus on appropriately coding missing values as '.' missing. The data coded with '.' missings could be later re-analyzed with multiple imputation, if desired.

## Chapter 4

4.1  Stata also has `drop` and `keep` commands that allow for excluding variables and cases from the data file with an `if` qualifier. However, when reading large data sets like the NSFH into Stata it is helpful to read only a portion of the cases on the `use` command because of memory limitations.

4.2  Another method is explicitly to change the end of command delimiter to the same symbol—a semicolon—used by SAS. Programmers who switch frequently back and forth between SAS and Stata sometimes like this approach. To do so, we type `#delimit ;` After this, Stata treats all words from a command name to the semicolon as part of one command. The command `#delimit cr` can be issued after the long line to switch back to using the end of the line (*cr* stands for carriage return which is the code that marks the end of the line to the computer, although not visible to us in most text editors or word processors).

4.3  Stata's Version 11, released in summer 2009, includes syntax highlighting in the Do-File Editor.

4.4  Four days overlap in both survey years: June 10–13. Because the year of the interview was not directly coded, there is no way to assign definitely them to 1996 or 1997. Our coding treats them all as 1996. For a manuscript, we could look at how sensitive the results are to assigning them to 1996 or 1997, or randomly assign them to one year, but we will use 1996 for the chapter exercises. Given that there are only five cases interviewed on these days, and the interviews differed by just one year, it is unlikely that the results would be sensitive to this coding.

4.5  We could similarly look at the sensitivity of assuming that these mail interviews were completed in 1996 or 1997, or randomly assign these cases to 1996 or 1997. Because about 10 percent of interviews (121) were completed by mail, this coding might have more of an effect on results than the overlap in interview days between interview years, although again the one-year difference in the years likely attenuates any effect.

## Chapter 5

5.1  That is, first substitute $\hat{Y}_i$ for $\hat{\beta}_0 + \hat{\beta}_1 X_i$ in Equation 5.3 (i.e., $Y_i = \hat{Y}_i + \hat{\varepsilon}_i$). Then, subtract $\hat{Y}_i$ from both sides ($Y_i - \hat{Y}_i = \hat{\varepsilon}_i$).

5.2 For illustrative purposes, we first trimmed about 1 percent of cases with the most extreme values of hours (over 140 hours per work). Then, after regressing hours of chores on number of children, we kept a random set of 10 cases within each level of number of children whose residuals fell within half a standard deviation of the fitted line. We also omitted about 30 cases who reported more than four children in the household.

5.3 Recall that the empirical rule tells us that in a normal distribution about 95 percent of the values lie within two standard deviations of the mean, about 68 percent fall within one standard deviation of the mean, and nearly 100 percent fall within three standard deviations of the mean.

5.4 This can be seen, for example, by solving Equation 5.6 for $\bar{Y}$. Adding $\hat{\beta}_1 \bar{X}$ to both sides results in $\hat{\beta}_0 + \hat{\beta}_1 \bar{X} = \bar{Y}$.

5.5 As you learned in introductory statistics $\hat{\sigma}_Y = \sqrt{\Sigma(Y_i - \bar{Y})^2/(n-1)}$. In this sample $\hat{\sigma}_Y = \sqrt{5371.92/(50-1)} = 10.47$ (the numerator is calculated by squaring the values in the third column and then summing the results). Note that because we drew cases with small residuals to create the stylized sample, the difference between the conditional and unconditional standard deviations is larger in this set of 50 cases than in the full NSFH sample.

5.6 Recall that the central limit theorem states that the sum of identically distributed random variables approaches a normal distribution as the sample size increases. This convergence to the normal distribution happens even if the original variables are not normally distributed, although larger sample sizes are required as the original variables' distributions deviate more from the normal.

5.7 Stata's commands can be abbreviated to the first three letters (e.g., `reg` for `regress`). We use the full command because the full commands are usually easier for students to understand as they start to learn the Stata syntax.

5.8 The abbreviation `_cons` in Stata stands for *constant*. The intercept is sometimes referred to as the constant because its value is included in the predicted value for every case (i.e., it is a "constant") in contrast to the slope whose contribution varies depending on the level of $X$.

5.9 SAS and Stata differ in rounding the values. This is especially important in our case with *g2earn*, because of the small size of the coefficient estimate. In this case, Stata leaves more significant digits on the estimate, where significant digits are values that contribute to the meaning of the value (ignoring leading and trailing zeros). As we discuss below, it is possible to rescale a variable before estimating the model to eliminate the leading decimals and assure the presentation of more significant digits in the output. It is also important to take care when rounding coefficient estimates for the prediction equation and presenting results in papers: presenting fewer digits is easier to read but keeping more digits will reduce rounding error in calculations. We can always read the *t*-value off the output, and we will see in later chapters how to ask SAS and Stata to make predictions for us (thus reducing rounding errors).

5.10 The DF column in SAS does not list the degrees of freedom for the *t*-test. We will learn in Chapter 6 how these DF correspond to related $F$-tests. Also note that the computer calculates the $t$-value based on the stored values of the coefficient and standard error. The coefficient estimate and its standard error, listed in the output, have been rounded to a smaller number of digits. Thus, a $t$-value hand calculated based on the listed coefficient estimate and standard error differs slightly from the computer's calculation of the $t$-value. To minimize rounding error in values reported in a manuscript, it is preferable to use the statistic as calculated by the statistical software.

5.11   To fully examine the latter idea, we might want to measure the child's educational attainment and occupation in addition to earnings. And, ideally, we might measure professional networks and personal jobseeking strategies.

5.12   To minimize rounding error, we used in our calculation the most significant digits available across the packages (from Stata for the point estimate and from SAS for the standard error).

5.13   Based on Equation 5.1, we should also subtract the mean value to calculate a standardized variable. For example, $newY_i = (1/\hat{\sigma}_Y)(Y_i - \bar{Y})$. But, subtracting a constant, like the mean, will not affect the estimate of the slope.

5.14   This result is explicit in SAS's correlation output, which lists the same correlation value above and below the diagonal (see again Display B.5.7).

5.15   In the full NSFH sample, the completely standardized coefficient estimate is smaller, 0.25, just above Cohen's cutoff for a small effect, but still larger than the effect for *g1miles* and *g2earn* in our distance example.

## Chapter 6

6.1   We might also use the indicator variable approach considered in Chapter 7 and Chapter 9 to distinguish various levels of educational attainment. This approach would be especially important if we thought that the effect of one more year of high school differed from the effect of one more year of college.

6.2   We excluded cases missing any of the variables used for the miles examples in this chapter, thus the sample size in Display B.6.1 is slightly smaller than the sample size in Chapter 5 ($n = 5,475$). The standard deviations are thus slightly different than those used in Chapter 5.

6.3   As in Chapter 5, we might want to instead semistandardize the coefficient for earnings rescaled to $10,000 units. Then, we would have $10,000 * 0.001022/645.0074 = 0.016$. Thus, a $10,000 increase in earnings is associated with the adult living about 0.016 of a standard deviation further from the mother.

6.4   The abbreviations are used differently in different subfields. For example, SSE is sometimes used to abbreviate Explained Sum of Squares which we use instead here for Sum of Squared Errors. In the SAS output, TSS is listed as "Corrected Total," SSE is the "Error Sum of Squares," and MSS is the "Model Sum of Squares."

6.5   Although some researchers use the overall $F$-test in a fashion similar to the Bonferroni-type adjustment (i.e., not examining individual coefficients unless the model $F$ is significant), having a nonsignificant $F$-test when individual $t$-statistics are significant is unusual in well-specified models. The more common importance of the model $F$-test is in identifying multicollinearity. As we'll see in Chapter 11, a hallmark of multicollinearity is the nonintuitive situation in which the model $F$ test is significant while none of the individual $t$-tests for the slopes is significant.

6.6   That is: $(n-1) - (n-2) = n - 1 - n + 2 = (n - n) + (2 - 1) = 1$

6.7   SAS and Stata round the $t$-value to two decimals. If you instead calculate the $t$-value directly using the listed coefficient and standard error, you can match the $F$-value more exactly. $6.383609 / 0.435193 = 14.668455$ and $14.668455 * 14.668455 = 215.16$.

6.8   That is: $(3{,}116 - 1) - (3{,}116 - 3) = 3{,}116 - 1 - 3{,}116 + 3 = 3 - 1 = 2$.

6.9   For the numerator, $(3{,}116 - 2) - (3{,}116 - 3) = 3{,}116 - 2 - 3{,}116 + 3 = 1$

6.10  Notice that Stata's sum of squares tables in Display B.6.1 and Display B.6.8 list the residual and total sum of squares in scientific notation because of their large size. We can read the values directly from SAS's output, or we can calculate them more precisely in the Stata output by multiplying the MS column value in the residual row (e.g., 412,112.921 in Display B.6.1) by the residual degrees of freedom (e.g., 5,472 in Display B.6.1) although this introduces some rounding error. We will also see later in the text how to ask SAS and Stata to calculate the $F$-value directly for us, to avoid rounding error.

6.11  That is: $(5{,}475 - 3) - (5{,}475 - 7) = 5{,}475 - 3 - 5{,}475 + 7 = 4$.

6.12  We will later learn how to use these test commands to test other null hypotheses, such as that the coefficients for two variables are equivalent.

6.13  It is useful to remember that a Pearson correlation between two variables can be squared to determine how much variation the variables share. It is also instructive to note that a correlation of .50 reflects just 25 percent shared variation. It is not until the correlation exceeds .70 that the two variables share 50 percent variation $(.7 * .7 = .49)$.

6.14  The $R$-squared for this stylized sample is larger than the $R$-squared found in Display B.6.4 for the full NSFH sample of $n = 3{,}116$ $(r = 0.0646)$ because we chose the stylized sample of 50 to include observations that fell relatively close to the regression line.

6.15  In fact, an effect size measure for multiple regression is based on $R$-squared (Cohen 1992).

6.16  Presentation of confidence intervals for coefficients has become the norm in some fields, such as public health. In Chapter 7, we present an example of the presentation of confidence intervals. As noted earlier, presenting confidence intervals is useful because they make explicit that the population coefficient is estimated with uncertainty.

## Chapter 7

7.1   In general, there are usually multiple ways to perform data management tasks, such as data creation. One person's code may be simpler than another's for a particular task, but as long as both accomplish the objective, both are fine. As a student, it is important to use the approach that you find easiest to understand initially. Then, you can move on to more compact syntax. As a student and scientist, it is also important that your code is easy for you to proof and for others to understand. Sometimes longer, less elegant, syntax can be easier to check for errors or to be understood by others (or by you a few months down the road). At times, SAS or Stata will also have an approach to a specific task not offered in the other package. For example, in Stata, it is possible to create dummy variables using the tabulate command combined with a generate option. For example, "`tabulate M484 if M484<97, gen(M484)`" creates dummy variables for each category of the original variable, assigning missing values to '.' missing. Similarly, `xi i.M484` does the same, although missing values must be dealt with separately and the smallest variable is used as reference by default (the command `char M484[omit] 2` can be used to instead exclude the second category).

7.2 SAS has syntax that allows us to type the *if* qualifier just once, referred to as `if-then-do`. Statements that follow the `do` are executed only for cases that meet the expression, until a closing end statement. For example, we could define our dummies as follows:

```
if M484<97 then do;
  aframer=(M484=1);
  white= (M484=2);
  mexamer=(M484=3);
  amind= (M484=7);
  other= ((M484>=4 & M484<=6) | M484=8 | M484=9);
end;
```

7.3 The result in Equation 7.3 will be clear to some students, recognizing that the two $\beta_0$ terms in the equation $(\beta_0 + \beta_1) - (\beta_0)$ cancel out, because one is positive and one is negative. To show why this is so, we can rewrite the equation as follows:

$$(\beta_0 + \beta_1) - (\beta_0) = \beta_0 + \beta_1 - \beta_0 = \beta_0 - \beta_0 + \beta_1 = \beta_1$$

Using the distributive property of multiplication, we first multiply through to remove the parentheses, then collect terms (put the two $\beta_0$ terms side by side), and then subtract one $\beta_0$ from the other. To put this another way, the $\beta_0$ term is common to both expected values, so when we subtract one expected value from the other, we are left with the term that is not common to both: $\beta_1$.

7.4 More specifically, the OLS results are analogous to the *t*-test assuming equal variance on the outcome across the groups. This is analogous to the homoskedasticity assumption in OLS regression. We will discuss in Chapter 11 how to test this assumption and, if needed, attempt to make the data better conform to this assumption in OLS.

7.5 Like the analogy of the two-sample *t*-test to a regression with a single dummy variable, Analysis of Variance (ANOVA) and Analysis of Covariance (ANCOVA) conduct hypothesis tests that parallel those of regression models with sets of dummies and interval-level controls. Whereas interpretation of regression models focuses on the coefficients for individual variables, ANOVA and ANCOVA focus on partitioning the variance, often for sets of variables at a time. The sums of squares decomposition that we discussed in Chapter 6 is fundamental to ANOVA.

7.6 As noted, other strategies for coding the categories allow for different interpretations. See Endnote 7.1.

7.7 What if we were interested in a contrast between an included group and all other groups combined? Then, we could include just one dummy variable to indicate that group, and exclude the rest. For example, if we were interested in contrasting whites versus all other groups combined, then we would estimate the model: $Y_i = \beta_0 + \beta_3 D_{3i} + \varepsilon_i$. By grouping American Indians and Mexican Americans with African Americans as a single reference category, this model implies that the means for these omitted groups do not differ. This could be tested by estimating the initial model: $Y_i = \beta_0 + \beta_1 D_{1i} + \beta_2 D_{2i} + \beta_3 D_{3i} + \varepsilon_i$. If the coefficient estimates for the dummy variables $D_1$ and $D_2$ were insignificant, we would have empirical evidence that the means for these groups do not differ, which could be used to justify re-estimating the model with both $D_1$ and $D_2$ excluded. Researchers sometimes use this approach to simplify the presentation of their results, referring to it as "collapsing together" categories. Sometimes the approach is used to collapse together categories with small sample sizes, although ideally the approach is supported by conceptual reasons or prior empirical results suggesting that the other groups would not differ from one another (i.e., if the sample sizes of the subgroups are small, then the standard errors will be relatively large, and it will be difficult to reject the null hypothesis that the means for the subgroups are equivalent).

7.8 This rationale implies a directional hypothesis (whites less than American Indians; whites less than Mexican Americans). Although we will use two-tailed tests throughout this chapter, as we discussed in Chapter 5, if you have strong à priori rationale for a directional hypothesis, and your results are consistent with your hypothesis, then you can calculate the one-tailed $p$-values by dividing the $p$-values reported by SAS and Stata in half.

7.9 We speculated in the text that the other groups might live closer to their mothers than would whites, implying a one-sided test. In the bottom panel of Display B.7.7, the coefficient estimates for Mexican Americans and African Americans are negative, consistent with the hypothesis that they live closer to their mothers than do whites. Thus, to calculate the one-sided $p$-value, we would divide the listed $p$-value in half ($0.197/2 = 0.0985$ for *mexamer* and $0.022/2 = 0.011$ for *aframer*). In contrast, the coefficient estimate for American Indians is positive, inconsistent with our hypothesis. Thus, to calculate the one-sided $p$-value for *amind* we should subtract half the listed $p$-value from one ($[1-(.433/2)] = .7835$). Doing so does not change any of our conclusions with a 5 percent alpha level, although the difference between Mexican Americans and whites is now a trend or marginal significance (just below the cutoff often used for marginal significance of $\alpha = 0.10$).

7.10 In each row, we simplified the algebra for the difference, by first multiplying the $-1$ through the parentheses using the distributive property of multiplication (e.g., $-(\beta_0 + \beta_2)$ becomes $-\beta_0 - \beta_2$). And, the positively and negatively signed $\beta_0$ terms cancel out.

7.11 We might have used theory and prior research to hypothesize in advance that women would live closer to their mothers than men. If so, then our coefficient estimate would be consistent with our hypothesis, and thus, to calculate the one-sided $p$-value, we would divide the $p$-value listed by SAS and Stata in half. For *female*, the two-sided $p$-value in Display B.7.12 is 0.072. Thus, the one-sided $p$-value would be $0.072/2 = 0.036$.

7.12 If we had hypothesized in advance that African Americans would live closer to their mothers than other groups, then the positive coefficient estimates in Display B.7.12 would be consistent with our hypotheses and we would divided SAS and Stata's two-sided $p$-values in half (making the contrasts with *amind* and *white* marginally significant, since $0.150/2 = 0.075$ and $0.164/2 = 0.082$). We also know from the `test` commands in Display B.7.9 that the other contrasts, among the included groups, are not significant (even for one-sided hypotheses). Had any been significant, a common way to indicate such contrasts would be to put superscript letters on pairs of categories that do (or do not) have coefficient estimates that differ significantly from each other. Literature Excerpt 7.2 provides an example of using such subscripts.

7.13 As we will discuss in Chapter 8, it is possible to make the intercept meaningful by altering what a zero on the predictor variable represents, a technique often referred to as *centering* the predictor.

7.14 The authors' inclusion of only the coefficient estimates and $p$-values conserves space, and space is always a premium in journals; however, we will recommend providing standard-errors in tables (because the standard errors can help us to identify problems with our models, which we will discuss in Chapter 11).

7.15 These outcome variables could also be treated as ordinal. For example, each value of educational expectations is linked to an explicit level of education, and whether the differences between those levels are equal is not known (e.g., is moving from not finishing high school to graduating from high school equivalent to moving from completing a few years of college to graduating from a four-year college?). The roadmap in Chapter 12 includes advanced models specifically designed for this kind of ordinal outcome.

7.16 The authors do not tell us whether these are one-tailed or two-tailed $p$-values, although most likely they are two-tailed. Although some of the authors' arguments suggest directional alternative hypotheses, they have generally laid out two mechanisms (family resources and family instability) with somewhat different predictions.

7.17 The authors also related the family types to the child's likelihood of having graduated high school and enrolled in college by Wave 2. Importantly, these models also reveal that children in cohabiting families are less likely than one or more other groups to graduate from high school and to enroll in college. These strengthen the findings we discuss in the text of this chapter because they are based on the child's own report, rather than the mother's reports. The maternal reports could be picking up effects of family type on mother's perceptions of the child, rather than actual child effects.

7.18 That is, $(0.58/10) * (2/2) = 1.16/20$. Similarly, $(0.37/10) * (2/2) = 0.74/20$ and $(0.80/10) * (2/2) = 1.60/20$.

7.19 That is, $(0.08/10) * (10/10) = 0.80/100$. And, $(1.36/10) > (1/10)$.

## Chapter 8

8.1 All of the Excel spreadsheets from this chapter are on the textbook web site *http://www.routledge.com/textbooks/9780415991544*.

8.2 Because of our focus on gender differences, it is useful to have in mind that the NSFH asked respondents to report the time they spent on household tasks in nine different categories, covering stereotypically male and female chores. The items, from Self-Enumerated Questionnaire #1, are: (1) Preparing meals, (2) Washing dishes and cleaning up after meals, (3) Cleaning house, (4) Outdoor and other household maintenance tasks (lawn and yard work, household repair, painting, etc.), (5) Shopping for groceries and other household goods, (6) Washing, ironing, mending, (7) Paying bills and keeping financial records, (8) Automobile maintenance and repair, and (9) Driving other household members to work, school, or other activities.

8.3 We do not code cohabiting partners in this example, although it would be possible to do so in the NSFH. Indeed, when we have done so, we found that women who are cohabiting have statistically equivalent average hours of chores as single women.

8.4 It is also possible to respecify the model using dummies for the four cells in the cross-tabulation of the two dummies. We can think of the four cells as a four-category variable capturing joint characteristics across the two dummy variables. We will need to create new dummy indicators for three of the cells, and allow one cell to be the reference category. For example, we might specify:

$$Y_i = \beta_0^+ + \beta_1^+ D_{1i}^+ + \beta_1^+ D_{2i}^+ + \beta_3^+ D_{3i}^+ + \varepsilon_i$$

where $D_{1i}^+$ indicates being married and male, $D_{2i}^+$ indicates being unmarried and female, and $D_{3i}^+$ indicates being married and female. The reference group is unmarried males. The results will be equivalent to the results for including two dummies and their product, although the interpretation will differ. Each coefficient now captures the difference in means between the group with each pair of characteristics, and the reference category. To test whether the effect of one variable differs depending on the level of the other variable, would require tests of equality of coefficients. For the effect of being married to be the same within gender we would test whether $(\beta_3^+ - \beta_2^+) = \beta_1^+$. For the effect of being female to be the same within marital status we would test whether $(\beta_3^+ - \beta_1^+) = \beta_2^+$. These two equations are equivalent. Both test whether $\beta_3^+ = \beta_1^+ + \beta_2^+$. This tests whether the sum of the effect of having just one of the characteristics (married but not female, $D_{1i}^+$; female but not married, $D_{2i}^+$) relative to neither characteristic (the reference category of not married and not female) equals the effect of having both characteristics simultaneously (being married and female, $D_{3i}^+$), relative to neither characteristic (the reference category of not married and not female). Because we typically want to first test this equality constraint, it is more common in the social sciences to use the product term approach than to use the explicit coding for the joint characteristics across the categorical variables.

8.5 In a real world project, rather than being simple means, the predicted values would take advantage of multiple regression and be adjusted for other variables in the model. We show how to make such predictions, with controls, in Display H.8.11.

8.6 These are about the 10th and 90th percentiles for women. For men, the 10th and 90th percentiles are about 30 and 60, although at least 5 percent of men work 15 or fewer hours per week

8.7 As we have emphasized previously, SAS and Stata conduct a two-sided test by default. In this case, the sign of the product term is negative, but its absolute value of 2.87 is greater than the cutoff of 1.96. Above, we laid out a directional hypothesis, expecting that the sign on *hrwork* would be negative for both genders, but of larger magnitude (more negative) for women than men. The sign on *hrwork* and on the product term *fem_hrwork* are both negative, and thus consistent with these directional hypotheses. Thus, to conduct a one-sided hypothesis test, we would divide the *p*-values in half (doing so has little practical effect in our case, however, because the *p*-values are both less than 0.001).

8.8 It is preferable to have SAS and Stata make the calculations that will be presented in a manuscript, to avoid rounding error (and other human error). We could also use the full estimates in Display B.8.6 in our hand calculation (rather than rounding them to two significant digits as we did in Equation 8.3). This would give $-0.1243169 - 0.1256042 = -.2499211$.

8.9 Again, it is preferable to have SAS and Stata make the calculations that will be presented in a manuscript, to avoid rounding error (and other human error). If we need to make hand calculations to present in a paper, we could also use the full estimates in Display B.8.6 (rather than rounding them to two significant digits as we did in Equation 8.3). For 10 hours of paid work, this would give $16.99004 - 10 * 0.1256042 = 15.73$.

8.10 The standard deviation of *hrwork* is similar within gender, at 14.08 for men and 14.17 for women.

8.11 Clearly, for a publication, we would introduce additional variables to the model. For example, our rationale for marital status and number of children may extend also to the number and/or characteristics of other adults in the household.

8.12 We present just the `lincom` results for simplicity, now that we have fully examined all three approaches. We could also use the other approaches of recentering each variable and re-estimating the regression, using the `test` command, or requesting the variance–covariance matrix of coefficients.

8.13 The footnote on the front page which notes that some of the data for the study (e.g., percentage of Evangelical Protestants in a county) came from the American Religion Data Archive (now the Association of Religion Data Archives; *http://www.thearda.com/*), an important resource for students interested in religion. The authors combine these data with other sources, primarily the 2000 US Census.

8.14 The authors do not report how fully sex segregation and percentage women in the labor force vary and covary across counties. If the empirical distribution of sex segregation does not vary completely, from zero to one, within one or more of the conditions (45 percent, 55 percent and 65 percent in the labor force), then the predictions would be out of range. It is always important to check the univariate range and bivariate distribution of predictors when specifying and interpreting interactions. If these are constrained, using larger existing data sources may be more appropriate for testing a substantive interaction. For example, family and neighborhood income tend to be highly correlated, especially in localized studies. Large national samples may be better suited to testing interactions between these variables.

## Chapter 9

9.1 Other bases for the logs are of course possible, but in our regression models we will always use the natural log.

9.2 Each of the data sets is available publicly, most of them through ICPSR. Cornwell, Laumann and Schumm (2008) use the National Social Life and Health Aging Project (*http://www.norc.org/nshap*). Willson (2003) uses the Mature Women cohort of the National Longitudinal Surveys of Labor Market Experience (*http://www.bls.gov/nls/*). Pritchett and Summers (1996) data come from the Penn World Tables (*http://pwt.econ.upenn.edu/php_site/pwt_index.php*) and the World Bank (*http://www.worldbank.org/*; see the paper for additional notes about their use of these data). Glauber (2007) uses the National Longitudinal Survey of Youth (NLSY 1979; *http://www.bls.gov/nls/*).

9.3 This formula is derived by taking the derivative of Equation 9.1, setting the result equal to zero, and solving for $X$; that is, $\beta_1 + 2\beta_2 X = 0$ leads to $X = -\beta_1/2\beta_2$.

9.4 This formula is the derivative of Equation 9.1.

9.5 Recall that the exponential is often referred to as the anti-log and $\exp[\ln(X)] = X$.

9.6 Generally, $b^{(m+n)} = b^m b^n$. For exponentials, $e^{(m+n)} = e^m e^n$ or equivalently $\exp(m+n) = \exp(m)\exp(n)$.

9.7 In our simple model, we do not include the adult's own education. Because of intergenerational correlations in educational attainment, the basic association between mothers' education and distance may be picking up some effects of the adult's education.

9.8 An alternative hypothesis, we will see some evidence for below, is that larger families may be more family oriented and may stay close together.

9.9 As we noted above, the linear model is nested in the quadratic model. Thus, they can be compared with the general linear $F$-test. There is only one restriction in the null hypothesis for this test, and the square of the $t$-value for the quadratic term is equivalent to the F value for this general linear $F$-test.

9.10 Six and 16 are the 5th and 95th percentiles for *glyrschl* and thus contain most of the observations (the full range of *glyrschl* is 0 to 17).

9.11 In the table below, we round the values to at least three significant digits to make it readable. This introduces some rounding error. In practice, you can use Excel, SAS, or Stata to make calculations, but this table emphasizes the accounting of the transformations.

## Chapter 10

10.1  This is difficult to understand fully based on the information provided in the articles, although the direction of effect of education is opposite of the authors' expectations, and differs from the bivariate associations shown in their Table 2. And, the coefficient on missing income is significant in the multiple regression but not bivariate associations.

10.2  This outcome is really ordinal, rather than interval. In the roadmap in Chapter 12, we discuss alternative models designed explicitly for ordinal outcomes.

## Chapter 11

11.1  Although scholars (including the original developer of DFFIT) have tried to change the name (Welsch 1986), it is commonly used (sometimes as DFITS rather than DFFITS).

11.2  Although DFFITS is sometimes presented in absolute value in the formula, so that it only takes on positive values, both SAS and Stata calculate DFFITS with the sign indicated. Thus, the absolute value should be taken before comparing with the approximate cutoff.

11.3  In Stata, graphs can be easily saved using the command `graph export`. For example, we could type `graph export box_g1miles.wmf` to export the graph created by `graph box g1miles`.

11.4  These graphs can also be saved with the `graph export` command in Stata. The `jitter` option can improve readability for large data sets (we used the option, `jitter(17)` for the graphs shown in Figure 11.3).

11.5  Recall that `ods` stands for output delivery system and we also use an `ods` command to save the output to a rich text file.

11.6  Because the DFBETAS are specific to individual predictors, we also created separate dummy indicators of whether a case was extreme on the overall measures (hat diagonal, studentized residual, Cook's distance, and DFFITS) and whether a case was extreme on the DFB for mother's years of schooling (either the linear *DFB_g1yrschl* or quadratic *DFB_sqg1yrschl* terms) and whether a case was extreme on the DFB for any of the number of sisters' dummies (*DFB_g2numsis4*, *DFB_g2numsis56*, or *DFB_g2numsis7p*). Results were similar, so we focus on a measure that combines these three.

11.7  This can be done directly in Stata, since the diagnostic variables are added to the original data set. In SAS, the diagnostics must be merged into the original data file using the following command: `data influence3; merge miles influence2; run;` where *influence2* is the datafile we created in Display B.11.1. Note that this adds a new data step to our batch program. The SAS system refers to the most recently created data file by default. This has been fine in earlier batch programs, because we had only one data step in the batch program. But, when we have multiple data steps in a batch program, we can use the `data=` option on procedures to refer to the desired data set (e.g., `proc reg data=miles;` or `proc reg data=influence3;`).

11.8 The predictions for mother's years of schooling were made by substituting in the indicated value and holding the rest of the variables constant at their means. The resulting predicted value in log $Y$ units was converted to natural units by taking the exponential and calculating the adjustment factor, using the process shown in Chapter 9. The predicted values for number of sisters were similarly calculated by substituting in the relevant pattern of zeros and ones on the dummies and holding the remaining variables constant at their means, and then converting the predicted log $Y$ value to $Y$ units.

11.9 We rewrote these formulae using the conventions introduced in Chapter 10 for calculating direct and indirect effects, using 1 for the outcome, 2 for the first predictor, and 3 for the second predictor.

11.10 For example, suppose the correlation taken to six digits, rather than rounded to four digits, is 0.990556. Then,

$$VIF = \frac{1}{(1 - 0.990556)} = 105.89$$

11.11 Many courses in psychometric approaches are available. Check your local universities as well as resources like summer courses at the University of Michigan (*http://www.icpsr.umich.edu/sumprog/*) and the University of Kansas (*http://www.quant.ku.edu/StatsCamps/overview.html*) and likely at your local university.

## Chapter 12

12.1 It is not entirely clear from the text, however, whether the coefficients in the table for the *Child characteristics* are from the reduced model (without *Risk factors and child–mother interaction* controlled) or the full model (with *Risk factors and child–mother interaction* controlled). Her text suggests that they are from the reduced model (p. 215). With only one set of coefficients and standard errors presented for each outcome, we also cannot see how one set of coefficients and standard errors change when the other set of variables are controlled.

12.2 In fact, Li-Grining reports this partially standardized result for gender in the text (p. 215). The text also clarifies that the variable, labeled *Gender*, is coded a "1" for boys and "0" for girls (pp. 213, 215).

12.3 Note that the exact $t$-values that Vaisey reports are important in his case because his sample size has just 50 observations.

## Appendix F

F.1 Stata's Version 11, released in summer 2009, now includes .pdf files of all manuals.

## Appendix H

H.1 To rename the worksheet, right click on the tab named "Sheet 1" and choose "Rename." To delete the other two worksheets, right click on their tabs, and choose delete.

H.2 If you replicate your models in both SAS and Stata, you can always select a rectangle directly in the SAS windows. If you only use Stata and open the results .log in Microsoft Office, we recommend switching the Font to *Courier New* style of *9 point* size. Doing so will make it easier to view the results on the screen, without wrapping. A powerful text editor, like TextPad (*www.textpad.com*) or UltraEdit (*www.ultraedit.com*), is worth the investment if you regularly use Stata.

# BIBLIOGRAPHY

Acock, Alan C. 2005. "SAS, Stata, SPSS: A comparison." *Journal of Marriage and the Family* 67: 1093–5.

Agresti, Alan. 2002. *Categorical Data Analysis* (2nd Edition). New York: Wiley.

Agresti, Alan. 2007. *An Introduction to Categorical Data Analysis* (2nd Edition). Hoboken: Wiley.

Allison, Paul D. 2001. *Missing Data*. Thousand Oaks: Sage.

Allison, Paul D. 2005. *Fixed Effects Regression Methods for Longitudinal Data Using SAS*. Cary, NC: SAS Institute.

Allison, Paul D. 2009. *Statistical Horizons*. Available at *http://www.statisticalhorizons.com/*.

American Sociological Association. 2007. *American Sociological Association Style Guide*. Washington DC: Author.

American Psychological Association. 2009. *Publication Manual of the American Psychological Association*. Washington DC: Author.

Andrich, David. 1988. *Rasch Models for Measurement*. Thousand Oaks: Sage.

Baron, Reuben M. and David A. Kenny. 1986. "The Moderator–Mediator Variable Distinction In Social Psychological Research: Conceptual, Strategic, and Statistical Considerations." *Journal of Personality and Social Psychology*, 51(6): 1173–82.

Belsey, David A., Edwin Kuh, and Roy E. Welsch. 1980. *Regression Diagnostics: Identifying Influential Data and Sources of Collinearity*. New York: Wiley.

Berk, Richard A. 2004. *Regression Analysis: A Constructive Critique*. Thousand Oaks: Sage.

Bickel, Robert. 2007. *Multilevel Analysis for Applied Research: It's Just Regression!* New York: Guilford.

Bland, J. Martin and Douglas G. Altman. 1995. "Multiple Significance Tests: The Bonferroni Method." *British Medical Journal*, 310: 170.

Blossfeld, Hans-Peter and Götz Rohwer. 2002. *Techniques of Event History Modeling* (2nd Edition). New York: Lawrence Erlbaum Associates.

Bollen, Ken A. 1989. *Structural Equation Models with Latent Variables*. New York: Wiley.

Box G.E.P. and D.R. Cox. 1964. "An Analysis of Transformations." *Journal of the Royal Statistical Society, Series B*, 26: 211–52.

Brumbaugh, Stacey M., Laura A. Sanchez, Steven L. Nock, and James D. Wright. 2008. "Attitudes Toward Gay Marriage in States Undergoing Marriage Law Transformation." *Journal of Marriage and Family*, 70: 345–59.

Burnham, Kenneth P. and David R. Anderson. 2002. *Model Selection and Multi-Model Inference*. New York: Springer-Verlag.

Bushway, Shawn, Brian D. Johnson, and Lee Ann Slocum. 2007. "Is the Magic Still There? The Use of the Heckman Two-Stage Correction for Selection Bias in Criminology." *Journal of Quantitative Criminology*, 23: 151–78.

Chatterjee, Samprit and Ali S. Hadi. 1986. "Influential Observations, High Leverage Points, and Outliers in Linear Regression." *Statistical Science*, 1(3): 379–93.

Chatterjee, Sangit and Mustafa Yilmaz. 1992. "A Review of Regression Diagnostics for Behavioral Research." *Applied Psychological Measurement*, 16: 209–27.

Chow, Gregory C. 1960. "Tests of Equality between Sets of Coefficients in Two Linear Regressions." *Econometrica*, 28(3): 591–605.

Clark, William A.V. and Suzanne Davies Withers. 2007. "Family Migration and Mobility Sequences in the United States: Spatial Mobility in the Context of the Life Course." *Demographic Research*, 17: 591–622.

Cohen, Jacob. 1969. *Statistical Power Analysis for the Behavioral Sciences*. New York: Academic Press.

Cohen, Jacob. 1992. "A Power Primer." *Psychological Bulletin*, 112(1): 155–9.

Cooney, Teresa M. and Peter Uhlenberg. 1992. "Support from Parents Over the Life Course: The Adult Child's Perspective." *Social Forces*, 71: 63–84.

Cornwell, Benjamin, Edward O. Laumann, and L. Philip Schumm. 2008. "The Social Connectedness of Older Adults: A National Profile." *American Sociological Review*, 73: 185–203.

de Leeuw, Jan and Richard Berk. 2004. *Introduction to the Series, Advanced Quantitative Techniques in the Social Sciences*. Thousand Oaks: Sage.

Dooley, David and Joann Prause. 2005. "Birth Weight and Mothers' Adverse Employment Change." *Journal of Health and Social Behavior*, 46: 141–55.

Downs, George W. and David M. Rocke. 1979. "Interpreting Heteroscedasticity." *American Journal of Political Science*, 23(4): 816–28.

DuMouchel, William H. and Greg J. Duncan. 1983. "Using Sample Survey Weights in Multiple Regression Analyses of Stratified Samples." *Journal of the American Statistical Assocation*, 78(383): 535–43.

Duncan, Greg J., Chantelle J. Dowsett, Amy Claessens, Katherine Magnuson, Aletha C. Huston, Pamela Klebanov, Linda S. Pagani, Leon Feinstein, Mimi Engel, Jeanne Brooks-Gunn, Holly Sexton, Kathryn Duckworth, and Crista Japel. 2007. School readiness and later achievement. *Developmental Psychology*, 43: 1428–46.

Fairweather, James S. 2005. "Beyond the Rhetoric: Trends in the Relative Value of Teaching and Research in Faculty Salaries." *Journal of Higher Education*, 76: 401–22.

Fox, John. 1991. *Regression Diagnostics*. Newbury Park: Sage.

Fox, John. 2008. *Applied Regression Analysis and Generalized Linear Models* (2nd Edition). Los Angeles: Sage.

Freese, Jeremy. 2007. "Replication Standards for Quantitative Social Science: Why Not Sociology?" *Sociological Methods and Research*, 36: 153–72.

Ghilagaber, Gebrenegus. 2004. "Another Look at Chow's Test for the Equality of Two Heteroscedastic Regression Models." *Quality and Quantity*, 38: 81–93.

Giordano, Peggy C., Monica A. Longmore, and Wendy D. Manning. 2006. "Gender and the Meanings of Adolescent Romantic Relationships: A Focus on Boys." *American Sociological Review*, 71: 260–87.

Glauber, Rebecca. 2007. "Marriage and the Motherhood Wage Penalty among African Americans, Hispanics, and Whites." *Journal of Marriage and Family*, 69: 951–61.

Greene, William H. 2008. *Econometric Analysis*. Saddle River, NJ: Pearson-Prentice Hall.

Grossbard-Shechtman, Shoshana. 1993. *On the Economics of Marriage: A Theory of Marriage, Labor, and Divorce*. Boulder: Westview Press.

Gujarati, Damodar N. 1970a. "Use of Dummy Variables in Testing for Equality between Sets of Coefficients in Two Linear Regressions: A Note." *American Statistician*, 24(1): 50–2.

Gujarati, Damodar N. 1970b. "Use of Dummy Variables in Testing for Equality between Sets of Coefficients in Two Linear Regressions: A Generalization." *American Statistician*, 24(5): 18–22.

Gujarati, Damodar N. 2003. *Basic Econometrics*. Boston: McGraw-Hill.

Harrington, Donna. 2008. *Confirmatory Factor Analysis*. Oxford: Oxford University Press.

Hayes, Andrew F. and Li Cai. 2007. "Using Heteroskedasticity-Consistent Standard Error Estimates in OLS Regression: An Introduction and Software Implementation." *Behavior Research Methods*, 39: 709–22. See *www.comm.ohio-state.edu/ahayes/SPSS% 20programs/HCSEp.htm* for downloadable macros.

Heckman, James J. 1979. "Sample Selection Bias as a Specification Error." *Econometrica*, 47: 153–61.

Hilbe, Joseph. 1996. "Windows File Conversion Software." *The American Statistician*, 50: 268–70.

Hoaglin, David C. and Peter J. Kempthorne. 1986. "Comment." *Statistical Science*, 1(3): 408–12.

House, James S., James M. LaRocco, and John R.P. French, Jr. 1982. "Response to Schaefer." *Journal of Health and Social Behavior*, 23: 98–101.

Huberty, Carl J. 2002. "A History of Effect Size Indices." *Educational and Psychological Measurement*, 62: 227–40.

Huffman, Matt L. 1999. "Who's in Charge? Organizational Influences on Women's Representation in Managerial Positions." *Social Science Quarterly*, 80: 738–56.

Iacobucci, Dawn. 2008. *Mediation Analysis*. Thousand Oaks: Sage.

ICPSR 2009. *Summer Program in Quantitative Methods of Social Research*. Available at *www.icpsr.umich.edu/sumprog*.

Jaccard, James and Robert Turrisi. 2003. *Interaction Effects in Multiple Regression*. Thousand Oaks: Sage.

Jann, Ben. 2007. "Making regression tables simplified." *The Stata Journal*, 7(2): 227–44.

Jarrell, Stephen B. and T.D. Stanley. 2004. "Declining Bias and Gender Wage Discrimination? A Meta-Regression Analysis." *Journal of Human Resources*, 39: 828–38.

Kalleberg, Arne L., David Knoke, and Peter V. Marsden. 2001. "National Organizations Survey (NOS), 1996–1997." Available at *www.icpsr.org*.

Kalleberg, Arne L., David Knoke, Peter V. Marsden, and Joe L. Spaeth. 1994. "National Organizations Survey (NOS), 1991." Available at *www.icpsr.org*.

Kolenikov, Stanislav 2001. "Review of Stata 7." *Journal of Applied Econometrics*, 16: 637–46.

Korn, Edward L. and Barry I. Graubard. 1995. "Examples of Differing Weighted and Unweighted Estimates from a Sample Survey." *The American Statistician*, 49: 291–5.

Kutner, Michael H., Christopher J. Nachtsheim, and John Neter. 2004. *Applied Linear Regression Models*, (4th Edition). New York: McGraw-Hill.

LaRocco, James M., James S. House, and John R. P. French Jr. 1980. "Social Support, Occupational Stress, and Health." *Journal of Health and Social Behavior*, 21: 202–18.

Lawton, Leora, Merril Silverstein, and Vern Bengtson. 1994. "Affection, Social Contact, and Geographic Distance Between Adult Children and Their Parents." *Journal of Marriage and Family*, 56: 57–68.

Li-Grining, Christine P. 2007. "Effortful Control among Low-Income Preschoolers in Three Cities: Stability, Change, and Individual Differences." *Developmental Psychology*, 43(1), 208–21.

Long, J. Scott. 1983. *Confirmatory Factor Analysis: A Preface to LISREL*. Thousand Oaks: Sage.

Long, J. Scott. 1997. *Regression Models for Categorical and Limited Dependent Variables*. Thousand Oaks: Sage.

Long, J. Scott. 2007. "SPost: Postestimation analysis with Stata." Downloaded May 30 2008 from *www.indiana.edu/~jslsoc/spost.htm*.

Long, J. Scott. 2009. *The Workflow of Data Analysis Using Stata: Principles and Practice For Effective Data Management and Analysis*. College Station, TX: Stata Press.

Long, J. Scott and Laurie H. Ervin. 2000. "Using Heteroscedasticity Consistent Standard Errors in the Linear Regression Model." *The American Statistician*, 54: 217–24.

Long, J. Scott and Jeremy Freese. 2006. *Regression Models for Categorical Dependent Variables Using Stata* (2nd Edition). College Station, TX: Stata Press.

Lyons, Christopher J. 2007. Community (Dis)Organization and Racially Motivated Crime. *American Journal of Sociology*, 113(3), 815–63.

MacKinnon, David. 2008. *Introduction to Statistical Mediation Analysis*. New York: Erlbaum.

Maddala, G.S. 1983. *Limited-Dependent and Qualitative Variables in Econometrics*. New York: Cambridge University Press.

Magdol, Lynn and Diane R. Bessel. 2003. "Social Capital, Social Currency, and Portable Assets: The Impact of Residential Mobility on Exchanges of Social Support." *Personal Relationships*, 10: 149–69.

Marsden, Peter V., Arne L. Kalleberg, and Cynthia R. Cook. 1993. "Gender Differences in Organizational Commitment: Influences of Work Positions and Family Roles." *Work and Occupations*, 20: 368–90.

McCartney, Kathleen and Robert Rosenthal. 2000. "Effect Size, Practical Importance, and Social Policy for Children." *Child Development*, 71: 173–80.

McVeigh, Rory and Julianna M. Sobolewski. 2007. "Red Counties, Blue Counties, and Occupational Segregation by Sex and Race." *American Journal of Sociology*, 113: 446–506.

Michielin, Francesca and Clara H. Mulder. 2007. "Geographical Distances between Adult Children and Their Parents in the Netherlands." *Demographic Research*, 17: 655–78.

Morales, Leo S., Peter Guitierrez, and Jose J. Escarce. 2005. "Demographic and Socioeconomic Factors Associated with Blood Lead Levels among American Children and Adolescents in the United States." *Public Health Reports*, 120(4): 448–54.

Mueser, Peter R., Kenneth Troske, and Alexey Gorislavsky. 2007. Using State Administrative Data to Measure Program Performance." *Review of Economics and Statistics*, 89(4): 761–83.

Mulder, Clara H. 2007. "The Family Context and Residential Choice: A Challenge for New Research." *Population, Space and Place*, 13: 265–78.

National Institute of Health. 2003. Final NIH Statement on Sharing Research Data. Available at *http://grants.nih.gov/grants/guide/notice-files/NOT-OD-03-032.html*.

National Institute of Health. 2007. NIH Data Sharing Policy. Available at *http://grants.nih.gov/grants/policy/data_sharing/*.

O'Brien, Robert M. 2007. "A Caution Regarding Rules of Thumb for Variance Inflation Factors." *Quality and Quantity*, 41: 673–90.

Ostini, Remo and Michael L. Nering. 2006. *Polytomous Item Response Theory Models*. Thousand Oaks: Sage.

Pinquart, Martin. 2003. "Loneliness in Married, Widowed, Divorced, and Never-Married Older Adults." *Journal of Social and Personal Relationships*, 20: 31–53.

Primary Respondent Self-Enumerated Schedule. Available at *ftp://elaine.ssc.wisc.edu/pub/nsfh/i1sc.001*.

Pritchett, Lant and Lawrence H. Summers. 1996. "Wealthier is Healthier." *Journal of Human Resources*, 31: 841–68.

Rabe-Hesketh, Sophia and Anders Skrondal. 2005. *Multilevel and Longitudinal. Modeling Using Stata*. College Station, TX: Stata Press.

Raftery, Adrian E. 1995. "Bayesian model selection in Social Research." *Sociological Methodology*, 25: 111–63.

Raley, R. Kelly, Michelle L. Frisco and Elizabeth Wildsmith. 2005. "Maternal Cohabitation and Educational Success." *Sociology of Education*, 78: 155.

Raudenbush, Stephen W. and Anthony S. Bryk. 2002. *Hierarchical Linear Models: Applications and Data Analysis Methods* (2nd Edition). Thousand Oaks: Sage.

Reiter, Jerome P., Elaine L. Zanutto, and Larry W. Hunter. 2005. "Analytical Modeling in Complex Surveys of Work Practices." *Industrial and Labor Relations Review*, 59: 82–100.

Research Triangle Institute. 2009. *SUDAAN*. Available at *www.rti.org/SUDAAN/*.

Reskin, Barbara and Debra Bran McBrier. 2000. Why not Ascription? Organizations' Employment of Male and Female Managers. *American Sociological Review*, 65: 210–33.

Roan, Carol L. and R. Kelly Raley. 1996. "Intergenerational Coresident and Contact: A Longitudinal Analysis of Adult Children's Response to their Mother's Widowhood." *Journal of Marriage and the Family*, 58(3), 708–17.

Rogerson, Peter A., Richard H. Weng, and Ge Lin. 1993. "The Spatial Separation of Parents and their Adult Children." *Annals of the Association of American Geographers*, 83: 656–71.

Sampson, Robert J. and John H. Laub. 1995. *Crime in the Making: Pathways and Turning Points Through Life*. Boston: Harvard University Press.

SAS. 2009. *Courses and Schedules*. Available at *http://support.sas.com/training/us/*.

Sastry, Jaya and Catherine E. Ross. 1998. "Asian Ethnicity and Sense of Personal Control." *Social Psychology Quarterly*, 61(2): 110.

Schaefer, Catherine. 1982. "Shoring Up the 'Buffer' of Social Support." *Journal of Health and Social Behavior*, 23: 96–8.

Schochet, Peter Z. 2008. *Technical Methods Report: Guidelines for Multiple Testing in Impact Evaluations*. Washington, DC: National Center for Education Evaluation and Regional Assistance, Institute of Education Sciences, US Department of Education.

Scientific Software International. 2009. *LISREL 8.8*. Available at *www.ssicentral.com*.

Seidman, David. 1976. "On Choosing Between Linear and Log-Linear Models" *The Journal of Politics*, 38: 461–6.

Self-Enumerated Questionnaire Skip Map. Available at *ftp://elaine.ssc.wisc.edu/pub/nsfh/ T1se.pdf*.

Shaffer, Juliet Popper. 1995. "Multiple Hypothesis Testing." *Annual Review of Psychology*, 46: 561–84.

Sharkey, Patrick. 2008. "The Intergenerational Transmission of Context." *American Journal of Sociology*, 113: 931–69.

Smith, Tom W., Arne L. Kalleberg, Peter V. Marsden. 2005. "National Organizations Survey (NOS), 2002." Available at *www.icpsr.org*.

Stata. 2009a. *Which Stata is Right for Me?* Available at *www.stata.com/products/ whichstata.html*.

Stata. 2009b. *Training*. Available at *www.stata.com/training/*.

Stavig, Gordon R. 1977. "The Semistandardized Regression Coefficient." *Multivariate Behavioral Research*, 12: 255–8.

Stigler, Stephen M. 1986. *The History of Statistics: The Measurement of Uncertainty Before 1900*. Cambridge, MA: Harvard University Press.

Stolzenberg, Ross M. and Daniel A. Relles. 1997. "Tools for Intuition about Sample Selection Bias and Its Correction." *American Sociological Review*, 62(3): 494–507.

Suits, Daniel B. 1957. "Use of Dummy Variables in Regression Equations." *Journal of the American Statistical Association*, 52(280): 548–51.

Sweet, James, Larry Bumpass and Vaughn Coll 1988. The Design and Content of the National Survey of Families and Households. *NSFH Working Paper No. 1*. Available at *www.ssc.wisc.edu/cde/nsfhup/nsfh1.pdf*.

Tukey, John W. 1977. *Exploratory Data Analysis*. Reading: Addison Wesley.

UCLA Statistical Computing Group. 2009. *Overview of the Statistical Consulting Group*. Available at *www.ats.ucla.edu/stat/overview.htm*.

University of Kansas. 2009. *KU Summer Institute in Statistics*. Available at *http://quant.ku.edu/ StatsCamps/overview.html*.

US Bureau of Labor Statistics. 2008. *Consumer Price Index for All Urban Consumers (CPI-U).* Available at *ftp://ftp.bls.gov/pub/special.requests/cpi/cpiai.txt.*

US Department of Justice. 2007. *Hate Crime by Jurisdiction.* Available at *www.fbi.gov/ucr/hc2007/jurisdiction.htm.*

Vaisey, Stephen. 2007. "Structure, Culture, and Community: The Search for Belonging in 50 Urban Communes. *American Sociological Review,* 72: 851–73.

Vavra, Janet K. 2002. "Preservation: Maintaining Information for the Future." ICPSR Bulletin, XXII(3): 1–8.

Weakliem, David. 2004. "Introduction to the Special Issue on Model Selection." *Sociological Methods and Research,* 33: 167–87.

Webster's Revised Unabridged Dictionary. 2008. Dummy. Retrieved June 2, 2008, from Dictionary.com website: Available at *http://dictionary.reference.com/browse/dummy.*

Webster's Revised Unabridged Dictionary. 2009. Mediate. Retrieved June 2, 2009, from Dictionary.com website: Available at *http://dictionary.reference.com/browse/mediate.*

Webster's Revised Unabridged Dictionary. 2009. Moderate. Retrieved June 2, 2009, from Dictionary.com website: Available at *http://dictionary.reference.com/browse/moderate.*

Welsch, Roy E. 1986. "Comment." *Statistical Science,* 1(3): 403–5.

West, Candace and Don. H. Zimmerman. 1987. "Doing Gender" *Gender and Society,* 1(2): 125–51.

Wheaton, Blair. 1985. "Models for the Stress-Buffering Functions of Coping Resources." *Journal of Health and Social Behavior,* 26: 352–64.

Willson, Andrea E. 2003. "Race and Women's Income Trajectories: Employment, Marriage, and Income Security over the Life Course." *Social Problems,* 50: 87–110.

Wolf, Douglas A., Vicki Freedman, and Beth J. Soldo. 1997. "The Division of Family Labor: Care for Elderly Parents." *The Journals of Gerontology,* Series B, 52B: 102–9.

Wooldridge, Jeffrey M. 2003. *Introductory Econometrics: A Modern Approach.* Mason, OH: Thomson.

Wooldridge, Jeffrey M. 2009. *Introductory Econometrics: A Modern Approach* (4th Edition). Mason, OH: South-Western.

Wright, Sewall. 1921. "Correlation and Causation." *Journal of Agricultural Research,* 20: 557–85.

Yamaguchi, Kazuo. 1991. *Event History Analysis.* Newbury Park: Sage.

# GLOSSARY/INDEX

**Bonferroni adjustment:** A conservative adjustment for the increasing risk of making a Type I error when multiple hypothesis tests are conducted; the alpha level for each individual test is calculated by dividing the overall alpha level by the total number of tests being conducted   166

*Caseid*   40–1

**Case:** Participant in a study (such as a person, family, school, business, state, or country)   40

**Case-Sensitive:** The interpretation of a word differs depending on the combination of uppercase and lowercase letters used (e.g., use differs from Use differs from USE); Stata is case-sensitive; SAS is not case-sensitive   74

Censored dependent variables   407

**Chow Test:** Tests whether any coefficients differ in a regression model estimated separately for two or more subgroups (e.g., could test whether the intercept and/or slope differs in a regression of income on education estimated separately for men and for women)   277–86

**Clustered:** Data in which some sample members share a common grouping, such as relatives clustered in families, families clustered in neighborhoods, students clustered in schools, or cities clustered within states; clustering also exists in longitudinal data, when multiple time points are clustered within individuals   405, 409

**Codebook:** A description of the contents of a raw data file   54, 56–7

Cohen's *d*   139–43

**Command Window:** The Stata window in which commands can be interactively typed   67–8

Comments   85

Common cause, *see* **Confounder**

**Completely Standardized Regression Coefficient:** Regression coefficient interpreted in standardized units of the predictor and the outcome   134–6, 139–43, 163, 189

**Complex Samples:** Samples that are not simple random samples; often oversamples some subgroups and/or include clusters (because initial sampling units are drawn, and then sample members are drawn from those units; for example, cities might be drawn and then public schools sampled from those cities)   410

**Conditional Distribution:** Distribution of the outcome variable within levels of the predictor variable; the center of the conditional distribution is the conditional mean; the spread of the conditional distribution is captured by the conditional variance (the square root of which is the conditional standard deviation)   103–4, 112

**Conditional Effect:** The estimated association between one predictor and the outcome at a particular level of another predictor   158, 261–5, 269–73, 282–5, 288–96